Mathematical Programming
for
Industrial Engineers

INDUSTRIAL ENGINEERING

A Series of Reference Books and Textbooks

1. Optimization Algorithms for Networks and Graphs, *Edward Minieka*
2. Operations Research Support Methodology, *edited by Albert G. Holzman*
3. MOST Work Measurement Systems, *Kjell B. Zandin*
4. Optimization of Systems Reliability, *Frank A. Tillman, Ching-Lai Hwang, and Way Kuo*
5. Managing Work-In-Process Inventory, *Kenneth Kivenko*
6. Mathematical Programming for Operations Researchers and Computer Scientists, *edited by Albert G. Holzman*
7. Practical Quality Management in the Chemical Process Industry, *Morton E. Bader*
8. Quality Assurance in Research and Development, *George W. Roberts*
9. Computer-Aided Facilities Planning, *H. Lee Hales*
10. Quality Control, Reliability, and Engineering Design, *Balbir S. Dhillon*
11. Engineering Maintenance Management, *Benjamin W. Niebel*
12. Manufacturing Planning: Key to Improving Industrial Productivity, *Kelvin F. Cross*
13. Microcomputer-Aided Maintenance Management, *Kishan Bagadia*
14. Integrating Productivity and Quality Management, *Johnson Aimie Edosomwan*
15. Materials Handling, *Robert M. Eastman*
16. In-Process Quality Control for Manufacturing, *William E. Barkman*
17. MOST Work Measurement Systems: Second Edition, Revised and Expanded, *Kjell B. Zandin*
18. Engineering Maintenance Management: Second Edition, Revised and Expanded, *Benjamin W. Niebel*
19. Integrating Productivity and Quality Management: Second Edition, Revised and Expanded, *Johnson Aimie Edosomwan*
20. Mathematical Programming for Industrial Engineers, *edited by Mordecai Avriel and Boaz Golany*

Additional Volumes in Preparation

Mathematical Programming
for
Industrial Engineers

edited by

Mordecai Avriel
Boaz Golany

*Technion—Israel Institute
of Technology
Haifa, Israel*

Marcel Dekker, Inc.　　　New York•Basel•Hong Kong

Library of Congress Cataloging-in-Publication Data

Mathematical programming for industrial engineers / edited by Mordecai Avriel, Boaz Golany.
    p. cm. — (Industrial engineering ; v. 20)
  Includes bibliographical references and index.
  ISBN: 0-8247-9620-9 (alk. paper)
  1. Programming (Mathematics) 2. Industrial engineering—Mathematics. I. Avriel, M. II. Golany, B. III. Series.
T57.7.M317 1996
519.7'002467—dc20
                                              96-5923
                                                CIP

The publisher offers discounts on this book when ordered in bulk quantities. For more information, write to Special Sales/Professional Marketing at the address below.

This book is printed on acid-free paper.

Copyright © 1996 by MARCEL DEKKER, INC. All Rights Reserved.

Neither this book nor any part may be reproduced or transmitted in any form or by any means, electronic or mechanical, including photocopying, microfilming, and recording, or by any information storage and retrieval system, without permission in writing from the publisher.

MARCEL DEKKER, INC.
270 Madison Avenue, New York, New York 10016

Current printing (last digit):
10 9 8 7 6 5 4 3 2 1

PRINTED IN THE UNITED STATES OF AMERICA

To Or and Shir Wolf

—M. A.

To Yannai, Inbal, and Tomer Golany

—B. G.

Preface

Mathematical programming consists of models that describe complex real-world problems and algorithms to find optimal solutions to the problems represented by the models. Optimization theory—the science of finding the best possible solution for a given problem—is implemented through several approaches, such as simulation and marginal analysis, but in the last 30 years mathematical programming has become the key approach to implementing optimization theory for real-world problems. Its impact is felt in diverse areas such as transportation, telecommunications, and military operations analysis, and, in particular, it has played an important role in industrial engineering. Mathematical programming models are found in every industrial engineering topic, and in many cases they constitute the lion's share of the quantitative aspects of these areas. Anyone who wishes to learn about the underlying theory of mathematical programming subjects that have significant application areas in industrial engineering, and about the application areas themselves, must browse today through several books, since there is no single source that combines sound theoretical principles of mathematical programming with industrial engineering applications. The purpose of writing this book was to bridge the gap between texts on the theory of mathematical programming and industrial engineering books that describe application areas based on mathematical programming. Our choice of mathematical programming topics to be covered was motivated by their proven relevance to industrial engineering applications. For each mathematical programming topic a concise, but rigorous, derivation and description are given, followed by a review of some relevant application areas. Special attention is given to providing a good balance between the theoretical and the applied portions of each topic. Since some of the application areas rely on more than one mathematical programming topic, they are covered from several points of view and appropriate cross-references between the topics are supplied.

The book can serve as a required or recommended textbook as it has in the study programs offered by our Faculty of Industrial Engineering and Management at Technion—Israel Institute of Technology. The book can be used as a textbook in an Optimization Methods course, offered to senior-year undergraduate and first-year graduate students in industrial engineering. Students in this course will have taken an introductory course in operations research, as well as a basic production management course in industrial engineering. Individual chapters can further serve as recommended readings and references in other courses of the industrial engineering program: The linear programming chapter (Chapter 2) serves as a reference text in Deterministic Models in Operations Research (a first course in operations research). The integer programming and networks chapters (Chapters 3 and 4) are well suited as reference material in the courses Integer Programming and Linear and Combinatorial Programming (also senior-year undergraduate elective and first-year graduate core courses). The dynamic programming, nonlinear programming, and multiple objective chapters (Chapters 5, 6, and 7) can serve as a textbook in a miniseries of courses known as Systems Analysis 1 and 2 (a pair of graduate courses), in which students learn how to apply mathematical programming techniques in formulating and solving real-world problems arising in industry. The integer programming, nonlinear programming, and stochastic programming chapters (Chapters 3, 6, and 8) can also be used in a second course in finance, Investments and Portfolio Management. The heuristics chapter (Chapter 9) is an attractive reference for a course known as Scheduling and Sequencing, an elective industrial engineering course offered to students who have taken the basic production management courses.

We were fortunate to work with an excellent team of authors who contributed chapters. John Birge, Eric Denardo, James Evans, Fred Glover, Leon Lasdon, Timothy Magee, John Mulvey, Michael Pinedo, John Plummer, David Simchi-Levi, Al Soyster, Mario Tabucanon, José Ventura, and Allan Waren are all well-known scholars in their respective areas of expertise. We are grateful to all of them for the time and energy they have put into this book. Special thanks are also due to Russell Dekker, Editor-in-Chief, to Walter Brownfield, Production Editor, and the staff at Marcel Dekker, Inc., for their continuous support of our efforts, and to Eva Gaster for her assistance in compiling and typing the index of the book. Partial support was provided by the Fund for the Advancement of Research at Technion.

Mordecai Avriel
Boaz Golany

PERSONAL ACKNOWLEDGMENT

It is with mixed feelings of gratitude and sadness that I recognize two persons from whom I have learned a great deal of mathematical programming and industrial engineering and who have unfortunately passed away during the three years in which this book was written: first, my mentor, Professor Abraham Charnes, one of the founding fathers of Operations Research, who as my Ph.D. supervisor opened the doors of this profession to me; and second, Professor Yaakov Roll, a leader in developing industrial engineering education and practice in Israel, who was both my teacher and my friend. I wish to express special gratitude to Professor G. Kozmetsky, the former director of the IC^2 Institute at the University of Texas and to the Institute's staff, who provided me with excellent facilities and intellectual stimulation in working on the early drafts of the manuscript.

Boaz Golany

Contents

Preface		v
Contributors		xiii

1 Introduction — 1
Mordecai Avriel and Boaz Golany

  1 The Role of Mathematical Programming in Industrial Engineering — 3
  2 New Opportunities for Mathematical Programming Models — 4
  3 Categories of Mathematical Programming Models — 5
  4 Topics in Industrial Engineering — 8
  5 The Plan of This Book — 8
     References — 9

2 Linear Programming — 11
Allen L. Soyster and José A. Ventura

  1 Introduction — 11
  2 Early History and Origins of Linear Programming — 20
  3 Model Types and Applications — 24
  4 Graphical Method and Geometric Insights — 42
  5 Foundations of Linear Programming — 49
  6 Simplex Method — 58
  7 Duality and Sensitivity Analysis — 70
  8 Interior Point Methods — 82
  9 Computer-Aided Model Building and Data Management — 93
  10 Beyond the Model — 109
     Appendix A.1: Employment of Industrial Engineers in the United States — 114
     Appendix A.2: Data for Production–Distribution Model — 115
     References — 116

3 Integer Programming — 123
Timothy M. Magee and Fred Glover

1	Integer Programming Models	125
2	Cutting Planes and Valid Inequalities	149
3	Branch and Bound Procedures	169
4	Lagrangean Relaxation	209
5	Applications	257
6	Conclusions	260
	References	261

4 Graph Theory and Networks — 271
James R. Evans

1	Introduction	271
2	Spanning Tree Problems	276
3	Shortest-Path Problems	282
4	Minimum-Cost Flow Problems	286
5	Arc Routing Problems	291
6	Facility Layout	298
7	Cellular Manufacturing	300
8	Production Lot Sizing	302
9	Employee Scheduling	303
10	Snow and Ice Control	305
11	Final Remarks	306
	References	306

5 Dynamic Programming — 307
Eric V. Denardo

1	Introduction	307
2	Recursions	308
3	The Simplest Dynamic Program	316
4	Language of Dynamic Programming	322
5	Formulating Dynamic Programs	325
6	The Dynamic Lot Size Model	331
7	Capacity Expansion with Known Requirements	337
8	Decision Trees	342
9	Decision Analysis	348
10	The Markov Decision Model	361
11	Inventory Control: Optimal (s, S) Policies	375
12	Conclusion	380
	References	382

Contents   xi

6 Nonlinear Programming — 385
Leon Lasdon, John Plummer, and Allan D. Waren

    1 Introduction — 385
    2 Unconstrained Problems — 399
    3 Constrained Problems — Optimality Conditions — 409
    4 Constrained Optimization: Algorithms — 415
    5 Applications of Nonlinear Programming — 447
    References — 483

7 Multiobjective Programming for Industrial Engineers — 487
Mario T. Tabucanon

    1 Introduction — 487
    2 Theoretical Concepts and Approaches — 488
    3 Techniques — 491
    4 Applications — 501
    5 Concluding Remarks — 540
    References — 540

8 Stochastic Programming — 543
John R. Birge and John M. Mulvey

    1 Introduction — 543
    2 Financial Planning and the Value of the Stochastic Solution — 544
    3 Modeling Stochastic Parameters — 550
    4 Discrete Bounding Approximations — 552
    5 Generating Scenarios — 554
    6 Solution Procedures for Stochastic Programs — 555
    7 Industrial Engineering Applications — 559
    8 Conclusions — 569
    References — 570

9 Heuristic Methods — 575
Michael Pinedo and David Simchi-Levi

    1 Introduction — 575
    2 Heuristics — 581
    3 Routing Applications — 594
    4 Scheduling Applications — 599
    5 Location and Layout Applications — 613
    6 Discussion — 615
    References — 615

Author Index — 619
Subject Index — 627

Contributors

Mordecai Avriel Faculty of Industrial Engineering and Management, Technion—Israel Institute of Technology, Haifa, Israel

John R. Birge Department of Industrial and Operations Engineering, The University of Michigan, Ann Arbor, Michigan

Eric V. Denardo Professor of Operations Research, Yale University, New Haven, Connecticut

James R. Evans Department of Qualitative Analysis and Operations Management, University of Cincinnati, Cincinnati, Ohio

Fred Glover School of Business, University of Colorado, Boulder, Colorado

Boaz Golany Faculty of Industrial Engineering and Management, Technion—Israel Institute of Technology, Haifa, Israel

Leon Lasdon Department of Management Science and Information Systems, The University of Texas at Austin, Austin, Texas

Timothy M. Magee Center for Advanced Decision Support for Water and Environmental Systems, Department of Civil Engineering, University of Colorado, Boulder, Colorado

John M. Mulvey Department of Civil Engineering and Operations Research, Princeton University, Princeton, New Jersey

Michael Pinedo Department of Industrial Engineering and Operations Research, Columbia University, New York, New York

John Plummer Department of Computer Information Systems and Quantitative Methods, Southwest Texas State University, San Marcos, Texas

David Simchi-Levi Department of Industrial Engineering and Management Sciences, Northwestern University, Evanston, Illinois

Allen L. Soyster Department of Industrial and Manufacturing Engineering, College of Engineering, The Pennsylvania State University, University Park, Pennsylvania

Mario T. Tabucanon School of Advanced Technologies, Asian Institute of Technology, Bangkok, Thailand

José A. Ventura Department of Industrial and Manufacturing Engineering, College of Engineering, The Pennsylvania State University, University Park, Pennsylvania

Allan D. Waren Department of Computer Information Science, Emeritus, Cleveland State University, Cleveland, Ohio

Mathematical Programming
for
Industrial Engineers

1

Introduction

Mordecai Avriel and Boaz Golany
Technion—Israel Institute of Technology, Haifa, Israel

Industrial engineering is a relatively young profession that traces its roots to the early part of the 20th century. Many of the first industrial engineers were trained in other areas, mainly mechanical engineering, and became interested in the interaction that took place on the shop floor among machines, workers, and other means and resources used in production processes. Although the industrial revolution has been in full swing for several decades, it has taken a relatively long time for people involved in the production processes to grasp the need to establish a new discipline that will be devoted to analyzing the complex organization of the production resources and the means by which they can be utilized efficiently and effectively.

Industrial engineering (IE) was not taught as a separate area in most universities until the second half of the century, and even today there are major universities that do not have a separate IE department. The first industrial engineers used mostly crude and simple techniques to tackle the problems they faced. In the early days, the focus of IE was primarily on the individual workstation level and on developing methods for measuring and improving the performance of individual workers. For example, the most advanced manufacturing facilities of that time were the assembly lines of the automobile industry. These gave rise to methods such as the man-machine chart, which was aimed at reducing idle time and accelerating the production rate. This early focus was still dominating the IE research community in the early 1950s, when the first volumes of the *Journal of Industrial Engineering* were published. Typical papers in the first two volumes (1952–1953) dealt mainly with time studies and work-factor analysis.

At the same time IE has attracted the attention of researchers from a variety of other areas of science and, most notably, from the newly emerging areas of operations research and management science. These professions, born in the stormy days of World War II and conceived for the purpose of assisting the Allied military effort, focused much of their initial energy and interest on solving problems of production, inventory, and distribution of war material. As soon as the war ended, the accumulated expertise and knowledge in these disciplines was directed into a variety of civilian fields, including those related to IE. The postwar era has witnessed unprecedented growth of industrial activities around the globe. The traditional conflicts over the control and usage of natural resources were quickly replaced by global competition in manufacturing and marketing of technology. The need to maintain a competitive edge meant that organizations had to place higher priorities on industrial engineering activities, such as how to plan operations in the presence of competitive pressures, how to select and obtain appropriate technology, and how to incorporate it into the operations and utilize it in efficient and reliable ways. Thus, in the mid-1950s IE began evolving in several different directions. Spurred by successful applications of operations research during the wartime and the advent of the digital computer, IE has undergone significant changes. Evidence for the evolution in IE can be found in a paper published by Salveson (1954) in the *Journal of Industrial Engineering* entitled "Mathematical Methods of Management Programming." In this paper, Salveson promoted the use of mathematical models to improve productivity. He said, "the value of the conceptual model lies in the frame of reference it offers for analyzing and testing rules, methods and procedures in the prototype system in relation to their efficiency in achieving optimal managerial decisions in planning and scheduling." In another early paper in the same journal, Emerson (1953) presented a table in which he matched IE applications with relevant mathematical techniques. In one case he wrote, "lot size for inventories, purchases and production . . . need to be solved simultaneously to obtain optimum values [with the techniques of] simultaneous equations, determinants and matrix algebra."

More recently, research in IE is confronting two challenging problems. First, the systems to be modeled for the purpose of improving or optimizing their performance have typically become very complex, involving many diverse elements with a significant amount of uncertainty. Second, these systems are highly dynamic, evolving rapidly over time as new technologies are introduced at an ever-increasing rate. All this has led to the broadening of IE areas of interest. Modern industrial engineers are searching for ways to increase productivity and improve quality in all areas of business, industry, and government. The traditional role of industrial engineers in the manufacturing sector has been extended to include service

Introduction

industries such as health care, education, and transportation. In some cases, the focus is still on the individual workplace; in others, the focus has shifted to the production system as a whole—how to make its components act in concert so as to increase the overall system productivity. In all cases, a paradigm of IE is that productivity can be quantified and improved, either in a continuous manner or in leaps, to support the changes that are necessary for any business to succeed.

The challenges that reality has put in front of the research community cannot go unnoticed by university departments and individuals charged with the task of teaching and training future industrial engineers. Universities are continuously required to update their curricula so as to adapt to the growing demands of the marketplace. In particular, IE departments face a need to improve the quality and quantity of technical courses that equip the students with the mathematical programming tools that enable them later to confront successfully the challenges in their future workplace.

1. THE ROLE OF MATHEMATICAL PROGRAMMING IN INDUSTRIAL ENGINEERING

One can summarize the IE tasks in the contemporary industrial environment as integrating quantitative and qualitative data, using engineering techniques and information systems to improve the production of goods and services. Mathematical programming touches on almost every facet of the engineering toolbox carried by the industrial engineer. Starting with the "bread and butter" of IE, the planning and control of production and inventory, one finds industrial engineers applying material requirement planning (MRP) or more advanced versions of it (MRP-II, hierarchical production planning, etc.). Many of these tools embody linear programming optimization as part of the algorithm. Other production and inventory situations are best served through dynamic programming applications. In the area of facility planning analysis, industrial engineers apply linear and integer programming models, and facility location is one of the many IE areas in which multiobjective programming may be necessary due to the multiple criteria that characterize such problems. In addressing staffing and scheduling problems, industrial engineers often use integer programming or network techniques. Real-time monitoring and control systems for production processes typically require nonlinear or stochastic programming. In all of these areas, when an exact solution to the problem at hand is difficult to obtain, recourse is sought by applying one of the many heuristic (HE) methods developed in mathematical programming. A somewhat representative array of state-of-the-art mathematical programming applications to various industrial settings can be found in Burkard et al. (1995) and Ciriani and Leachman (1993). Other, conventional references can be found in Hillier and Lieberman (1990), Holzman (1981) and Williams (1985).

These are just a few of the IE areas affected by mathematical programming. Of course, other quantitative techniques, such as statistical analysis and simulation, not addressed in this book, are also valuable tools of IE. It is only through systematic design and integration of the appropriate tools and data that industrial engineers can indeed make a difference in their workplace.

2. NEW OPPORTUNITIES FOR MATHEMATICAL PROGRAMMING MODELS

The introduction of personal computers in the early 1980s triggered a wave of revolutionary advances in the power of computing available to all. This caused a profound change in the practicality and availability of mathematical programming models. Techniques that in the past required mainframe computers and, therefore, were limited in their use to large organizations, are now available on personal computers. As the price of hardware and software products continues to slide down, more and more potential users who were excluded before are now offered new capabilities. Two developments in particular have contributed the most to the dramatic increase in the availability of mathematical programming models on PCs. These are the spreadsheet optimizers and the modeling language systems.

2.1. Spreadsheet Optimizers

These tools allow the user to define and solve mathematical programming models within the environment of a spreadsheet application. Perhaps the most popular of these optimizers today is the Solver in the Microsoft Excel spreadsheet. For a user familiar with mathematical programming concepts and the use of spreadsheets it requires very little training, since it is mostly menu driven. The user defines the variables, constraints, and the objective function by selecting cells in the spreadsheet and then invokes the Solver, which also offers some flexibility in defining algorithmic parameters. Although spreadsheet optimizers are usually limited to relatively small problems, they are inexpensive and can be helpful in a small business setting to provide quick and efficient solutions for real applications.

2.2. Modeling Language Systems

These systems allow the user to enter a mathematical programming model in a straightforward manner, writing symbols almost the same way as one would write them on paper; define the data parameters; select which solver to use; and run the model. One example of modeling languages is the

Introduction

general algebraic modeling system (GAMS); see Brooke et al. (1992). GAMS was developed by a team of researchers working for the World Bank in the early 1980s. Its development was motivated by the frustration of the team with the available mathematical programming software at that time. Their mission involved the formulation and execution of many large-scale linear programming models. Running the models on a mainframe proved to be boring but demanding at the same time, as changes in either the data or the model formulation were difficult to implement and test. Consequently, easily avoidable errors were generated. GAMS enables a user, even one who is not very familiar with the algorithms on which the various solvers are based, to have access to the full power of mathematical programming. Users can rely on a library of existing models, communicate easily with others, and construct many "what if" scenarios for further evaluation. The system has since been implemented on PCs and is now available on a number of hardware platforms. Similar systems soon followed, for example, AMPL, MPL, Xpress-MP, and others; see the survey by Nash (1995).

Another development that triggers new opportunities for applied mathematical programming is the significant global increase in automation of various production processes since the 1980s. This trend has manifested itself in many ways: starting at individual workstations where robots replace manual operations, through flexible manufacturing systems (FMS) — a group of computer-controlled machines connected by an automatic system that transport material, products, and tools — to the operations of automatic warehouses that rely on an automatic storage and retrieval system (AS/RS), finally reaching, in some cases, the status of the "factory of the 21st century," where the entire plant system is automatically driven. With increased automation, the tools of mathematical programming are becoming more relevant. First, the setup cost involved in installing these systems is large and there is a need to utilize the invested capital to its maximal potential. Second, there is a reduction in the number of constraints that affect the system. Machines and robots can work 24 hours a day, they can be placed in a dark and noisy environment, their production rate is more stable, etc. Many mathematical programming techniques came "ahead of their time" in the sense that they have provided exact solutions to situations that were characterized by large human-related variability (e.g., scheduling). As the degree of automation increases, some of these techniques gain relevance.

3. CATEGORIES OF MATHEMATICAL PROGRAMMING MODELS

Mathematical programming in its simplest form deals with the problem of maximizing a real-valued function, such as the profit resulting from operating a system (or minimizing a real-valued function, such as operating cost), subject to constraints, such as capacities or operating conditions, expressed as equations or inequalities. Mathematically, this can be represented as

(MP)   Maximize (minimize) $f(x)$ (the objective function)
subject to (the constraints)
$$g_i(x) \le 0, \quad i = 1, \ldots, m$$
$$h_k(x) = 0, \quad k = 1, \ldots, p$$

where x is an n-vector of real variables and f, g_i, h_k are real-valued functions depending on the vector variable x. Notice that this formulation is very general and many special cases of it will be presented throughout the following chapters.

If the functions above are all linear functions in the variables, then the mathematical program is called a *linear program* (LP). Chronologically, the first formulation and solution procedures of mathematical programs were those of linear programs. Optimal production planning is an example that can usually be formulated as a linear program. Linear programming is still the best known and most popular branch of mathematical programming. From a mathematical perspective, elegant theorems were derived that explain the existence and characterization of optimal solutions of linear programs. From an application-oriented perspective, extremely powerful solution techniques are available that can tackle large-scale problems with thousands of variables and more.

Many real-world applications cannot be modeled as linear programs. One example of such applications is the case in which the variables are restricted to be (nonnegative) integers, rather than real numbers. Modifying the problem (MP) above by requiring that the variable take on (nonnegative) integer values (that is, $x = 0, 1, 2, \ldots$), whereas the functions f, g_i, h_k are linear in the variables, results in an *integer program* (IP). In a special case of integer programs, called a *0-1 integer program*, or *binary program*, the variables are restricted to be either 0 or 1 (binary variables). Selecting the best projects among many alternative ones, and subject to constraints, is an example that is formulated as a 0-1 integer program. Another generalization of a linear program is to a *nonlinear program* (NLP). In a nonlinear program the variables are real valued, but some (or all) of the functions, f, g_i, h_k are nonlinear in the vector variable x. Many industrial processes (e.g., chemical processes) are nonlinear in the operating variables (e.g., temperature, pressure). Formulating the optimization problem of such processes often results in a nonlinear program.

Consider now an optimization problem in which there are several, often conflicting, objective functions. For example, a firm may be interested in maximizing its profits from a production process and at the same

Introduction

time in minimizing the environmental damage resulting from the process. Further assume that both the profits and the environmental damage can be formulated in terms of the operating variables and are subject to the same technological constraints. The optimal solutions of the mathematical programs formulated first with one of the objective functions alone and, second, with the other one alone look very different from each other. To obtain a "compromise" solution that would take into account both objectives requires a special formulation, called *multiobjective programming* (MOP).

A mathematical program consists of variables and parameters. The values of the variables are determined by the solution of the mathematical program, whereas the values of the parameters are fixed by the modeler. For example, in a linear objective function, given by

$$f(x) = c_1 x_1 + c_2 x_2 + \cdots + c_n x_n$$

the $x_1, x_2, \ldots, x_n$ are the variables and the $c_1, c_2, \ldots, c_n$ are the parameters. Suppose now that the objective above represents some future production cost function and the c_j are unit costs of raw materials. In many realistic cases these future unit costs are not known with certainty at the time of planning the production, and they should be treated as random variables. A mathematical program in which some (or all) of the parameters appearing in the objective function and constraints are random needs special formulation and solution techniques. Such a program is called a *stochastic program* (SP).

Another branch of mathematical programs is characterized by variables (the vector x) that can represent flow in networks and constraints that describe various conditions and attributes of the network formulation. Such programs fall into the category of *network programming* (NP) (or, more generally, into the area of graph theory). These programs are particularly important in situations involving discrete entities (e.g., manpower scheduling) and in scenarios in which material needs to be transported among different facilities.

Certain types of optimization problems can be represented as stagewise (sequential) decision processes. A mathematical program that describes the states under which the system can operate, the decision stages, and the rules that govern the transition from stage to stage is called a *dynamic program* (DP).

For some mathematical programs it is very difficult to find an optimal solution. In such case a heuristic method (HE) may be useful in finding a "good" solution.

All of these programming categories, and many more, are discussed in this book.

4. TOPICS IN INDUSTRIAL ENGINEERING

The chapters of this book are organized by mathematical programming categories. Each chapter refers to various IE subjects that are of interest for both practitioners and researchers. Most of these subjects are related to production and inventory management. Specifically, the areas of production planning, scheduling, forecasting, and inventory control provide many examples and cases studies used throughout the book. The list below brings together IE application areas and mathematical programming categories that can be found in this book:

Production (and aggregate) planning — LP, IP, NLP, NE, MOP, DP, SP
Product mix (and project selection) — LP, MOP
Capacity planning — DP, SP
Blending and mixing — LP, NLP
Cutting stock — LP, IP
Sequencing and scheduling — IP, MOP, NE, HE
Staffing and manpower planning — LP, DP, SP
Transportation and transshipment — LP, DP, NE, HE
Facility layout and location — LP, NE, HE
Forecasting — NLP
Inventory planning — LP, NLP, DP, SP
Inventory control (stochastic) — DP
Project management — NE, DP
Maintenance decisions — MOP
Cellular manufacturing — NE
Financial planning — NLP, SP, DP

As can be seen in the list, many IE topics are covered in the book by more than one mathematical programming category. This gives the reader an excellent opportunity to compare different approaches to similar planning and control scenarios, learn about the different ways to formulate them, and realize the potentially different outcomes that each method is capable of providing.

5. THE PLAN OF THIS BOOK

This is a textbook that might be used in senior-level courses on production-inventory systems or graduate-level courses in the same general area, and also a reference book on mathematical programming categories. The book can be best understood by readers who have some engineering or other quantitative background and perhaps have taken an introductory course in operations research/management science. The authors of the following chapters review the main structure of knowledge in the various mathematical programming categories, point out adequate references for further reading, and move into areas of application that are of specific interest to industrial engineers.

We have deliberately included more material than is necessary in any given one-quarter or one-semester course. We expect that instructors using this book will choose the chapters and sections that are most relevant to their course subjects.

There are eight substantive chapters in this book. Chapter 2 by Soyster and Ventura presents linear programming, by far the most practiced tool of mathematical programming. The chapter lists nine major areas of application of LP to industrial engineering before launching into a detailed description of the methodology. The chapter contains a section on implementation in which many useful hints are provided to potential LP users. Chapter 3 by Magee and Glover deals with integer and 0-1 programming. It provides a very extensive review of state-of-the-art IP techniques, algorithms, and software. Chapter 4 by Evans succinctly describes the broad areas of networks and graph theory and their applications to IE. It succeeds in presenting the relevant mathematical programming building blocks in this area as well as some important IE applications in a compact, yet comprehensive way. Chapter 5 by Denardo contains a unique, thought-provoking presentation of dynamic programming. Readers who have some degree of familiarity with DP will also find it useful to learn this new means of presentation as well as diving into the more technical sections in this chapter. Chapter 6 by Lasdon, Plummer, and Waren provides a well-polished presentation of nonlinear programming techniques and algorithms. It further dwells on implementation issues, devoting significant sections to practical guidelines for NLP software implementation. Chapter 7 by Tabucanon is dedicated to multiobjective programming and its applications to IE. The chapter reviews current techniques and describes several case studies that are presented in full — from the formulation stage through the analysis of results. Chapter 8 by Birge and Mulvey describes stochastic programming. It presents the most important model structures and algorithms available today in this area, as well as some IE applications (mainly in financial planning). Chapter 9 by Pinedo and Simchi-Levi spans a range of known heuristics that have proved themselves in numerous IE applications.

REFERENCES

Brooke, A., Kendrick, D., Meeraus, A., *GAMS — A User Guide*. The Scientific Press, 1992.

Burkard, R. E., Ibaraki, T., Queyranne, M., *Mathematics of Industrial Systems*, Annals of Operations Research. Baltzer Science Publishers, 1995.

Ciriani, T. A., Leachman, R. C., *Optimization in Industry: Mathematical Programming and Modeling Techniques in Practice*. Wiley, New York, 1993.

Emerson, H. P. "A Mathematical Foundation for Industrial Engineering," *Journal of Industrial Engineering*, Vol. 4, pp. 14–18, 1953.

Hillier, F. S., Lieberman, G. J., *Introduction to Mathematical Programming*. McGraw-Hill, New York, 1990.

Holzman, A. G., *Mathematical Programming for Operations Researchers and Computer Scientists*. Marcel Dekker, New York, 1981.

Nash, S. G., "Software Survey," *OR/MS Today*, Vol. 22, No. 2, pp. 60–71, 1995.

Salveson, M. E., "Mathematical Methods for Management Programming," *Journal of Industrial Engineering*, Vol. 5, pp. 9–15, 1954.

Williams, H. P., *Model Building in Mathematical Programming*. Wiley, New York, 1985.

2

Linear Programming

Allen L. Soyster and José A. Ventura
The Pennsylvania State University, University Park, Pennsylvania

1. INTRODUCTION

1.1. Overview

It is no accident that linear programming is a subject that is taught in virtually all industrial engineering (IE) departments in the United States and throughout the world. Since the late 1950s, it is a subject that has flourished in an IE environment. As the focus of industrial engineering broadened from the improvement of the individual workplace to coordination and optimization of the activities spanning departments, plant sites, and entire companies, more sophisticated tools were required. Formalized methods were needed to compare, evaluate, and optimize systems and procedures involving hundreds of products and the individuals, materials, and machines required to make them. Similarly, tactics were needed for the most effective deployment of hundreds of trucks and airplanes to best service a company's customers and passengers. Since the late 1950s, the methodology of linear programming has been used successfully to save companies many millions of dollars. This chapter highlights these successes, provides the reader with an understanding of the methodology, and concludes with some insights into how to implement this methodology.

1.2. A Prototype Problem

To help understand the nature of linear programming, consider a particular application for a company in the distribution business. To simplify matters,

Table 1 Unit Transportation Cost for Prototype Example

	Pittsburgh	Atlanta	Houston	Denver	San Francisco
New York	$2	$3	$4	$5	$8
Chicago	$2	$4	$3	$3	$7
Los Angeles	$8	$6	$4	$3	$2

suppose the company distributes a single product and maintains three warehouses, one in New York, one in Chicago, and one in Los Angeles. Furthermore, suppose that there are five major distribution points that stock the product, namely Pittsburgh, Atlanta, Houston, Denver, and San Francisco. The question is: "how many units should be shipped each month from each warehouse to each distribution depot?" That this should be the question is based on the facts that

Minimizing the distribution cost is important, and
There are limitations on the availability of the product at each warehouse and specification of how many units each distribution depot needs.

To add some specificity to this illustrative example, consider Table 1, in which the cost of shipping a unit from each warehouse to each distribution depot is shown. Furthermore, the monthly supply at each warehouse and monthly demand at each distribution depot are given in Figure 1. Note that New York has a supply of 200 units per month to distribute and Pittsburgh requires a monthly shipment of 100 units. Overall, there are 1000 units of capacity at the three warehouses and 1000 units of demand at the five distribution depots.

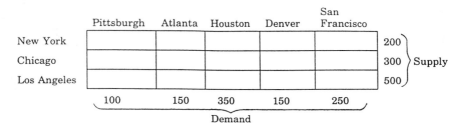

Figure 1 Supply/demand for prototype example.

Linear Programming

At this level of the discussion, the problem to be answered is how the "empty boxes" in Figure 1 should be completed. In particular, the 200 units of supply at New York need to be somehow distributed across the five distribution depots, and likewise for the 300 units at Chicago and 500 units at Los Angeles. However, the distribution must also result in an allocation so that Pittsburgh receives 100 units, Atlanta receives 150 units, etc. One possible solution to the problem is illustrated in Figure 2. In this specification, the total monthly shipping cost is $30,500, which is obtained by multiplying the shipping costs in Table 1 with the quantities in Figure 2. This example illustrates the basic nature of linear programming. In this problem there is an

Objective Function (here, minimize the shipping costs) in the presence of
Constraints (here, the availability of supply and the specified demand)

It should be clear that there are many different ways to distribute the product across the distribution depots, and we seek the alternative that minimizes the total overall monthly cost. One needs a methodology that can find the best alternative. The methodology should be applicable not just to this problem (a problem with three warehouses and five distribution depots) but also to many problems with any number of warehouses and depots. Furthermore, the methodology should be applicable to other related problems with the generic form of optimizing some objective, subject to certain constraints.

Next, we construct a mathematical model of this distribution problem. Mathematical models are routinely used by engineers to help understand or predict or optimize some system. In any model one must

Define *decision variables* (unknowns),
State formal relationships among variables, and ultimately
Solve for the unknowns.

	Pittsburgh	Atlanta	Houston	Denver	San Francisco
New York	50	50	100		
Chicago	50	100	50	100	
Los Angeles			200	50	250

Figure 2 Example solution for prototype example.

The same process is true for linear programing models. First, we define our decision variables. For this example, let x_{ij} represent the number of units that will be shipped from warehouse i to depot j. In this manner, we define the following set of 15 variables:

$x_{11}, x_{12}, x_{13}, x_{14}, x_{15}$ (shipments from New York)
$x_{21}, x_{22}, x_{23}, x_{24}, x_{25}$ (shipments from Chicago)
$x_{31}, x_{32}, x_{33}, x_{34}, x_{35}$ (shipments from Los Angeles)

The objective is to minimize the following *linear function* (which has units of dollars), where the coefficients are obtained from Table 1:

$$2x_{11} + 3x_{12} + 4x_{13} + 5x_{14} + 8x_{15} + 2x_{21} + 4x_{22} + 3x_{23} + 3x_{24} + 7x_{25} + 8x_{31} + 6x_{32} + 4x_{33} + 3x_{34} + 2x_{35}$$

where the variables are subject to both supply and demand constraints as follows:

$$\begin{aligned} x_{11} + x_{12} + x_{13} + x_{14} + x_{15} &= 200 \\ x_{21} + x_{22} + x_{23} + x_{24} + x_{25} &= 300 \\ x_{31} + x_{32} + x_{33} + x_{34} + x_{35} &= 500 \end{aligned} \quad \text{supply}$$

and

$$\begin{aligned} x_{11} + x_{21} + x_{31} &= 100 \\ x_{12} + x_{22} + x_{32} &= 150 \\ x_{13} + x_{23} + x_{33} &= 350 \\ x_{14} + x_{24} + x_{34} &= 150 \\ x_{15} + x_{25} + x_{35} &= 250 \end{aligned} \quad \text{demand}$$

$$x_{ij} \geq 0, \quad \begin{aligned} i &= 1, 2, \ldots, 5 \\ j &= 1, 2, 3 \end{aligned} \quad \text{nonnegativity}$$

The first supply constraint (New York) specifies that the 200 units available must be distributed to the five different depots. The demand constraints ensure that each destination receives its particular requirement. Finally, all the decision variables are restricted to be nonnegative, which is almost always the case in linear programs, because the variables represent tangible entities like the number of refrigerators to be built, the amount of money to invest in a new project, or the number of square feet to allocate to the welding department. In this example, the nonnegativity requirement eliminates the possibility of $x_{24} = -50$, which would (presumably) mean shipping 50 units "backward" from depot 4 to warehouse 2.

Altogether, the linear programming model for this prototype example is as shown on p. 15:

Linear Programming

$$\begin{aligned}
\text{Minimize} \quad & 2x_{11} + 3x_{12} + 4x_{13} + 5x_{14} + 8x_{15} + 2x_{21} + 4x_{22} + 3x_{23} + 3x_{24} + 7x_{25} + 8x_{31} + 6x_{32} + 4x_{33} + 3x_{34} + 2x_{35} & (1.1)\\
\text{subject to} \quad & x_{11} + x_{12} + x_{13} + x_{14} + x_{15} = 200 \\
& \qquad\qquad x_{21} + x_{22} + x_{23} + x_{24} + x_{25} = 300 \\
& \qquad\qquad\qquad\qquad\qquad\qquad\qquad\qquad x_{31} + x_{32} + x_{33} + x_{34} + x_{35} = 500 \\
& x_{11} \qquad\qquad\qquad\qquad + x_{21} \qquad\qquad\qquad\qquad + x_{31} = 100 \\
& \quad x_{12} \qquad\qquad\qquad\qquad + x_{22} \qquad\qquad\qquad\qquad + x_{32} = 150 \\
& \qquad x_{13} \qquad\qquad\qquad\qquad + x_{23} \qquad\qquad\qquad\qquad + x_{33} = 350 \\
& \qquad\quad x_{14} \qquad\qquad\qquad\qquad + x_{24} \qquad\qquad\qquad\qquad + x_{34} = 150 \\
& \qquad\qquad x_{15} \qquad\qquad\qquad\qquad + x_{25} \qquad\qquad\qquad\qquad + x_{35} = 250 \\
& x_{ij} \geq 0, \quad i = 1, 2, 3, \; j = 1, \ldots, 5 & (1.2)
\end{aligned}$$

The mathematical model on p. 15 is a *linear program*. It has two parts: an objective function (1.1) and a set of constraints (1.2). There are many alternative ways to distribute the product from the three warehouses to the five distribution depots, but we seek the alternative that minimizes the total transportation cost. This problem has 15 variables, 8 linear constraints, and a linear objective function. In fact, the linearity specification is the reason why it is called a linear programming problem. For this example, the optimal solution is $x_{11} = 50$, $x_{12} = 150$, $x_{21} = 50$, $x_{23} = 250$, $x_{33} = 100$, $x_{34} = 150$, $x_{35} = 250$, and all other variables equal to zero. Such a solution results in an objective function value (monthly shipping cost) of $27,500.

In summary, the problem statement embodied by (1.1) and (1.2) is a linear program. It is called a program because, once a solution is obtained, the company can implement a distribution plan or program; i.e., the terms program and plan, in our context, should be considered synonyms. This particular linear program is an example of a special class of models called *transportation problems*. These are characterized by a set of origins (in our case warehouses) and destinations (in our case distribution depots) and some "product" must be transported from these origins to the destinations in a least-cost manner. Later in this chapter several applications of this model are provided.

1.3. Linear Programs: What They Are and Are Not

A simplifying feature of linear programs is the fact that the basic building blocks of such models are familiar things — lines and planes. A line in two-dimensional space is the locus of points (x_1, x_2) satisfying a relationship such as

$$2x_1 - 3x_2 = 8 \tag{1.3}$$

and a linear inequality is derived from this relationship by replacing the equality (1.3) with an inequality ($\leq$ or $\geq$). This would specify a set of points that lie on one side of the line or the other. The same ideas apply to higher-dimensional spaces, and all linear programs are defined in terms of linear equalities and inequalities. Consider the following linear program:

$$\begin{aligned}
\text{Maximize} \quad z = 3x_1 &- x_2 + 2x_3 \\
5x_1 + 6x_2 &+ x_3 = 18 \\
3x_1 + 2x_2 &+ 4x_3 \leq 8 \\
x_1 &\geq 0 \\
x_2 &\geq 0 \\
x_3 &\geq 0
\end{aligned} \tag{1.4}$$

Linear Programming

The objective in this linear program is to choose values for the three unknowns x_1, x_2, and x_3 that maximize the linear expression

$$z = 3x_1 - x_2 + 2x_3$$

In fact, one can view this problem as one of choosing feasible values for (x_1, x_2, x_3) that result in the largest possible value of z. Different values of z, of course, correspond to a family of parallel hyperplanes. The allowable choices for (x_1, x_2, x_3) must satisfy the constraints of the linear program (1.4). These constraints include one equality and four inequalities. The region that satisfies all five of these constraints is called the *feasible region*. The problem, then, is to choose a point in the feasible region that maximizes z, the value of the objective function. It should be emphasized that all linear programs are of this form. Some problems may contain thousands of variables and thousands of constraints, but they are all characterized by the use of the basic building blocks of linear equalities and inequalities.

Before proceeding further it is important to understand fully the implications and the restrictions on the type of problems amenable to modeling by linear programs such as (1.4). We first emphasize that such models contain no specifications about *integrality* of the values of the variables. For the linear program (1.4), the optimal solution is $x_1 = 1.50$, $x_2 = 1.75$, and $x_3 = 0$. The nonintegrality of the optimal solution for linear programs may limit their usefulness in some potential applications. In general, if integrality is required, one needs to apply different solution methodologies that are discussed in the integer programming chapter in this book. Roughly speaking, an integer programming problem is specified in the same form as (1.4) with the added restriction that all variables must be integer valued. However, this restriction increases the complexity of the solution procedures and required computing time by orders of magnitude.

Another characteristic of linear programs is the inherent *proportionality* of the variables. This simply means that if one doubles the value of a variable, this variable's contribution to the objective function and the consumption of resources (as part of the constraints) will also double. Hence, in applications that include processes or activities with increasing returns to scale, this proportionality characteristic may invalidate the use of a linear programming model. For example, if the objective function in (1.4) is changed to

$$z = 3x_1^2 - x_2 + 2x_3 \qquad (1.5)$$

then doubling x_1 would lead to a fourfold increase in the objective value. Equation (1.5) is an example of a nonlinear objective function. Problems with nonlinearities (in the objective function, constraints, or both) are termed *nonlinear programs* and are also orders of magnitude more difficult to solve than linear programs.

Another characteristic of linear programs is their deterministic nature. All the data coefficients in the linear programming model are known with *certainty*. In (1.4), the coefficient of x_1 in the objective function is precisely and exactly 3. Suppose that it is known only that this coefficient lies in the interval [2.5, 3.5]. What if the right-hand side of the first constraint can assume the values 9, 10, and 11 with probabilities 1/4, 1/2, and 1/4? These are extensions that preclude the direct use of linear programming. These extensions generate another class of problems termed *stochastic programs* and are treated elsewhere in this book.

These discussions should sharpen the reader's insight into the applicability of and restrictions on using linear programs to model and solve real problems. The bottom line is that linear programs are linear in both the objective function and constraints. What we have seen is that in certain applications there may be a specific need for

>Integrality of variables, or
>Nonlinearities in relationships, or
>Uncertainty in the data.

In these cases, linear programming approaches may not be appropriate and other methodologies may be needed. Nonetheless, it is important to understand that other modeling approaches and allied solution techniques such as integer, nonlinear, and stochastic programming are heavily influenced by the structure, theory, and algorithms aligned with linear programming.

1.4. Guide to the Reader

The organization of this chapter about linear programming features three major parts:

>I. Concepts, History, and Applications
>II. Solution and Analysis
>III. Implementation

The chapter is intended to serve the needs of a diverse set of readers. Part I, Concepts, History, and Applications, provides an overview of what linear programming is, where it came from and how it has been applied. The present subsection should answer, in broad terms, what a linear program is, its potential usefulness, and its limitations. The next section, Early History and Origins of Linear Programming, chronicles the early history of linear programming, its early expectations, and its early pioneers. The early his-

Linear Programming

tory of operations research and that of linear programming are remarkably intertwined, growing in stature and importance together. But industrial engineering and its early pioneers also contributed to the emergence and importance of linear programming. The final section in part I is entitled Model Types and Applications. The set of applications described in this section is specifically chosen to be of direct interest to industrial engineers employed in their profession worldwide. The applications described in this subsection represent the major categories in which the U.S. Bureau of Labor classifies employment sectors for industrial engineers (e.g., transportation equipment, textiles, food, public utilities). This set of applications are also "real" applications. For all the application areas, we first provide a generic, prototype model and, subsequently, review several real applications of this generic model that have been reported in the literature.

Part II, Solution and Analysis, provides a review of the main mathematical constructs of solution algorithms, sensitivity analysis, and duality as well as a separate, self-contained section on interior point methods. Part II is similar to other textbooks about linear programming. On the other hand, part III, Implementation, focuses on how linear programming theory is actually applied in a computing framework and how models with thousands of variables and constraints can be formulated, verified, and solved. Often, computer implementation is not addressed in textbooks in serious manner nor does it receive the attention it deserves. In our experience, the main reason why the technology of linear programming is not as widely used as it could be is related to the management of data and its integration into the model. This point of view is emphasized in a 1994 editorial, "Software Tools for Mathematical Programming" [47]:

> There are several reasons why these methods . . . are not widely used in real decision processes. One of them is that much work is needed to get the model set-up and the data collected and compiled into the appropriate order to solve it with the right algorithm. Once created, the model must be tested, analyzed, modified, solved, translated into different forms; once solved the results must be reported into a user-specified format. Often too much time and effort must be invested to manage the model, . . . the main limiting factor . . . is no longer the solution procedures, but the *model management*, i.e. the creation, modification, documentation, transformation of models, survey, views, analysis of results, result specification.

Various viewpoints on formulating, validating, and, in general, managing the necessary data are discussed and contrasted in the section entitled Computer-Aided Model Building and Data Management. Another section, Beyond the Model, concludes part III with advice from individuals who have been there—individuals who have successfully constructed linear programming models to assist their businesses in a decision-making environment.

So what final advice do we provide about reading this chapter? Here is our advice: The three parts can, for the most part, be read independently of one another.

> Part I, Concepts, History, and Applications, is for the entrepreneur and manager. Even if you know little about models and their optimization, you will learn and may be surprised at the wide array of ordinary business functions that can be addressed and improved with this technology. Also, current and past students of linear programming will be interested in learning about the origins of linear programming and the contributions and viewpoints of some of the early pioneers.
>
> Part II, Solution and Analysis, is like most classical texts and can be used either to learn the material for the first time (we hope) or to provide a review for someone needing a refresher. In addition to discussions of topics like the simplex algorithm and duality, we have provided a "user-friendly" description of recently developed interior point solution methods.
>
> Part III, Implementation, is important for anyone who is serious about using linear programs to solve real problems in their company. We believe that the section will help you understand and make some judicious choices about software and needed computer skills.

2. EARLY HISTORY AND ORIGINS OF LINEAR PROGRAMMING

2.1. Origins in Operations Research

In 1991 George Dantzig published a short paper on the history of linear programming [42]. This is a wonderful paper in that Dantzig engages the reader in a fashion that makes you feel you were there when various events happened in the 1940s and 1950s. Dantzig begins by describing his job responsibilities at the Pentagon during World War II. The following paragraph, excerpted from reference 42, describes the type of environment that led to the idea of linear programming:

> My own contributions grew out of my World War II experience in the Pentagon. During the war period (1941-45), I had become an

Linear Programming

expert on programming-planning methods using desk calculators. In 1946 I was Mathematical Advisor to the US Air Force Comptroller in the Pentagon. . . . My Pentagon colleagues, D. Hitchcock and M. Wood, challenged me to see what I could do to mechanize the planning process. I was asked to find a way to more rapidly compute a time-staged deployment, training and logistical supply program. In those days 'mechanizing' planning meant using analog devices or punch-card equipment. There were no electronic computers.

But there was something still missing in this project to "mechanize the planning process." Dantzig continues,

The *activity analysis* model I formulated would be described today as a time-staged, dynamic linear program with a staircase matrix structure. *Initially there was no objective function*; broad goals were never stated explicitly in those days because practical planners simply had no way to implement such a concept. Noncomputability was the chief reason, I believe, for the total lack of interest in optimization prior to 1947.

The idea of something like an "objective function" seemed to be foreign at the time. In fact, according to Dantzig, one of his three major contributions to linear programming was "replacing ground rules for selecting good plans by general objective functions." It is interesting that Dantzig ranks this achievement in the same category as the "invention of the simplex," which he pinpoints as occurring in the summer of 1947.

The actual name "linear program" evolved from Dantzig's first paper, which was entitled "Programming in a Linear Structure" [13]. In 1948 Dantzig related that the famous economist Tjalling Koopmans recommended the shortened name *linear program*, where the name program stemmed from a military term about "plans and schedules for training and supply."

In the early 1950s, the National Bureau of Standards had begun work on computational issues involved with linear programming [43]. Two new societies were emerging: Operations Research Society of America and The Institute of Management Science. At the first national meeting of the Operations Research Society of America, J. F. Magee [62] presented a paper entitled "Linear Programming in Production Scheduling." The early volumes of the *Journal of Operations Research* included papers by Charnes, Cooper, and Farr [11] on product mix and process selection applications and by Flood [24] on the transportation model for scheduling tanker routes. One of the first references (1952) to a proposed application of linear programming is a "blending problem" by Charnes, Cooper, and Mellon [12]. Linear programming was also receiving notoriety in the new journal *Management Science*. The first issue included a work by Charnes and Cooper [10] that explained the operation of the simplex method in the context of a transportation problem.

But what about the early impact of linear programming in business and industry? It is fair to say that in the 1950s this new technology did not yet strongly affect industrial practices. One must remember that the digital computer was still in its infancy and few businesses were prepared to actually solve such problems. But in 1955, a conference was held at the University of Pennsylvania entitled "The Current Use of Linear Programming in American Business" (see [87]). Participants at this conference discussed (or knew about) 20 examples of the industrial use of linear programming. In 1956 one of the first papers on linear programming applications, written by someone from industry (Salveson at General Electric), was presented at the Second Annual Meeting of the Institute of Management Science [82]. In this paper, Salveson develops a linear programming model for machine loading in one of GE's sheet metal shops. This application incorporated "20 heavy presses . . . into 6 categories . . . 24 different parts . . . with the number of operations ranging from 2 to 13 . . . ." In the next year, 1957, a survey of applications of linear programming in the oil industry was published [29].

During this early formative period in the development of linear programming, the limitation of computational resources was a significant constraint that clearly affected widespread use of the methodology. As a benchmark, Orden [75] reported that a 10×10 transportation problem in 1951 required 3 minutes to solve on the SEAC (Standards Eastern Automatic Computer). A 10×50 transportation problem required 80 minutes. But things were changing rapidly. Smith [87] related that by 1955, cost-competitive computer codes for executing the simplex algorithm (described in Section 6) existed on the IBM 701, 702, 650, and UNIVAC computers. By the late 1950s the technology of linear programming and its potential usefulness in business and industry were becoming widely known.

During the next several decades the development and use of linear programming flourished in several disciplines. Orden [75] suggests that "intellectual bonds" of linear programming exist in areas beyond operations research to include economics, mathematics, and computer science. The fact is that linear programming has found a home in many intellectual pursuits and disciplines. And the discipline continues to be one of the most important operations research tools. In 1989, Harpell, Lane, and Mansour [37] reported the results of a 10-year study in which educators and practitioners of operations research were asked to rate the importance of various operations research tools. Educators rated linear programming as the most important methodology in the two most recent surveys in [37], and practitioners ranked linear programming among their top three most important methodologies in all the surveys in this 10-year period. The study and use of linear programming are now inextricably woven into the operations research discipline.

2.2. Early Development in Industrial Engineering

In 1953 Emerson [20] published a paper in the *Journal of Industrial Engineering* entitled "A Mathematical Foundation for Industrial Engineering," which referenced the use of simultaneous linear equations in solving industrial engineering problems, but linear programming was not explicitly mentioned. The first reference to linear programming in the *Journal of Industrial Engineering* was in a paper by Salveson [81] (1954) entitled "Mathematical Methods in Management Programming." (Curiously, however, Salveson did not use the term linear programming; instead, he referred to the term "management programming.") The paper suggested the following uses for linear programming by industrial engineers:

1) . . . determine the optimal shop load . . .
2) . . . determine optimal amount of overtime . . .
3) . . . determine optimal mix . . .
4) . . . determine amount of rerouting . . .
5) . . . determine level of inventory . . .

In 1954 Malcolm published his paper [63] in the *Journal of Industrial Engineering* advocating the "widening of Industrial Engineering." Included in this paper was a suggestion that systems planning may be performed by a technique known as "linear programming." Malcolm says,

> The technique is promising for use in production planning, but will require considerable development and refinement before becoming an acceptable technique.

Over the next few years the *Journal of Industrial Engineering* featured a series of papers and viewpoints that demonstrated interest in and the potential of linear programming in industrial engineering. In 1956 Thompson [93] provided some first-hand testimony about the potential use of linear programming at Argus Cameras for space planning and layout. McGarrah [67] (1956) gave the industrial engineering community some insight into the use of linear programming for managing aggregate employment levels. In 1957 Beckwith and Vasani [4] published a paper on the use of the "assignment problem" in industrial engineering. Erickson and Randolph [21] showed how the "transportation model" could be used for selecting material handling equipment. In 1959, Greene, Chatto, Hicks, and Cox [34] described an application of a "product mix" formulation for the packing industry. But the first industrial application of linear programming was described in the *Journal of Industrial Engineering* in 1961. Metzger and Schwarzbek [70] describe how linear programing was used at General Motors Central Foundry to minimize the cost of raw materials in charging a cupola.

The introduction of operations research techniques such as linear programming into the curricula in industrial engineering departments during the 1950s was not rapid, but several strong editorials in the *Journal of Industrial Engineering* (see Lehrer [58] and Gryna [35]) in the mid-1950s advocated substantial changes in industrial engineering curricula to accommodate these new tools. Today, linear programming is a standard topic in virtually every industrial engineering department in the world.

3. MODEL TYPES AND APPLICATIONS

3.1. Overview

In this section we review eight particular linear programming models that have a special relationship with industrial engineering. These applications reflect the actual jobs and tasks performed by industrial engineers. According to the 1990 Bureau of Labor Statistics, there are 133,000 industrial engineers employed in the United States (and maybe several times more worldwide). The majority of industrial engineers, approximately 77%, are employed in manufacturing industries. Appendix A.1 in this chapter provides a summary of employment of industrial engineers in the various economic sectors. The Bureau of Labor Statistics also estimates that in a "moderate" growth scenario the employment of industrial engineers should increase by about 20% in the next 15 years. The new jobs will be about evenly split between the manufacturing and service sectors; i.e., the service sector will grow at a faster rate.

This employment information provides an appropriate rationale for choosing and highlighting applications of linear programming for industrial engineers. The remainder of this section is organized around a simple theme in which an illustrative, prototype linear programming model is presented first. Then, some typical applications that specifically pertain to the industries employing industrial engineers are cited. In all cases, only applications that are identified with specific companies are included, e.g., Citgo in the [Chemical] sector and General Motors in the [Transportation Equipment] sector. These applications can serve readers as a point of departure in evaluating how other organizations in their field have utilized linear programming and what kind of successes, problems, and ultimate benefits have ensued. The eight generic models are as follows:

1. Transportation and Assignment Models
2. Transshipment and Shortest Path Models
3. Staffing Models
4. Multiperiod Production/Inventory Models
5. Product Mix and Process Selection Models
6. Blending and Mixing Models
7. Cutting Stock Models
8. Multifacility Location Models

3.2. Transportation and Assignment Models

In Section 1.2, the concept of a transportation model is described. A product must be distributed from a set of origins, $i \in I$, to a set of destinations, $j \in J$, and the per unit transportation cost for transporting a unit from origin i to destination j is c_{ij}. Such a model can be viewed as a network with I origins and J destinations as illustrated in Figure 3.

A special case of this model, termed the assignment model, is one in which the number of origins equals the number of destinations and the supply at each origin and demand at each destination are unity. This specialization is termed the assignment problem because it can be used to *assign*, for example, n workers to n jobs. As an example, consider the case in which there are three workers and three jobs. Figure 4 shows a matrix in which the entries represent the hours required for worker i to complete job j. Denote by $x_{ij} = 1$ the assignment of worker i to job j, and $x_{ij} = 0$ otherwise. Then, a linear programming formulation of this model is

$$\begin{aligned}
\text{Minimize} \quad & 2x_{11} + x_{12} + 3x_{13} + x_{21} + 2x_{22} + 4x_{23} + 5x_{31} + 2x_{32} + 2x_{33} \\
\text{subject to} \quad & x_{11} + x_{12} + x_{13} = 1 \\
& x_{21} + x_{22} + x_{23} = 1 \\
& x_{31} + x_{32} + x_{33} = 1 \\
& x_{11} + x_{21} + x_{31} = 1 \\
& x_{12} + x_{22} + x_{32} = 1 \\
& x_{13} + x_{23} + x_{33} = 1 \\
& x_{ij} \geq 0, \quad i = 1, 2, 3; j = 1, 2, 3
\end{aligned}$$

An important characteristic of this linear programming model (which is not obvious) is that in the optimal solution the decision variables automatically assume a value of either zero or one. Hence, it is sufficient simply to require $x_{ij} \geq 0$ as in any ordinary linear program.

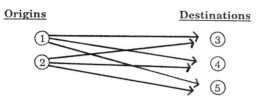

Figure 3 Transportation network.

The assignment problem and the more general class of transportation models boast a long history of applications. One of the first applications of linear programming (1954) focused on scheduling a military tanker fleet and was modeled as a transportation problem [24]. Flood [24] says,

> The object of the analysis was to determine optimum routes and schedules for a fleet of tankers carrying bulk petroleum products for military use throughout the world.

Since this earlier time, the transportation model has been used to solve problems in many sectors. An application at McDonnell Douglas [Transportation Equipment] [88] dealing with fuel management generated annual cost savings of 5 to 6%. An application of this model to production planning for the Olean Tile Company [Chemical] [59] reduced distribution costs by about $500,000 per year. In 1989 a variant of the transportation model was used by Barber Ellis Fine Papers [Paper] [7] to assist in the distribution of about 10,000 different products. The assignment model has been used by American Airlines [Transportation] [86] for scheduling crewmen for retraining (saving $250,000 per year). And in 1993 Sandia National Laboratories [Government] reported the use of the transportation model for disposing of nuclear wastes [79].

Worker \ Job	1	2	3
1	2	1	3
2	1	2	4
3	5	2	2

Figure 4 3 × 3 Assignment problem.

Linear Programming

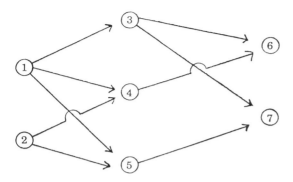

Figure 5 Transshipment model.

3.3. Transshipment and Shortest Path Models

The transportation model, shown as a network in Figure 3, depicts the flow of "product" from origin 1 to the three destinations 3, 4, and 5. An extension of this model is one in which there can be *intermediate destinations* as illustrated in Figure 5. In Figure 5, nodes 3, 4, and 5 could represent intermediate destinations and nodes 6 and 7 could represent final destinations. In a manner analogous to the transportation model, let x_{ij} represent the quantity of the product flowing from node i to node j. A node is one of the junctions $\{1, 2, 3, 4, 5, 6, 7\}$ represented in Figure 5, and c_{ij} would represent the cost of transporting a single unit from node i to node j. In effect, we define a variable x_{ij} for each arc in the network. For this example, assume that supplies s_1 and s_2 are available at nodes 1 and 2 and demands d_6 and d_7 are required at nodes 6 and 7. Here, nodes 3, 4, and 5 are transshipment nodes, with no net supply or demand.

The constraints of the linear programming formulation in *detached coefficient form* (because the variable names are omitted and only the coefficients show) are shown in Figure 6. Note that an equation is defined

x_{13}	x_{14}	x_{15}	x_{23}	x_{24}	x_{25}	x_{36}	x_{37}	x_{46}	x_{47}	x_{56}	x_{57}	
1	1	1										$= s_1$
			1	1	1							$= s_2$
-1			-1			1	1					$= 0$
	-1			-1				1	1			$= 0$
		-1			-1					1	1	$= 0$
						-1		-1		-1		$= -d_6$
							-1		-1		-1	$= -d_7$

$x_{ij} \geq 0;\ i=1,\ldots,5;\ j=1,\ldots,7.$

Figure 6 Transshipment coefficient matrix.

for each node in the network and this equation represents a conservation of flow. Flows into a node are specified as negative and flows leaving a node are considered positive. This structure results in a coefficient matrix in which each column contains exactly one positive 1 and one negative 1, because each arc (represented by variable x_{ij}) must leave some node and enter some other node. The objective is to determine the arc flows that satisfy the demand at destinations 6 and 7 using supplies at sources 1 and 2, and minimize the following objective function:

$$c_{11}x_{13} + c_{14}x_{14} + c_{15}x_{15} + \cdots + c_{56}x_{56} + c_{57}x_{57}$$

It is also noteworthy to point out that if one would "solve" this transshipment model (12 variables and 7 constraints), the optimal values of $\{x_{ij}\}$ would automatically be integers (if the supplies, $\{s_i\}$, and demands, $\{d_j\}$, are also integers).

An interesting special case of the transshipment model is the so-called PERT/CPM network. This is a network that is commonly used to represent a project with several activities and one in which certain activities must be completed before others can be started. The network illustrated in Figure 7 (this example is from Schrage [83]) represents some of the ordinary activities in building a house. Each arc represents a particular activity, and the number associated with each arc represents the number of days required to complete the activity.

The minimum time required to complete the project can be determined from a property of the graph. This minimum time corresponds to the *longest path* from node 1 to node 9. This longest path is termed the *critical path*. The critical path can be determined from the solution of a transshipment model. One formulates such a model in exactly the same fashion described and illustrated in Figures 5 and 6. The model would have

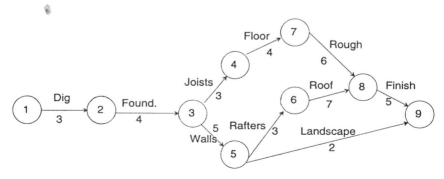

Figure 7 Activities in the construction of a house.

Linear Programming

10 variables (arcs) and 9 equations (nodes). This special application is accommodated in a transshipment model in the manner in which the right-hand sides are specified. For this problem, the right-hand sides of the first and last constraints are

$$x_{12} = 1$$
$$-x_{59} - x_{89} = -1$$

and in all other (seven constraints), fix the right-hand sides at zero; e.g., at node 5 one would have

$$-x_{35} + x_{57} + x_{59} = 0$$

The objective for this PERT/CPM application is to maximize

$$3x_{12} + 4x_{23} + 3x_{34} + 5x_{35} + 4x_{46} + 3x_{57} + 2x_{59} + 6x_{68} + 7x_{78} + 5x_{89}$$

Now observe that any path from node 1 to node 9, is specified by fixing some variables at 1 and all other variables equal to zero; e.g., the path 1-2-3-5-9 is specified by $x_{12} = x_{23} = x_{35} = x_{59} = 1$ and all other variables are set equal to zero. The elegance of this network formulation is as follows: *the shortest possible time to complete the project is the longest path from node 1 to node 9*. In this example, the optimal solution is

$$x_{12}^* = x_{23}^* = x_{35}^* = x_{56}^* = x_{68}^* = x_{89}^* = 1, \text{ other } x_{ij}^* = 0$$

This solution has an objective function value of 27, which represents the minimal time required to complete the project, and the critical path is 1-2-3-5-6-8-9.

A major user of transshipment-type models has been and continues to be the petroleum industry. Klingman et al. [56] describe a supply/distribution/marketing model for Citgo [Chemical] that "addresses simultaneously most of the short-term planning and operational issues associated with supply, distribution and marketing of refined petroleum products." For one product, unleaded gasoline, a transshipment submodel included 3000 nodes and 15,000 arcs. Citgo estimated that the use of this tool saved approximately $2.5 million annually. Other applications of these network-based techniques in the [Chemical] sector include a supply and distribution model at Amoco [68] and a logistics model at British Petroleum [84]. Harrison [38] describes how the Ballyclough Cooperative Creamery [Food] in Ireland used the transshipment model to improve its distribution efficiency by reducing distance traveled by 11% and saving IRf 1.5 million annually.

3.4. Staffing Models

One of the more common responsibilities of an industrial engineer, in manufacturing as well as the service sector, is personnel staffing. Consider a business (e.g., a hospital, factory, toll booth on a turnpike, or a utility) that operates 24 hours a day. In these settings, different periods of the day may require different numbers of workers. As an example, suppose that the operation of the intensive care wing of a hospital requires the numbers of nurses shown in Figure 8 to be on duty. Let x_t represent the number of nurses who begin their shift in period t and work for 8 hours. The objective is to minimize the number of nurses required. In this illustrative example the linear program for this illustrative example would be

$$\begin{align} \text{Minimize } & x_1 + x_2 + x_3 + x_4 + x_5 + x_6 \\ & x_1 + x_6 \geq 8 \\ & x_1 + x_2 \geq 10 \\ & x_2 + x_3 \geq 12 \\ & x_3 + x_4 \geq 10 \\ & x_4 + x_5 \geq 8 \\ & x_5 + x_6 \geq 6 \\ & x_j \geq 0, \quad j = 1, \ldots, 6 \end{align} \tag{3.1}$$

Note that there is one constraint for each time period; e.g., in period 2, nurses who begin their shift in either period 1 or period 2 will be on duty. In this example, we would also like to have the optimal solution be integer valued (because x_t represents the number of nurses). In fact, this will be the case for the linear program (3.1). An optimal solution for (3.1) (there are more than one) is

$$x_1^* = 0, x_2^* = 8, x_3^* = 4, x_4^* = 8, x_5^* = 2, x_6^* = 6$$

which requires 28 nurses. However, if the special constraint structure in (3.1) is changed by adding other constraints, this integrality property of (3.1) will no longer hold. For example, suppose that one adds an additional

SHIFT	(1)	(2)	(3)	(4)	(5)	(6)
TIME	12 AM-4	4 - 8 AM	8-12 Noon	12 Noon-4	4-8 PM	8-12 Midnight
Number	8	10	12	10	8	6

Figure 8 Nurse staffing requirements.

constraint to (3.1) specifying that at least 33% of the nurses begin work in shift 3. Furthermore, suppose that it is required that at least one nurse reports at the beginning of each shift. These additional restrictions require the following constraints:

$$x_3 \geq 0.33(x_1 + x_2 + x_3 + x_4 + x_5 + x_6)$$

and

$$x_j \geq 1, \quad j = 1, 2, \ldots, 6 \tag{3.2}$$

If the seven constraints in (3.2) are added to the linear program (3.1), a new optimal solution is

$$x_1^* = 7.358$$
$$x_2^* = 2.642$$
$$x_3^* = 9.358$$
$$x_4^* = 1.000$$
$$x_5^* = 7.000$$
$$x_6^* = 1.000$$

Hence, 7.358 nurses are to report to work in shift 1 (and 28.358 are needed in total). Obviously, there is a problem here. This example illustrates the types of difficulties that result from the lack of integrality in some applications of linear programming. But this example also illustrates the capabilities associated with *integer programming*, the subject of another chapter in this book. Using methods described in that chapter, one can constrain each variable to be integer valued, and when these methods are applied to this example problem, an optimal integer solution can be obtained that requires 29 nurses specified by

$$x_1^* = 7, \, x_2^* = 3, \, x_3^* = 10, \, x_4^* = 3, \, x_5^* = 5, \, x_6^* = 1$$

One successful application of *staffing models* is reported by the Transport Company of Montreal [Government] [5]. This public transit company employs 3000 bus drivers and nearly 1000 subway operators and ticket collectors and must address large variations in required service levels. A linear programming-based model saved the transit authority nearly $4 million annually. In another application, scheduling of police in San Francisco [Government] [92], linear programming models were used in an "integer application" as part of a heuristic strategy. The improvement in staff assignments resulted in $11 million annual cost savings and a 20% response time improvement. United Airlines [Transportation] [45] reported a $6 million annual savings in developing work schedules for 4000 employees using linear programming and other network optimization techniques. In this application, 30-minute intervals over a 7-day period were used to generate monthly schedules and linear programming was used as the scheduling module. Another application of staff planning was reported by the Financial Services Group of Canada Systems Group [Finance] [97]. Due to a surge in the demand for processing of retirement savings, this group implemented a modest-size linear programming–based manpower planning model (202 constraints and 226 variables). The application resulted in $300,000 annual savings. Another model of modest size has been used to schedule telephone operators for off-track betting in Australia with annual savings estimated at $350,000 [101]. This model has also been used extensively in the [Health] sector. In [76], a goal-programming approach to nurse scheduling is applied at St. Luke's Medical Center in Phoenix, Arizona.

3.5. Multiperiod Production Scheduling and Inventory Management Models

Some of the earliest applications of linear programming focused on the optimization of production schedules and minimization of inventory costs. As early as the 1950s, linear programming models were being constructed for this application, e.g., see [62]. In this class of models, one can relate production decisions in a time period t with inventory levels in periods $t + 1, t + 2, \ldots$. This application of linear programming, in a real sense, was one of the first efforts of industrial engineers to view the operations of the factory as a system. This system's viewpoint is apparent in the fact that there are two distinct classes of variables: production decisions and inventory levels. These variables are linked together in a series of relationships that relate production in a given period t with incoming inventory I_{t-1} (carried forward from the previous period) and ending inventory I_t (carried into the next period). If x_t represents the amount of production in period t, the general production-inventory relationship is

$$x_t + I_{t-1} = d_t + I_t$$

where d_t is the demand in period $t = 1, 2, \ldots, T$ and T is the length of the planning horizon. A simple three-period example of this model with per unit production costs of $2, $3, and $4 in periods $t = 1, 2, 3$, inventory holding costs of $1 per period, and demands of 10, 20, and 25 units in the three-period horizon is as follows:

Linear Programming

$$\begin{align}
\min \quad & 2x_1 + 3x_2 + 4x_3 + 1I_1 + 1I_2 + 1I_3 \\
& x_1 \qquad\qquad\qquad - I_1 \qquad\qquad\qquad = 10 \\
& \qquad x_2 \qquad\qquad + I_1 - I_2 \qquad\qquad = 20 \\
& \qquad\qquad x_3 \qquad\qquad + I_2 - I_3 = 25 \\
& x_j \geq 0, \quad j = 1, 2, 3; \quad I_t \geq 0, \quad t = 1, 2, 3
\end{align}$$

Note that production scheduled in period 1 can be "brought forward" to satisfy demand in all future periods.

There are many ways to enhance this model further; some of these enhancements are:

> Include production capacity constraints and inventory storage constraints.
> Expand the model to include a transportation component to account for distribution to markets.
> Expand the model to include multiple products.
> Expand the model to account for the costs accompanying changing production rates from one period to the next.

These enhancements suggest that the basic production/inventory model can be tailored to suit the individual needs and constraints of most firms. In fact, there is little doubt that this has been one of the leading areas of application of linear programming by industrial engineers.

Klingman et al. [54] describe an application of this type of model to W. R. Grace [Chemical]. This model includes over 40,000 variables that characterize the production-distribution processes of its phosphate-based products. There are 12 monthly time periods in the model, and its focus is on annual planning and control of over $500 million of working capital. Another application in the [Chemical] sector is reported by Owens-Corning [74]. This multiperiod, multiproduct model included over 10,000 variables and resulted in scheduling efficiencies saving over $50,000 annually. Tatung Wire and Cable (Industrial Machinery) [28] describe the benefits of the use of this type of linear programming model for reducing "layoffs, set-ups and costs."

Two earlier applications are also noteworthy. The Chessie System [Transportation] [8] in 1980 described an application of this model that increased annual profits by $2 million. The model included manpower requirements as well as time-based production decisions and totaled over 3500 variables. In [52], the experience of Kelly-Springfield Tire and Rubber [Industrial Products] with operations research in general, and linear programming in particular, is reviewed. Of particular interest is the production planning system, which addresses "the determination of the number of molds needed to cure each product at each factory during each day of the planning period." The Kelly-Springfield linear programming model included about 1000 variables and 300 constraints and its use was estimated to generate annual savings of $4.2 million.

3.6. Product Mix and Process Selection Models

Consider the rather common situation in which a firm must select a mix of products to produce and, for each product, select a process for manufacture. This may be the most frequent application of linear programing by industrial engineers. For clarity, we present an illustrative example in a single-period context, but the same formulation could be integrated into the multiperiod context as described in the previous application. This model, in its simplest form, is characterized by the data shown in Figure 9.

Note that product 1 has two alternative manufacturing sequences (denoted by a and b), product 2 has only one alternative, and product 3 can be manufactured by two alternative processes (a and b). The problem is to determine how much of each product should be manufactured by each process. If the unit profits for products, 1, 2, and 3 are $4, $3, and $6, respectively, and $x_{1a}, x_{1b}, x_2, x_{3a}, x_{3b}$ represent the quantity of each product to produce by the alternative processes, the linear programming formulation is

$$\begin{aligned}
\text{Maximize} \quad & 4x_{1a} + 4x_{1b} + 3x_2 + 6x_{3a} + 6x_{3b} \\
\text{subject to} \quad & 2x_{1a} \phantom{+ 4x_{1b}} + 2x_2 + 3x_{3a} + 2x_{3b} \le 100 \quad (\text{machine A}) \\
& \phantom{2x_{1a} +} 4x_{1b} + 7x_2 + 2x_{3a} + x_{3b} \le 200 \quad (\text{machine B}) \\
& 6x_{1a} + 5x_{1b} + x_2 + 5x_{3a} + 9x_{3b} \le 250 \quad (\text{machine C}) \\
& x_{1a}, x_{1b}, x_2, x_{3a}, x_{3b} \ge 0
\end{aligned}$$

Note that each variable represents the simultaneous choice of a product and manufacturing sequence. If there are several hundred products and several alternative process selections for each product, the number of variables could reach the thousands. Again, note that this model could also be enhanced by incorporating the downstream distribution of the finished products to markets.

	PRODUCT					
	1a	1b	2	3a	3b	Capacity
Machine A	2 hrs	0	2	3	2	100 hours
Machine B	0	4	7	2	1	200 hours
Machine B	6	5	1	5	9	250 hours

Figure 9 Process selection alternatives.

Linear Programming

Applications of linear programming abound in this area, even in the nonmanufacturing sector. In the [Transportation] sector, North American Van Lines [1] reported using linear programming to plan fleet configuration and reduce its inventory of tractors by $3 million. Applications in the [Finance] sector are many also. The Central Carolina Bank and Trust Company [2] has developed a product mix model for optimizing the choice of assets and liabilities on its balance sheet.

There are also many applications of this model in the [Government] sector. Might [71] outlines how the U.S. Air Force uses linear programming to determine "procurement budget for different aircraft and munitions" for models that include over 250,000 variables. Another large-scale application is maintained by the U.S. Department of Agriculture (USDA) Forest Service [Government]. In [51], a model denoted Forest Planning (FORPLAN) is described that optimizes the management of timber and uses a product mix type of orientation. Similar efforts are reported in [64] by the Forest Research Institute of New Zealand [Government]. Other applications exist in the [Health] sector. The American Hospital Supply Corporation reported an application of a type of product mix problem in the acquisition of biological heart valves that resulted in annual savings exceeding $1.5 million [41].

3.7. Blending and Mixing Models

Consider the case in which there are n raw materials that must be combined in certain quantities so that the resultant mix contains constituent ingredients within certain ranges. In particular, suppose that a type of iron to be blended requires the following minimum and maximum alloying contents:

	Carbon	Silicon	Manganese
min	3%	4%	0%
max	4%	5%	1%

Furthermore, there are four raw materials available to be blended. The compositions of the raw materials is shown in Table 2.

A batch of 1000 pounds of this specially blended iron is required. The question is how many pounds of each raw material should be included in the charge to minimize the total cost. Denote by x_i the number of pounds to be used of each of the four materials, i.e., pig iron 1, pig iron 2, alloy 1, and alloy 2. Note that the total quantity of carbon (in pounds) in the blend would be $0.02x_1 + 0.05x_2$. Hence, the ratio

$$\frac{0.02x_1 + 0.05x_2}{x_1 + x_2 + x_3 + x_4}$$

Table 2 Alloying Parameters and Costs

	PRODUCT			
Raw Material	Carbon	Silicon	Manganese	Cost/lb.
Pig Iron #1	2%	1%	0	$.05
Pig Iron #2	5%	0	1%	$.04
Alloy 1	0	60%	40%	$.25
Alloy 2	0	20%	80%	$.35

represents the fraction of carbon in the overall charge. The constraint that at least 3% of the iron is carbon would mean

$$\frac{0.02x_1 + 0.05x_2}{x_1 + x_2 + x_3 + x_4} \geq 0.03$$

and such a restriction is converted to a linear inequality as follows:

$$-0.01x_1 + 0.02x_2 - 0.03x_3 - 0.03x_4 \geq 0$$

Proceeding in this manner, a linear programming formulation to minimize the cost of a 1000-pound blend, subject to the restrictions on the constituents, is

$$\begin{array}{ll}
\text{Minimize} & 0.05x_1 + 0.04x_2 + 0.25x_3 + 0.35x_4 \\
& -0.01x_1 + 0.02x_2 - 0.03x_3 - 0.03x_4 \geq 0 \quad \text{carbon} \\
& -0.02x_1 + 0.01x_2 - 0.04x_3 - 0.04x_4 \leq 0 \\
& -0.03x_1 + 0.56x_3 + 0.16x_4 \geq 0 \quad \text{silicon} \\
& -0.04x_1 + 0.55x_3 + 0.15x_4 \leq 0 \\
& -0.01x_1 + 0.39x_3 + 0.79x_4 \geq 0 \quad \text{manganese} \\
& x_1 + x_2 + x_3 + x_4 = 1000 \quad \text{demand} \\
& x_j \geq 0, \quad j = 1, \ldots, 4
\end{array}$$

One of the early industrial engineering applications of blending occurred at the Central Foundry of General Motors [Transportation Equipment] [70]. In this early application, the model included only 30 variables and 22 constraints. As one would expect, the [Chemical] sector has been a prominent user of blending- and mixing-type models. Klingman et al. [55] describe how linear programming has helped management control costs at Citgo. In this application a linear program was used to model a refinery and, according to Klingman, "is now used routinely to provide critical

Linear Programming

decision information in . . . crude selection, refinery run levels, product component production levels, feedstock selection, ..." Citgo has estimated that this and related applications have saved $70 million annually. Another type of blending problem concerns the determination of a diet, i.e., how to meet minimal dietary levels (protein, carbohydrates, vitamins, etc.) at least cost. A recent application of this model has been formulated for the United Rescue Mission of Knoxville, Tennessee [Health] [44]. Here, a PC based-solution can be obtained in less than 1 minute, a necessary characteristic for implementation in such a small business environment.

3.8. Cutting Stock Models

The cutting stock problem is one that is typical of industries which must transform standard sizes of raw materials, usually obtained from vendors, into sizes appropriate for their own products. The acquisition and shearing of rolls of steel sheet or paper is an example. The following illustrative example is adapted from [89]. Coils of paper can be acquired in widths of 10 and 20 feet. In a particular time period, the demand for paper rolls is given by

Width	Length
5 ft	10,000 ft
7 ft	30,000 ft
9 ft	20,000 ft

The 10- and 20-foot rolls must be slit into 5-, 7-, and 9-foot rolls to meet these demands; e.g., one needs 10,000 feet of the 5-foot width. The essence of the linear programming formulation for this problem is the identification of alternative patterns (which will be the variables) for slitting the 10- and 20-foot rolls. For example, the 10-foot rolls could be slit in the following three alternative ways:

1. | ← 5 → | ← 5 → |
2. | ← 7 → | ←3→ |
3. | ← 9 → | 1 |

In the first method, there is no waste, but the second and third patterns generate a "residual roll" of waste which is 3 feet and 1 foot, respectively. The alternative patterns and resulting waste are summarized in Table 3.

For this application, define two sets of variables x_{11}, x_{12}, x_{13} and x_{21}, x_{22}, x_{23}, x_{24}, x_{25}, x_{26}. Variable x_{11} represents the number of linear feet of the

Table 3 Alternative Cutting Patterns

Width	10 Foot Patterns			20 Foot Patterns					
	#1	#2	#3	#1	#2	#3	#4	#5	#6
5 ft rolls	2	0	0	4	2	2	1	0	0
7 ft rolls	0	1	0	0	1	0	2	1	0
9 ft rolls	0	0	1	0	0	1	0	1	2
Waste (ft)	0	3	1	0	3	1	1	4	2

10-foot roll, which is to be slit according to pattern 1. The other eight variables are defined in the same way. Finally, let c_1 and c_2 represent the cost per foot of the 10- and 20-foot rolls, respectively. A linear programing formulation of this model is

$$\begin{aligned}
\text{Minimize} \quad & c_1 x_{11} + c_1 x_{12} + c_1 x_{13} + c_2 x_{21} + c_2 x_{22} + c_2 x_{23} + c_2 x_{24} + c_2 x_{25} + c_2 x_{26} \\
\text{subject to:} \quad & 2x_{11} \phantom{+ x_{12} + x_{13}} + 4x_{21} + 2x_{22} + 2x_{23} + x_{24} \phantom{+ x_{25} + 2x_{26}} \geq 10{,}000 \\
& x_{12} \phantom{+ x_{13} + 4x_{21}} + x_{22} \phantom{+ 2x_{23}} + 2x_{24} + x_{25} \phantom{+ 2x_{26}} \geq 30{,}000 \\
& \phantom{2x_{12} +}x_{13} \phantom{+ 4x_{21} + 2x_{22}} + x_{23} \phantom{+ 2x_{24}} + x_{25} + 2x_{26} \geq 20{,}000 \\
& x_{ij} \geq 0, \quad i = 1, 2, \quad j = 1, 2, \ldots, 7
\end{aligned}$$

The three constraints represent the demand (feet) for the 5-, 7-, and 9-foot widths. Note that the objective coefficients for x_{11}, x_{12}, and x_{13} are the same, simply the cost per foot of a 10-foot roll. In this formulation the objective is to minimize the overall cost of meeting the demand.

The cutting stock problem has received much attention from the beginnings of linear programming. The classic reference on the use of linear programming for this application is the 1961 paper by Gilmore and Gomory [31]. Many applications have been implemented in the [Textile] and [Food] sectors. Blue Bell [Textiles] [19] described how one of its subsidiaries (Wrangler jeans) used this model to help reduce required inventory levels by $115 million and fabric waste by $1 million. An application in the [Food] sector has been developed for AFFCO Exports Ltd. of New Zealand [99]. In this application the choice of cutting patterns are optimized for the carving of animal carcasses. The authors say, "For the lamb carcass, the optimum solution selects 50 or so of 318 patterns included in the linear programming model." There are numerous other applications in [Textile] and [Food] sectors. A survey of applications in the clothing industry is provided in [23]. Also, see [36] for a taxonomy of problem types and solution procedures for this model.

3.9. Multifacility Location Models

Another application of linear programming in industrial engineering involves the location of several new facilities in relation to existing facilities. As an example, suppose that three existing facilities are located at the coordinates $P_1 = (10, 30)$, $P_2 = (20\ 10)$, and $P_3 = (40, 20)$. Two new facilities are to be located with respect to the existing facilities. The locations of new facilities are represented by the coordinates $X_1 = (x_1, y_1)$ and $X_2 = (x_2, y_2)$. Figure 10 graphically illustrates this problem, including possible locations for the new facilities.

The objective of the problem is to minimize the annual cost of travel between the two new facilities and between each new facility and each existing facility. The cost of travel between any two facilities is assumed to be proportional to the rectilinear distance between the corresponding locations. The rectilinear distances between the two new facilities and new facility $X_i = (x_i, y_i)$ and existing facility $P_j = (a_j, b_j)$ are, respectively,

$$d(X_1, X_2) = |x_1 - x_2| + |y_1 - y_2|$$
$$d(X_i, P_j) = |x_i - a_j| + |y_i - b_j|$$

A constant of proportionality between a pair of facilities represents the annual volume of material or the number of trips between the two facilities. In this example, assume that the constant of proportionality between the two new facilities is $v_{12} = 20$, and the constants of proportionality between

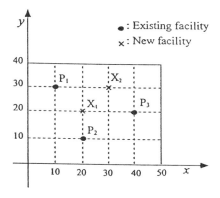

Figure 10 Possible solution to the layout problem.

new and exiting facilities are provided by matrix $\mathbf{W} = [w_{ij}]$, where w_{ij} is the constant of proportionality between new facility i and existing facility j. Let the 2×3 matrix $\mathbf{W}$ be as follows:

$$\mathbf{W} = \begin{pmatrix} 20 & 10 & 0 \\ 40 & 0 & 50 \end{pmatrix}$$

The objective for this location problem involves the determination of the values of (x_1, y_1) and (x_2, y_2) that minimize the weighted sum of rectilinear travel distances; hence, the problem can be formulated as

$$\begin{aligned} \text{Minimize } f(x_1, x_2, y_1, y_2) = \ & 20(|x_1 - x_2| + |y_1 - y_2|) \\ & + 20(|x_1 - 10| + |y_1 - 30|) \\ & + 10(|x_1 - 20| + |y_1 - 10|) \\ & + 40(|x_2 - 10| + |y_2 - 30|) \\ & + 50(|x_2 - 40| + |y_2 - 20|) \end{aligned}$$

A key to solving location problems with rectilinear distances is that they can be decomposed into two subproblems. In this example, note that the objective function can be decomposed into two functions:

$$f(x_1, x_2, y_1, y_2) = f_x(x_1, x_2) + f_y(y_1, y_2)$$

where

$$\begin{aligned} f_x(x_1, x_2) = \ & 20|x_1 - x_2| + 20|x_1 - 10| \\ & + 10|x_1 - 20| + 40|x_2 - 10| + 50|x_2 - 40| \\ f_y(y_1, y_2) = \ & 20|y_1 - y_2| + 20|y_1 - 30| + 10|y_1 - 10| \\ & + 40|y_2 - 30| + 50|y_2 - 20| \end{aligned}$$

In this manner, one can formulate the problem of minimizing the annual cost of travel in the x direction separately from the same problem in the y direction. Moreover, both problems can be recast as linear programs, as we shall see. However, because of the absolute values, the problem of minimizing the annual cost of travel in the x direction is the following nonlinear program:

$$\text{Minimize } f_x(x_1, x_2) \tag{3.3}$$

To convert this nonlinear program into a linear one, for each absolute value in the objective function, one needs to introduce two new nonnegative variables, replace the absolute value by a linear component, and define a new equality constraint. For the absolute value corresponding to the two new facilities, let p_{12} be the amount by which x_1 is to the right of x_2 and q_{12} the amount by which x_1 is to the left of x_2. If x_1 is to the right of x_2, then

Linear Programming

$$p_{12} = x_1 - x_2 = |x_1 - x_2| \quad \text{and} \quad q_{12} = 0 \tag{3.4}$$

Similarly, if x_1 is to the left of x_2, then

$$q_{12} = x_2 - x_1 = |x_1 - x_2| \quad \text{and} \quad p_{12} = 0 \tag{3.5}$$

Thus, (3.4) and (3.5) imply

$$|x_1 - x_2| = p_{12} + q_{12} \quad \text{and} \quad x_1 - x_{12} = p_{12} - q_{12} \tag{3.6}$$

Similarly, for the absolute value corresponding to new facility i and existing facility j, let r_{ij} and s_{ij} represent the amount by which x_i is to the right or the left, respectively, of a_j. Thus,

$$|x_1 - x_2| = p_{12} + q_{12} \quad \text{and} \quad x_1 - x_2 = p_{12} - q_{12}$$

Using these new linear relations and constraints, the problem of minimizing $f_x(x_1, x_2)$ becomes the following linear program:

Minimize $20(p_{12} + q_{12}) + 20(r_{11} + s_{11}) + 10(r_{12} + s_{12}) + 40(r_{21} + s_{21}) + 50(r_{23} + s_{23})$
subject to

$$\begin{aligned}
x_1 - x_2 - p_{12} + q_{12} &= 0 \\
x_1 \quad\quad\quad - r_{11} + s_{11} &= 10 \\
x_1 \quad\quad\quad\quad\quad - r_{12} + s_{12} &= 20 \\
x_2 \quad\quad\quad\quad\quad\quad\quad - r_{21} + s_{21} &= 10 \\
x_2 \quad\quad\quad\quad\quad\quad\quad\quad\quad - r_{23} + s_{23} &= 20
\end{aligned}$$

$$x_1, x_2, p_{12}, q_{12} \geq 0 \tag{3.7}$$
$$r_{ij}, s_{ij} \geq 0, \quad i = 1, 2, j = 1, 2, 3$$

Note that nonnegativity conditions on x_1 and x_2 can be imposed in this problem because all existing facilities have nonnegative x coordinates.

In this example, the optimal solution is

$$(x_1^*, y_1^*) = (20, 30)$$
$$(x_2^*, y_2^*) = (10, 30)$$
$$f(x_1^*, x_2^*, y_1^*, y_2^*) = 2600$$

where the values for (x_1^*, x_2^*) are obtained by solving problem (3.7), and the values for (y_1^*, y_2^*) are computed by solving the corresponding linear program in the y direction. This optimal solution illustrates a result that is generally true: there exists at least one optimal solution in which each new facility coordinate coincides with some existing facility coordinate.

A typical example of a multifacility location problem would be the location of new distribution centers in a manufacturing system with products flowing between each of the new distribution centers and any number of existing manufacturing plants, distribution centers, and demand points, as well as among the new distribution centers. Other examples would be the location of new retail stores with respect to existing distribution centers and market points, or the location of new departments in a manufacturing plant with respect to existing departments.

Facility location textbooks, such as those by Francis et al. [26] and Love et al. [61], provide a comprehensive introduction to mathematical models and quantitative methods for various types of location problems. One successful application of location theory was the study of the Emergency Medical Service (EMS) system in Austin, Texas [Government] [18] to determine what services should be delivered, by whom, via what types of equipment, and sited at which locations. The resulting plan called for a system quite different from that in operation: four advanced life-support and eight basic life-support vehicles were to operate from 2 EMS-only stations and 10 shared-use fire stations. The proposed plan saved $3.4 million in construction costs and $1.2 million per year in operating cost. The average response time was reduced despite an upsurge in calls for service. Other applications of location models include the relocation of the Red Cross Blood Donor Clinic and Transfusion Center in Quebec, Canada [Government] [77] and the location of a high-energy physics laboratory in Texas [Government] [94].

4. GRAPHICAL METHOD AND GEOMETRIC INSIGHTS

Linear programs with only two variables can be solved graphically. Although the graphical approach cannot be used to solve practical problems, it provides valuable insights into properties of the optimal solutions of linear programs for the general case. The graphical solution method and important special cases of linear programs are presented in this section.

4.1. Graphical Method

The graphical procedure is illustrated through an example of a product mix problem similar to the one presented in Section 3.6. A manufacturer has at its disposal fixed amounts of three different resources: labor, machine 1, and machine 2. The weekly availability of these resources is 180, 150, and 160 hours, respectively. The manufacturer uses the resources to produce two different products, A and B. The relevant production data are provided in Table 4. The objective of this problem is to maximize the total weekly profit subject to the limited resources available.

The unknown activities to be determined are the number of units to be produced for each product type. Let

Linear Programming

Table 4 Production Requirements and Profit Data

Product	A	B
Labor (hrs/unit)	4	3
Machine 1 (hrs/unit)	2	3
Machine 2 (hrs/unit)	4	2
Profit ($/unit)	9	12

x_1 = units of product A to be produced
x_2 = units of product B to be produced

The optimal product mix is the solution of the following linear program:

$$\begin{aligned}
\text{Maximize} \quad & z = 9x_1 + 12x_2 \\
\text{subject to} \quad & 4x_1 + 3x_2 \leq 180 \quad \text{(labor)} \\
& 2x_1 + 3x_2 \leq 150 \quad \text{(machine 1)} \\
& 4x_1 + 2x_2 \leq 160 \quad \text{(machine 2)} \\
& x_1, x_2 \geq 0
\end{aligned} \qquad (4.1)$$

This product mix example can be represented graphically in a two-dimensional coordinate system by defining the horizonal and vertical axes, respectively, as the units of products A and B to be produced. The nonnegativity constraints mean that only solutions in the first quadrant are feasible.

As shown in Section 3, a linear program may have equality and inequality constraints. Graphically, an equality constraint is plotted as a straight line and an inequality constraint is represented by a *halfspace*. A halfspace is bounded by the straight line obtained when the constraint is considered as an equality. In this example, the three resource constraints are represented by three halfspaces. The intersection of the first quadrant and the three constraints define the *feasible region*, which is also defined as the set of all feasible solutions. The feasible region is a *polyhedral set*, i.e., the intersection of a finite number of halfspaces. Formal definitions of halfspaces and polyhedral sets are provided later in Section 5.

A simple way to represent an inequality constraint graphically is to find the intercept on each axis of the line corresponding to the linear equation obtained by replacing the inequality sign by an equality sign. For example, the corresponding linear equation for the first constraint is

$$4x_1 + 3x_2 = 180$$

The intercepts are obtained by setting $x_1 = 0$, then $x_2 = 180/3 = 60$, and when $x_2 = 0$, $x_1 = 180/4 = 45$. Thus, the intercepts are (0, 60), and (45, 0). For this constraint, the *feasible halfspace* is the one including the origin, since (0, 0) satisfies the constraint. By a similar procedure, one can plot the remaining two constraints. The resulting feasible region is shown in Figure 11.

The *optimal solution* of the product mix problem is the feasible solution with the largest objective function value, z. For a particular value of z, the objective function is represented by the line

$$z = 9x_1 + 12x_2$$

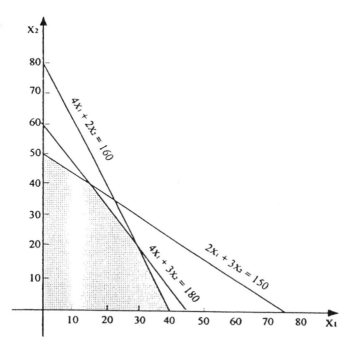

Figure 11 Feasible region for problem (4.1).

Linear Programming

All the points on this line have the same objective function value. In Figure 12, the lines corresponding to $z = 300$ and $z = 450$ are graphed. The two lines are parallel and, in this case, the line farthest from the origin has the largest objective function value. Note that both lines include segments that are feasible. This suggests that it is possible to generate other lines with larger objective function values by moving in a perpendicular direction farther from the origin. Proceeding in this manner, note that the optimal solution occurs at the point $(x_1^*, x_2^*) = (15, 40)$, a point in the feasible region through which the line $z^* = 615$ passes. Here, the optimal solution is the intersection point of the two linear equations obtained from the first two resource constraints. This intersection point is called *optimal extreme point* and defined as the solution to the following system of equations:

$$4x_1 + 3x_2 = 180$$
$$2x_1 + 3x_2 = 150$$

This example illustrates a fundamental property. If a linear program has an optimal solution, there always exists an *extreme point* that is optimal. In the plane, an extreme point is a feasible point defined by the intersection of at least two linear equations corresponding to equality or inequality constraints. Therefore, an alternative way to find an optimal

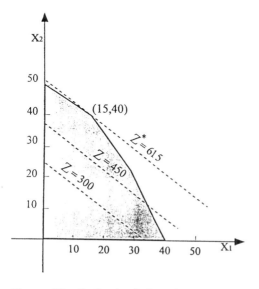

Figure 12 Optimal solution of problem (4.1).

solution is to determine the objective function value of all the extreme points. The feasible region of the linear program (4.1) contains the following set of extreme points. (A formal definition of an extreme point is provided in Section 5.)

$$E = \{(0,0), (40,0), (20,30), (15,40), (0,50)\}$$

Note that if one knew only the points in E, it would be possible to generate the entire feasible region. This observation is the key to the general solution algorithm presented in Section 6.

4.2. Special Cases

A linear program may have more than one optimal solution. But the number of optimal solutions is either one (a unique solution) or infinite. To illustrate the latter case for the product mix example, if the objective function is changed to

$$z = 8x_1 + 12x_2$$

then the number of optimal solutions becomes infinite. The set of optimal solutions would be defined by the line segment bounded by the extreme points (0, 50) and (15, 40). In this case, the linear program has *multiple or alternate optima* because the lines defined by the objective function for a given value of z and the second constraint are parallel. The new optimal objective function value would be $z^* = 600$. This case is shown in Figure 13.

The feasible region of this product mix problem is *bounded*. In other words, the distance from the origin to any feasible point is bounded and cannot become arbitrarily large. This is not always the case. If the inequality signs in this product mix example are changed from less than or equal to greater than or equal to, that is, if the problem is changed to

$$\begin{aligned}
\text{Maximize} \quad & z = 9x_1 + 12x_2 \\
\text{subject to} \quad & 4x_1 + 3x_2 \geq 180 \\
& 2x_1 + 3x_2 \geq 150 \\
& 4x_1 + 2x_2 \geq 160 \\
& x_1, x_2 \geq 0
\end{aligned} \quad (4.2)$$

then the feasible region becomes *unbounded*. If the feasible region of a linear program is unbounded, it contains at least one *extreme direction*. The extreme directions are parallel to the lines corresponding to the constraints defining the unboundedness of the feasible region. The sets of extreme points and extreme directions completely define an unbounded feasible region. For problem (4.2), these sets are (see Figure 14):

Linear Programming

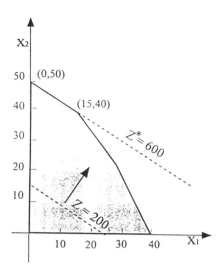

Figure 13 Multiple optimal solutions.

$$E = \{(75, 0), (22.5, 35), (0, 80)\} \quad \text{(extreme points)}$$
$$D = \{(1, 0), (0, 1)\} \quad \text{(extreme directions)}$$

If a problem has an unbounded feasible region, the optimal objective function value may be bounded or unbounded. A solution is unbounded if, for any value of z, it is always possible to find a feasible solution with a better objective function. The modified product mix problem (4.2) has an *unbounded solution*. If the optimal objective function value is unbounded, there must exist at least one extreme direction, called an *ascent extreme direction*, whose dot product with the cost vector results in a positive number, e.g.,

$$(9, 12)\begin{pmatrix} 1 \\ 0 \end{pmatrix} = 9 > 0$$

and

$$(9, 12)\begin{pmatrix} 0 \\ 1 \end{pmatrix} = 12 > 0$$

In this example, both extreme directions are ascent extreme directions.

Finally, note that even though the feasible region is unbounded, this does not necessarily mean that the optimal objective function is un-

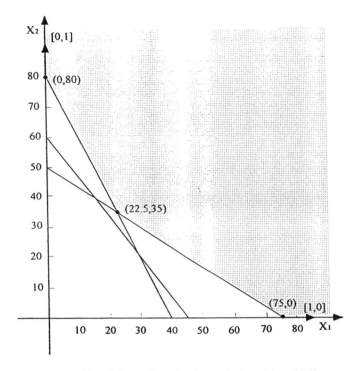

Figure 14 Feasible region of unbounded problem (4.2).

bounded. If the objective of (4.2) is changed from maximization to minimization, i.e.,

$$\text{Minimize} \quad z = 9x_1 + 12x_2$$
$$\text{subject to} \quad 4x_1 + 3x_2 \geq 180$$
$$2x_1 + 3x_2 \geq 150$$
$$4x_1 + 2x_2 \geq 160$$
$$x_1, x_2 \geq 0$$

then the optimal solution becomes (22.5, 35); i.e., the optimal objective function value is bounded.

If the feasible region of a linear program is empty, then the linear program is said to be *infeasible*. Note that if we add the constraint

$$x_1 \geq 45$$

Linear Programming   49

to the linear program (4.1), the resulting feasible region is empty and the problem becomes infeasible. This situation is illustrated in Figure 15, where the two disjoint sets represent the set of points satisfying the original constraints and the halfspace corresponding to the new constraint. Typically, real-world linear problems are neither unbounded nor infeasible. If either of these cases occurs, it is necessary to review the formulation of the problem.

From the geometric insights gained in the presentation of the graphical approach, one should observe some general properties of linear programs. If the linear program is neither infeasible nor unbounded, an optimal solution always occurs at an extreme point of the feasible region. Moreover, if the linear program is unbounded, there must exist an extreme direction that is ascent. These conclusions will be exploited in the simplex method (presented in Section 6), which is a method for solving linear programs that iterates from an extreme point to an improved extreme point until an optimal solution is found.

5. FOUNDATIONS OF LINEAR PROGRAMMING

In this section we introduce the standard form of a linear program and provide the necessary geometric and algebraic foundations to understand the linear programming algorithms presented later in this chapter. Impor-

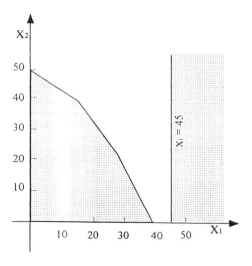

Figure 15 Empty feasible region of modified problem (4.1).

tant notation and terminology that will be used throughout the chapter is also introduced.

5.1. Standard Form

The simplex method and interior point methods for solving linear programming problems require the linear program to be in standard form. Given that $x_1, x_2, \ldots, x_n$ are the decision variables, the standard form of a linear program is

$$
\begin{aligned}
\text{Maximize} \quad & z = c_1 x_1 + c_2 x_2 + \cdots + c_n x_n \\
\text{subject to} \quad & a_{11} x_1 + a_{12} x_2 + \cdots + a_{1n} x_n = b_1 \\
& a_{21} x_1 + a_{22} x_2 + \cdots + a_{2n} x_n = b_2 \\
& \quad\quad\quad\quad\quad\quad \vdots \\
& a_{m1} x_1 + a_{m2} x_2 + \cdots + a_{mn} x_n = b_m \\
& x_1, x_2, \ldots, x_n \geq 0
\end{aligned}
$$

where $c_1, c_2, \ldots, c_n$ are the objective function (profit) coefficients, a_{ij}, $i = 1, 2, \ldots, m$ and $j = 1, 2, \ldots, n$, are the constraint (technological) coefficients, and $b_1, b_2, \ldots, b_m$ are the right-hand side (RHS) coefficients. In addition to the m equality constraints, the decision variables are restricted to be nonnegative. Without loss of generality, the RHS coefficients are also assumed to be nonnegative. If an RHS coefficient is negative, it is possible to multiply both sides of the corresponding equation by -1 and the RHS coefficient becomes positive. Let

$$
\mathbf{x} = \begin{pmatrix} x_1 \\ \vdots \\ x_n \end{pmatrix}, \quad \mathbf{c} = \begin{pmatrix} c_1 \\ \vdots \\ c_n \end{pmatrix}, \quad \text{and} \quad \mathbf{b} = \begin{pmatrix} b_1 \\ \vdots \\ b_m \end{pmatrix}
$$

be column vectors of sizes n, n, and m, respectively, and

$$
\mathbf{A} = \begin{pmatrix} a_{11} & \cdots & a_{1n} \\ \vdots & \ddots & \vdots \\ a_{m1} & \cdots & a_{mn} \end{pmatrix}
$$

be the $m \times n$ constraint matrix. The superscript T represents the transpose of a vector or matrix. Also, let $\mathbf{0}$ be the zero column vector in R^n. Then, the above linear program can be written in matrix-vector form as follows:

$$
\begin{aligned}
\text{Minimize} \quad & z = \mathbf{c}^T \mathbf{x} \\
\text{subject to} \quad & \mathbf{A}\mathbf{x} = \mathbf{b} \\
& \mathbf{x} \geq \mathbf{0}
\end{aligned} \tag{5.1}
$$

Linear Programming

It is always possible to convert a linear program to standard form. If a constraint is of the type $\leq$, it is possible to add a nonnegative slack variable to the left-hand side to obtain an equivalent equality constraint. If the constraint is $\geq$, an equality is obtained by subtracting a nonnegative surplus variable from the left-hand side. In general, the variables can be nonpositive or unrestricted. An unrestricted variable x_j can be replaced by the difference of two nonnegative variables, and a nonpositive variable can be replaced by a nonnegative variable. Consider the following example:

$$\text{Maximize} \quad z = 9x_1 + 2x_2 + 5x_3$$
$$\text{subject to} \quad 4x_1 + 3x_2 + 6x_3 \leq 50$$
$$x_1 + 2x_2 - 3x_3 \geq 8 \qquad (5.2)$$
$$2x_1 - 4x_2 + x_3 = 5$$
$$x_1 \geq 0,\ x_2 \leq 0,\ x_3 \text{ unrestricted}$$

To convert the first and second constraints to standard form, one adds (nonnegative) slack variable x_4 and (nonnegative) surplus variable x_5 to the corresponding constraints; that is,

$$4x_1 + 3x_2 + 6x_3 + x_4 \qquad = 50$$
$$x_1 + 2x_2 - 3x_3 \qquad - x_5 = 8$$

Next, variable x_2 is replaced by $x_2' = -x_2$, and variables x_3^+ and x_3^- are introduced and x_3 is replaced by $x_3 = x_3^+ - x_3^-$. All three new variables are nonnegative. The resulting linear program is

$$\text{Maximize} \quad z = 9x_1 - 2x_2' + 5x_3^+ - 5x_3^-$$
$$\text{subject to} \quad 4x_1 - 3x_2' + 6x_3^+ - 6x_3^- + x_4 \qquad = 50$$
$$x_1 - 2x_2' - 3x_3^+ + 3x_3^- \qquad - x_5 = 8$$
$$2x_1 + 4x_2' + x_3^+ + x_3^- \qquad = 5$$
$$x_1, x_2', x_3^+, x_3^-, x_4, x_5 \geq 0$$

5.2. Hyperplanes, Halfspaces, and Polyhedral Sets

In this subsection we present several definitions and geometric constructs that provide the foundations for the linear programming solution methods. Given an n-dimensional vector $\mathbf{a}$ and a scalar α, a *hyperplane* in R^n is the set defined by

$$H = \{\mathbf{x} \in R^n | \mathbf{a}^T\mathbf{x} = \alpha\}$$

A hyperplane is a line in R^2 and a plane in R^3. A hyperplane in R^n defines two *closed halfspaces* (see an example in Figure 16):

$$H_L = \{\mathbf{x} \in R^n | \mathbf{a}^T\mathbf{x} \leq \alpha\}$$

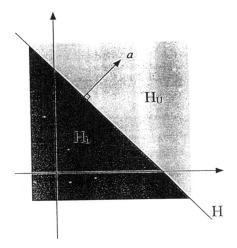

Figure 16 Example of a hyperplane and two closed halfspaces.

$$H_U = \{x \in R^n | a^T x \geq \alpha\}$$

These two halfspaces intersect at the hyperplane H. If one removes H from H_L and H_U, two *open halfspaces* are obtained:

$$H'_L = \{x \in R^n | a^T x < \alpha\}$$
$$H'_U = \{x \in R^n | a^T x > \alpha\}$$

A *polyhedral set* in R^n is a set defined by the intersection of a finite number of closed halfspaces. Thus, the set

$$S_1 = \{x \in R^n | Ax \leq b, x \geq 0\}$$

where A is an $m \times n$ matrix, $b \in R^m$, and 0 is the zero vector in R^n, is a polyhedral set defined by m halfspaces corresponding to $Ax \leq b$ and n halfspaces corresponding to $x \geq 0$. If the polyhedral set is nonempty and bounded, it is called a *polytope*.

In a linear program in standard form, the constraint set

$$S_2 = \{x \in R^n | Ax = b, x \geq 0\}$$

is also a polyhedral set defined by $2m + n$ halfspaces corresponding to $Ax \leq b$, $Ax \geq b$, and $x \geq 0$. Equivalently, S_2 is defined by m hyperplanes corresponding to $Ax = b$ and n halfspaces corresponding to $x \geq 0$.

The (*relative*) *interior* of a polyhedral set is obtained by maintaining the equality relations and replacing the inequality relations by strictly inequality relations. In other words, the interior of a polyhedral set is defined by the intersections of hyperplanes and open halfspaces. The interiors of S_1 and S_2 are defined by

$$I(S_1) = \{\mathbf{x} \in R^n : \mathbf{Ax} < \mathbf{b}, \mathbf{x} > \mathbf{0}\}$$
$$I(S_2) = \{\mathbf{x} \in R^n : \mathbf{Ax} = \mathbf{b}, \mathbf{x} > \mathbf{0}\}$$

5.3. Affine Sets, Convex Sets, and Simplices

Consider r vectors (points) $\mathbf{x}^1, \mathbf{x}^2, \ldots, \mathbf{x}^r$ in R^n, and r scalars $\lambda_1, \lambda_2, \ldots, \lambda_r$. The expression

$$\lambda_1 \mathbf{x}^1 + \lambda_2 \mathbf{x}^2 + \cdots + \lambda_r \mathbf{x}^r \tag{5.3}$$

is termed a *linear combination* of the r vectors. The set of vectors $\mathbf{x}^1, \mathbf{x}^2, \ldots, \mathbf{x}^r$ is called *linearly dependent* if there exist some scalars $\lambda_1, \lambda_2, \ldots, \lambda_r$, not all zero, such that

$$\lambda_1 \mathbf{x}^1 + \lambda_2 \mathbf{x}^2 + \cdots + \lambda_r \mathbf{x}^r = \mathbf{0} \tag{5.4}$$

where $\mathbf{0}$ is the n-vector of zeros. If (5.4) holds only when $\lambda_1 = \lambda_2 = \cdots = \lambda_r = 0$, the set of r vectors is *linear independent*.

The linear combination (5.3) is called *affine* when

$$\lambda_1 + \lambda_2 + \cdots + \lambda_r = 1$$

In addition, when

$$\lambda_1 + \lambda_2 + \cdots + \lambda_r = 1$$
$$\lambda_1, \lambda_2, \ldots, \lambda_r \geq 0$$

the linear combination becomes a *convex combination*. Geometrically, the set of all affine combinations of two distinct points $\mathbf{x}^1, \mathbf{x}^2 \in R^n$ is the unique straight line defined by these points. The set of all convex combinations of $\mathbf{x}^1$ and $\mathbf{x}^2$ is the line segment bounded by these two points.

A set $S \subset R^n$ is called *affine* if it contains every affine combination of any two points $\mathbf{x}^1, \mathbf{x}^2 \in S$. The set S is said to be *convex* if it contains every convex combination of any two points $\mathbf{x}^1, \mathbf{x}^2 \in S$. Figure 17 shows some examples of convex and nonconvex sets. The definitions imply that an affine set is also convex, but a convex set is not necessarily affine. For example, a hyperplane is both affine and convex, but a halfspace is only convex. Note that the intersection of affine sets is an affine set, and the intersection of convex sets is a convex set. Therefore, since polyhedral sets are defined by the intersection of halfspaces, and halfspaces are convex, polyhedral sets are also convex.

The following are two useful types of polyhedral sets. The *convex hull*

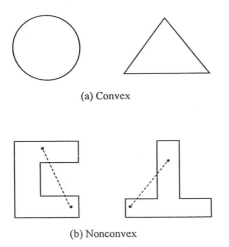

Figure 17 Examples of convex and nonconvex sets.

of a set of points $x^1, x^2, \ldots, x^r$ in R^n is the set containing all the convex combinations of these points. It can also be defined as the smallest convex set containing the points. For example, a polytope is a convex hull defined by its extreme points. A *simplex* of dimension r is the convex hull of a set of $r + 1$ *noncoplanar* points in R^r; that is, the $r + 1$ points cannot lie in the same hyperplane in R^r. For example, a line segment is a simplex for $r = 1$, and a triangle and a tetrahedron are simplices for $r = 2$ and $r = 3$, respectively.

5.4. Basic Solutions and Extreme Points

Consider the system of linear equations from the linear program (5.1):

$$\mathbf{Ax} = \mathbf{b}$$

We assume that $m \leq n$ and the *rank* of $\mathbf{A}$ is m, which means that the set of row vectors from $\mathbf{A}$ is linearly independent or, equivalently, that there exists at least one set of m linearly independent columns in $\mathbf{A}$. Thus, we can rearrange the columns of $\mathbf{A}$ and partition $\mathbf{A}$ as follows:

$$\mathbf{A} = [\mathbf{B} | \mathbf{N}]$$

where $\mathbf{B}$ is an $m \times m$ matrix with rank m; that is, $\mathbf{B}$ is *nonsingular*, and $\mathbf{N}$ is an $m \times (n - m)$ matrix. $\mathbf{B}$ is called the *basis* or *basic matrix*, because its rows or columns form a basis for R^m. The vector $\mathbf{x}$ can be divided in the same manner as $\mathbf{A}$; that is,

Linear Programming

$$x = \begin{pmatrix} x_B \\ x_N \end{pmatrix}$$

The vectors x_B and x_N are referred to as *basic and nonbasic vectors*, respectively. Then the systems of linear equations can be written as

$$Bx_B + Nx_N = b$$

and, since **B** is nonsingular, the basic vector can be written as a function of the nonbasic vector, i.e.,

$$x_B = B^{-1}b - B^{-1}Nx_N$$

In the special case $m = n$, the vector x_N is empty and x_B has a unique solution. Otherwise, for every set of values of x_N, there is a unique solution for x_B. If one sets $x_N = 0$, then $x_B = B^{-1}b$ and the vector

$$x = \begin{pmatrix} x_B \\ x_N \end{pmatrix} = \begin{pmatrix} B^{-1}b \\ 0 \end{pmatrix}$$

becomes a *basic solution*. Furthermore, if $B^{-1}b \geq 0$, then **x** is a *basic feasible solution* to the linear program (5.1). Since **A** contains n columns and the rank of **A** is m, there are at most $n!/[m!(n-m)!]$ ways of choosing the m columns of **B**. Thus, the maximum number of basic solutions is $n!/[m!(n-m)!]$.

Next, the concepts of basic feasible solutions and extreme points of a linear program are linked together. Let $S = \{x \in R^n | Ax = b, x \geq 0\}$ be the polyhedral set defined by the constraints in (5.1). A point **x** is an *extreme point* of S if it cannot be written as a convex combination of two distinct points of S. In other words, **x** cannot lie in the interior of the line segment defined by two distinct points of S. It can be shown (see [3]) that the extreme points occur only at the intersection of n linearly independent hyperplanes that form the boundary of S. Hence, an extreme point $x \in S$ lies on the m linearly independent hyperplanes defined by $Ax = b$, and $n - m$ hyperplanes defined by the corresponding $x_i = 0$. This means that **x** is also a basic feasible solution in which the nonbasic variables correspond to the variables defining the $n - m$ hyperplanes $x_i = 0$, and the remaining components of **x** are the basic variables. Conversely, it can be shown that every basic feasible solution corresponds to an extreme point. Finally, two distinct extreme points x^1 and x^2 of S are *adjacent* if the line segment joining them in an *edge* of the convex set (see Figure 18). Equivalently, two extreme points are adjacent if their corresponding basic feasible solutions contain $m - 1$ common basic variables.

The following example illustrates the relationship between extreme points and basic feasible solutions. Let S be defined by the following set of constraints:

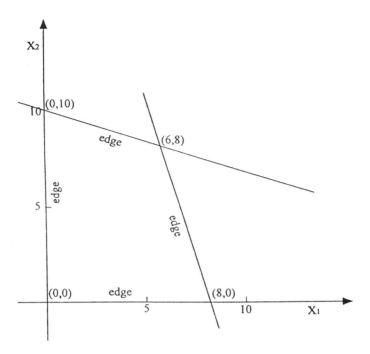

Figure 18 Feasible region and extreme points of the constraint set (5.4).

$$\begin{aligned} x_1 + 3x_2 + x_3 &= 30 \\ 4x_1 + x_2 + x_4 &= 32 \\ x_1, x_2, x_3, x_4 &\geq 0 \end{aligned} \quad (5.5)$$

In this polyhedral set, $n = 4$ and $m = 2$. Therefore, there are at most $4!/(2! \cdot 2!) = 6$ possible basic solutions. Table 5 shows all the possible bases, the corresponding basic solutions of (5.5), and their type. The four basic feasible solutions are also shown in Figure 18.

5.5. Nondegeneracy

The correspondence between basic feasible solutions and extreme points, in general, is not one-to-one. Each basic feasible solution corresponds to a unique extreme point, but each extreme point may correspond to more than one basic feasible solution. As an example, consider the set S defined by

Linear Programming

Table 5 Polyhedral Set Defined by the Constraints in (5.4)

Basis	Inverse of basis	Basic variables	Nonbasic variables	Type of basic solution
$\begin{pmatrix} 1 & 3 \\ 4 & 1 \end{pmatrix}$	$\begin{pmatrix} -1/11 & 3/11 \\ 4/11 & -1/11 \end{pmatrix}$	$x_1 = 6$ $x_2 = 8$	$x_3 = 0$ $x_4 = 0$	Feasible
$\begin{pmatrix} 1 & 1 \\ 4 & 0 \end{pmatrix}$	$\begin{pmatrix} 0 & 1/4 \\ 1 & -1/4 \end{pmatrix}$	$x_1 = 8$ $x_3 = 22$	$x_2 = 0$ $x_4 = 0$	Feasible
$\begin{pmatrix} 1 & 0 \\ 4 & 1 \end{pmatrix}$	$\begin{pmatrix} 1 & 0 \\ -4 & 1 \end{pmatrix}$	$x_1 = 30$ $x_4 = -88$	$x_2 = 0$ $x_3 = 0$	Infeasible
$\begin{pmatrix} 3 & 1 \\ 1 & 0 \end{pmatrix}$	$\begin{pmatrix} 0 & 1 \\ 1 & -3 \end{pmatrix}$	$x_2 = 32$ $x_3 = -66$	$x_1 = 0$ $x_4 = 0$	Infeasible
$\begin{pmatrix} 3 & 0 \\ 1 & 1 \end{pmatrix}$	$\begin{pmatrix} 1/3 & 0 \\ -1/3 & 1 \end{pmatrix}$	$x_2 = 10$ $x_4 = 22$	$x_1 = 0$ $x_3 = 0$	Feasible
$\begin{pmatrix} 1 & 0 \\ 0 & 1 \end{pmatrix}$	$\begin{pmatrix} 1 & 0 \\ 0 & 1 \end{pmatrix}$	$x_3 = 30$ $x_4 = 32$	$x_1 = 0$ $x_2 = 0$	Feasible

$$\begin{aligned}
x_1 + x_2 + x_3 &= 20 \\
x_1 + 2x_2 + x_4 &= 30 \\
2x_1 + x_2 + x_5 &= 30 \\
x_1, x_2, x_3, x_4, x_5 &\geq 0
\end{aligned} \quad (5.6)$$

Figure 19 displays the feasible region S. The four extreme points of S are

$$\begin{aligned}
\mathbf{x}^1 &= (15, 0, 5, 15, 0)^T \\
\mathbf{x}^2 &= (0, 0, 20, 30, 30)^T \\
\mathbf{x}^3 &= (0, 15, 5, 0, 15)^T \\
\mathbf{x}^4 &= (10, 10, 0, 0, 0)^T
\end{aligned}$$

The first three extreme points correspond to unique basic feasible solutions in which the basic variables coincide with the unique sets of three positive components. Extreme point $\mathbf{x}^4$ corresponds to three different basic feasible solutions. The reason is that this extreme point is defined by only two positive variables, x_1, and x_2, which are also basic. The third basic variable can be any of the other three variables when the variable is set equal to zero. In Figure 19 note that $\mathbf{x}^4$ is at the intersection of the three constraints, which implies that the corresponding slack variables are zero.

In a linear program in standard form with n variables and m constraints, a basic feasible solution is called *degenerate* if the number of positive basic variables is at most $m - 1$. If the number of positive basic

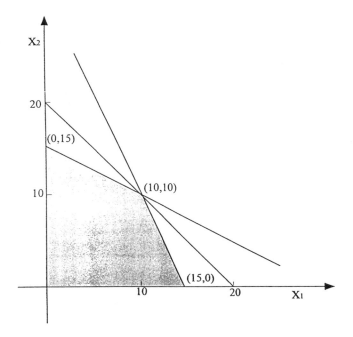

Figure 19 Feasible region and extreme points of the constraint set (5.6).

variables is exactly m, then basic solution is *nondegenerate*. It will be seen later that degenerate basic feasible solutions can create problems in the convergence of the simplex method.

6. SIMPLEX METHOD

6.1. Overview

The simplex method is an iterative procedure for generating and examining different extreme points (basic feasible solutions) of a linear program, each one improving the current value of the objective function until an optimal solution is found.

The simplex method starts at an extreme point, often the origin, and systematically moves to adjacent extreme points. The selection of the next extreme point is based on the potential improvement of the objective function. Because the same extreme point cannot be repeated twice and the maximum number of extreme points (basic feasible solutions) is bounded by $n![m!(n - m)!]$, where n and m are the numbers of variables and con-

Linear Programming

straints of the linear program in standard form, respectively, the method terminates with an optimal extreme point in a finite number of iterations.

First, an algebraic description of the simplex method is presented below. Then, the steps of the algorithm and the *simplex tableau* are introduced. The tableau is a useful tool for executing the simplex method manually. Next, the case of degenerate basic solutions is discussed and two procedures for identifying an *initial basic feasible solution* are provided. Finally, the special cases of linear programs introduced in Section 4.2 are discussed in the context of the simplex tableau.

6.2. Algebra of the Simplex Method

Consider a linear program in standard form:

Maximize $z = c^T x$

subject to $Ax = b$

$x \geq 0$

Let x be a basic feasible solution. Then x can be denoted by

$$x = \begin{pmatrix} x_B \\ x_N \end{pmatrix}$$

where $x_B \geq 0$ is the *m-vector of basic variables*, and $x_N = 0$ is the $(n - m)$-*vector of nonbasic variables*. The partition of x defines a similar partition for A and c; that is,

$$A = [B | N]$$

and

$$c = \begin{pmatrix} c_B \\ c_N \end{pmatrix}$$

Then,

$$x_B = B^{-1}b = \hat{b}$$

and

$$z = c_B^T x_B = c_B^T \hat{b}$$

The linear program can be equivalently stated as

Maximize $z = c_B^T x_B + c_N^T x_N$

subject to $Bx_B + Nx_N = b$

$x_B \geq 0, x_N \geq 0$

Then the basic vector in explicit form is

$$\mathbf{x}_B = \mathbf{B}^{-1}\mathbf{b} - \mathbf{B}^{-1}\mathbf{N}\mathbf{x}_N$$
$$= \hat{\mathbf{b}} - \mathbf{B}^{-1}\mathbf{N}\mathbf{x}_N$$

and the objective function can be written as a function of the nonbasic variables,

$$z = \mathbf{c}_B^T(\hat{\mathbf{b}} - \mathbf{B}^{-1}\mathbf{N}\mathbf{x}_N) + \mathbf{c}_N^T\mathbf{x}_N$$
$$= \mathbf{c}_B^T\hat{\mathbf{b}} - (\mathbf{c}_B^T\mathbf{B}^{-1}\mathbf{N} - \mathbf{c}_N^T)\mathbf{x}_N$$

In the above expression, the vector of coefficients multiplying $\mathbf{x}_N$ is called the *vector of nonbasic reduced costs*. The reduced cost of nonbasic variable x_j is defined as follows:

$$z_j - c_j = \mathbf{c}_B^T\mathbf{B}^{-1}\mathbf{a}_j - c_j$$

where $\mathbf{a}_j$ is the column of $\mathbf{A}$ corresponding to x_j, and $z_j = \mathbf{c}_B^T\mathbf{B}^{-1}\mathbf{a}_j$.

The reduced cost coefficient $z_j - c_j$ represents the decrease in the objective function value to generate one unit of x_j. Therefore, if the vector of nonbasic reduced costs is nonnegative (i.e., $z_j - c_j \geq 0$, for all nonbasic x_j), then the current basic feasible solution is optimal. If not, one would need to choose a nonbasic variable with negative reduced cost, say x_k, to enter the basis. Normally, the variable with the most negative reduced cost is selected, which corresponds to the steepest increase of the objective function value. As the nonbasic variable x_k increases, all the basic variables may change. The vector

$$\hat{\mathbf{a}}_k = \mathbf{B}^{-1}\mathbf{a}_k$$

indicates the decrement of the original basic vector as x_k increases by one unit. A new basic feasible solution is generated if x_k is increased until a basic variable becomes zero. This occurs when x_k is set to

$$\theta = \min\{\theta_i: \text{ for all } \hat{a}_{ik} > 0\}$$

where $\theta_i = x_{Bi}/\hat{a}_{ik}$, and $\hat{a}_{ik}$ is the ith element of $\hat{\mathbf{a}}_k$. Assume the minimum occurs for $i = r$, that is, $\theta = \theta_{Br}$. Thus, x_k is the nonbasic variable entering the basis, and x_r is the basic variable leaving the basis.

The new basic feasible solution and objective function value are

$$(x_{Bi})_{\text{new}} = \hat{b}_i - \hat{a}_{ik}\theta, \quad \text{for all } i \neq k$$
$$(x_{Br})_{\text{new}} = x_k = \theta$$
$$z_{\text{new}} = \mathbf{c}_B^T\hat{\mathbf{b}} - (\mathbf{c}_B^T\mathbf{B}^{-1}\mathbf{a}_k - c_k)\theta$$
$$= \mathbf{c}_B^T\hat{\mathbf{b}} - (z_k - c_k)\theta$$

Linear Programming

Since $(z_k - c_k)$ is negative and θ is nonnegative, the new objective function value could never decrease. Moreover, if θ is positive, the new objective function value would strictly increase. If θ is zero, the current basis is degenerate and the objective function value does not change.

The new basic feasible solution becomes the current solution, the new basis is obtained from **B** by replacing the column corresponding to x_{B_r} by $\mathbf{a}_k$, and the method continues with a new iteration.

6.2. Steps of the Simplex Method

Step 1. Formulate the linear program in standard form.

Step 2. Select an initial basic feasible solution and calculate the value of the objective function.

Step 3. Select a nonbasic variable to enter the basis that improves the objective function value. If the objective value cannot be improved, stop; the current basic feasible solution is optimal.

Step 4. Select a current basic variable to leave the basis in such a manner that the new basic solution is feasible, i.e., satisfies the nonnegativity constraints.

Step 5. Perform the change of basis by pivoting. The new basic solution becomes the current solution. Evaluate the objective function value, and go to step 3.

The pivoting process required in step 5 is explained in detail in the subsection below.

6.3. Simplex Tableau

The simplex tableau is used to represent the current basic feasible solution, determine the incoming nonbasic variable and leaving basic variable, and perform the pivoting step to generate a new basic solution. The structure of the simplex tableau is shown in Table 6. The top row includes the vector of

Table 6 Simplex Tableau

Basic	$\mathbf{x}_B$	$\mathbf{x}_N$	Solution
z	0	$\mathbf{c}_B^T \mathbf{B}^{-1} \mathbf{N} - \mathbf{c}_N$	$\mathbf{c}_B^T \mathbf{B}^{-1} \mathbf{b}$
$\mathbf{x}_B$	I	$\mathbf{B}^{-1} \mathbf{N}$	$\mathbf{B}^{-1} \mathbf{b}$

decision variables rearranged in basic and nonbasic components (in practice the vector is not rearranged). Under the basic variables, one has the zero vector and the identity matrix. The nonbasic reduced costs and the updated coefficients of the nonbasic matrix (nonbasic matrix premultiplied by the inverse of B) are beneath the nonbasic variables.

Now, through an illustrative example, we show how to set up and update the simplex tableau. Consider the product mix example of Section 4.1. Three slack variables are used to formulate the linear program in standard form:

$$\begin{aligned} \text{Maximize} \quad & z = 9x_1 + 12x_2 \\ \text{subject to} \quad & 4x_1 + 3x_2 + x_3 = 180 \\ & 2x_1 + 3x_2 + x_4 = 150 \\ & 4x_1 + 2x_2 + x_5 = 160 \\ & x_1, x_2, x_3, x_4, x_5 \geq 0 \end{aligned} \quad (6.1)$$

In this problem, $m = 3$ and $n = 5$, since the RHS vector is positive, the three slack variables are selected to form the initial basic feasible solution. Thus, the nonbasic variables, x_1 and x_2, are set to zero, and the values of the basic variables are

$$x_3 = 180$$
$$x_4 = 150$$
$$x_5 = 160$$

Notice that the columns corresponding to the three slack variables form the identity matrix. Thus, the initial basis is $\mathbf{B} = \mathbf{I}$. This choice for a basic feasible solution is reflected in the initial tableau (see Table 7). The

Table 7 Initial Tableau

Basic	x_1	x_2	x_3	x_4	x_5	Solution	θ_i
z	-9	-12	0	0	0	0	
x_3	4	3	1	0	0	180	60
x_4	2	③	0	1	0	150	50
x_5	4	2	0	0	1	160	80

Linear Programming

nonbasic reduced costs are the negative of the objective function coefficients, the objective function value is 0, and the coefficient matrix does not need to be updated.

The most negative reduced cost is $z_2 - c_2 = -12$. Thus, x_2 is selected to enter the basis, and column $\mathbf{a}_2$ is called the *pivot column*. Based on the coefficients of $\mathbf{a}_2$, the ratios θ_i are computed to determine the value that x_2 should assume to generate the new basic solution. The ratios θ_i are given in the RHS column next to the tableau. The minimum ratio is $\theta_2 = 50$, corresponding to the second basic variable, $x_{B_2} = x_4$, which is selected to leave the basis. The row corresponding to the second basic variable is called the *pivot row*, and the coefficient $a_{22} = 3$ is called the *pivot element*.

The change of basis is performed by pivoting. This involves a series of row operations to transform the reduced cost $z_2 - c_2$ to 0 and the pivot column to the second elementary vector, i.e., $\hat{\mathbf{a}}_2 = \mathbf{e}_2 = (0, 1, 0)^T$. The new row 2 is obtained by dividing row 2 by 3. The new row 0 (row of reduced costs) is the result of adding 12 times new row 2 to row 0. The new row 1 is obtained by subtracting three times new row 2 from row 1. Finally, the new row 3 is obtained by subtracting two times new row 2 from row 3. The results of these calculations are shown in the new tableau in Table 8. In this tableau, the basic variable x_4 has been replaced by x_2, and the objective function value has increased from 0 to 600.

Because there still is a negative reduced cost in the new tableau, another iteration is required. In this iteration, x_1 enters the basis and x_3 leaves. The resulting tableau is shown in Table 9. This tableau is optimal. Therefore, the optimal solution is

$$x_1^* = 15, \ x_2^* = 40, \ x_3^* = 20, \ z^* = 615$$

Table 8 Second Tableau

Basic	x_1	x_2	x_3	x_4	x_5	Solution	θ_i
z	-1	0	0	4	0	600	
x_3	②	0	1	-1	0	30	15
x_2	2/3	1	0	1/3	0	50	75
x_5	8/3	0	0	-2/3	1	60	45/2

Table 9 Third and Optimal Tableau

Basic	x_1	x_2	x_3	x_4	x_5	Solution
z	0	0	1/2	7/2	0	615
x_1	1	0	1/2	-1/2	0	15
x_2	0	1	-1/3	2/3	0	40
x_5	0	0	-4/3	2/3	1	20

6.4. Degeneracy

As indicated in Section 5, a basic solution of a linear program may have some basic variables equal to zero. Such a basic solution is said to be degenerate. In this situation, the simplex method may cycle among different degenerate basic solutions, which correspond to the same extreme point.

A simple way of avoiding cycling is to apply Bland's procedure [6] to select the variables to enter and leave the basis at each iteration of the simplex method. Bland's rules are:

1. Select as the entering variable that nonbasic variable with the smallest subscript among all the nonbasic variables with negative reduced cost.
2. Select the leaving variable as the one with the minimum θ_i. In case of a tie, select the variable with the minimum θ_i and the smallest subscript in the basis.

Although there are other methods available to avoid cycling, such as the perturbation method of Charnes [9] and the lexicographic ordering of Dantzig et al. [16], Bland's procedure is the simplest to use.

6.5. Finding an Initial Basic Feasible Solution

If the RHS vector of a linear program (with all less than or equal to constraints) is nonnegative, one can use the slack variables to form an initial basic feasible solution. If this is not the case, a starting basic feasible solution may not be obvious. In general, one does not use the original decision variables in the initial basis because it may be too difficult to find a combination that yields a feasible solution.

Linear Programming

Consider the linear program (5.2) in standard form (where the variables have been renumbered for simplicity),

$$\text{Maximize } z = 9x_1 - 2x_2 + 5x_3 - 5x_4$$
$$\text{subject to}$$
$$4x_1 - 3x_2 + 6x_3 - 6x_4 + x_5 \qquad\qquad = 50$$
$$x_1 - 2x_2 - 3x_3 + 3x_4 \qquad - x_6 = 8 \qquad (6.2)$$
$$2x_1 + 4x_2 + x_3 - x_4 \qquad\qquad = 5$$
$$x_1, x_2, x_3, x_4, x_5, x_6 \geq 0$$

In this example, slack variable x_5 could be put in the basis with initial value $x_5 = 50$. However, including surplus variable x_6 in the basis would yield an infeasible solution because its initial value would be $x_6 = -8$. In addition, there are neither slack nor surplus variables in the third constraint because it is an equality constraint.

An initial basic solution can always be constructed by using all available slack variables and, for each constraint without a slack variable, an *artificial variable* can be defined and included in the basis. The artificial variables do not have any meaningful interpretation and must be somehow subsequently eliminated from the basis by penalizing them in the objective function, so that there is no artificial variable with a positive value remaining in the basis in the final tableau. In this example, two nonnegative artificial variables, say x_7 and x_8, are added to the last two constraints so that the system becomes

$$4x_1 - 3x_2 + 6x_3 - 6x_4 + x_5 \qquad\qquad = 50$$
$$x_1 - 2x_2 - 3x_3 + 3x_4 \qquad - x_6 + x_7 \qquad = 8$$
$$2x_1 + 4x_2 + x_3 - x_4 \qquad\qquad + x_8 = 5$$

There are two common methods for eliminating the artificial variables from the basis. The main difference between the methods is the way in which the objective function is formulated. For a maximization problem, these methods are:

> *Two-phase method*: the objective of phase I is to eliminate the artificial variables from the basis. This can be achieved by maximizing the negative sum (i.e., minimizing the sum) of the artificial variables. Thus, one solves the problem with the objective function coefficients being 0 for the decision variables and -1 for the artificial variables. In the above example, the objective function of phase I would be
>
> $$\text{Maximize } z = -x_7 - x_8$$
>
> The objective of phase II reverts to solving the original problem. Therefore, one deletes the artificial variables from the problem,

uses the basic feasible solution generated in phase I, and proceeds with the simplex method using the original objective function.

Big-M method: the objective of this method is to force the artificial variables out of the basis by penalizing them in the objective function. One uses an objective function with the original cost coefficients for the decision variables and assigns a very large negative profit, $-M$, to the artificial variables (M is a very large positive number). For the example problem, the objective function of this method would be

Maximize $z = 9x_1 - 2x_2 + 5x_3 - 5x_4 - Mx_7 - Mx_8$

The first method is computationally more stable because one does not have to compute with numbers of different magnitudes, which would be the case when using the big-M method.

The two-phase method is illustrated by solving problem (6.2). The initial and successive tableaus for phase I are shown in Tables 10–12. The initial tableau, which is also optimal, for phase II is shown in Table 13. Note that the first two rows of Tables 10 and 13 include the negative of the vector of objective function coefficients and the vector of reduced costs. The first row can be transformed to the second one by performing row operations to obtain zero nonbasic reduced costs.

6.6. Special Cases of a Linear Program

As discussed in Section 4.2, a linear program can have a unique optimal solution, alternate optima, or can be infeasible or unbounded. The illustra-

Table 10 Phase I—Initial Tableau

Basic	x_1	x_2	x_3	x_4	x_5	x_6	x_7	x_8	Solution	θ_i
z	0	0	0	0	0	0	1	1	0	
	-3	-2	2	-2	0	1	0	0	-13	
x_5	4	-3	6	-6	1	0	0	0	50	25/2
x_7	1	-2	-3	3	0	-1	1	0	8	8
x_8	②	4	1	-1	0	0	0	1	5	5/2

Linear Programming

Table 11 Phase I—Second Tableau

Basic	x_1	x_2	x_3	x_4	x_5	x_6	x_7	x_8	Solution	θ_i
z	0	4	7/2	-7/2	0	1	0	3/2	-11/2	
x_5	0	-11	4	-4	1	0	0	-2	40	
x_7	0	-4	-7/2	(7/2)	0	-1	1	-1/2	11/2	11/7
x_1	1	2	1/2	-1/2	0	0	0	1/2	5/2	

tive example in subsection 6.3 has a unique optimal solution. The case of unique optimal solution is characterized in an optimal tableau by the following two conditions:

All the artificial variables are equal to 0.
All the nonbasic reduced costs are positive.

The other three cases are discussed below.

6.6.1. Alternate or Multiple Optima

Consider the example of alternate optima used in Section 4.2. The problem in standard form is

Table 12 Phase I—Third Tableau

Basic	x_1	x_2	x_3	x_4	x_5	x_6	x_7	x_8	Solution
z	0	0	0	0	0	0	1	1	0
x_5	0	-109/7	0	0	1	-8/7	8/7	-18/7	324/7
x_7	0	-8/7	-1	1	0	2/7	2/7	-1/7	11/7
x_1	1	10/7	0	0	0	-1/7	1/7	3/7	23/7

Table 13 Phase II—First and Optimal Tableau

Basic	x_1	x_2	x_3	x_4	x_5	x_6	Solution
z	-9	2	-5	5	0	0	0
	0	144/7	0	0	0	1/7	152/7
x_5	0	-109/7	0	0	1	-8/7	324/7
x_4	0	-8/7	-1	1	0	-2/7	11/7
x_1	1	10/7	0	0	0	-1/7	23/7

Maximize $z = 8x_1 + 12x_2$
subject to
$$4x_1 + 3x_2 + x_3 = 180$$
$$2x_1 + 3x_2 + x_4 = 150$$
$$4x_1 + 2x_2 + x_5 = 160$$
$$x_1, x_2, x_3, x_4, x_5 \geq 0$$

An optimal simplex tableau is shown in Table 14. If, in the optimal tableau of a linear program, there is at least one nonbasic reduced cost equal to 0, the problem has alternate optima. In this example, $z_3 - c_3 = 0$, which means that, if x_3 enters the basis, the objective function value will not

Table 14 Optimal Tableau with Alternative Optima—I

Basic	x_1	x_2	x_3	x_4	x_5	Solution
z	0	0	0	4	0	600
x_1	1	0	(1/2)	-1/2	0	15
x_2	0	1	-1/3	2/3	0	40
x_5	0	0	-4/3	2/3	1	20

Linear Programming

Table 15 Optimal Tableau with Alternative Optima—II

Basic	x_1	x_2	x_3	x_4	x_5	Solution
z	0	0	0	4	0	600
x_3	2	0	1	-1	0	30
x_2	2/3	1	0	1/3	0	50
x_5	8/3	0	0	-2/3	1	60

change. Table 15 shows the other optimal tableau obtained by entering x_3 and removing x_1 from the basis. These two optimal basic solutions (extreme points) given in Tables 14 and 15 define a line segment and all the points on this segment are optimal. The set of optimal solutions is characterized by all *convex combinations* of these two optimal extreme points.

6.6.2. Infeasible Solution

An infeasible solution occurs when the feasible region is empty. This case arises when the value of at least one artificial variable cannot be reduced to 0. In particular, if at the end of phase I of the two-phase method, an artificial variable is still in the basis with a positive value, then there is no feasible solution.

6.6.3. Unbounded Problem

A linear program is unbounded if, at a given iteration of the simplex method, there is a nonbasic variable with negative reduced cost and all the elements of its column are nonpositive. Consider the unbounded problem of Section 4.2 in standard form:

$$\text{Maximize} \quad z = 9x_1 + 12x_2$$
$$\text{subject to} \quad 4x_1 + 3x_2 - x_3 = 180$$
$$2x_1 + 3x_2 - x_4 = 150$$
$$4x_1 + 2x_2 - x_5 = 160$$
$$x_1, x_2, x_3, x_4, x_5 \geq 0$$

Three artificial variables, one for each constraint, need to be added to generate an initial basic solution. Tables 16 and 17 show respectively the

Table 16 Phase I—Initial Tableau for Unbounded Solution

Basic	x_1	x_2	x_3	x_4	x_5	x_6	x_7	x_8	Solution	θ_i
	0	0	0	0	0	1	1	1	0	
z	-10	-8	1	1	1	0	0	0	-490	
x_6	4	3	-1	0	0	1	0	0	180	45
x_7	2	3	0	-1	0	0	1	0	150	75
x_8	④	2	0	0	-1	0	0	1	160	40

initial and final tableaus of phase I. Then, the first and second tableaus of phase II are illustrated in Tables 18 and 19, respectively. Note that all the elements of the column corresponding to x_5 in Table 19, including the reduced cost, are negative. This implies that the solution value is unbounded.

7. DUALITY AND SENSITIVITY ANALYSIS

7.1. Overview

When one solves a linear program, the usual purpose is to determine the particular values of the variables that, subject to the constraints, optimize the objective function. This is what we mean by saying that we "solve" the

Table 17 Phase I—Second Tableau for Unbounded Solution

Basic	x_1	x_2	x_3	x_4	x_5	x_6	x_7	x_8	Solution	θ_i
z	0	0	0	0	0	1	1	1	0	
x_2	0	1	0	-1/2	1/4	0	1/2	-1/4	180	45
x_3	0	0	1	-1/2	-3/4	-1	1/2	3/4	150	75
x_1	1	0	0	1/4	-3/8	0	-1/4	3/8	160	40

Table 18 Phase II—Initial Tableau for Unbounded Solution

Basic	x_1	x_2	x_3	x_4	x_5	Solution	
z	-9	-12	0	0	0	0	
	0	0	0	-15/4	-3/8	1245/2	θ_i
x_2	0	1	0	-1/2	1/4	35	
x_3	0	0	1	-1/2	-3/4	15	
x_1	1	0	0	(1/4)	-3/8	45/2	90

problem. In practice, however, solving the problem may require investigating and, possibly, solving a whole family of related problems. This can be illustrated with the example introduced in subsection 4.1. This is a product mix problem with two products and three constraints and, for convenience, the formulation is replicated below:

$$\text{Maximize } 9x_1 + 12x_2$$
$$4x_1 + 3x_2 \leq 180 \quad \text{(labor)}$$
$$2x_1 + 3x_2 \leq 150 \quad \text{(machine 1)}$$
$$4x_1 + 2x_2 \leq 160 \quad \text{(machine 2)}$$
$$x_1, x_2 \geq 0$$

Table 19 Phase II—Second Tableau for Unbounded Solution

Basic	x_1	x_2	x_3	x_4	x_5	Solution
z	15	0	0	0	-6	960
x_2	2	1	0	0	-1/2	80
x_3	2	0	1	0	-3/2	60
x_4	4	0	0	1	-3/2	90

Consider the following set of possibilities and questions related to this example:

> Suppose that additional hours of machine 1 can be acquired at a cost of $3/hour.
> what happens if the actual labor hours availability are only 170 units instead of 180?
> Due to a changing market, product 1 profit is only $8/unit.
> Suppose that a new product is being developed that will require 4 hours of labor, 2 hours on machine 1 and 5 hours on machine 2. What is the minimum selling price to make the product profitable? These are the type of questions that may be routinely faced by industrial engineers and the management of a company producing a mix of manufactured products. Can the linear program provide answers and valuable insight to these and related questions? The answer is a resounding *yes*.

One of the truly amazing and insightful by-products of formulating problems in a linear programming framework is the wealth of information that is provided in addition to the actual optimal solution. In particular, once the optimal solution is found, one also knows the marginal value of additional resources, i.e., how the optimal value of the objective function will change if the availability of machine 1 is increased from 150 to 151 hours. In this example, (as we will subsequently show) these marginal values are $0.50, $3.50, and $0, respectively. Hence, an additional hour of machine 1 would result in an increase in the objective value of the optimal solution from $615 to $618.50. If an additional hour could be acquired for $3, then do it! The second question also addresses what would occur given a change in the resource base. If the material availability is reduced from 180 to 170, does this mean that one must resolve the problem from the beginning? Here the answer is *no*. In fact, all the questions posed here, plus many others, can be answered by performing some postoptimality procedures utilizing the optimal simplex tableau. In this section we highlight some of these insights and postoptimality procedures. However, we refer the reader to [3], [22], and [48] for additional details and a more complete presentation of sensitivity and other postoptimality procedures related to linear programming.

7.2. The Dual Problem

For the product mix example, suppose one seeks a set of "prices" for the three resources (labor, machine 1 hours, machine 2 hours) so that the owners of the firm would neither want to acquire additional resources at

Linear Programming

these prices nor want to sell their existing resources at these prices; i.e., these prices establish a sort of an internal equilibrium for the firm. These prices would be the marginal value *to the firm* of its resources. These equilibrium prices are obtained from what is called a *dual program*. Every linear program possesses one. This dual program is, itself, another linear program whose objective is to determine explicitly these marginal or equilibrium prices.

Consider a linear program written in inequality form as

(P) Maximize $\mathbf{c}^T\mathbf{x}$

$\mathbf{Ax} \leq \mathbf{b}$

$\mathbf{x} \geq \mathbf{0}$.

We will call this the *primal* problem because it is stated first. The dual linear program utilizes the same three data sets, $\mathbf{c}$, $\mathbf{A}$, and $\mathbf{b}$, but in a different fashion. For problem (P), let the $\mathbf{A}$ matrix be $m \times n$, that is, m constraints and n variables. Then, the dual linear program associated with (P), and denoted (D) for dual, is

(D) Minimize $\mathbf{w}^T\mathbf{b}$

$\mathbf{w}^T\mathbf{A} \geq \mathbf{c}$

$\mathbf{w} \geq \mathbf{0}$

where $\mathbf{w}$ is an $m \times 1$ vector. Note that (D) contains m variables and n constraints, whereas (P) contains n variables and m constraints. The constraints in the dual problem are generated from the columns in the primal problem. As an example, the first column in the product mix problem in this section is

Coefficients	Dual variable
9 (objective)	
4 (labor constraint)	w_1
2 (machine 1)	w_2
4 (machine 2)	w_3

A dual variable, w_i, is associated with each of the three constraints. The first dual constraint is

$$4w_1 + 2w_2 + 4w_3 \geq 9 \tag{7.1}$$

i.e., the column of coefficients, weighted by the dual variables w_1, w_2, w_3, is constrained to be greater than or equal to the objective coefficient for the column. Note that if the dual variables w_l are interpreted as the marginal value of the three resources (180 hours of labor, 150 hours of machine 1, 160 hours of machine 2), then (7.1) specifies that the total marginal value of product 1 must be at least equal to the profit for product 1. Finally, the objective coefficients for the dual problem are the RHS coefficients of the primal problem. In this matter, the dual linear program for this product mix problem is

$$\begin{aligned}
\text{Minimize} \quad & 180w_1 + 150w_2 + 160w_3 \\
\text{subject to} \quad & 4w_1 + 2w_2 + 4w_3 \geq 9 \\
& 3w_1 + 3w_2 + 2w_3 \geq 12 \\
& w_1, w_2, w_3 \geq 0
\end{aligned}$$

The optimal solution to the dual linear program is

$$w_1^* = 0.50, \; w_2^* = 3.50, \text{ and } w_3^* = 0$$

and these values are the equilibrium prices or marginal values for the primal problem. For example, $w_3^* = 0$ and this means that there is no value of additional hours on machine 2. This makes sense, for in the optimal solution of the primal problem one does not even utilize all the existing 160 hours; 20 are left over. Note also that the optimal value of the objective function in this dual program is 615, which is the same as the optimal value of the objective function of the primal. This will always be the case. In fact, the intimate relationship between the primal and dual is even more remarkable. One does not even need to explicitly solve the dual linear program to determine its optimal solution. The optimal values of the dual program appear explicitly in the optimal tableau of the primal problem. Observe that the values $w_1^* = 0.50$, $w_2^* = 3.50$, and $w_3^* = 0$ appear in the reduced cost row (under x_3, x_4, and x_5) of the optimal tableau shown in Table 9. The optimal value of the dual program is obtained as a by-product in solving the primal!

7.3. Properties of Dual Linear Programs

In this section we highlight some of the more important aspects of the theory and properties of a pair of dual linear programs. To begin, we first note that if **w** is any feasible solution to the dual program (D) and **x** is any feasible solution to the primal problem (P), then

$$\mathbf{w}^T \mathbf{b} \geq \mathbf{c}^T \mathbf{x} \tag{7.2}$$

Linear Programming

This follows from two observations:

1. If **w** is a feasible solution to (D), then

$$\mathbf{w}^T \mathbf{A} \geq \mathbf{c}^T \tag{7.3}$$
$$\mathbf{w} \geq \mathbf{0}$$

2. If **x** is a feasible solution to (P), then

$$\mathbf{A}\mathbf{x} \leq \mathbf{b} \tag{7.4}$$
$$\mathbf{x} \geq \mathbf{0}$$

Since $\mathbf{x} \geq \mathbf{0}$, multiplying the vector inequality (7.2) by **x** yields the scalar inequality

$$\mathbf{w}^T \mathbf{A}\mathbf{x} \geq \mathbf{c}^T \mathbf{x} \tag{7.5}$$

Similarly, since $\mathbf{w} \geq \mathbf{0}$, multiplying the vector inequality (7.3) by $\mathbf{w}^T$ yields the scalar inequality

$$\mathbf{w}^T \mathbf{A}\mathbf{x} \leq \mathbf{w}^T \mathbf{b} \tag{7.6}$$

Combining (7.4) and (7.5) yields (7.2).

Relationship (7.2) is termed the *weak duality theorem* of linear programming. It simply says that the objective value of any feasible solution to (D) is an upper bound to the objective value of any feasible solution to (P). There is also a *strong duality theorem*, which we now describe. Suppose a pair of dual feasible solutions (**x**, **w**) [meaning that **x** is feasible for (P) and **w** is feasible for (D)] satisfy the following relationships:

$$\begin{aligned}
\mathbf{A}\mathbf{x} &\leq \mathbf{b} \\
\mathbf{x} &\geq \mathbf{0} \\
\mathbf{w}^T \mathbf{A} &\geq \mathbf{c}^T \\
\mathbf{w} &\geq \mathbf{0}
\end{aligned} \tag{7.7}$$

$$\mathbf{w}^T(\mathbf{A}\mathbf{x} - \mathbf{b}) = 0 \tag{7.9}$$
$$(\mathbf{w}^T \mathbf{A} - \mathbf{c}^T)\mathbf{x} = 0 \tag{7.10}$$

Relationships (7.7) and (7.8) simply state that **x** and **w** are feasible for (P) and (D). Relationships (7.9) and (7.10) are called *complementarity slackness* conditions. They state that if in the optimal solution, a given inequality is satisfied as a strict inequality, then the corresponding dual variable (price) associated with that constraint must be zero. This is more easily seen if (7.9) is written in a less compact form, i.e.,

$$w_1(\mathbf{a}^1\mathbf{x} - b_1)$$
$$+ w_2(\mathbf{a}^2\mathbf{x} - b_2)$$
$$\cdot$$
$$\cdot$$
$$\cdot$$
$$+ w_m(\mathbf{a}^m\mathbf{x} - b_m) = 0$$

where $\mathbf{a}^i$ is the ith row of $\mathbf{A}$ (the transpose sign has been omitted). As an example, in the product mix problem, the third constraint in the optimal solution is satisfied as a strict inequality, i.e.,

$$4x_1^* + 2x_2^* < 160$$

which means that $w_3^* = 0$. If a pair of dual feasible solutions $(\mathbf{x}, \mathbf{w})$ satisfy (7.7), (7.8), (7.9), and (7.10) then $\mathbf{x}$ is optimal for (P) and $\mathbf{w}$ is optimal for (D) and this is the *strong duality theorem*. That such is the case follows directly from (7.9) and (7.10), which show that

$$\mathbf{w}^T\mathbf{b} = \mathbf{c}^T\mathbf{x} \tag{7.11}$$

Relationship (7.11), combined with the weak duality theorem, means that no primal solution objective value can be larger than $\mathbf{w}^T\mathbf{b}$. But $\mathbf{c}^T\mathbf{x}$ exactly equals $\mathbf{w}^T\mathbf{b}$. Hence, it follows that $\mathbf{c}^T\mathbf{x}$ must be optimal for (P). A similar argument shows that $\mathbf{w}$ must be optimal for (D).

To complete this analysis, we must address the case in which one of the dual linear programs is either unbounded or infeasible. What happens here? Consider a very simple, one-variable problem as follows:

(P) Maximize x
$-x \leq 1$
$x \geq 0$

This problem is unbounded because x can be made arbitrarily large and still remain feasible. The dual program is

(D) Minimize w
$-w \geq 1$
$w \geq 0$

This dual program is obviously infeasible.
Another example is

(P) Maximize x
$0x \leq -1$
$x \geq 0$

Linear Programming

which is obviously infeasible. Here the dual program is

(D) Minimize $-w$

$$0w \geq 1$$
$$w \geq 0$$

which is also infeasible. These examples exhaust all the possibilities. In fact, any pair of dual linear programs will assume one of the following four states:

Primal	Dual
(1) Feasible	Feasible
(2) Unbounded	Infeasible
(3) Infeasible	Unbounded
(4) Infeasible	Infeasible

Furthermore, the strong duality theorem shows that in case (1) the values of the pair of feasible solutions will be equal. In this case

$$z^* = \mathbf{c}^T\mathbf{x}^* = \mathbf{w}^{*T}\mathbf{b} \tag{7.12}$$

Relationship (7.12) makes it obvious that w_i^*, the ith dual variable, measures the rate of change of z^* with respect to b_i, i.e.,

$$\frac{\partial z^*}{\partial b_i} = w_i^*$$

To conclude this analysis, we show that the duality theory associated with the pair (P) and (D) is sufficient to handle all the different forms of linear programs. For example, if one has a primal constraint

$$\mathbf{a}^T\mathbf{x} \geq b$$

then it can be converted into a less than or equal to constraint by multiplying by -1. Similarly,

$$\mathbf{a}^T\mathbf{x} = b$$

is equivalent to

$$\mathbf{a}^T\mathbf{x} \geq b$$

and

$$\mathbf{a}^T\mathbf{x} \leq b$$

Hence, the form (P) is sufficient to handle all possible situations. But it is instructive to consider explicitly a linear program in standard form:

$$\text{Maximize} \quad c^T x$$
$$Ax = b \quad (7.13)$$
$$x \geq 0$$

This problem (P) is equivalent to

$$(P) \quad \text{Maximize} \quad c^T x$$
$$Ax \leq b \quad (w_1)$$
$$-Ax \leq -b \quad (w_2)$$
$$x \geq 0$$

and its dual linear program would be

$$\text{Minimize} \quad w_1^T b - w_2^T b$$
$$w_1^T A - w_2^T A \geq c$$
$$w_1, w_2 \geq 0$$

Now replace the two nonnegative vectors, w_1 and w_2, with the single vector

$$w = w_1 - w_2$$

Note that w is not constrained to be nonnegative. Substituting w for $w_1 - w_2$ yields

$$\text{Minimize} \quad w^T b$$
$$w^T A \geq c \quad (7.14)$$
$$w \text{ arbitrary (in sign)}$$

This shows that if some constraint in the primal is an equality, the corresponding dual variable is arbitrary in sign. Hence, the dual linear program for (7.13) is (7.14).

7.4. Sensitivity Analysis

The previous section shows how one can determine "equilibrium prices" for the RHS resources and, in doing so, ascertain whether acquiring or selling off some of the resources will increase the overall value of the objective function. For example, these equilibrium prices are sometimes used to justify the acquisition of new equipment and other scarce resources (see [57]). This same theme carries over into this section. One may want to know how to determine whether a new product can be produced at a profit or if a certain method improvement for a process may change an unprofitable

Linear Programming

product into a profitable one. These and other issues focus on the *sensitivity* of the optimal solution to changes in the data, namely the cost vector **c**, the resource vector **b**, and the technological coefficients a_{ij} in the **A** matrix. For example, in the product mix illustration given earlier in this section, the optimal solution is $x_1^* = 15$ and $x_2^* = 40$. How will this optimal solution change if c_1 is increased from \$9 to \$10 or if b_1 is decreased from 180 to 170? Moreover, can one determine the "new" solution without resorting to solving the revised problem from scratch?

All these and related questions focus on how **x***, an optimal solution to

Maximize $\mathbf{c}^T\mathbf{x}$

$\mathbf{Ax} \leq \mathbf{b}$

$\mathbf{x} \geq 0$

changes as a function of changes in **c**, **A**, and **b**. There is a well-defined set of procedures and algorithms for performing this in a systematic manner [3, 48]. In [3], the sensitivity procedures are presented in the following framework.

Changes in RHS vector **b**
Changes in **c** vector
   Nonbasic coefficients
   basic coefficients
Changes in **A** matrix
   Nonbasic columns
   Basic columns
Addition of a new activity (column)
Addition of a new constraint

In all cases, the focus is on determining the new optimal solution, subject to changes in the data, *without resorting to starting over*.

Here, we will highlight some of these procedures by illustrating the analysis in the context of our product mix example. (We emphasize that, in practice, most computer codes have options to assist the user with the various sensitivity procedures.) The information needed for performing most of the sensitivity procedures is extracted from the optimal simplex tableau, namely

   w* (optimal dual solution)
   $\mathbf{B}^{-1}$ (optimal basis inverse)

To begin, suppose that the RHS is changed from **b** to $\mathbf{\bar{b}}$. One may believe that such a change would be complicated, because one is dealing with a set of simultaneous linear equations and if the right-hand sides are changed one might expect that all the variables would need to change. But things may not be that complicated. The current optimal solution, $\mathbf{x}^*$, is a basic solution, i.e., $\mathbf{x}^* = (\mathbf{x}_B^*, \mathbf{x}_N^*)$ and

$$\mathbf{x}_B^* = \mathbf{B}^{-1}\mathbf{b}$$
$$\mathbf{x}_N^* = \mathbf{0}$$

Hence, if the right-hand side $\mathbf{b}$ is replaced by $\bar{\mathbf{b}}$, the new basic solution would be

$$(\mathbf{x}_B)_{new} = \mathbf{B}^{-1}\bar{\mathbf{b}}$$
$$\mathbf{x}_N = \mathbf{0}$$

And if $(\mathbf{x}_B)_{new} \geq \mathbf{0}$, it will also be optimal! That this is the case follows from the fact that the right-hand side vector $\mathbf{b}$ does not affect the optimality conditions $(x_j - c_j)$ in the optimal tableau. For the example in this section, one has

$$\mathbf{B}^{-1} = \begin{pmatrix} \frac{1}{2} & -\frac{1}{2} & 0 \\ -\frac{1}{3} & \frac{2}{3} & 0 \\ -\frac{4}{3} & \frac{2}{3} & 1 \end{pmatrix}$$

If the RHS vector is changed from $\mathbf{b}$ to $\bar{\mathbf{b}}$, i.e.,

$$\mathbf{b} = \begin{pmatrix} 180 \\ 150 \\ 160 \end{pmatrix} \longrightarrow \bar{\mathbf{b}} = \begin{pmatrix} 170 \\ 150 \\ 160 \end{pmatrix}$$

then the new basic solution is

$$(\mathbf{x}_B)_{new} = \mathbf{B}^{-1}\bar{\mathbf{b}} = \begin{pmatrix} 10 \\ 43.3 \\ 33.3 \end{pmatrix}$$

Here, $x_1^* = 10$, $x_2^* = 43.3$, and $x_3^* = 33.3$ are optimal because $(\mathbf{x}_B)_{new} \geq \mathbf{0}$. In fact,

$$\mathbf{x}_B = \mathbf{B}^{-1}\bar{\mathbf{b}}$$

is optimal for any RHS $\mathbf{b}$ as long as $(\mathbf{x}_B)_{new} > \mathbf{0}$.

Next, consider the situation in which a new product is being evaluated for possible adoption. To illustrate, consider the data for a new product for the product mix example of this section. Suppose for this new product that the marketing department is projecting a selling price of $8/unit and the industrial engineering staff estimates that each unit will require 4 hours of labor, 2 hours on machine 1, and 5 hours on machine 2. Will this new product be profitable and, hence, generate a different product mix? Furthermore, to answer this question, does one need to resolve completely a new product mix problem as formulated in (7.15), where x_3 represents the quantity of the new product to be produced?

$$\text{Maximize } 9x_1 + 12x_2 + 8x_3$$
$$4x_1 + 3x_2 + 4x_3 \leq 180$$
$$2x_1 + 3x_2 + 2x_3 \leq 150 \quad (7.15)$$
$$4x_1 + 2x_2 + 5x_3 \leq 160$$
$$x_j \geq 0$$

The answer is no. The value of $z_3 - c_3$ can be determined directly from the knowledge of the optimal dual solution w^*. In this case

$$z_3 - c_3 = \mathbf{w}^{*T} \begin{pmatrix} 4 \\ 2 \\ 5 \end{pmatrix} - 8 = 1$$

Because $z_3 - c_3 > 0$, the presence of this variable would not affect the optimality of the current solution. Otherwise, a new column $\mathbf{B}^{-1}\mathbf{a}_3$ would be added to the tableau (with $z_3 - c_3$ in the index row) and the simplex iterations would be restarted starting from the current (near-optimal) solution.

Other sensitivity procedures are similar to the ones described here and require varying degrees of sophistication and effort. The reader is referred to [3] or [48] for a thorough presentation of all the various possibilities. In addition, these references address the topic of *parametric* programming. Here, one considers questions similar to those posed in this section but in a *continuous* context. For example, rather than determining optimal solutions for just two specific objective vectors $\mathbf{c}_1$ and $\mathbf{c}_2$, one may be interested in a whole family of optimal solutions, $\mathbf{x}(\lambda)$, which are optimal for the family of objective vectors $\mathbf{c}(\lambda)$ defined as

$$\mathbf{c}(\lambda) = \lambda \mathbf{c}_1 + (1 - \lambda)\mathbf{c}_2, \quad 0 \leq \lambda \leq 1$$

The determination of such a family of optimal solutions, indexed by the *parameter* λ, is called parametric programming.

8. INTERIOR POINT METHODS

This section introduces the concept of *computational complexity* and presents a new class of linear programming algorithms, called *interior point methods*. Whereas the simplex method iterates from extreme point to extreme point, along the boundary of the feasible region, interior point methods move through its interior. From a theoretical point of view, interior point methods have been shown to be more efficient than the simplex method in the worst case. The measurements of the computational efficiency of linear programming algorithms are discussed in the next section. The rest of the chapter is devoted to a simple interior point method called the *affine scaling algorithm*.

8.1. Computational Complexity in Linear Programming

The *worst-case analysis* of an algorithm measures the computational efficiency of an algorithm in problem solving under the worst-case scenario. In terms of the simplex method, the worst-case scenario is one in which every extreme point is examined before finding the optimal solution. The concept of computational complexity was introduced to classify algorithms in terms of their computational effort under the worst-case analysis. The computational effort of an algorithm is usually measured as a function of the number of elementary operations, such as additions, multiplications, and comparisons, that need to be performed to solve a particular problem and the size of the problem's input data, i.e., number of bits (binary digits) necessary to represent the input data in binary form. The input data of a linear program in standard form are completely characterized by the number of variables (n), number of equality constraints (m), and the coefficients of the objective function ($\mathbf{c}$), constraint matrix ($\mathbf{A}$), and RHS vector ($\mathbf{b}$). In the binary system, it takes $\lceil \log_2(1 + z) \rceil$ bits, where $\lceil x \rceil$ is the rounded-up integer value of x, to represent a positive integer z. In general, an additional bit is necessary to represent the sign of an integer. Assuming that the input data of the linear program in standard form are all integers (possibly converted from rational data to this form), the number of bits necessary to store the input data is

$$L = (1 + \lceil \log_2(1 + m) \rceil) + (1 + \lceil \log_2(1 + n) \rceil)$$
$$+ \sum_{i=1}^{m} \sum_{j=1}^{n} (1 + \lceil \log_2(1 + |a_{ij}|) \rceil) + \sum_{i=1}^{m} (1 + \lceil \log_2(1 + |b_i|) \rceil)$$
$$+ \sum_{i=1}^{n} (1 + \lceil \log_2(1 + |c_i|) \rceil)$$

The complexity of an algorithm for linear programming can be written as a function of the triplet (n, m, L), say $f(n, m, L)$. When the com-

Linear Programming

plexity function $f(n, m, L)$ of an algorithm is a polynomial function of n, m, and L, the algorithm is said to be of *polynomial complexity*. Otherwise, the algorithm is a *nonpolynomial-time algorithm*.

The computational complexity of the simplex method depends on the number of iterations required to find an optimal solution, and the number of elementary operations performed at each iteration, in the worst case. Klee and Minty [53] have produced a class of problems defined by m equality constraints and $n = 2m$ nonnegative variables, for which the simplex algorithm requires $2^m - 1$ iterations, traversing all the extreme points of the feasible region. Therefore, although the simplex method works well in practice, it is a nonpolynomial-time algorithm.

In 1984 Karmarkar [49] introduced a polynomial-time algorithm for linear programming. This new algorithm not only has better complexity than the simplex method but also is reported to be superior to the simplex method in some practical applications, especially for certain classes of large-scale problems [50].

The projective scaling method of Karmarkar is radically different from the simplex method. The search direction of the former is through the interior of the feasible region, whereas the latter search is restricted to the boundary of the feasible region. The projective method considers a linear program over a simplex with zero optimal objective function value. It moves through the interior of a simplex defined in the null space of the constraint matrix, by transforming the space at each iteration so that the current solution is in the center of the simplex.

Since Karmarkar's projective scaling algorithm was introduced, a flurry of research activities in interior point methods have been undertaken. Figure 20 illustrates the conceptual difference between a path followed by the simplex method and a path followed by an interior point method. Among many variants of Karmarkar's original algorithm, the affine scaling method attracted much attention, because it allows users to solve linear programs in standard form without requiring the special simplex structure. The basic idea of the affine scaling algorithm was developed by the Soviet mathematician Dikin [17] in 1967. Because of its simplicity and practicality, this method is explained in detail below. Polynomial complexity of the affine scaling method, however, has not yet been proved.

8.2. Preliminaries of the (Primal) Affine Scaling Algorithm

Consider a linear program in standard form:

$$\text{Maximize} \quad z = \mathbf{c}^T\mathbf{x}$$
$$\text{subject to} \quad \mathbf{A}\mathbf{x} = \mathbf{b} \quad (8.1)$$
$$\mathbf{x} \geq \mathbf{0}$$

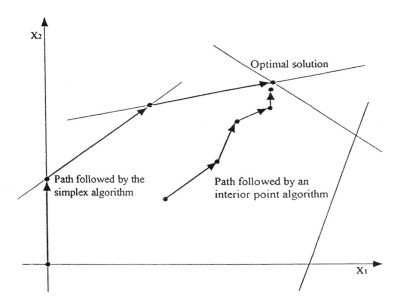

Figure 20 Simplex method versus interior point method.

where $\mathbf{A}$ is an $m \times n$ full rank matrix. The *interior* of the feasible region of (8.1), $S = \{\mathbf{x} \mid \mathbf{A}\mathbf{x} = \mathbf{b}, \mathbf{x} \geq \mathbf{0}\}$, is

$$I(S) = \{\mathbf{x} \mid \mathbf{A}\mathbf{x} = \mathbf{b}, \mathbf{x} > \mathbf{0}\}$$

A vector $\mathbf{x}$ is called an *interior point* of the linear program if $\mathbf{x} \in I(S)$. In the remainder of this section, it is assumed that $I(S) \neq \phi$.

For an interior point method the initial solution is critical. As shown in Figure 21, if one starts at $\mathbf{x}^0$, a point near the center of S, and moves in the direction of $\mathbf{c}$, the *gradient of the objective function*, one can take a large step toward the optimal solution. However, if one starts at $\bar{\mathbf{x}}^0$, an interior point close to the boundary of S, the size of the step toward the optimum will be short. The major problem with using the direction of the gradient is that the gradient is the same no matter what point is chosen. As the goal is to maximize the objective function, one could take a large step in the direction $\mathbf{c}$, which may result in a nonoptimal point on the boundary. Once on the boundary, the method could become the simplex method.

To circumvent this problem with the step size, Karmarkar had a simple but ingenious idea. Do not proceed all the way to the boundary. Proceed in the direction of the gradient but stop while the point is still in the interior.

Linear Programming

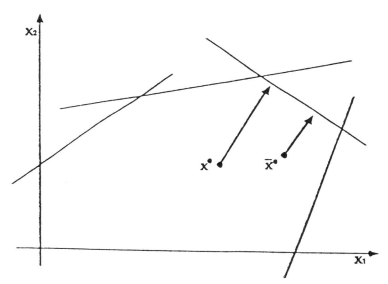

Figure 21 Differences in possible improvements.

From this new point, he performed a projective transformation that transformed the point from being close to the boundary of the original simplex to being the center of a new simplex. From this new center, it is then possible to take another large step toward the optimal solution. This concept of *projective transformation* is illustrated in Figure 22.

It is possible to implement Karmarkar's original idea of moving a near boundary point to the middle of the feasible region in several ways. A simple and practical method is doing an *affine transformation*. An affine transformation is just a rescaling process. Before we present the affine scaling algorithm, we need to define an affine transformation. Let $\mathbf{x}^k = (x_1^k, x_2^k, \ldots, x_n^k)^T$ be an interior point. Then we define the following diagonal matrix:

$$\mathbf{D}_k = \text{diag}(\mathbf{x}^k) = \begin{pmatrix} x_1^k & 0 & \cdots & 0 \\ 0 & x_2^k & \cdots & 0 \\ \vdots & \vdots & \vdots & \vdots & \vdots \\ 0 & 0 & \cdots & x_n^k \end{pmatrix}$$

Since $\mathbf{x}^k \in I(S)$, all components of $\mathbf{x}^k$ are positive. This implies that the matrix $\mathbf{D}_k$ is nonsingular with an inverse matrix $\mathbf{D}_k^{-1}$, which is also diagonal:

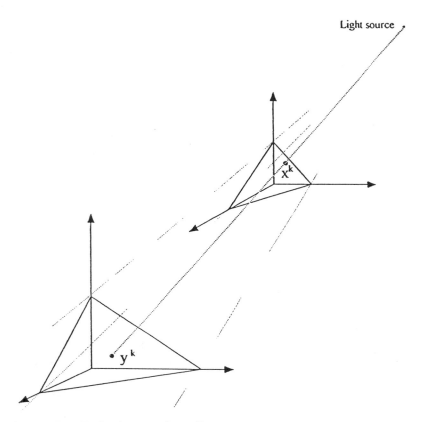

Figure 22 Projective transformation.

$$\mathbf{D}_k^{-1} = \begin{pmatrix} \frac{1}{x_1^k} & 0 & \cdots & 0 \\ 0 & \frac{1}{x_2^k} & \cdots & 0 \\ \vdots & \vdots & \vdots & \vdots \\ 0 & 0 & \cdots & \frac{1}{x_n^k} \end{pmatrix}$$

The affine transformation with respect to $\mathbf{x}^k$ is defined as $\mathbf{y} = \mathbf{D}_k^{-1}\mathbf{x}$. This transformation simply rescales each component x_i of $\mathbf{x}$ by dividing it by x_i^k. Geometrically, this is a linear transformation. The feasible point $\mathbf{x}^k$

Linear Programming

is mapped to the point $\mathbf{e} = (1, 1, \ldots, 1)^T$, the n-vector of ones (see Figure 23).

In the transformed y-space, the standard linear program (8.1) becomes

Maximize $z = (\mathbf{c}^k)^T \mathbf{y}$
subject to $\mathbf{A}_k \mathbf{y} = \mathbf{b}$
$\mathbf{y} \geq \mathbf{0}$

where $\mathbf{c}^k = \mathbf{D}_k \mathbf{c}$ and $\mathbf{A}_k = \mathbf{D}_k \mathbf{A}$.

Since the image of $\mathbf{x}^k$ is $\mathbf{e}$, one can move from $\mathbf{e}$ along a direction $\mathbf{d}^k$ that lies in the *null space* of $\mathbf{A}_k$ to maintain feasibility. The null space of $\mathbf{A}_k$ is the set

$$N(\mathbf{A}_k) = \{\mathbf{d} \mid \mathbf{A}_k \mathbf{d} = \mathbf{0}\}$$

Then the new point

$$\mathbf{y}^{k+1} = \mathbf{e} + \alpha^k \mathbf{d}^k$$

will remain interior if one selects an appropriate step size $\alpha^k > 0$. The inverse image of $\mathbf{y}^{k+1}$ is

$$\mathbf{x}^{k+1} = \mathbf{D}_k \mathbf{y}^{k+1}$$

which is interior to the original linear programming problem.

Since we have a maximization problem, $\mathbf{d}^k$ can be the *steepest ascent direction*, which is the direction of the greatest increase in objective function value. Therefore, one can set $\mathbf{d}^k$ equal to the projection of $\mathbf{c}^k$ into the

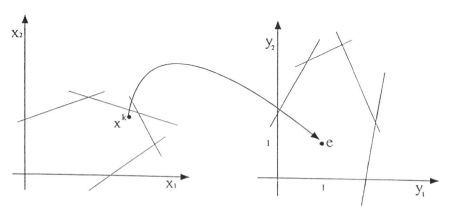

Figure 23 Affine transformation.

null space of $\mathbf{A}_k$. In order to do so, we need to define the projection matrix for the null space of $\mathbf{A}_k$. The *null space projection matrix* [3, 48] is

$$\mathbf{P}_k = \mathbf{I} - \mathbf{A}_k^T(\mathbf{A}_k\mathbf{A}_k^T)^{-1}\mathbf{A}_k$$
$$= \mathbf{I} - \mathbf{D}_k\mathbf{A}^T(\mathbf{A}\mathbf{D}_k^2\mathbf{A}^T)^{-1}\mathbf{A}\mathbf{D}_k$$

and the search direction becomes

$$\mathbf{d}^k = \mathbf{P}_k\mathbf{c}^k = [\mathbf{I} - \mathbf{D}_k\mathbf{A}^T(\mathbf{A}\mathbf{D}_k^2\mathbf{A}^T)^{-1}\mathbf{A}\mathbf{D}_k]\mathbf{c}^k$$

Now, one chooses the appropriate step size, α^k, such that

$$\mathbf{y}^{k+1} = \mathbf{e} + \alpha^k\mathbf{d}^k > \mathbf{0}$$

If $\mathbf{d}^k \geq \mathbf{0}$, then α^k can be set arbitrarily large, maintaining feasibility. In this case, the problem becomes unbounded. If at least one component of $\mathbf{d}^k$ is negative, then α^k can be selected as follows:

$$\alpha^k = \min\{\alpha/(-d_i^k)\,|\,d_i^k < 0,\, i = 1, \ldots, n\}$$

where $0 < \alpha < 1$. When α is close to 1, the current solution $\mathbf{x}^k$ moves closer to the boundary. Next, one has to map $\mathbf{y}^{k+1}$ back to the original solution space to determine the new improved solution $\mathbf{x}^{k+1}$, that is,

$$\mathbf{x}^{k+1} = \mathbf{D}_k\mathbf{y}^{k+1}$$
$$= \mathbf{x}^k + \alpha^k\mathbf{D}_k\mathbf{d}^k$$

Since $\mathbf{d}^k = \mathbf{P}_k\mathbf{c}^k$ and $\mathbf{c}^k = \mathbf{D}_k\mathbf{c}$, it can be shown that

$$\mathbf{x}^{k+1} = \mathbf{x}^k - \alpha^k\mathbf{D}_k^2[\mathbf{c} - \mathbf{A}^T\mathbf{w}^k]$$

where

$$\mathbf{w}^k = (\mathbf{A}\mathbf{D}_k^2\mathbf{A}^T)^{-1}\mathbf{A}\mathbf{D}_k^2\mathbf{c}$$

If $\mathbf{x}^k$ were an extreme point, the preceding expression could be reduced to

$$\mathbf{w}^k = (\mathbf{B}^T)^{-1}\mathbf{c}_B$$

Hence, $\mathbf{w}^k$ is called the vector of dual estimates associated to $\mathbf{x}^k$ in the affine scaling algorithm. Moreover, in this case, the vector

$$\mathbf{r}^k = \mathbf{c} - \mathbf{A}^T\mathbf{w}^k$$

reduces to

$$\mathbf{r}^k = \mathbf{c} - \mathbf{A}^T(\mathbf{B}^T)^{-1}\mathbf{c}_B$$

Hence, $\mathbf{r}^k$ is called the reduced cost vector associated with $\mathbf{x}^k$.

Note that $\mathbf{w}^k$ can be found by solving the following system of linear equations:

Linear Programming

$$(AD_k^2A^T)w^k = (AD_k^2c)$$

which can be done without finding the inverse of matrix $(AD_k^2A^T)$. In addition,

$$\begin{aligned}c^Tx^{k+1} &= c^Tx^k + \alpha^k c^T D_k d^k \\ &= c^Tx^k + \alpha^k (c^k)^T d^k \\ &= c^Tx^k + \alpha^k (d^k)^T d^k \\ &= c^Tx^k + \alpha^k \|d^k\|^2\end{aligned}$$

which implies that x^{k+1} is an improved solution as long as $d^k \neq 0$.

8.3. (Primal) Affine Scaling Algorithm

The specific steps of the affine scaling algorithm are given below.

Step 1: Let $x^0 \in I(S)$, $\alpha \in (0, 1)$, and set $k = 0$.

Step 2: Let $D_k = \text{diag}(x^k)$, and compute

$$(AD_k^2A^T) = (AD_k)(AD_k)^T$$

and

$$(AD_k^2c) = (AD_k)D_kc.$$

Step 3: Solve the linear system of equations

$$(AD_k^2A^T)w^k = (AD_k^2c)$$

Step 4: Compute

$$r^k = c - A^Tw^k$$

and

$$d^k = D_k r^k$$

Step 5: If $d^k > 0$, the linear program is unbounded. Otherwise, compute

$$\alpha^k = \min\{\alpha/(-d_i^k) \mid d_i^k < 0, i = 1, \ldots, n\}$$

and

$$x^{k+1} = x^k + \alpha^k D_k d^k$$

Step 6: Check for convergence. If the convergence criterion is satisfied, x^{k+1} is the optimal solution, and stop. Otherwise, increase k by 1, and go to step 2.

The value α is a scalar that is used to compute the step size α^k to guarantee that the new iterate stays in the interior of the feasible region. Generally, $\alpha = 0.98$ works well in practice, but it can be any value strictly between 0 and 1.

If in step 6 of the algorithm $\mathbf{d}^k = \mathbf{0}$, then we stop because the current iterate is optimal. However, this would occur only in the first iteration of a special linear program with $\mathbf{c} = \mathbf{0}$. This is the only case in which an interior solution (and any feasible solution) is optimal. Because typically $\mathbf{c} \neq \mathbf{0}$, the algorithm must have a mechanism to terminate when the current iterate is close enough to an optimal solution. A *purification scheme* can be used to determine an optimal basic feasible solution (extreme point) from the current iterate [90, 91]. Purification schemes are beyond the scope of this chapter and will not be discussed. Now we present two simple rules that may be used in step 6 as termination criteria. These rules are:

1. Small decrease in objective function value: the algorithm stops when the relative change of objective function value in two consecutive iterations is less than some prescribed tolerance, i.e.,

$$\frac{|\mathbf{c}^T\mathbf{x}^{k+1} - \mathbf{c}^T\mathbf{x}^k|}{(|\mathbf{c}^T\mathbf{x}^k| + \delta)} \leq \varepsilon$$

   where δ is a small positive number that ensures numerical stability in the computation, and ε is the prescribed tolerance.

2. Difference between primal and dual solutions: the sequences $\{\mathbf{x}^k\}_k$ and $\{\mathbf{w}^k\}_k$ converge to optimal primal and dual feasible solutions, respectively. Then, when the relative difference between primal and dual objective values is within certain tolerance, the algorithm stops:

$$\frac{|\mathbf{c}^T\mathbf{x}^k - \mathbf{b}^T\mathbf{w}^k|}{(|\mathbf{c}^T\mathbf{x}^k| + \delta)} \leq \varepsilon$$

The following example illustrates the steps of the affine scaling algorithm:

$$\text{Maximize} \quad z = -x_1 + 4x_2$$
$$\text{subject to} \quad -x_1 + 2x_2 + x_3 \quad\quad\quad = 30 \quad\quad (8.2)$$
$$\quad\quad\quad\quad\quad x_1 \quad\quad\quad\quad + x_4 = 30$$
$$x_1, x_2, x_3, x_4 \geq 0$$

In this problem,

$$\mathbf{c} = (-1, 4, 0, 0)^T$$

$$\mathbf{A} = \begin{pmatrix} -1 & 2 & 1 & 0 \\ 1 & 0 & 0 & 1 \end{pmatrix}$$

and

Linear Programming

$$\mathbf{b} = (30, 30)^T$$

In step 1, set $\alpha = 0.98$, and let the initial feasible interior solution and objective function value be

$$\mathbf{x}^0 = (4, 10, 14, 26)^T$$

and

$$z^0 = 36$$

In addition, $k = 0$. In step 2,

$$D_0 = \begin{pmatrix} 4 & 0 & 0 & 0 \\ 0 & 10 & 0 & 0 \\ 0 & 0 & 14 & 0 \\ 0 & 0 & 0 & 26 \end{pmatrix}$$

In step 3,

$$\mathbf{w}^0 = (\mathbf{A}\mathbf{D}_0^2\mathbf{A}^T)^{-1}\mathbf{A}\mathbf{D}_0^2\mathbf{c} = (1.333535, 0.007712)^T$$

In step 4,

$$\mathbf{r}^0 = \mathbf{c} - \mathbf{A}^T\mathbf{w}^0 = (0.325823, 1.332930, -1.333535, -0.007712)^T$$

and

$$\mathbf{d}^0 = \mathbf{D}_0\mathbf{r}^0 = (1.303293, 13.329301, -18.669489, -0.200507)^T$$

In step 5,

$$\alpha_0 = \min\{\alpha/(-d_i^0) \mid d_i^0 < 0, i = 1, \ldots, 4\}$$
$$= 0.98/18.669489 = 0.052492$$

and the new feasible interior solution and objective function value are

$$\mathbf{x}^1 = \mathbf{x}^0 + \alpha^0\mathbf{D}_0\mathbf{d}^0$$
$$= (4.273650, 16.996825, 0.280000, 25.726350)^T$$

and

$$z^1 = 63.713650$$

The algorithm continues until a feasible interior solution sufficiently close to the optimal solution is obtained. The optimal solution is

$$\mathbf{x}^* = (30, 30, 0, ,0)^T \quad \text{and} \quad z^* = 90$$

8.4. Finding an Initial Interior Feasible Solution

The big-M and two-phase methods used to obtain an initial basic feasible solution for the simplex algorithm can be adapted for finding an initial interior feasible solution. These methods are briefly discussed below.

In the big-M method, one needs to add an artificial variable, say x_a, with a very large negative cost $-M$ to the original problem in standard form, and make $\mathbf{x} = \mathbf{e}$ and $x_a = 1$ become an initial interior feasible solution. The modified linear program for the big-M method is

$$\text{Maximize} \quad z = \mathbf{c}^T\mathbf{x} - Mx_a$$
$$\text{subject to} \quad \mathbf{Ax} + (\mathbf{b} - \mathbf{Ae})x_a = \mathbf{b}$$
$$\mathbf{x} \geq \mathbf{0}, \quad x_a \geq 0$$

Similarly, in the two-phase method, $\mathbf{x} = \mathbf{e}$ and $x_a = 1$ is an interior feasible solution in the following phase I problem:

$$\text{Minimize} \quad z = x_a$$
$$\text{subject to} \quad \mathbf{Ax} + (\mathbf{b} - \mathbf{Ae})x_a = \mathbf{b}$$
$$\mathbf{x} \geq \mathbf{0}, \quad x_a \geq 0$$

In both methods, x_a will become the blocking variable after few iterations and, by setting the step size parameter $\alpha = 1$, instead of $\alpha = 0.98$, a strictly positive solution for $\mathbf{x}$ will be generated and x_a will become zero. Then x_a can be dropped and the positive solution can be used along with the original objective function to restart the affine scaling algorithm.

Consider the linear program (8.2). Let

$$(\mathbf{b} - \mathbf{Ae}) = \begin{pmatrix} 30 \\ 30 \end{pmatrix} - \begin{pmatrix} -1 & 2 & 1 & 0 \\ 1 & 0 & 0 & 1 \end{pmatrix} \begin{pmatrix} 1 \\ 1 \\ 1 \\ 1 \end{pmatrix} = \begin{pmatrix} 28 \\ 28 \end{pmatrix}$$

Then, the modified linear program for the big-M method is

$$\text{Maximize} \quad z = -x_1 + 4x_2 \qquad\qquad - Mx_5$$
$$-x_1 + 2x_2 + x_3 \qquad\quad + 28x_5 = 30$$
$$x_1 \qquad\qquad\qquad + x_4 + 28x_5 = 30$$
$$x_1, x_2, x_3, x_4, x_5 \geq 0$$

where x_5 is the artificial variable. An interior feasible solution to the modified linear program is

$$\begin{pmatrix} x_1 \\ x_2 \\ x_3 \\ x_4 \\ x_5 \end{pmatrix} = \begin{pmatrix} 1 \\ 1 \\ 1 \\ 1 \\ 1 \end{pmatrix}$$

9. COMPUTER-AIDED MODEL BUILDING AND DATA MANAGEMENT

9.1. Overview and Need

In many texts, the study of linear programming terminates after the simplex algorithm is fully developed and postoptimality issues are discussed. In essence, once the mathematical processes for solving and analyzing

$$\text{Maximize} \quad \mathbf{c}^T\mathbf{x}$$
$$\mathbf{Ax} = \mathbf{b}$$
$$\mathbf{x} \geq \mathbf{0}$$

are developed, the instructors and authors presume the job is done. In many cases, this is entirely appropriate. The study of linear programming at most schools is a one-semester course. The subject is quite robust and serves as a precursor to enhancements involving nonlinearities, integralities, and uncertainties. On the other hand, actual applications of linear programming to determine shipping schedules for next month, inventory acquisition plans, or staffing decisions require a process for prescribing actual numbers for the generic vectors and matrices $\mathbf{c}$, $\mathbf{b}$, and $\mathbf{A}$. Then these numbers are processed by an optimizer (like the simplex method) to obtain optimal values for the decision variables. Finally, the optimal solution needs to be organized in a fashion so that management can "understand" the solution and take appropriate action.

This overall process may be best understood in the context of the transportation model presented earlier. The transportation model, in general, is characterized by the algebraic form:

$$\text{Minimize} \quad \sum_{i \in I} \sum_{j \in J} c_{ij} x_{ij}$$
$$\sum_{j \in J} x_{ij} \leq b_i \quad i \in I \quad (9.1)$$
$$\sum_{i \in I} x_{ij} \geq d_j \quad j \in J$$
$$x_{ij} \geq 0$$

A single "product" is being transported from a set of origins I to a set of destinations J at a cost per unit of c_{ij}. The quantity shipped from i to j is represented by the variable x_{ij}. The parameters b_i and d_j represent, respectively, the supply and demand.

This mathematical statement fully describes the transportation model in an unambiguous, mathematical fashion, in particular, an algebraic fashion. We emphasize that the linear program (1.1) and (1.2) presented in subsection 1.2 is a particular realization of the transportation model, i.e., one in which the parameters c_{ij}, b_i, and d_j are given specific numeric values. This contrast between an abstract model like (9.1) and data to support a specific realization of the model and the resulting linear program like (1.1) and (1.2) is clearly and elegantly described in Fourer et al. [25]. These authors distinguish between an *abstract model*, *data for a model*, and a *resulting linear program*. In the context of the transportation model, the abstract model is (9.1), the *data* are provided in a set of tables like

Table SUPPLY	
New York	200
Chicago	300
Los Angeles	500

Table DEMAND	
Pittsburgh	100
Atlanta	150
Houston	350
Denver	150
San Francisco	250

Table SHIPPING COSTS

	Pittsburgh	Atlanta	Houston	Denver	San Francisco
New York	$2	3	4	5	8
Chicago	2	4	3	3	7
Los Angeles	8	6	4	3	2

and the *resulting linear program* is (1.1) and (1.2).

The basic question is how one efficiently generates the linear program specified by (1.1) and (1.2), which is then subsequently processed and solved by an optimizer (like the simplex). Furthermore, it is important to realize that these processes have both a human and a computer component, and hence there is a real issue about using processes that are both "computer friendly" and "people friendly." In addition, the overall scheme needs to be robust in the sense that a resulting linear program like (1.1) and (1.2) should be "data driven." In particular, if a different set of suppliers and availabilities is specified, e.g.,

Linear Programming

Table SUPPLY

New York	200
Chicago	100
Los Angeles	500
Boston	100
Miami	200

then only this one table should need to be modified and a new linear programming realization should be generated (with relative ease) by reexecuting the *same* set of computer instructions.

This discussion leads us to address four particular issues:

1. How does an abstract model like (9.1) become transformed into a computer model?
2. How are data arranged and organized to facilitate the transformation of an abstract model into a particular linear program?
3. How is the particular linear program processed to obtain an optimal solution?
4. How does one ensure that mistakes were not made in processes 1, 2, and 3?

First, note that issue 3 has already been addressed in part II, Solution and Analysis. But this issue is explicitly included here because it helps illustrate the fact that both people and computers are involved in these four issues and the overall system must somehow accommodate the needs and limitations of both. For example, the simplex algorithm is a process of iteratively replacing a single vector in a basis with another different vector. Hence, from the viewpoint of the simplex algorithm, it would be advantageous if the data are presented in the form of a set of vectors. On the other hand, algebraic models are almost always presented in the form of equations and inequalities, a language understood by all engineers. Engineers are also quite familiar with iterative processes and indirect references to data sets. For example, suppose one needs to calculate the dot product of two vectors **a** and **b**. If **a** and **b** are

$$\mathbf{a} = \begin{pmatrix} 6 \\ -1 \\ 3 \end{pmatrix}, \quad \mathbf{b} = \begin{pmatrix} 5 \\ 2 \\ -6 \end{pmatrix}$$

then one might explicitly calculate the dot product by performing five binary operations (three multiplications and two additions), i.e.,

$$6 \cdot 5 + (-1) \cdot 2 + 3 \cdot (-6) = 10$$

Such a process is feasible if the vectors contain just three components, but suppose the vectors contain 3000 components! In such a case, one would alternatively consider a recursive set of steps such as

```
      SUM = 0
      DO 10   I = 1,N
      PRODUCT = A(I) * B(I)
      SUM = SUM + PRODUCT
   10 CONTINUE
      PRINT SUM.
```

Here, the user provides a data structure in the form of arrays that is referenced indirectly. No one would ever consider computing the dot product any other way. The same contrast is true in the construction of linear programs. The earlier transportation model (1.1) and (1.2) illustrates the point. This problem contains 15 variables, 8 constraints, and 53 nonzero coefficients. For such a problem one could explicitly construct an equation like

$$x_{11} + x_{12} + x_{13} + x_{14} + x_{15} = 200$$

by "entering" the sequence of keystrokes $(1,1,1,1,1, =, 200)$ into a program (computer) and somehow identifying the 1's with the coefficients of the variables $x_{11}, x_{12}, x_{13}, x_{14}, x_{15}$. Because there are only eight equations, such a strategy is obviously feasible (and is often used in teaching students to solve problems with a computer and is illustrative of the various spreadsheets that incorporate linear programming software). But if the transportation model contained 1000 origins and 3000 destinations, obviously some better process is required, and this observation sets the stage for addressing issues 1 and 2 outlined in this subsection.

9.2. Illustrative Example

In this subsection we develop a simple, single-period model for optimizing the production and distribution activities for a firm that

> Produces two products,
> At two different plants, and
> Services two different markets.

This model is used subsequently to facilitate a comparison among various approaches to computer-aided model building. Let x_{ij} be the quantity of product i ($i = 1, 2$) produced at plant j ($j = 1, 2$) and y_{ijk} the quantity of product i produced at plant j and shipped to market k ($k = 1, 2$). Define three types of constraints.

Linear Programming

1. Capacity
   (a) Machine availabilities (machines A and B)
   (b) Raw materials
2. Production-distribution balance
3. Demand

The linear programming model for this problem is depicted in Figure 24. In this figure, we have provided a complete linear programming realization and explicitly depicted numerical realizations of the coefficients. The reason for providing a picture of the complete model is to emphasize the *pattern* for this model. Note that the various columns and rows show a regularity and predictability. *In general, the vast majority of real applications of linear programming models are characterized by a few simple, recurring patterns of nonzero coefficients and, in short, this is why computerized procedures for generating realizations of linear programs are efficient and feasible.* All the various computer-aided model building and data management systems are firmly entrenched in a data-driven environment. The data that are used to support the realization of a linear program reside in a set of data tables. These data tables are used as input to a model building language, which essentially replaces the abstractly defined coefficients such as c_i, a_{ij}, and b_j with actual numbers. This linear programming realization is then presented to the optimizer for solution. The optimal solution so obtained could be vector of possibly 10,000 or 20,000 numbers (or more). But herein lies another problem: who could or would want to sort through and use such a vector (list) to assist in arriving at decisions? Management and other users should be provided a people-friendly report that captures the important results from the optimal solution vector.

The basic elements of any computer-aided model building and data management system for linear programming models are summarized in Figure 25.

A reasonable question to ask in relation to Figure 25 is, "where is the abstract model defined?" The answer is in two parts: (1) it resides in the actual computer code within the model building block, and (2) it resides in the very structure of the data tables. For example, the computer code may specify instructions to include a construct like

$$x_t + I_{t-1} - I_t = d_t, \quad t = 1, \ldots, T$$

and the data tables could specify an actual number for T and numeric values for $\{d_t\}$ as well as generate "names" for variables $\{x_t\}$ and $\{I_t\}$. We now illustrate many of these ideas by showing how the linear program depicted in Figure 24 can be constructed.

To begin, consider the concept of a data table. This is not different,

OBJECTIVE		PRODUCTION VARIABLES				DISTRIBUTION VARIABLES								RESOURCES
		15	20	20	22	2	3	4	4	3	5	1	3	
		x_{11}	x_{21}	x_{12}	x_{22}	y_{111}	y_{112}	y_{211}	y_{212}	y_{121}	y_{122}	y_{221}	y_{222}	
Machine Capacity	Plant #1 A	2	1											≤ 100 (hours)
	Plant #1 B	3	5											≤ 200
	Plant #2 A			3	2									≤ 250
	Plant #2 B			4	4									≤ 100
Raw Materials	Plant #1	10	12											≤ 1000 (lbs)
	Plant #2			15	20									≤ 1500
Production-Distribution Balance	Plant #1	-1				1	1							= 0
	Plant #1		-1					1	1					= 0
	Plant #2			-1						1	1			= 0
	Plant #2				-1							1	1	= 0
Demand	Market #1					1				1				= 110 (units)
	Market #1							1				1		= 150
	Market #2						1				1			= 200
	Market #2								1				1	= 160

Figure 24 Production-distribution model.

Linear Programming

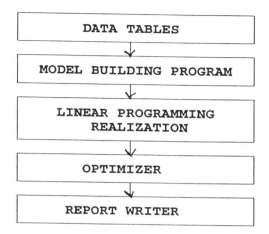

Figure 25 Elements of a computer-aided linear programming model.

in fact, from the ordinary notion of an "array" of data. As an example, consider the following data table:

Table DEMAND

	N	S
CHR	110	200
TBL	150	160

This table provides the capability of the model building language to specify the right-hand side of the demand constraints, where we now refer to markets 1 and 2 as market North and market South. For example, the North demand for Chairs is referred to by the number

DEMAND (CHR, N)

in the same fashion as one ordinarily refers to entries in a matrix; i.e., CHR refers to a row and N to a column. Another useful table would be

Table PRODCOST

	1	2
CHR	$15	$20
TBL	20	22

which represents the production costs for chairs and tables at the two different plants 1 and 2. The objective coefficient for the variable x_{11} could then be referred to by

PRODCOST (CHR, 1).

In addition, the algebraic variable x_{11} could be replaced with a user-friendly literal by using the row and column names from the data tables. The variable x_{11}, by itself, connotes nothing about chairs or plants. Alternatively, one could replace the reference to x_{11} with the literal

CHR1

by compounding the names of row 1 and column 1 from Table PROD-COST. The names of the distribution variables $\{y_{ijk}$ could be generated by using (and compounding) the row and column names from both Tables DEMAND and PRODCOST. The reasons for representing variables in large-scale models with mnemonic literals rather than algebraic symbols and subscripts are twofold:

Simpler debugging processes
Facilitation of report generation

When one is building a linear programming model, a routine debugging activity is for the modeler to "view" constraints generated by the software. If such a view is presented to the modeler in the form

$$2x_{29,2062} + 7x_{371,5162} - y_{1062} - z_{105,37} = 100$$

then the modeler would need to make several references to a "dictionary" of variable subscripts to verify that the constraint is valid. The subscripted x, y, and z variables, by themselves, give little information (to a human) about the constraint validity. The alternative being suggested here is to construct variable names using a sequence of letters and numbers that follow a precise pattern. As a more robust example, one might specify a family of variables in a form such as

$$(xxx)(y)(z) \qquad (9.2)$$

where xxx = product type
       y = originating plant
       z = market destination.

Hence, the variable name CHR1N could represent the number of chairs (CHR) to be produced at plant (1), destined for the market North. Using this naming convention, the first Demand constraint of Figure 25 would become

CHR1N + CHR2N ≥ 110.

Linear Programming

Such a constraint can be validated by a programmer without reference to a dictionary of subscripts, because all the needed information is explicit in the variable names. Variable names like (9.2) also greatly simplify aggregation of the optimal solution values into useful subsets, like the collection of all variables with *xxx* = CHR (which would represent the total of all chairs produced). The task of constructing a "report writer" is simplified by orders of magnitude if the variable (and row) names are deftly defined.

For the example in this section, one possible design of data tables and row/column names that could be used by a model building program to generate the linear programming realization depicted in Figure 24 is provided in Appendix A.2. Some characteristics of this design include:

- Consistency of names of rows and columns in the various tables (because these names will be used to refer to appropriate entries for retrieving coefficients).
- Consistency of units among all entries in a given table (we did not attempt to minimize the number of tables by mixing units such as raw material and machine hours).
- Tables are named in an obvious mnemonic fashion that is understandable by people (because people will be responsible for the integrity and validity of the standards used in the tables).

Finally, suppose that in this example we wanted to add a new product (say a "hutch," which we shorten to HUT) to be produced in plant 1. To generate a new linear programming realization to account for including this expansion of the problem, one would make the following modifications to the data tables in Appendix A.2:

Add a new column to MACHSTD1.
Add a new column to RAWSTD.
Add a new row to PRODCOST (with only one entry).
Add a new row to DIS1COST.
Add a new row to DEMAND.

Then, one executes the same computer program and a new linear program will be realized. No changes whatsoever in the actual computer code are required. The new model is realized completely through changes in the data tables.

9.3. Basic Approaches

Over the past three decades many approaches and computer languages have evolved for assisting industrial engineers in transforming a model and data into a linear program. A particularly insightful overview and comparison of these approaches is provided by Greenberg and Murphy [33]. These authors classify the approaches and computer languages in three general categories:

1. Process-oriented structures (OMNI)
2. Algebraic (GAMS)
3. Block structures (MathPro)

(The names in parentheses represent examples of commercial computer codes representative of the particular class.) In this section we provide the reader with a basic idea of the nature and usefulness of each of these approaches. However, specific details and instructions about specific computer software are not addressed. The example in the previous subsection (9.2) is used to highlight the fundamental strengths of each approach as well as provide contrasts to the advantages and disadvantages of the other methods.

Consider first the process-oriented approach. In this approach one views the linear program as a *collection of columns*. For example, the variable x_{11}, which we now denote as CHR1, is represented as a column with five nonzero entries. In a language like OMNI, an engineer would generate a column CHR1 with row entries as follows:

	CHR1
COST	15.
MACHA1	2.
MACHB1	3.
RAW1	10.
PLT1CHR	−1.

The names of rows, in general, would be obtained from the row and column names of the data tables (or, in some cases, even the names of the data tables themselves). The rows are named in a manner that facilitates people understanding; for example, MACHA1 is the name of the row for addressing the capacity of machine A in plant 1. Similar columns (and right-hand sides) are generated for all variables in the model. These columns are then delivered to a problem solver, like the simplex, for model optimization. This strategy is termed process oriented because the columns often represent real, physical processes. For example, CHR1 is a variable that represents the "process" of making chairs at plant 1. A chair requires 2 hours of machine A time, 3 hours of machine B time, and 10 pounds of raw material. Each chair also generates an input (-1 coefficient) to the PLT1CHR balance equation and $15 for the COST row.

There is a subtle but rather important advantage of this process-

Linear Programming

oriented approach. Business and industry typically organize themselves around processes. For example, there are probably a manager and a work force focused on making chairs in plant 1 and a similar group in plant 2. In this sense, the data and even the organization chart for the company may nicely coincide with this view of the linear programming model, i.e., as a collection of columns. Also, the idea of organizing the data as a collection of columns matches the general form of the input required by the simplex algorithm. In fact, the IBM format, which was established in the 1960s for data input to a simplex optimizer (MPS, MPSX), specifically requires that the data to be organized in a column format.

The second general approach for transforming a model and data into a linear program is algebraic in nature and is typified by (among others) the commercial code general algebraic modeling system (GAMS) [30]. In contrast to the process-oriented column focus, the algebraic approach focuses on the generation of rows. In many ways, this is a much more natural method for transforming an abstract model into a computer file, especially for engineers. Engineers deal with equations and inequalities on a daily basis. Most engineers would view the linear program in Figure 24 as a collection of equations and inequalities rather than a collection of columns.

The basic idea in the algebraic approach is that the algebraic relations in the abstract model should also be represented in the computer language in an algebraic fashion. This view is best illustrated through our example. The abstract model of Figure 24 would be specified as follows:

$$\text{Minimize} \sum_{i \in I} \sum_{j \in J} c_{ij} x_{ij} + \sum_{i \in I} \sum_{j \in J} \sum_{k \in K} d_{ijk} y_{ijk} \qquad (9.3)$$

subject to

(machine capacity) $\quad \sum_{i \in I} a_{im} x_{ij} \leq b_m, \quad m \in M, j \in J$

(raw material) $\quad \sum_{i \in I} f_{ij} x_{ij} \leq g_j, \quad j \in J$

(production-distribution) $\quad -x_{ij} + \sum_{k \in K} y_{ijk} = 0, \quad i \in I, j \in J$

(demand) $\quad \sum_{j \in J} y_{ijk} \geq d_{ik}, \quad i \in I, k \in K$

$\quad x_{ij}, y_{ijk} \geq 0$

In this algebraic approach, the specification of the constraints in the computer model should "look like" the abstract, algebraic representation. To illustrate this, we will describe some of the basic ideas of the GAMS language. In GAMS, in addition to data tables, one also defines *classes*. In this example, define the classes (or sets) as

$$I = \{CHR, TBL\} \quad \text{(products)}$$
$$J = \{1, 2\} \quad \text{(plants)}$$
$$M = \{A, B\} \quad \text{(machines)}.$$

In a manner analogous to the use of subscripts for ordinary variables like x_i or y_{ij}, one defines the variable X over classes. The *declaration*

X(I, J)

means that a class of variables (X) is defined so that a variable exists to represent each pair of elements in the sets I and J. For example, X(CHR,2) is one such variable. Constraints are constructed by appropriately combining coefficients, obtained from data tables, with variable names. Certain keywords are reserved for assisting in this process. For example, the keywork SUM is used to sum all the terms in an expression. The GAMS statement

$$\text{SUM}[I, \text{MACHSTD1}(A, I)*X(I, 1)] \tag{9.4}$$

specifies that the terms

MACHSTD1(A, I)*X(I, 1)

be summed for all elements in the set I; that is, (9.4) literally means

$$\sum_{I = CHR, TBL} \text{MACHSTD1}(A, I)*X(I, 1).$$

To understand how a class of constraints can be generated in GAMS, consider the first two constraints (machine capacities for plant 1) in Figure 24. These constraints, using the variable names generated by the tables and classes in our example, are

2X(CHR, 1) + 1 X(TBL, 1) ≤ 100   (machine A)
3X(CHR, 1) + 5 X(TBL, 1) ≤ 200   (machine B)

These two constraints can be generated in GAMS with the statement

MACH(M, 1)...SUM[I, MACHSTD1(M, I)*X(I, 1)] = L
= MACHAVL(M, 1).

Note that a constraint [named MACH(M,1)] is generated for each element in the set M. The presence of the set M in such an expression defines an implicit "do-loop" (which can be imbedded within other such loops). The reader is referred to [30] for details of these and other matters.

An obvious advantage of the algebraic approach is the direct one-to-one relationship between the rows in the abstract model and the computer

Linear Programming

code. Greenberg and Murphy [33] say, "Many model builders from academic backgrounds think readily in algebra, while people with process industry backgrounds think in terms of processes."

Is there an approach that integrates advantages of both of these preceding viewpoints? In fact, the so-called block structure approach attempts just this. The idea here is that any linear program is composed of several "blocks" of data (within the **A** matrix) and each of these blocks is characterized by a few simple patterns. For the linear program (9.3), these patterns are shown in Figure 26. In Figure 26, note that each of the five blocks and the RHSs has a unique and simple pattern. For example, block (4) is simply the identity matrix multiplied by -1. In the block structure approach, one takes advantage of these relative simple patterns in the construction of the model. In doing so, of course, one is simultaneously generating equations and columns.

The methodologies used in this block structure, not surprisingly, utilize basic procedures derived from both the process-oriented and algebraic approaches. This block structure approach is also relatively new. In 1987 Welch [98] described the practitioner's approach to modeling (PAM) as a modeling language "in which the practitioner describes an LP matrix with a set of two dimensional tables." MathPro [66] is a block structure methodology for building linear programming models and is composed of a set of screens specifically designed for interactive use on a PC. As such, this approach is somewhat more difficult to characterize and the reader is referred to [66] for further details.

In summary, all three approaches described here have the common objective of assembling numbers into the objective vector **c**, the technology matrix **A**, and the RHS vector **b**. From a symbolic point of view, the process-oriented approach accomplishes this by generating columns as illus-

	PRODUCTION VARIABLES	DISTRIBUTION VARIABLES	RHS
Machine Capacity	(1)		
Raw Materials	(2)		Supply
Production/ Distribution	(3)	(4)	0
Demand		(5)	Demand

Figure 26 Block patterns for illustrative problem

trated in Figure 27. In the algebraic approach, one focuses on generating rows (where $\mathbf{a}^i$ represents row i of the $\mathbf{A}$ matrix) as suggested in Figure 28.

9.4. Testing and Debugging

Any linear programming solution method will terminate in one of three possible states:

1. Optimal
2. Infeasible
3. Unbounded

Consider the case in which one just obtained the first "run" of a 5000 row × 10,000 column production-distribution model of a company and the optimizer says the problem is infeasible. What does one do? To begin, one must somehow decide whether the infeasibility occurred because of trying to accomplish something beyond the capability of plant and equipment, or whether mistakes were generated in the construction of the constraints, or whether signs were reversed or maybe numbers 10, 20, and 50 in a table should have been 100, 200, and 500. In this latter case, there could be a myriad of possible errors ranging from erroneously entered data to errors in the computer code. This section is about what to do when infeasibilities occur.

Typically, a optimizer will specify particular constraints that cannot be satisfied, and this is usually a good place to begin searching for problems. However, infeasibility is a property of the overall model and, in attempting to obtain feasibility, the optimizer may choose to satisfy some constraints and not others. In any case, our experience suggests that one should first verify that the structure of a model (what variables appear in

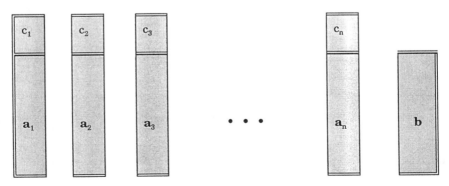

Figure 27 Building blocks in a process-oriented approach.

Linear Programming

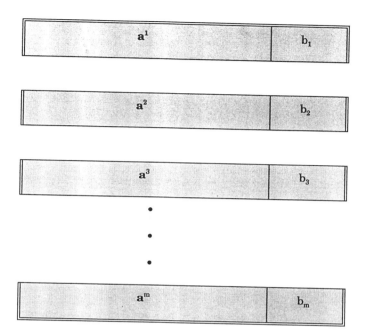

Figure 28 Building blocks in algebraic approach.

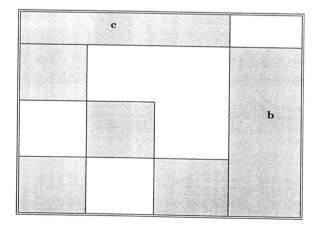

Figure 29 Building blocks in block structure approach.

what rows) actually reflects what one presumes has been generated. In essence, look at "pictures" of the various submatrices of the overall problem. If things appear as they should be (from a qualitative viewpoint), then a more detailed perusal of the size (and sign) of coefficients and right-hand sides should be checked. Were data from a table or set of tables processed as expected? Several "hand calculations" must often be done to verify the integrity of the code.

If a linear program terminates in infeasibility or unboundedness, Greenberg says that *mechanical* failures have occurred [32]. What to do when mechanical failures occur is often simpler than what to do when (and if) one obtains an optimal solution. If infeasibility or unboundedness occurs, one knows there is a problem and one knows what needs to be done — revise, check, or update the computer code and database and then seek optimality! But when an optimal solution is finally obtained, how does one know that the resultant solution is an accurate reflection of the model's intent? In a model with 5 constraints and 10 variables, integrity can be ensured by a simple inspection of each and every constraint. Solution integrity is not so obvious if the model is of size $5000 \times 10,000$.

With the experience gained in implementing several large-scale linear program models over several years [39, 49, 73, 78], four processes emerge that we recommend for ensuring the integrity of any large-scale linear programming model. The four processes are:

1. Verification of the desired model structure
2. Corroboration of data (units, size, current)
3. Comparison of model solution with current real-world activity
4. Scenario testing

These processes may involve some or all of the following individuals:

Model builders
Process and data resource personnel
Model users

The first process, verification of the desired model structure, is the responsibility of the model builders. Here, one needs to verify that the constraints and coefficients being generated by the computer code do, in fact, match one's specifications. The need for such a check in any computer routine is obvious. The task of corroboration of data is not so obvious. In large-scale models there may be a multitude of processes in which even the very definition of the units in the model is fuzzy. Moreover, data can become obsolete quickly. Here, process experts and data resource personnel of the company or firm must be given the opportunity to verify the database is accurate and valid. The third process, comparison of model solution with current real-world activity, is often the most important. Here, the model

Linear Programming

109

builders and model users compare the results of the model with what is currently happening in the company. If the company is currently building 500 refrigerators a month and the optimal solution to a linear programming product mix model specifies 0 refrigerators per month, then either there must be a huge "glitch" in the model or there is a startling indication that current operations and profitability of the company are far from optimal. However, such differences, when they occur, usually indicate that the model does not yet fully characterize the constraints and profitability of the firm's alternatives. Even more important, this process of comparing model results with current activities is important to ensure that the decision variables within the model are consistent with what the firm measures, counts, or monitors. If one cannot perform a reasonable, one-to-one accounting between the model output and firm's current operating system (production counts, labor hours, raw material usage, shipping records, etc.), then one may be optimizing the wrong model. As an example of how the variables of a model should be consistent with what is going on in the real world, we refer to the application to police scheduling [92] of Section 3.4. Here the authors say, "Prior to implementation, the San Francisco Police Department tested the system against manual schedules using the existing . . . scheduling strategy." In this application, the variables in the model corresponded the type of decisions being used by the San Francisco Police Department at that time.

Finally, there is an important and insightful process that the model builder can carry out alone, i.e., scenario testing. Here, one perturbs the database and determines whether the model solution changes in an expected manner. This is similar to what Greenberg [32] calls a "deep validity" test. For example, consider an application in a production and inventory planning context. If one increased the inventory holding costs, one would expect that in the new optimal solution the total inventory carried forward would decrease. If the inventory levels do not change no matter what the inventory costs are, there is probably some problem. Similarly, if the production standard (hours/part) is increased for some process, the operating level of this process should decrease or at least remain at its earlier level. The actual scenarios that one may consider depend, of course, on the nature of the model. However, if the model becomes a truly successful part of the business, there will be no shortage of scenarios suggested by interested parties.

10. BEYOND THE MODEL

10.1. An Early Application

In 1966, one of the authors was involved in an application of linear programming for the Stauffer Chemical Company in which weekly production schedules for several hundred products were generated. This particular application was certainly a successful one and, in retrospect, it is clear now that we addressed two important issues (without particularly realizing it) that are often overlooked in the implementation of linear programming:

> In a successful application, the optimizing step is only one part of the total effort required. Data acquisition, data management, interpretation of the solution, and report generation are essential activities.
> Involvement of the users and support from management are absolutely essential.

We were solving, on a weekly basis, a production scheduling problem with 220 constraints and 310 variables, a problem of reasonable size at that time. And, in those days, software for matrix generation and report writing was not widely available. But a problem with 200 constraints and 300 variables could still be processed in several hours. One prepared a deck of punched cards (one for each nonzero coefficient), which were then read onto a magnetic tape unit. The optimizing step required several minutes and the output was simply a printout of the 310 optimal values of the variables. These 310 numbers were then interpreted and manually converted into next week's shop floor production schedule.

This linear programming model (which was strongly influenced by Manne's paper [65]) was particularly simple to assemble. The model was designed so that weekly runs required one to update only the objective coefficients (**c** vector) and RHS (**b** vector). The same **A** matrix could be used week after week. In our paper [57], we say

> The clerical effort required to implement the model is minimal. Only the requirements vector . . . and objective function would ever have to be changed from one period to the next. . . . Unlike (our) model, the Manne model (with 100 products, 10 machines and four time periods) . . . would require 30,000 constraint coefficients to have to be computed for each application.

In fact, our 1966 model was explicitly designed, in part, to minimize clerical preparation time and the management of our input data. Actually, a major use of our time occurred *after* the optimal solution was obtained. This optimal solution was "transformed" into next week's production schedule using the same process routing sheets and other shop floor data that were currently in use. No additional training nor new forms and procedures were required.

Another vitally important factor in this application was the support and involvement of the users and the management at the plant. Management ownership of this new scheduling process was firmly entrenched from

Linear Programming

the beginning. The support was there because we addressed management's problem: generate a feasible schedule for meeting next week's demand.

10.2. Voice of Experience

In this section, we highlight some viewpoints expressed by practitioners who have been involved in linear programming applications. Our focus is on the overall context of the application and what the "voice of experience" says that really matters. In what follows we excerpt some particular viewpoints expressed by the practitioners in the survey of applications presented in Section 3.

One important theme, which arises with great regularity, is that management of data and the overall turnaround time for executing a study are major factors in one's overall success. In the application concerning the production of heart valves [41] reported earlier, the authors say,

> Two major points should be interjected here:
> - Data gathering, as described, is not cheap.
> - The linear programming model runs are only a minor part of the total effort.

In one of the applications dealing with forestry management in New Zealand [64], the authors say, "While computer runtime for any specific model is relevant, what is more important is the time it takes to carry out a total exercise."

As early as 20 years ago, the theme of data management was already being voiced as a critical issue. In a linear programming application in the petroleum industry in 1975, Wiig [100] described the company's approach to data as follows:

> A considerable database is required for operation of the planning model. The model is maintained regularly, and is tied to the company's general processing system. . . . key elements are refinery capacities, manufacturing costs, . . . pipeline tariffs, . . . terminal inventories.

Another often overlooked aspect of a successful linear programming application is how the model results are transformed from a set of optimal primal and dual variables (x, w) into information meaningful to decision makers. In our 1966 application at the Stauffer Chemical plant discussed earlier, we converted (with some assistance from company personnel) the optimal linear programming solution onto existing scheduling forms. Today, such laborious tasks can be made part of the overall system. Software (called report writers) exists that can give management the optimal solution in a user friendly fashion. Such report writers are designed to collect efficiently various subsets of variables (like the number of refrigerators to be scheduled in shift 3 at plants A, B, and C in February) and present the results in a simple, clear format. In the application of linear programming at North American Van Lines reported earlier [1], the authors describe three different output reports:

> *Fleet Development Reports* include forecasted (1) fleet sizes, . . . (2) class schedules . . . (3) inventory levels.
> *Tractor Sales and Warranty Reports* reflect net contribution . . . book values . . .
> *Used Truck Allowances* . . . inventory wholesale prices . . . get-ready costs . . . .

At United Airlines [45], a discussion of their report writer includes tasks such as "produces monthly shift schedules and places them in an interactive database where they can be accessed by schedulers. In addition, it produces a variety of coverage and cost reports associated with each shift schedule."

The second issue, and likely the most important for a successful application of linear programming, is the need for strong involvement of the users and support from management. It is, indeed, difficult to underestimate this point and the literature abounds with such testimony. Mehring and Gutterman [68] describe a linear programming-based effort to optimize the supply and distribution of products at Amoco (U.K.) Limited. This is an excellent paper about the potential use of linear programing in the petroleum industry. However, the model was never actually used. The authors' say:

> We attribute Amoco U.K.'s failure to use (the model) largely to the characteristics of the environment and the implementation process. Obstacles to substantial user involvement and management support arose early in this project and, as history now suggests, were never dealt with satisfactorily.

A paper on the use of linear programming for scheduling nurses reinforces the point. Franz and Miller [27] say:

> As attractive as the arguments may be for the advantages of the resident scheduling model, the implementation effort must be regarded as a failure at this time. . . . This study reinforces the need for an internal organizational champion or change agent to sell an approach to decision makers. Clearly, any project met with indifference from top management is doomed unless there are insiders to advocate and support the project.

Linear Programming

In contrast, the support and involvement of management are always stressed in successful applications. The application at CITGO [55] resulted in a $70 million annual savings. The authors say, "The success of the project . . . is attributable to several factors. Foremost was the support of top management. . . . A second factor of almost equal importance was the enthusiastic and dedicated support of operational managers."

The notion of user involvement and management support is now sometimes expressed in other terminology, namely "satisfying the customer." The customer, in our context, may want to minimize costs, or meet production requirements, or reduce inventory, or, in the final analysis, make a profit. Linear programming is a means, not an end itself, as eloquently stated by Woolsey [102]:

> . . . reasons for failure in the field of mathematical programming have nothing to do with modeling or technical expertise . . . reasons are . . . failure because the modeler does not realize that the method is a means to an end rather than the end itself.

As a concluding comment for this section, we include a short excerpt from a paper by Lockett [60] reviewing the current status of linear programming and its applications in 1985. We believe one of his concluding paragraphs is just as applicable today as nearly 10 years ago:

> Many of the authors mention that linear programming should be seen as only part of a "process". Nearly all the excellent examples of real applications discussed earlier are presented in this fashion. The authors understand the organizational environment within which the model will have to operate, and this is usually far more important and difficult to deal with than the mathematics. It is this recognition of the role of models within the overall process which gives us hope for the future. If we are to use the power of linear programming to the full, its limitations must also be accepted. Therefore, one of the main barriers will be researchers themselves who do not spend enough time on understanding the process under observation. More time spent on this, and less on the mathematics of modelling, would be very rewarding. When discussing the reasons for failure the literature comes to the same type of conclusions, e.g., "lack of user commitment, lack of user understanding, and high organizational resistance to change." In order to introduce models we have to make changes within organizations, i.e., there is a process to go through. It is seeing the problems in their true organizational context that can bring great rewards.

APPENDIX A.1: EMPLOYMENT OF INDUSTRIAL ENGINEERS IN THE UNITED STATES*

I. *Manufacturing*				77%
1. Durable goods			65%	
a. Transportation equipment (cars, planes)	25%			
b. Electronic/electrical	10%			
c. Industrial machinery	10%			
d. Other	20%			
2. Nondurable Goods			12%	
a. Chemical/rubber	5%			
b. Textile/apparel	3%			
c. Food	2%			
d. Paper	1%			
e. Other	1%			
II. *Service*				23%
a. Transportation/public utilities	6%			
b. Engineering/management services	5%			
c. Government	3%			
d. Financial	2%			
e. Other (health, leisure, etc.)	7%			
Total				100%

Source: 1990 Bureau of Labor statistics

Linear Programming

APPENDIX A.2: DATA FOR PRODUCTION–DISTRIBUTION MODEL

1. Production Data

(Plant 1)

Table MACHSTD1

	CHR	TBL
A	2	1
B	1	5

(Plant 2)

Table MACHSTD1

	CHR	TBL
A	3	2
B	4	4

2. Materials and Machine Availability Data

Table RAWSTD

	CHR	TBL
1	10	12
2	15	20

Table MACHAVL

	1	2
A	100	250
B	200	100

3. Production Cost Data

Table PRODCOST

	1	2
CHR	15	20
TBL	20	22

4. Distribution Cost Data

(Plant 1)

Table DIS1COST

	N	S
CHR	2	3
TBL	3	4

(Plant 2)

Table DIS2COST

	N	S
CHR	3	1
TBL	5	3

5. Demand Data

Table DEMAND

	N	S
CHR	110	200
TBL	150	160

REFERENCES

1. Avramovich, D., Cook, T. M., Langston, G. D., Sutherland, F., "A Decision Support System for Fleet Management: A Linear Programming Approach," *Interfaces*, Vol. 12, pp. 1-9, 1982.
2. Balbirer, S. D., Shaw, D., "An Application of Linear Programming to Bank Financial Planning," *Interfaces*, Vol. 11, pp. 77-83, 1981.
3. Bazaraa, M. S., Jarvis, J. J., Sherali, H. D., *Linear Programming and Network Flows*, 2nd ed. Wiley, New York, 1990.
4. Beckwith, R. E., Vaswani, R., "The Assignment Problem—A Special Case of Linear Programming," *Journal of Industrial Engineering*, Vol. 8, pp. 167-172, 1957.
5. Blais, J., LaMont, J., Rousseau, J., "The HASTUS Vehicle and Manpower Scheduling System at the Société de Transport de la Communaute urbaine de Montréal," *Interfaces*, Vol. 20, pp. 26-42, 1990.
6. Bland, R. G., "New Finite Pivoting Rules for the Simplex Method," *Mathematics of Operations Research*, Vol. 2, pp. 103-107, 1977.
7. Bookbinder, J. H., McAuley, P. T., Schulte, J., "Inventory and Transportation Planning in the Distribution of Fine Papers," *Journal of Research Society of America*, Vol. 40, No. 2, pp. 155-166, 1989.
8. Brosch, L., Buck, R., Sparrow, W., "Boxcars, Linear Programming and Sleeping Kitten," *Interfaces*, Vol. 10, pp. 53-61, 1980.
9. Charnes, A., "Optimality and Degeneracy in Linear Programming," *Econometrica*, Vol. 20, pp. 160-170, 1952.
10. Charnes, A., Cooper, W. W., "The Stepping Stone Method of Explaining Linear Programing Calculations in Transportation Problems," *Management Science*, Vol. 1, pp. 49-69, 1954.
11. Charnes, A., Cooper, W. W., Farr, D., "Linear Programming and Profit Preference Scheduling in a Manufacturing Firm," *Journal of Operations Research Society of America*. Vol. 1, pp. 114-129, 1953.
12. Charnes, A., Cooper, W. W., Mellon, B., "Blending Aviation Gasolines," *Econometrica*, Vol. 20, pp. 135-159, 1952.
13. Dantzig, G. B., "Programming in a Linear Structure," USAF Comptroller Report, Washington, DC, 1948.
14. Dantzig, G. B., "Maximization of a Linear Function of Variables Subject to

Linear Inequalities," in *Activity Analysis of Production and Allocation*, ed. T. C. Koopmans, pp. 339-347. Wiley, New York, 1951.
15. Dantzig, G. B., *Linear Programming and Extensions*. Princeton University Press, Princeton, NJ, 1963.
16. Dantzig, G. B., Orden, A., Wolfe, P., "The Generalized Simplex Method for Minimizing a Linear Form Under Linear Inequality Restraints," *Pacific Journal of Mathematics*, Vol. 5, pp. 183-195, 1955.
17. Dikin, I. I., "Iterative Solution of Problems of Linear and Quadratic Programming" (in Russian), *Doklady Akademiia Nauk USSR* Vol. 174, pp. 747-748; (English translation) *Soviet Mathematics Doklady*, Vol. 8, pp. 674-675, 1967.
18. Eaton, D. J., Daskin, M. S., Simmons, D., Bullock, W., Jansma, G., "Determining Emergency Medical Service Vehicle Deployment in Austin, Texas," *Interfaces*, Vol. 15, pp. 96-108, 1985.
19. Edwards, J. R., Wagner, H. M., Wood, W. P., "Blue Bell Trims Its Inventory," *Interfaces*, Vol. 15, pp. 34-52, 1985.
20. Emerson, H. P., "A Mathematical Foundation for Industrial Engineering," *Journal of Industrial Engineering*, Vol. 4, pp. 14-18, 1953.
21. Erickson, V., Randolph, P. H., "An Application of Linear Programming to the Assignment of Materials Handling Equipment," *Journal of Industrial Engineering*, Vol. 8, pp. 386-388, 1957.
22. Fang, S.C., Puthenpura, S., *Linear Optimization and Extensions: Theory and Algorithms*, Prentice-Hall, Englewood Cliffs, NJ, 1993.
23. Farley, A. A., "Mathematical Programming Models for Cutting-Stock Problems in the Clothing Industry," *Journal of Operations Research Society of America*, Vol. 39, pp. 41-53, 1988.
24. Flood, M. M., "Application of Transportation Theory to Scheduling a Military Tanker Fleet," *Journal of Operations Research*, Vol. 2, pp. 150-162, 1953.
25. Fourer, R., Gay, D. M., Kernighan, B. W., "A Modeling Language for Mathematical Programming," *Management Science*, Vol. 36, pp. 519-554, 1990.
26. Francis, R. L., McGinnis, L. F., White, J. A., *Facility Layout and Location: An Analytical Approach*, 2nd ed. Prentice-Hall, Englewood Cliffs, NJ, 1992.
27. Franz, L. S., Miller, J., "Scheduling Medical Residents to Rotations: Solving the Large-Scale Multiperiod Staff Assignment Problem," *Journal of Operations Research Society of America*, Vol. 41, pp. 269-279, 1993.
28. Garg, U., Tsai, C., "Modeling and Analysis of a Large Wire and Cable Plant Operation," *Interfaces*, Vol. 16, pp. 77-85, 1986.
29. Garvin, W. W., Crandall, H. W., Small, J. B., Spellman, R. A., "Applications of Linear Programming in the Oil Industry," *Management Science*, Vol. 3, pp. 407-430, 1957.
30. Brooke, A., Kendrick, D., Meeraus, A., *General Algebraic Modeling Systems (GAMS)*, Scientific Press, San Francisco, 1992.
31. Gilmore, P. C., Gomory, B. E., "A Linear Programming Approach to the

Cutting Stock Problem," *Journal of Operations Research Society of America*, Vol. 9, pp. 849–859, 1961.
32. Greenberg, H. J., "How to Analyze the Results of Linear Programs — Part 1: Preliminaries," *Interfaces*, Vol. 23, p. 56–67, 1993.
33. Greenberg, H. J., Murphy, F. H., "A Comparison of Mathematical Programming Modeling Systems," *Annals of Operations Research*, Vol. 38, pp. 177–238, 1992.
34. Greene, J. H., Chatto, L., Hicks, C. R., Cox, C. B., "Linear Programming in the Packing Industry," *Journal of Industrial Engineering*, Vol. 10, pp. 364–372, 1959.
35. Gryna, F. M., "Industrial Engineering and Operations Research," *Journal of Industrial Engineering*, Vol. 5, No. 12, p. 211, 1954.
36. Haessler, R. W., Sweeney, P. E., "Cutting Stock Problems and Solution Procedures," *European Journal of Operations Research*, Vol. 54, pp. 141–150, 1991.
37. Harpell, J. L., Lane, M. S., Mansour, A. H., "Operations Research in Practice: A Longitudinal Study," *Interfaces*, Vol. 19, pp. 65–74, 1989.
38. Harrison, H., "Management Science and Productivity Improvement in Irish Milk Cooperatives,"*Interfaces*, Vol. 16, No. 4, pp. 31–40, 1986.
39. Hibbard, W. R., Sherali, H. D., Soyster, A. L., Sousa, L. J., "Supply Prospects for U.S. Copper Industry: Alternative Scenarios: MIDAS II Computer Model," *Materials and Society*, Vol. 6, pp. 201–210, 1982.
40. Hibbard, W. R., Soyster, A. L., Gates, R. S., "A Disaggregated Supply Model of the U.S. Copper Industry Operating in an Aggregated World Econometric Supply/Demand System," *Materials and Society*, Vol. 4, pp. 261–284, 1980.
41. Hilal, S. S., Erikson, W., "Matching Supplies to Save Lives: Linear Programming the Production of Heart Valves," *Interfaces*, Vol. 11, pp. 48–56, 1981.
42. Lenstra, J. K., Rinnooy Kan, A. H. G., Schrijver, A., eds., *History of Mathematical Programming*. North-Holland, Amsterdam, 1991.
43. Hoffman, A. J., "Linear Programming at the National Bureau of Standards," in *History of Mathematical Programming*, eds. J. K. Lenstra et al. North-Holland, Amsterdam, 1991.
44. Holcomb, M. C., DePorter, E. L., "A Linear Programming Application Helps Feed the Homeless," *Computers and Industrial Engineering*, Vol. 19, pp. 548–552, 1990.
45. Holloran, T. J., "United Airlines Station Manpower Planning System," *Interfaces*, Vol. 16, pp. 39–50, 1986.
46. Hooker, J. N., "Karmarkar's Linear Programming Algorithm," *Interfaces* 16, pp. 75–90, 1986.
47. Hurlimann, T., "Software Tools for Mathematical Programming," Editorial, *European Journal of Operational Research*, Vol. 72, p. 213, 1994.
48. Ignizio, J. P., Cavalier, T. M., *Linear Programming*. Prentice-Hall, Englewood Cliffs, NJ, 1994.
49. Karmarkar, N., "A New Polynomial Time Algorithm for Linear Programming," *Combinatorica*, Vol. 4, pp. 373–395, 1984.

50. Karmarkar, N., Ramakrishnan, K. G., "Implementation and Computational Results of Karmarkar's Algorithm for Linear Programming," presented at the 13th Symposium on Mathematical Programming, Tokyo, 1988.
51. Kent, B., Bare, B., Field, R. C., Bradley, G., "Natural-Resource Land Management Planning Using Large-Scale Linear Programs: The USDA Forest Service experience with FORPLAN," *Journal of Operations Research Society of America*, Vol. 39, pp. 13–27, 1991.
52. King, R. H., Love, R., "Coordinating Decisions for Increased Profits," *Interfaces*, Vol. 10, pp. 4–19, 1980.
53. Klee, V., Minty, G., "How Good is the Simplex Algorithm?" in *Inequalities III*, ed. O. Shisha, pp. 159–175. Academic Press, New York, 1972.
54. Klingman, D., Mote, J., Phillips, N., "A Logistics Planning System at W. R. Grace," *Journal of Operations Research Society of America*, Vol. 36, pp. 811–822, 1988.
55. Klingman, D., Phillips, N., Steiger, D., Young, W., "The Successful Deployment of Management Science Throughout Citgo Petroleum Corporation," *Interfaces*, Vol. 17, pp. 4–25, 1987.
56. Klingman, D., Phillips, N., Steiger, D., Wirth, R., Rema, P., Krishnan, R., "An Optimization Based Integrated Short-Term Refined Petroleum Product Planning System," *Management Science*, Vol. 33, pp. 813–830, 1987.
57. Kortanek, K. O., Sodaro, D., Soyster, A. L., "Multi-Product Production Scheduling via Extreme Point Properties of Linear Programming," *Naval Research Logistics Quarterly*, Vol. 15, pp. 287–300, 1966.
58. Lehrer, R. N., " OR or IE?" *Journal of Industrial Engineering*, Vol. 4, p. 24, 1954.
59. Liberatore, M. J., Miller, T., "A Hierarchical Production Planning System," *Interfaces*, Vol. 15, pp. 1–13, 1985.
60. Lockett, G., "Applications of Mathematical Programming—Before, Now and After," *Journal of Operations Research Society*, Vol. 36, pp. 347–356, 1985.
61. Love, R. F., Morris, J. G., Wesolowsky, G. O., *Facilities Location: Models and Methods*. North-Holland, New York, 1988.
62. Magee, J. F., "Linear Programming in Production Scheduling," (First) Operations Research National Meeting, Washington, DC, 1952.
63. Malcolm, D. G., "Operations Research-Widening the Horizons of Industrial Engineering," *Journal of Industrial Engineering*, Vol. 5, pp. 3–6, 1954.
64. Manley, B., Threadgill, J., "LP Used for Valuation and Planning of New Zealand Plantation Forests," *Interfaces*, Vol. 21, pp. 66–79, 1991.
65. Manne, A. S., "Programming of Economic Lot Sizes," *Management Science*, Vol. 4, 1958.
66. *MathPro Usage Guide*. MathPro Inc., Washington, DC, 1990.
67. McGarrah, R. E., "Production Programming," *Journal of Industrial Engineering*, Vol. 7, pp. 263–270, 1956.
68. Mehring, J. S., Gutterman, M. M., "Supply and Distribution Planning Support for Amoco (U.K.) Limited," *Interfaces*, Vol. 20, 95–104, 1991.
69. Mehrotra, S., "On Finding a Vertex Solution using Interior Point Methods," *Linear Algebra and Its Applications*, Vol. 152, pp. 106–111, 191.

70. Metzger, R. W., Schwarzbek, R., "A Linear Programming Application to Cupola Charging," *Journal of Industrial Engineering*, Vol. 12, pp. 87–93, 1961.
71. Might, R. J., "Decision Support for Aircraft and Munitions Procurement," *Interfaces*, Vol. 17, pp. 55–63, 1987.
72. Morse, P. M., "Trends in Operations Research," *Journal of Operations Research Society of America*, Vol. 1, pp. 159–165, 1953.
73. Murphy, F. H., Soyster, A. L., *Economic Behavior of Electric Utilities*. Prentice-Hall, Englewood Cliffs, NJ, 1983.
74. Oliff, M., Burch, E., "Multiproduct Production Scheduling at Owens-Corning Fiberglass," *Interfaces*, Vol. 15, pp. 25–34, 1985.
75. Orden, A., "LP from the 40s to the 90s," *Interfaces*, Vol. 23, pp. 2–12, 1993.
76. Ozkarahan, I., Bailey, J., "Goal Programming Model Subsystem of a Flexible Nurse Scheduling Support System," *IIE Transactions*, Vol. 20, pp. 306–316, 1988.
77. Price, W. L., Turcotte, M., "Locating a Blood Bank," *Interfaces* 16, pp. 17–26, 1986.
78. Rapoport, L. A., Soyster, A. L., "LORENDAS: A Computer-Based System for Modeling of Long Range Energy Development and Supplies," *Natural Resources Forum*, Vol. 2, pp. 19–36, 1977.
79. Rautman, C. A., Reid, R. A., Ryder, E. E., "Scheduling the Disposal of Nuclear Waste Material in a Geologic Repository Using the Transportation Model," *Journal of Operations Research Society of America*, Vol. 41, pp. 459–469, 1993.
80. Salveson, M. E., "A Mathematical Theory of Production: Planning and Scheduling," *Journal of Industrial Engineering*, Vol. 4, pp. 3–6, 1953.
81. Salveson, M. E., "Mathematical Methods of Management Programming," *Journal of Industrial Engineering*, Vol. 5, pp. 9–15, 1954.
82. Salveson, M. E., "A Problem in Optimal Machine Loading," *Management Science*, Vol. 2, p. 232, 1956.
83. Schrage, L., *LINDO*. The Scientific Press, 1991.
84. Sear, T. N., "Logistics Planning in the Downstream Oil Industry," *Journal of Operational Research Society*, Vol. 44, pp. 9–17, 1993.
85. Shanno, D. F., "Computing Karmarkar Projection Quickly," *Mathematical Programming*, Vol. 41, pp. 61–71, 1988.
86. Shapiro, M., "Scheduling Crewmen for Recurrent Training," *Interfaces*, Vol. 11, pp. 1–8, 1981.
87. Smith, W. L., "Current Status of the Industrial Use of Linear Programming," *Management Science*, Vol. 2, pp. 156–158, 1956.
88. Stroup, J. S., Wollmer, R. D., "A Fuel Management Model for the Airline Industry," *Journal of Operations Research Society of America*, Vol. 40, pp. 229–237, 1992.
89. Taha, H. A., *Operations Research: An Introduction*. Macmillan, New York, 1971.
90. Tapia, R. A., Zhang, Y., "Cubically Convergent Method for Locating a

Nearby Vertex in Linear Programming," *Journal of Optimization Theory and Applications*, Vol. 67, pp. 217–225, 1990.
91. Tapia, R. A., Zhang, Y., "An Optimal-Basis Identification Technique for Interior Point Linear Programming Algorithms," *Linear Algebra and Its Applications*, Vol. 152, pp. 343–363, 1991.
92. Taylor, P. E., Huxley, S. J., "A Break from Tradition for the San Francisco Police: Patrol Officer Scheduling Using an Optimization-Based Decision Support System," *Interfaces*, Vol. 19, pp. 4–24, 1989.
93. Thompson, J. W., "The Industrial Engineer Uses Operations Research," *Journal of Industrial Engineering*, Vol. 7, pp. 101–160, 1956.
94. Thompson, R. G., Singleton, F. D., Thrall, R. M., Smith, B. A., "Comparative Site Evaluations for Locating a High-Energy Physics Lab in Texas," *Interfaces*, Vol. 16, pp. 35–49, 1986.
95. Todd, M. J., Ye, Y., "A Centered Projective Algorithm for Linear Programming," *Mathematics of Operations Research*, Vol. 15, pp. 508–529, 1990.
96. Ye, Y., "Recovering Optimal Basic Variables in Karmarkar's Polynomial Algorithm for Linear Programming," *Mathematics of Operations Research*, Vol. 15, pp. 564–572, 1990.
97. Von Lanzenauer, C. H., Harbrauer, E., Johnston, B., Shuttleworth, D., "RRSP Flood: LP to the Rescue," *Interfaces*, Vol. 17, pp. 27–33, 1987.
98. Welch, J. "PAM-A Practitioner's Guide to Modeling," *Management Science*, Vol. 33, pp. 610–625, 1987.
99. Whitaker, D., Cammell, S., "A Partitioned Cutting-Stock Problem Applied in the Meat Industry," *Journal of Operations Research Society of America*, Vol. 41, pp. 801–807, 1990.
100. Wiig, K. M., "Petroleum Product Distribution Planning System by Linear Programming," *Studies in Linear Programming*, eds. Salkin and Saha, North-Holland/American Elsevier, New York, 1975.
101. Wilson, E. J. G., Willis, R. J., "Scheduling of Telephone Operators – A Case Study," *Journal of the Society of Operational Research*, Vol. 34, pp. 991–997, 1983.
102. Woolsey, R. E. D., "A Novena to St. Jude, or Four Edifying Case Studies: In Mathematical Programming," *Interfaces*, Vol. 4, pp. 32–39, 1973.

3

Integer Programming

Timothy M. Magee and Fred Glover
University of Colorado, Boulder, Colorado

Integer programming (IP) is an extension of linear programming that encompasses a greatly enriched domain of practical applications by admitting an additional class of variables that are restricted to take on only integer values.

The modeling capability that results from using integer variables is surprisingly powerful. The most obvious use of integer variables is to model decisions that are integral by nature, such as the number of airplanes to purchase, the number of cargo ships to replace, and the number of employees to be assigned to a new organization.

However, a wide variety of problems that seem on the surface to have little or nothing to do with "integer-constrained linear optimization" can nevertheless be given an integer programming formulation. Indeed, the requirement of discreteness indirectly, if not directly, pervades many significant classes of problems and provides integer programming with application in an uncommonly wide variety of theoretical and practical disciplines. Production sequencing, job shop scheduling, logistics, plant location, assembly line balancing, mineral exploration, capital budgeting, resource allocation, and facilities planning constitute a few of the important industrial problem areas that frequently fall within the wider domain of integer programming.

Integer programming is not confined to industry, however; it also finds application in many other combinatorial optimization problems arising in engineering and scientific contexts. Computer design system reliability, prime implicant selection, signal coding, and energy storage system design all give rise to classes of problems with integer programming formulations.

Integer programming also has applications to economic analysis. The assumption of continuity underlying traditional economic theory is incompatible with the existence of indivisible plants, setup costs, and developmental expenditures. Conclusions reached by marginal analysis must, therefore, often be amended (or stated with appropriate qualification) to accommodate considerations that are the focus of integer programming.

Careful attention to modeling is essential to solving integer programs. Unlike linear programming formulations, integer programming formulations can easily be generated that are in principle correct but in practice unsolvable. Section 1 describes generic modeling techniques that have been used in a wide variety of applications.

Integer programming problems are both theoretically and practically more difficult to optimize than linear programming problems. No single solution methodology is appropriate for all problems. Frequently, tailoring a generic methodology to take advantage of the special structure of a particular problem is successful in producing good, and sometimes optimal, solutions. Thus, we focus on the general solution approaches: cutting planes (Section 2), branch and bound (Section 3), and lagrangean relaxation (Section 4).

The generic modeling techniques and solution methodologies presented here should be viewed not as an exhaustive summary of the integer programming landscape but rather as the main roads in that landscape. Many integer programs are accessible with these modeling techniques and methodologies in the sense of being solvable to optimality or near optimality in a practical time limit if the problems are not too large. In particular, mathematical programs arising from adding a small number of integer programming constructs to an essentially linear model are generally solvable.

However, the vast majority of integer programs also contain unique characteristics that can be exploited through a combination of clever modeling and specialized algorithms. In some cases, specialized methods can accelerate solution by one or even two orders of magnitude. At the same time, because of the potential complexity of integer programs, there remains a sizable collection for which "theoretically convergent" methods (that ensure optimal solutions in a finite amount of time) fail to obtain reasonably good solutions within many hours or even days of computer time. Models and methods that enable such problems to be solved much more effectively represent the frontiers of current research.

Integer Programming

Specific applications of integer programming in industrial engineering with references are presented in Section 5.

1. INTEGER PROGRAMMING MODELS

In this section we present a general mathematical formulation for integer programs. We also provide specific integer programming modeling techniques for extending linear programming to incorporate the following elements:

Integral quantities
Either-or decisions
Conditionally limiting the range of a continuous variable
Multiple choice constraints
Logical constraints
Some nonlinear functions

A linear programming problem in which all variables are additionally constrained to be integer is called a *pure* integer programming problem (PIP), and one in which only some of the variables are additionally constrained to be integer is called a *mixed* integer programming problem (MIP). Variables that are not integer constrained are called *continuous* variables. (Thus, in ordinary linear programming problems all variables are continuous.) The words "integer programming" often have reference to the pure integer problem, but we allow them also to refer to the more general mixed problem depending on the setting.

1.1. Mathematical Formulation of Integer Programs

A standard formulation of the linear mixed integer program is

$$\text{Maximize} \quad \sum_{j=1}^{n} c_j x_j + \sum_{k=1}^{p} h_k y_k$$

$$\text{subject to} \quad \sum_{j=1}^{n} A_{ij} x_j + \sum_{k=1}^{p} G_{ik} y_k \le b_k \quad \forall i$$

$$x_j, y_k \ge 0, \, x_j \text{ is integer}$$

or, in matrix notation,

$$\text{Maximize} \quad cx + hy$$
$$\text{subject to} \quad Ax + Gy \le b$$
$$x, y \ge 0, \, x \text{ is integer}$$

where x and y are the decision variable vectors and a problem instance is specified by the data (c, h, A, G, b).

Formulations that include minimization objectives and constraints of the form $Ax + Gy \geq b$ or $Ax + Gy = b$ can be stated equivalently in this standard form. The transformation of representation is identical to the transformation used for linear programming formulations.

1.2. Integral Quantities

A simple example of an integer program with integral quantities is that of a gardener who needs 107 pounds of fertilizer for his lawn and has the option of buying it either in 35-pound bags at \$14 each or 24-pound bags at \$12 each. His goal is to buy (at least) the 107 pounds he needs at the cheapest cost. Letting x_1 be the number of 35-pound bags he buys and x_2 be the number of 24-pound bags he buys, he seeks to

$$\begin{aligned}\text{Minimize} \quad & 14x_1 + 12x_2 \\ \text{subject to} \quad & 35x_1 + 24x_2 \geq 107 \\ & x_1, x_2 \geq 0 \text{ and integer}\end{aligned}$$

The "and integer" stipulation means that the gardener can't buy half or a third of a bag of fertilizer but must buy a whole bag or none at all. Without this stipulation the problem is an example of an ordinary linear programming problem.

To get a rudimentary grasp of how the integer restriction can affect the character of the problem, consider what would happen if the restriction were missing. The gardener would then observe that fertilizer in the 35-pound bags costs 40 cents a pound and fertilizer in the 24-pound bags costs 50 cents a pound and immediately perceive that his best policy would be to buy 3.06 of the 35-pound bags. However, in the presence of the integer restriction, his best policy is to buy only one of the 35-pound bags and buy three of the 24-pound bags. Clearly, his best policy has changed dramatically, a not at all unusual result of requiring the variables to be integer valued.

1.3. Rounding

In practice, the goal of obtaining integral quantities is usually not a sufficient motivation for using integer programming instead of linear programming if the quantities involved are large. Integral quantities are usually of interest only for small values of the variables; integer variables taking large values can often be approximated as continuous variables. If the optimization generates a fractional value for such a variable, rounding either up

Integer Programming

or down will not cause significant degradation in the solution for most applications. (Exceptions, however, do occur.) Uncertainty and modeling approximations usually have a greater impact on the usefulness of a solution.

In contrast, explicitly treating variables taking on large values as integer variables can create extremely difficult problems. Models with several large integer variables may have a large number of solutions close to the optimal solution and hence complicate the search for the optimal solution.

A small modification of the fertilizer example illustrates the point. Suppose the gardener has now become a farmer and purchases not 107 pounds of fertilizer but instead 1070 pounds. Removing the integer restriction leads to a best policy of purchasing 30.6 35-pounds bags. Buying either 30 or 31 bags is likely to be a good solution, within 2% of the linear programming solution. The figure of 1070 pounds is likely an estimate of the fertilizer required. A shortfall could be made up with later purchases. Any excess can be stored and used in the future.

In contrast, if the original gardener rounds up from 3.06 to four 35-pounds bags, he creates a 30% change from the linear programming solution. If rounding down to three 35-pound bags truly is infeasible, then integer programming would be justified for the gardener. As will be seen in subsequent sections, there are many IP problems in which the variables receive not only small values but only the values 0 and 1. For such problems, and in general for problems in which the integer variables take on small values, rounding can be dangerous even as an approximation strategy. See Glover and Sommer (1975). (In fact, an example exists of a "conditional transportation problem" that can be summarized by a 5 × 5 cost matrix and whose linear programming solution can be rounded in more than a million ways, none of which yields a feasible, let alone optimal, solution.)

1.4. Either-Or Decisions

Either-or decisions are extremely common. Examples include whether or not to build a new warehouse at a site, to buy a new machine, to run a third shift, to produce a new product, or to plant a new crop.

Our concern is with problems that contain several either-or decisions. Problems with a single important decision, such as whether or not to build a new dam, are usually best handled with other tools, such as decision analysis and simulation, which incorporate probabilistic factors influencing the decision. On the other hand, integer programming models involving more than a single variable can also be embedded in simulation and scenario analysis to handle elements of uncertainty. The domain called *discrete robust optimization* is dedicated to treating such issues.

Either-or decisions can be modeled with integer variables that are further restricted to be either 0 or 1, called *zero-one* variables or *binary* variables. All other integer variables are called *general* integer variables. Integer programs are classified by use of general integer variables: *general integer programming* implies that general integer variables are used, and *zero-one programming* implies that only zero-one variables are used.

Modeling with binary variables can be illustrated with the decision to build a warehouse at site j. The value of the variables x_j corresponds to the decision. When $x_j = 1$, the decision is to build at site j. When $x_j = 0$, the decision is to not build at site j.

1.4.1. The Knapsack Problem

One of the simplest integer programming problems that uses binary variables is the knapsack problem, so called because it can be interpreted as a problem of selecting a best set of items to go in a hiker's knapsack, given the value he attaches to these items and an upper limit on the amount of weight he can carry. Let c_j represent the value of item j, a_j the weight of item j, x_j the decision to pack item j, and b the upper limit on total weight. The integer programming formulation is

$$\text{Maximize} \quad \sum_j c_j x_j$$
$$\text{subject to} \quad \sum_j a_j x_j \leq b$$
$$x_j \in \{0, 1\}$$

For example, consider the problem

$$\text{Maximize} \quad 30x_1 + 17x_2 + 14x_3 + 11x_4 + 9x_5$$
$$\text{subject to} \quad 29x_1 + 20x_2 + 16x_3 + 12x_4 + 10x_5 \leq 37$$
$$x_j \in \{0, 1\}$$

If the integrality constraints are relaxed to produce an ordinary linear program, by replacing $x_j \in \{0, 1\}$ with $0 \leq x_j \leq 1$, the resulting problem is extremely easy to solve by ranking the variables according to the ratios, c_j/a_j, of the objective function coefficients to the constraint coefficients. For the example this would be

$$\frac{30}{29} \geq \frac{11}{12} \geq \frac{9}{10} \geq \frac{14}{16} \geq \frac{17}{20}$$

corresponding to the ordering x_1, x_4, x_5, x_3, x_2. Then the problem is solved by sequentially assigning each variable the largest possible value, given the values assigned to preceding variables. Thus, in the example we have

Integer Programming

$x_1 = 1$,   leaving $37 - 29 = 8$ pounds of remaining capacity

$x_4 = \dfrac{8}{12}$,   using the remaining capacity

$x_5 = 0$,

$x_3 = 0$,

$x_2 = 0$

This solution has a value of $30 + (8/12)11 = 37.33$. Rounding down yields a feasible integer solution $x = (1, 0, 0, 0, 0)$ with a value of 30. The best integer solution is $x = (0, 1, 1, 0, 0)$ with a value of $17 + 14 = 31$, which happens to be quite close in this instance, although significantly greater differences can arise. The earlier example problem of a gardener buying bags of fertilizer was also an example of a knapsack problem in general integer variables (not restricted to zero-one values). The essential feature of a knapsack problem is that all objective function coefficients and all constraint coefficients are nonnegative, when expressed either as a maximization problem with a "less than or equal to" constraint or a minimization problem with a "greater than or equal to" constraint. Problems consisting of only a single knapsack constraint are efficiently solved by specialized algorithms as a matter of practice, although theoretically such problems can be as hard as any integer programming problem.

Our main interest is in modeling and solving problems with several constraints. A frequently encountered generalization of the knapsack problem is the so-called multidimensional knapsack problem, whose binary form can be expressed by

Maximize cx

subject to $Ax \leq b$

$x \in \{0, 1\}$

where we specify c, A, and $b \geq 0$. An obvious and sometimes useful property of the multidimensional knapsack problem is that any linear programming solution can be rounded down to produce an integer solution. Practical applications of the multidimensional knapsack problem arise in capital budgeting, forestry, and cargo loading.

In Section 2 on cutting planes we discuss valid inequalities that can be added to problems with knapsack constraints to strengthen the formulation.

1.5. Indicator Variables

An important use of binary variables in IP formulations is as an indicator for the range of a continuous variable. The indicator variables can reflect, for example, fixed costs associated with a given range or logical constraints on the allowable ranges of continuous variables. We begin our discussion of indicator variables with their simplest and most common use, fixed charge structures.

1.5.1. The Fixed Charge Problem

The fixed charge problem is characterized by "one-shot" outlays (or setup costs) that are incurred in the process of starting or renewing a business venture. For example, a manager who is faced with deciding which of several machines to buy, automobile plants to build, oil wells to drill, or land areas to develop must account not only for the continuing costs of operation (once the projects are under way) but also for the initial fixed cost required to initiate the projects. This type of problem, especially in the presence of networks, is one of the most frequently occurring problems in practical applications.

To simply the discussion, we consider a simple two-dimensional model and omit subscripts:

$$\text{Minimize} \quad fx + cy$$
$$\text{subject to} \quad y \leq Mx$$
$$x, y \geq 0, \; x \text{ is integer}$$

The data are the fixed cost, f, the continuous cost, c, possibly zero, and the maximum value, M, of the continuous variable, y. when the continuous variable, y, is greater than zero, the binary variable, x, is forced to 1, and the fixed cost is incurred. The maximum value, M, of the continuous variable is an important part of fixed charge models; unnecessarily large values of M lead to computationally weak formulations.

Occasionally, fixed charge models also include a minimum activity level, m, for the continuous variable when $x = 1$. The minimum activity level is reflected in the constraint:

$$mx \leq y$$

If instead of a fixed charge being incurred a fixed benefit is acquired when y is nonzero, i.e., $f < 0$, then this constraint is necessary and m should be suitably large to form a tight model.

The fixed charge model can be generalized to an arbitrary model with one continuous variable and one binary variable. The constraints model a pair of operating ranges for each continuous variable. The binary variable indicates which range is in effect. Figure 1 illustrates a mixed-integer feasible region.

When $x = 0$, the feasible range for y is (m_0, M_0). In the figure, the range is (1, 4). For the fixed charge problem, this range is the single point,

Integer Programming

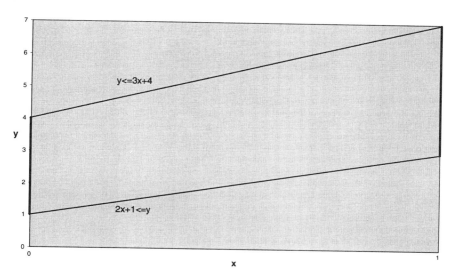

Figure 1 A mixed-integer feasible region.

$y = 0$. When $x = 1$, the feasible range for y is (m_1, M_1). In the figure, the range is (3, 7). We omitted the subscript 1 in our description of the fixed charge problem. The inequalities for the general model are

$$m_0 + (m_1 - m_0)x \leq y \leq M_0 + (M_1 - M_0)x$$

In the figure, the inequalities are

$$1 + 2x \leq y \leq 4 + 3x$$

The binary variable, x, may appear in other constraints as well to reflect, for example, logical restrictions (described later). The binary variable may also appear in the objective to reflect fixed costs related to the choice of operating range.

1.6. Multiple Alternative Constraints

A variety of "dichotomous" or "multiple alternative" situations can be accommodated by the introduction of appropriately defined binary variables. Problems to which such multiple alternative situations are relevant range from the design of a nuclear reactor complex (Glover et al., 1989) to the determination of demand reservoir locations for a water resource allocation project.

The common ingredient in these problems is a set of constraining relations:

$$\sum_j a_{ij} w_j \leq b_i \quad \forall i \in I$$

out of which at least q are required to be satisfied while the remaining $|I| - q$ may be satisfied or not. (The notation $|I|$ represents the number of elements in set I.) If we use x_i as an indicator variable for violating the constraints, we can express this condition with the following inequality:

$$\sum_i x_i \leq |I| - q$$

The final modeling step is to modify the constraints to include the indicator variables:

$$\sum_j a_{ij} w_j \leq b_i + (M_{1i} - b_i) x_i \quad \forall i \in I$$

M_{1i} must be chosen sufficiently large (larger than b_i) to relax constraint I when $x_i = 1$. Choosing M_{1i} any larger than this will unnecessarily weaken the constraint.

The last constraint can be viewed as an application of the general model with one continuous variable and one binary variable. Deriving this constraint in this way illustrates the general relationship between linear constraints and indicator variables.

Specifically, we define continuous variables equal to the level of each constraint:

$$y_i = \sum_j a_{ij} w_j$$

Next, we apply the generalized fixed charge constraint structure to the variable pair, x_i and y_i, and substitute the linear expression for y_i:

$$m_{0i} + (m_{1i} - m_{0i}) x_i \leq \sum_j a_{ij} w_j \leq M_{0i} + (M_{1i} - M_{0i}) x_i \quad \forall i \in I$$

Substituting $M_{0i} = b_i$ in the second inequality transforms the second inequality to the constraint suggested above. Deriving the constraint in this way generates a more general model, i.e., one including the first inequality. In addition, the relationship to the general model with one continuous variable and one binary variable may suggest further possibilities for improving a specific formulation. As with fixed charge models, an unnecessarily wide range of m and M values will lead to a computationally weak formulation.

Clearly, the last expression is useful beyond the special case of multiple alternative constraints. Utilizing m_0 and m_1 allows the integer variable

Integer Programming

to indicate a choice of ranges for any linear expression. Furthermore, the simple cardinality constraint limiting the number of unsatisfied constraints can be replaced creatively with logical conditions (or any other constraints) on the integer variables. Logical conditions can be used to reflect more subtle interaction of constraint ranges.

1.7. Nonconvex Regions

Nonconvex feasible regions, such as the one in Figure 2, may also be approached by adding integer variables which relax a subset of the constraints. In fact, multiple alternative constraints yield only a special case of a nonconvex feasible region.

One representation for a nonconvex region is a union of disjoint convex regions (separated by dashed lines in our example). Other representations are possible and computationally attractive depending on the structure of the nonconvex region.

The constraints for the individual regions are a building block in developing a global constraint set for the nonconvex region.

Region I is constrained by

$$-y_1 + y_2 \leq 5$$
$$y_1 + y_2 \leq 7$$
$$y_1 \leq 5$$

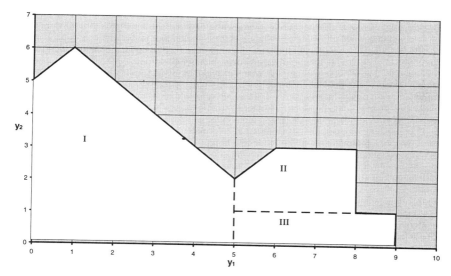

Figure 2 A nonconvex feasible region.

Region II is constrained by

$$-y_1 + y_2 \leq -3$$
$$5 \leq y_1 \leq 8$$
$$1 \leq y_2 \leq 3$$

Region III is constrained by

$$5 \leq y_1 \leq 9$$
$$y_2 \leq 1$$

We define an indicator variable, x_j, for each convex region and use them to form a global set of constraints for the nonconvex region. The linear expression in each of the above constraints forms the left-hand side of a constraint in the global set. The right-hand side of the global constraints is a function of the indicator variables and reflects the boundaries of the individual regions. Finally, we impose the constraint that exactly one of the regions must be selected.

$$-y_1 + y_2 \leq 5x_1 - 3x_2 - 4x_3$$
$$y_1 + y_2 \leq 7x_1 + 11x_2 + 10x_3$$
$$y_1 \leq 5x_1 + 8x_2 + 9x_3$$
$$y_1 \geq 5x_2 + 5x_3$$
$$y_2 \leq 6x_1 + 3x_2 + x_3$$
$$y_2 \geq x_2$$
$$x_1 + x_2 + x_3 = 1$$
$$y_1 \geq 0, \quad x_i \in \{0, 1\}$$

Nonconvex regions are geometrically similar to the integrality conditions for general integer programs. We highlight the main points of interest from polyhedral theory here.

1. For any set of constraints the convex hull of the feasible region is well defined.
2. The convex hull includes all feasible solutions and also includes infeasible solutions.
3. All linear objectives are optimized by extreme points of the convex hull.
4. The linear programming relaxation of any formulation of a nonconvex region must include the convex hull.

These observations lead to a criterion for a good formulation: the linear programming relaxation of a formulation of a nonconvex region should (ideally) be equivalent to the convex hull of the region.

Integer Programming

The above formulation could be simplified if this nonconvex region represented the entire problem. In this case, the convex hull of the region is sufficient. The convex hull of the region in Figure 2 is

$$-y_1 + y_2 \leq 5$$
$$3y_1 + 7y_2 \leq 45$$
$$2y_1 + y_2 \leq 19$$
$$y_1 \leq 9$$
$$y_i \geq 0$$

However, additional variables and constraints will require the previous formulation, with integer variables, to ensure that only feasible solutions are allowed. Points previously on the interior of the convex hull may now be extreme points of a higher-dimensional convex hull.

The following example of a one-dimensional nonconvex region illustrates these changes.

Nonconvex region: $0 \leq y_1 \leq 2.5 \cup 5 \leq y_1 \leq 8$
Convex hull: $0 \leq y_1 \leq 8$
Integer programming formulation:
$$5x_2 \leq y_1 \leq 2.5x_1 + 8x_2$$
$$x_1 + x_2 = 1$$
$$x_1, x_2 \in \{0, 1\}$$

which simplifies by using the second equation to eliminate x_1:

$$5x_2 \leq y_1 \leq 2.5 + 5.5x_2$$

Notice, that the integer programming formulation would be unnecessary if this was the entire problem: the convex hull would be sufficient for a linear objective in y_1 without additional constraints. Only the values 0 and 8 are possible solutions. However, the integer programming formulation will be necessary as additional variables are added to the formulation: the values 2.5, 5, and possibly others may be optimal.

For example, add a variable, y_2, and two constraints to the model:

$$y_1 + y_2 \leq 6$$
$$y_1 \geq y_2$$
$$y_2 \geq 0$$

The feasible regions (the two triangles with solid borders) are illustrated in Figure 3.

The new convex hull is composed of the constraints

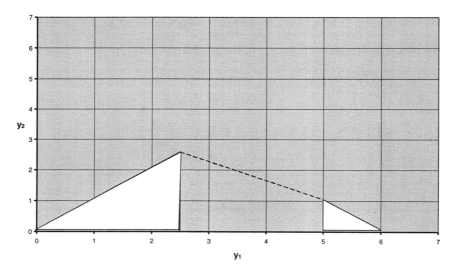

Figure 3 Nonconvex feasible region.

$$3y_1 + 5y_2 \leq 20$$
$$y_1 + y_2 \leq 6$$
$$y_1 \geq y_2$$
$$y_i \geq 0$$

The first constraint is new and joins the two regions. The constraint is represented by the dashed line in Figure 3.

The example illustrates two important aspects of the general case. first, the one-dimensional convex hull is an adequate formulation before, but not after, including y_2. Second, the integer variable, x_2, was not required to obtain the convex hull of y_1 and y_2. These properties hold in general.

If additional variables are added to the formulation, the convex hull above is inadequate to enforce the feasible region. One adequate formulation combines the one-dimensional MIP formulation and the new constraints:

$$5x_2 \leq y_1 \leq 2.5 + 5.5x_2$$
$$y_1 + y_2 \leq 6$$
$$y_1 \geq y_2$$
$$y_i \geq 0$$
$$x_2 \in \{0, 1\}$$

Integer Programming

However, stronger formulations can be created. The convex hull can be created using the same method as for the region in Figure 2, with one integer variable for each region. The constraint requiring one region to be chosen can be used to eliminate one integer variable, x_1. The convex hull is characterized by

$$5x_2 \leq y_1$$
$$y_1 + y_2 \leq 5 + x_2$$
$$y_1 \geq y_2 + 4x_2$$
$$y_1 \leq 2.5 + 3.5x_2$$
$$y_i \geq 0$$
$$x_2 \in \{0, 1\}$$

This three-dimensional hull implies the previous two-dimensional hull; the two-dimensional hull is a projection of the three-dimensional hull.

One reason we emphasize the importance of tight formulations is that there are many other ways of modeling nonconvex regions (Williams, 1993) that we will not present here. In selecting a method it is important to choose one that leads to a tight formulation and exploits the structure of the problem at hand.

A second reason is the philosophical viewpoint of integer programming as a modeling enhancement of linear programming; the integer programming feasible region is a subset of the linear programming feasible region. Integer variables restrict the solution to discrete values. In this section, we have generalized this notion: the integer variables restrict the solution to specific convex regions.

1.8. Disjunctive Constraints

Disjunctive constraints can be viewed as a special case of multiple alternative constraints and hence also as a special case of nonconvex regions. An effective modeling technique can be applied to disjunctive constraints that cannot be applied to arbitrary nonconvex regions.

The disjunctive constraint model specifies that at least one of the following constraints must be satisfied:

$$\sum_j a_{ij} y_j \leq b_i \quad \forall i \in I$$

We add the additional restriction that all $a_{ij} \geq 0$ and $y_j \geq 0$. These assumptions allow the y_j variables to be replicated for each constraint i, thus creating new variables we denote by y_{ij}. Binary variables, x_i, are also added to indicate whether particular constraints are enforced. If $x_i = 1$, then constraint i must be satisfied. If $x_i = 0$, then constraint i may or may not be satisfied. The new system becomes

$$\sum_j a_{ij} y_{ij} \leq b_i x_i \quad \forall i \in I$$

$$\sum_i y_{ij} = y_j \quad \forall j$$

The restriction that at least one constraint must be valid is enforced by

$$\sum_i x_i = 1$$

1.9. Logical Conditions

Logical conditions arise both directly and indirectly in mathematical programs. Examples of direct sources of logical conditions include:

> One of the projects X, Y, or Z must be funded.
> If an airline has flight legs A–B and B–C, then it competes in market A–C.
> If parts P_1, P_2, and P_3, are overhauled in period $t - 1$, then machine m can be used in period t.
> If the facility making part A and either of two facilities making part B operate in period t, then parts A and B can be assembled in period t.

Indirect sources of logical conditions include shared costs:

> If any transshipment takes place at a warehouse, then base level operating costs are incurred.
> If any part made on machine m is produced, then an operator must be present.
> If route R contains both the origin and destination for item I, then transfer costs are prevented.

We use X_j to represent a logical proposition and x_j to represent its associated binary variable. We set $x_j = 1$ when proposition X_j is true, and $x_j = 0$ when proposition X_j is false. The following symbols are used to represent logical relationships:

> $\vee$ means (inclusive) *or*
> $\wedge$ means *and*
> $\neg$ means *not*
> $\Rightarrow$ means *implies*
> $\Leftarrow$ means *is implied by*

Integer Programming

The *or* operator is explained first.

The proposition X_3 is equivalent to X_1 *or* X_2 is expressed by a pair of implications:

1. If X_1 *or* X_2 is true, then X_3 is true.
   $X_1 \lor X_2 \Rightarrow X_3$
2. If X_3 is true, then X_1 *or* X_2 is true.
   $X_1 \lor X_2 \Leftarrow X_3$

The first implication is represented by the pair of equations:

$$x_1 \leq x_3$$
$$x_2 \leq x_3$$

The second implication is represented by the equation:

$$x_1 + x_2 \geq x_3$$

The *not* operator translates more readily than the *or* operator: $\neg X_j$, is represented by $1-x_j$.

The *not* operator is used in De Morgan's laws to convert between *or* and *and* propositions:

$$\neg(X_1 \land X_2) = \neg X_1 \lor \neg X_2$$
$$\neg(X_1 \lor X_2) = \neg X_1 \land \neg X_2$$

The inequalities for the *and* operator are sufficiently straightforward that they could be stated without explanation as we did for the *or* operator. Instead, we use De Morgan's laws to illustrate the conversion process.

The proposition X_3 is equivalent to X_1 *and* X_2 is expressed by a pair of implications:

If X_1 *and* X_2 are true then X_3 is true:
$X_1 \land X_2 \Rightarrow X_3$
If X_3 is true then X_1 and X_2 are true:
$X_1 \land X_2 \Leftarrow X_3$

Applying the *not* operator to the first and implication yields:

$$\neg(X_1 \land X_2) \Leftarrow \neg X_3$$

Applying the first of De Morgan's laws to the left-hand side of the equation generates an *or* implication:

$$\neg X_1 \lor \neg X_2 \Leftarrow \neg X_3$$

Substituting for the variable representation for both the *not* operator and the *or* implication generates an inequality for the first and implication:

$$(1 - x_1) + (1 - x_2) \geq 1 - x_3$$

After rearranging terms, the first implication is represented by the simple inequality:

$$x_1 + x_2 \leq 1 + x_3$$

We repeat the procedure for the second implication.
Applying the *not* operator yields:

$$\neg(X_1 \wedge X_2) \Rightarrow \neg X_3$$

Applying De Morgan's laws yields:

$$\neg X_1 \vee \neg X_2 \Rightarrow \neg X_3$$

Substituting the variable representation:

$$1 - x_1 \leq 1 - x_3$$
$$1 - x_2 \leq 1 - x_3$$

After rearranging terms, the second implication is represented by the pair of equations:

$$x_1 \geq x_3$$
$$x_2 \geq x_3$$

The inequalities generated for the *or* operator and the *and* operator can be used to generate inequalities for other operators. We illustrate the procedure with the *exclusive or* operator which evaluates true when exactly one of its operands has the value true. The *exclusive or* operator is frequently abbreviated *XOR* and translated to *or* and *and* operators:

$$X_1 \text{XOR} X_2 = (X_1 \vee X_2) \wedge \neg(X_1 \wedge X_2)$$

We simplify this expression by substituting a new proposition for both of the expressions in parentheses:

$$X_1 \vee X_2 \Leftrightarrow X_3$$
$$X_1 \wedge X_2 \Leftrightarrow X_4$$

The overall expression can now be represented by X_5 in terms of the new propositions:

$$X_3 \wedge \neg X_4 \Leftrightarrow X_5$$

The process is completed by translating the individual expressions to inequalities as before.

$X_1 \vee X_2 \Leftrightarrow X_3$ becomes:

Integer Programming

$$x_1 \le x_3 \quad (1)$$
$$x_2 \le x_3 \quad (2)$$
$$x_1 + x_2 \ge x_3 \quad (3)$$

$X_1 \wedge X_2 \Leftrightarrow X_4$ becomes:

$$x_1 + x_2 \le 1 + x_4 \quad (4)$$
$$x_1 \ge x_4 \quad (5)$$
$$x_2 \ge x_4 \quad (6)$$

$X_3 \wedge \neg X_4 \Leftrightarrow X_5$ becomes

$$x_3 + (1 - x_4) \le 1 + x_5 \quad (7)$$
$$x_3 \ge x_5 \quad (8)$$
$$1 - x_4 \ge x_5 \quad (9)$$

The variables x_3 and x_4 are by-products of this mechanistic translation. The variables can be used elsewhere in a formulation or eliminated if their respective expressions do not appear elsewhere. Eliminating x_3 and x_4 yields the equivalent formulation:

$x_1 - x_2 \le x_5$	implied by inequalities (1), (6), and (7)
$x_2 - x_1 \le x_5$	implied by inequalities (2), (5), and (7)
$x_1 + x_2 \ge x_5$	implied by inequalities (3) and (8)
$x_1 + x_2 \le 2 - x_5$	implied by inequalities (4) and (9)

1.10. Nonlinear 0-1 Polynomials

The *and* operator frequently arises in integer programs in the form of a nonlinear term, $x_1 x_2$, where x_1 and x_2 are binary variables. Substituting x_3 for the nonlinear term and using the same inequalities as we did for the *and* operator removes the nonlinearity. Similar inequalities convert some more complex nonlinear terms to linear terms. For example, the term $x_1 x_2 x_3$ could be represented by the inequalities:

$$x_1 + x_2 + x_3 \le 2 + x_4$$
$$x_1 \ge x_4$$
$$x_2 \ge x_4$$
$$x_3 \ge x_4$$

Any powers in the nonlinear terms may be reduced to 1 because the variables are 0-1. For example:

$$x_1^2 x_2^3 = x_1 x_2$$

The general inequalities for a product of terms,

$$x_j \equiv \prod_{j \in J} x_j$$

are

$$\sum_{j \in J} x_j \leq x_j + |J| - 1$$
$$x_j \geq x_j \quad \forall j \in J$$

1.11. Specially Ordered Sets

Specially ordered sets (SOS) are useful supplements to integer variables for both easy modeling and efficient computation. Each SOS contains an ordered set of integer or continuous variables, λ_i. The sets are ordered by a *reference row* of weights.

Typically, two types of specially ordered sets are available in commercial mathematical programming packages. Specially ordered sets of type 1 (SOS1) have the restriction that *exactly one λ_i is nonzero*. These sets might be used, for example, to model a constraint to choose one of several competing projects that have some natural ordering. Specially ordered sets of type 2 (SOS2) have the restriction that *at most 2 adjacent λ_i are nonzero*. These sets are used in the next section to model nonlinear functions.

Sometimes, SOS are combined with the structure for convex combinations:

$$\sum_i \lambda_i = 1$$
$$0 \leq \lambda_i \leq 1$$

This restriction combined with SOS1 is also known as an SOS of type 3.

Branch and bound algorithms (described in Section 3) for integer programming easily generalize to include SOS.

1.12. Piecewise Linear Approximation of Separable Nonlinear Functions

A nonlinear function $f(y)$ where y is a vector is called separable if it can be written in the form

$$f(y) = \sum_i f_i(y_i)$$

where each function f_i is a function of a single variable y_i. If the functions

Integer Programming

are sufficiently well behaved, it may be possible to fit them with reasonable piecewise linear approximations without too much difficulty. In this context we shall for convenience drop the j subscript and understand $f(y)$ to designate one of the functions $f_j(y_j)$ where now y is a single variable rather than a vector. For example, such a function $f(y)$ and its linear approximation $z(y)$ might be as shown in Figure 4, where $f(y)$ is the curved line and $z(y)$ is the broken straight line.

We index the break points in the curve by i. The coordinates of point i are (y_i, z_i) and we define the incremental change about these points, $\Delta y_i = y_{i+1} - y_i$ and $\Delta z_i = z_{i+1} - z_i$. Two representations are frequently used for nonlinear curves, the λ-form and the δ-form, which we present in the following sections.

1.12.1. Caution in Applying IP to Nonlinear Functions

If a nonlinear optimization problem is convex, far more efficient methods may be used to solve the problem. An optimization problem is convex if all convex (concave) functions appear only with positive coefficients in a minimization (maximization) objective or on the left-hand side of a $\leq (\geq)$ constraint; the pressure is to reduce the convex (increase the concave) function. (Negative) coefficients reverse the sense of the objectives and con-

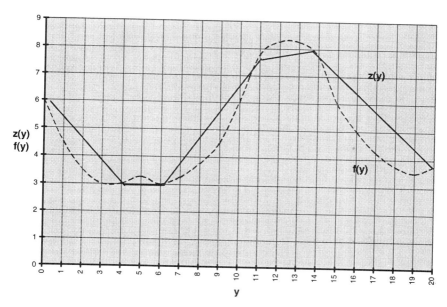

Figure 4 Piecewise linear approximation of a nonlinear function.

straints that lead to a convex optimization problem. Linear functions, which are both convex and concave, may appear in either objective sense, either inequality sense, and equations.

1.12.2. λ-Form

The λ-form selects a pair of adjacent points on the curve. The indicator variable for selecting point i is λ_i. The formulation is

$$z = \sum_i z_i \lambda_i$$
$$y = \sum_i y_i \lambda_i$$
$$\sum_i \lambda_i = 1$$
$$0 \leq \lambda_i \leq 1 \quad \forall i$$
$$\lambda_i \in \text{SOS2}$$

For example, suppose $f(y) = y^2$ and that we choose the integral values of y as the break points.

$$z = 0\lambda_0 + 1\lambda_1 + 4\lambda_2 + 9\lambda_3$$
$$y = 0\lambda_0 + 1\lambda_1 + 2\lambda_2 + 3\lambda_3$$
$$\lambda_0 + \lambda_1 + \lambda_2 + \lambda_3 = 1$$
$$0 \leq \lambda_i \leq 1 \quad \forall i$$
$$\lambda_i \in \text{SOS2}$$

The λ-form can also be implemented without using SOS sets. Removing SOS sets requires adding continuous variables w_i. The new formulation is

$$z = \sum_i z_i \lambda_i + \sum_i \Delta z_i w_i$$
$$y = \sum_i y_i \lambda_i + \sum_i \Delta y_i w_i$$
$$w_i \leq \lambda_i \quad \forall i$$
$$\sum_i \lambda_i = 1$$
$$\lambda_i \in \{0, 1\} \quad \forall i$$

The quadratic example becomes:

$$z = 0\lambda_0 + 1\lambda_1 + 4\lambda_2 + 9\lambda_3 + 1w_0 + 3w_1 + 5w_2$$
$$y = 0\lambda_0 + 1\lambda_1 + 2\lambda_2 + 3\lambda_3 + 1w_0 + 1w_1 + 1w_2$$
$$w_i \leq \lambda_i \quad \forall i$$
$$\lambda_0 + \lambda_1 + \lambda_2 + \lambda_3 = 1$$
$$\lambda_i \in \{0, 1\} \quad \forall i$$

Integer Programming

The λ-form can be valid even if z is not continuous. To do this, one selects the points i so that z is continuous on the half-open interval $[y_i - 1, y_i)$ and interprets $\lambda_i = 1$ to mean $y \in [y_i - 1, y_i)$, whereupon w_i is constrained to satisfy $\lambda_i - \varepsilon \geq w_i \geq 0$ for ε a small positive number.

It is also quite possible to make the representation valid when z is defined over disjoint intervals. In particular, if z is defined only at the points y_i, then one can simply represent z by $\Sigma_i z_i \lambda_i$, without any reference to the w_i variables.

The λ-form can be extended to nonseparable functions. We illustrate the extension when z is a function of two variables, x and y. The breakpoints now form a two-dimensional grid and the Δ variables are redefined:

$$\Delta x_{ij} = x_{i+1,j} - x_{i,j}$$
$$\Delta y_{ij} = x_{i,j+1} - x_{i,j}$$
$$\left(\frac{\Delta z}{\Delta v}\right)_{ij} = z_{i,j+1} - z_{i,j}$$
$$\left(\frac{\Delta z}{\Delta w}\right)_{ij} = z_{i+1,j} - z_{i,j}$$

and $\Delta y_i = y_{i+1} - y_i$ and $\Delta z_i = z_{i+1} - z_i$. The extension uses a discrete approximation to the partial derivative of z in both the x and y directions from each grid point.

$$z = \sum_{i,j} z_{ij}\lambda_{ij} + \sum_{i,j}\left(\frac{\Delta z}{\Delta v}\right)_{ij} v_{ij} + \sum_{i,j}\left(\frac{\Delta z}{\Delta w}\right)_{ij} w_{ij}$$
$$x = \sum_{i,j} x_{ij}\lambda_{ij} + \sum_{i,j} \Delta x_{ij} v_{ij}$$
$$y = \sum_{i,j} y_{ij}\lambda_{ij} + \sum_{i,j} \Delta y_{ij} w_{ij}$$
$$v_{ij} \leq \lambda_{ij} \quad \forall i,j$$
$$w_{ij} \leq \lambda_{ij} \quad \forall i,j$$
$$\sum_{i,j} \lambda_{ij} = 1$$
$$\lambda_{ij} \in \{0, 1\} \quad \forall i,j$$

Caution needs to be exercised in generating such piecewise linear approximations because they can expand a model quickly. The previous model generates 50 continuous variables, 25 integer variables, and over 50 constraints for a single function of two variables if each variables has five grid points.

1.12.3. δ-Form

The δ-form increments the values of y and z along the curve. The δ-form frequently has a computational advantage over the λ-form. The indicator variable for being *at or beyond* point i on the curve is δ_i. The formulation is

$$z = \sum_i \Delta z_i \delta_i$$
$$y = \sum_i \Delta y_i \delta_i$$
$$\delta_i \leq \delta_{i-1} \quad \forall i$$
$$0 \leq \delta_i \leq 1 \quad \forall i$$

at most one fractional δ_i

Some commercial packages accept the final condition in the same spirit as SOS.

The quadratic example becomes

$$z = 1\delta_1 + 3\delta_2 + 5\delta_3$$
$$y = 1\delta_1 + 1\delta_2 + 1\delta_3$$
$$1 \geq \delta_1 \geq \delta_2 \geq \delta_3 \geq 0$$

at most one fractional δ_1

The δ-form can also be implemented without the final condition. As when we removed SOS sets from the λ-form, we must introduce the continuous variables w_i. The new formulation is

$$z = \sum_i \Delta z_i w_i$$
$$y = \sum_i \Delta y_i w_i$$
$$\delta_i \leq w_i \leq \delta_{i-1} \quad \forall i$$
$$\delta_i \in \{0, 1\} \quad \forall i$$

The quadratic example becomes

$$z = 1w_1 + 3w_2 + 5w_3$$
$$y = 1w_1 + 1w_2 + 1w_3$$
$$1 \geq \delta_1 \geq w_1 \geq \delta_2 \geq w_2 \geq \delta_3 \geq w_3 \geq 0$$
$$\delta_1 \in \{0, 1\} \quad \forall i$$

In the following sections we consider general methods that can be applied to solve integer programming problems, focusing on those that have dominantly been used in practical applications to date. We also provide exercises to accompany some of these procedures, designed to provide a fuller understanding of the basic operation.

1.13. References

Some mathematical programming books have integer programming chapters of interest. For example, we suggest Bradley et al. (1977), Murty (1976), Shapiro (1979a), and Zionts (1974). Nemhauser (1994) highlights some recent developments in mathematical programming, including integer programming.

Integer Programming   **147**

Many other integer programming books exist. For a more comprehensive and theoretical treatment we recommend Nemhauser and Wolsey (1988). Others include Salkin and Mather (1989), Glover (1978), Balinski and Spielberg (1969), Papadimitriou and Steiglitz (1982), and Parker and Rardin (1988).

A good book on formulation for linear programming and integer programming is Williams (1993). Barnhart et al. (1993) describe the effect of one reformulation.

Garey and Johnson (1979) provide a comprehensive treatment of computational complexity including many problems that are modeled as integer programs.

Glover and Sommer (1975) describe the perils of rounding.

1.14. Exercises

1.1. (*The traveling salesman problem*) A salesman must call on customers in each of n cities, starting in one of the cities and returning to his starting point after he has visited each of the others exactly once. The salesman's objective is to determine an appropriate route (sequence) for visiting the cities so that he will minimize his total distance traveled. Formulate this problem as an integer program. (If one attempts to use only as many 0-1 variables as there are roads between pairs of cities and constrains each city to have exactly two of its roads on the salesman's route, what additional constraints are required to break "subtours" and make sure the route is connected?)

1.2. If the salesman does not have to return to his starting city after visiting all of the others, how can this be made into a special case of the problem of Exercise 1.1?

1.3. Show that the following machine scheduling problem is an instance of the traveling salesman problem. There are n jobs that must be sequentially processed on a machine, and the setup cost for each job depends on which job precedes it in the sequence. Determine an order for processing the jobs that minimizes total setup cost.

1.4. (*The Chinese postman problem*) In order to deliver the mail, a postman must travel every street of his route (at least once). Formulate this as an integer programming problem, under the assumption that the postman seeks to minimize the total distance traveled. (If the postman can determine a route that lets him travel each street exactly once, then this will be optimal. Such a route is called an "Euler tour" and exists for a connected network of streets if and only if an even number of streets meet at every intersection. The question of the existence of such a tour is raised in the famous Bridges of Koenigsberg puzzle.)

1.5. Accommodate the following "multiple alternative" constraints by the use of appropriately defined 0-1 variables.

$$\{x_1 \leq 1, x_2 \leq 1\} \quad \text{or} \quad \{x_1 \leq 1, 2 \leq x_2 \leq 3\}$$
$$\text{or} \quad \{x_1 + x_2 \leq 3, x_1 \geq 2\}$$

(Require at least one of these sets of constraints to hold. In this case, this will cause *exactly* one of the constraint sets to hold. Why?) Graph the alternative feasible regions.

1.6. Show that the following two mixed-integer programs are "equivalent" in the sense that x is feasible (optimal) for the first if and only if $y = U - x$ is feasible (optimal) for the second.

Minimize cx
subject to $Ax \geq b$
$U \geq x \geq 0$ and x_j integer for $j \in J$

Maximize cy
subject to $Ay \leq d$
$U \geq y \geq 0$ and y_j integer for $j \in J$

where $d = AU - b$ and U_j is integer for $j \in J$.

1.7. Using Exercise 1.6, show that the gardener's problem stated at the beginning of Section 1 is an instance of the knapsack problem. (Hint: show that redundant bounds can easily be found for the variables of the gardener's problem. Are these bounds also redundant when the gardener's problem is given the knapsack formulation?)

1.8. Show that the problem

Maximize $\sum d_j x_j$
subject to $\sum a_j x_j \leq b$
$U_j \geq x_j \geq 0$ and x_j integer for all j

(where b is a scalar) can always be made into a bounded variable knapsack problem regardless of the signs of the coefficients d_j and a_j. (Hint: Let $y_j = U_j - x_j$ if $d_j < 0$ and let $y_j = x_j$ if $d_j \geq 0$. Show that this gives a problem with all nonnegative coefficients in the objective function to be maximized. What can then be said about the optimal value of any y_j whose constraint coefficient is nonpositive?)

1.9. Show that if the cross-products $x_1 x_2$, $x_1 x_3$, $x_1 x_4$ are replaced by

0-1 variables x_{12}, x_{13}, x_{14}, as indicated in Section 1.10, then the lower bound constraints for x_{12}, x_{13}, and x_{14} can be replaced by

$$x_{12} + x_{13} + x_{14} \geq 3x_1 + x_2 + x_3 + x_4 - 3$$

Can you give a general rule for which this example constitutes a special case?

1.10. Show that if the cross-products $x_1x_2x_3x_4$, $x_1x_2x_3x_5$, $x_1x_2x_4x_5$ are replaced by 0-1 variables x_{1234}, x_{1235}, x_{1245}, then the lower bound constraints for these latter variables can be replaced by

$$x_{1234} + x_{1235} + x_{1245} \geq 3x_1 + 3x_2 + 2x_3 + 2x_4 + 2x_5 - 9$$

Can you give a general rule for which this example constitutes a special case?

1.11. Give two ways to replace a bounded integer variable by zero-one variables. What are the relative advantages and disadvantages of these replacements in terms of potential optimization algorithms?

1.12. Give an example of a discontinuous function and its representation with the λ-form of Section 1.12.

1.13. Illustrate by means of graphed examples that it is possible to construct a problem in two dimensions for which the linear programming solution is arbitrarily far from a feasible integer solution.

1.14. Construct an example in which no extreme point (corner point) of the feasible continuous region can be rounded to a feasible integer solution.

1.15. A well-known theorem of linear programing says that an optimal continuous solution, if one exists, can be found at one of the extreme points of the feasible region. (See Chapter 2 in this book.) Another theorem says that, for every extreme point, there exists a linear function that achieves its unique optimum over the convex feasible region at that extreme point. Show that, assuming you know optimal values for the 0-1 variables in advance, you can identify a linear objective function that will uniquely yield these values for the 0-1 variables. Is that enough to prove that an optimal solution to a 0-1 MIP problem can always be found at an extreme point?

2. CUTTING PLANES AND VALID INEQUALITIES

It has long been observed in integer programming that if one just knew *the right linear inequalities* to add to the underlying LP problem, then these linear inequalities would succeed in transforming the integer programming problem into an ordinary linear program; these inequalities reduce the feasible region of the LP relaxation until it equals the IP feasible region. This linear program could then be solved by the simplex method, yielding an optimal solution to the IP problem without paying any attention to the requirement that the variables assume integer values. In other words, the integer requirement would be satisfied automatically due to the addition of the appropriate linear inequalities.

Normally, the term cutting planes, or cuts, refers to constraints generated by a generic algorithm after solving an LP relaxation, while valid inequalities refer to constraints based on problem specific structure and are derived only from the formulation. The distinction in terminology is a minor one because both cuts and valid inequalities are typically added to an existing LP relaxation for the purpose of changing its optimal solution.

Section 1 provides an example of deriving valid inequalities for the specific problem of a knapsack problem. Like all valid inequalities, the inequalities remain valid if other constraints are added to the knapsack constraint. Section 2 describes Gomory's (1958, 1960, 1963a, 1963b) all-integer cuts and fractional cuts based on tableaus of the LP relaxation. Section 3 formally states the Chvátal-Gomory (Chvátal 1973) rounding method and a ranking of cuts based on the rounding method.

2.1. Tightening Knapsack Constraints for Zero-One Problems

In the general case, where both knapsack and nonknapsack constraints are present (i.e., where constraint coefficients can be of any sign), an approach of reformulating and tightening individual constraints is often an effective tactic for making a problem more easily solvable. Elaborate algorithms exist for automatically generating provably strong constraints. Rather than presenting these algorithms and the proof that they generate the strongest possible constraints, we present some generally effective methods and illustrate them with an example. Frequently, modelers can improve their formulation by "eyeballing" constraints and tightening them in an ad hoc manner that is difficult to duplicate, let alone surpass, with an automated procedure. We know of one case where constraints were provably weak in a theoretic sense and yet were indispensable in practice for tightening the formulation. See Applegate and Cook (1991).

In our knapsack example of Section 1.4, where all variable are zero-one, one reformulation is obvious. The value of the right-hand side can be reduced from 37 to 36: no integer solutions can achieve a value of 37. Our modified constraint is

$$29x_1 + 20x_2 + 16x_3 + 12x_4 + 10x_5 \leq 36$$

The modified constraint does not affect the integer programming solution

Integer Programming

but does create a linear programming relaxation with a smaller feasible region. In general, tightening by reducing the right-hand side is easy for constraints with a small number of variables, or more precisely a small number of coefficient values on the left-hand side of the constraint, but may be quite difficult when there are many different coefficient values. Consequently, other approaches are usually preferred.

A useful alternative is to add additional constraints rather than to replace the constraint. The additional constraints are implied by the original constraint and hence are *valid inequalities* to add to the model. The valid inequalities are intended to further tighten the linear programming relaxation. The strongest possible valid inequalities are called *facets*. In principle, any integer program can be solved as a linear program after generating all of its facets; the solution is guaranteed to be integer. In practice, there are almost always too many facets for this approach to succeed. However, in theory, only the subset of facets that would constrain the linear programming solution at optimality need be added. This subset is equal in number to the number of variables in the problem. Often, as an alternative to generating all facets, *separation algorithms* are used to generate valid inequalities that separate the linear programming relaxation from the set of feasible integer solutions. If a good and relatively small set of valid inequalities can be determined in advance, they can be added collectively prior to optimization.

In zero-one IP problems, any single constraint can be treated as a knapsack constraint, i.e., as a constraint with all nonnegative coefficients, by the device of replacing a variable x_j that has a negative coefficient by the variable $y_j = 1 - x_j$, called the complement of x_j. (That is, the identity $x_j = 1 - y_j$ is used to substitute $1 - y_j$ for x_j, which gives y_j a coefficient whose sign is opposite that of x_j.) For convenience in our following discussion, we will suppose such substitutions have already been made and we will continue to use the notation x_j to refer to the resulting variables.

A set, C, is *independent* if the solution, $x_j = 1$ for $j \in C$ and $x_j = 0$ for $j \notin C$, is feasible. If this solution is infeasible, C is a *dependent* set. For example, x_1, x_2, and x_3, in the knapsack problem above are a dependent set because their total weight, $29 + 20 + 16 = 65$, exceeds the available weight 37. The following valid inequality arises directly from the definition of a dependent set, C:

$$\sum_{j \in C} x_j \leq |C| - 1$$

which becomes

$$x_1 + x_2 + x_3 \leq 2$$

for our example.

A dependent set is *minimal* if all of its subsets are independent. The previous set is not minimal, because it has dependent subsets, such as the one formed by x_1 and x_2. We can strengthen the formulation by replacing the previous inequality with constraints generated by its minimal dependent sets. The constraints generated by all the minimal dependent sets of the knapsack example are

$$x_1 + x_2 \leq 1$$
$$x_1 + x_3 \leq 1$$
$$x_1 + x_4 \leq 1$$
$$x_1 + x_5 \leq 1$$
$$x_2 + x_3 + x_4 \leq 2$$
$$x_2 + x_3 + x_5 \leq 2$$
$$x_2 + x_4 + x_5 \leq 2$$
$$x_3 + x_4 + x_5 \leq 2$$

The last four constraints can be replaced by a stronger valid inequality. The *extension*, $E(C)$, of a minimal dependent set includes not only C but all variables with equal or larger knapsack weights:

$$E(C) = C \cup \{k \notin C : a_k \geq a_j \quad \forall j \in C\}$$

The minimal dependent set inequalities are strengthened by allowing the summation to range over the entire extension

$$\sum_{j \in E(C)} x_j \leq |C| - 1$$

We can further strengthen the inequality by stipulating that $E(C)$ also include all elements k such that

$$a_k - \text{Max}(a_j) + \sum_{j \in C} a_j > b$$
$$\quad\quad\quad j \in C$$

Using the extensions for our example we generate the following valid inequalities:

$$x_1 + x_2 + x_3 + x_4 \leq 2$$
$$x_1 + x_2 + x_3 + x_5 \leq 2$$
$$x_1 + x_2 + x_4 + x_5 \leq 2$$
$$x_1 + x_2 + x_3 + x_4 + x_5 \leq 2$$

In our example, the fourth inequality is obviously stronger than the first three and they can be omitted. In fact, the fourth inequality can be strengthened still further:

$$2x_1 + x_2 + x_3 + x_4 + x_5 \leq 2$$

Integer Programming

We have given an extended description of the knapsack problem not only for historical interest but also because, in spite of its apparent simplicity, this problem reappears in several guises and contexts in integer programming. (The problem of the gardener described in Section 1 is one form of the knapsack problem.) In fact, a number of solution techniques in integer programming can be interpreted as a direct extension of a "knapsack method" or involve the creation of "knapsack subproblems" whose solution provides valuable information for the solution of the original problem. This latter strategy is represented by the approach of creating *surrogate constraint relaxations* (Glover, 1975a), which generate new constraints as nonnegative linear combinations of other constraint inequalities (original constraints and valid inequalities derived from them) and typically replace the combined inequalities to generate a smaller problem. Surrogate constraint relaxations have produced highly effective methods for a variety of problems with special structures, including multidimensional knapsack problems and networks with side constraints. We discuss such approaches at greater length in Section 4.6.

2.2. Gomory Cuts

The cutting plane algorithm developed by Gomory (1958, 1969, 1963a, 1963b) was the first provably finite cutting plane algorithm proposed for solving integer programs. In spite of its guaranteed finiteness, however, its performance did not live up to the hopes generated by its emergence. Pure versions of Gomory's algorithm perform very poorly on some particular test problems that are easily solved by branch and bound. At the same time, the cuts remain of interest for two reasons. First, it is a general method and provides a framework to analyze problem specific cuts. Second, research indicates that combining cutting planes with branch and bound can be more efficient than using branch and bound in isolation. This combined technique is known as *branch and cut* (e.g., Padberg and Rinaldi, 1991; Balas et al., 1995; Nemhauser et al., 1994).

2.2.1. Graphical Illustration of Cutting

The rationale underlying a cutting plane procedure is illustrated in Figure 5. The "outer" region of Figure 5 represents the set of feasible solutions to the LP problem, and the "inner region" represents the "convex hull" of feasible integer solutions. (The grid points identify integer solutions.) This convex hull of feasible integer solutions clearly results by adding a set of additional inequalities to those defining the LP region. (That is, the inequalities of the convex hull are consistent with, and generally more restrictive than, the inequalities of the LP region.)

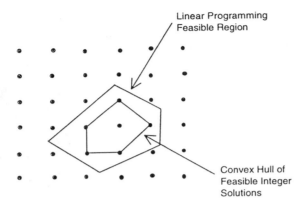

Figure 5 Convex hull of feasible integer solutions.

Because every extreme point of this convex hull is an integer solution, the simplex method will automatically obtain one of these points as an optimal solution (if the constraints defining the inner region are added to the problem) and thereby solve the integer programming problem. Consequently, it would appear highly desirable to be able to specify the inequalities that characterize this convex hull. As it turns out, this is exceedingly difficult to do, and no really effective computational procedure is known.

Fortunately, however, it is not really essential to identify the convex hull itself, because a much smaller and "less powerful" set of linear inequalities can work in its place. This is demonstrated in Figure 6. The problem

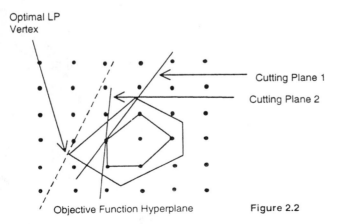

Figure 6 Cutting planes.

Integer Programming

depicted in this diagram is the same as in Figure 5. Here we have additionally identified the objective function hyperplane and the optimal LP vertex. The two lines called "Cutting Plane 1" and "Cutting Plane 2" represent two linear inequalities that have been added to the original LP constraints. Figure 6 discloses that these two inequalities, by themselves, succeed in converting the integer programming problem into a linear program. (The "inequality sense" of these two cutting planes is assumed to be "to the right"—i.e., points to the right are feasible and points to the left are infeasible.) In particular, the new LP solution obtained by adding these two inequalities is the integer point at the intersection of cutting planes 1 and 2—which is also, of course, the optimal solution to the integer programming problem. (The identity of the new LP optimum may be verified by shifting the objective function hyperplane parallel to itself.) Thus, even though the inequalities represented by the two cutting planes are less restrictive than others that define the convex hull of feasible integer solutions, they successfully transform the IP problem into a linear program. On the other hand, these inequalities constitute *supports* of the convex hull, and such supports may themselves be extremely difficult to obtain. Indeed, most inequalities that can be produced (with a reasonable amount of calculation) tend to be substantially weaker than these supports and succeed in "cutting off" only a fraction of the distance between the LP vertex and the optimal integer solution. (The term *cutting plane*—or *cut*—derives from this image of cutting off part of a region.) Thus, one may legitimately ask how it is ever possible to solve an IP problem by adding inequalities, because cutting planes that are not supports can never yield an integer solution.

The answer lies in applying the cutting approach *iteratively*. The iterative procedure first adds a cut inequality upon obtaining the LP optimum, then postoptimizes to obtain a new LP optimum that satisfies the cut. A cut that was not "visible" (i.e., readily susceptible to calculation) at the previous LP optimum is now calculated and added. Thus, progressing from one LP optimum to the next, deriving a cut at each step that was "unavailable" at the preceding optimum, a collection of cutting planes is finally amassed which ensures that the resulting "new LP optimum" will in fact be the solution to the integer program.

An illustration of this type of process is shown in Figure 7. Figure 7 is the same as Figure 6 except for the addition of cutting plane 0. Successive LP optima obtained upon adding the three cutting planes 0, 1, 2 in sequence are indicated by the circles. Thus, in particular, cutting plane 0 is assumed to be calculated from information available at the original LP optimum. The addition of this cut leads the new LP optimum just above and to the right of the original. Cutting plane 1 is then derived from the information

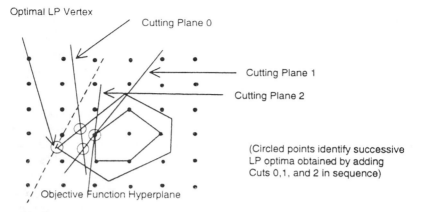

Figure 7 Separating cutting planes.

available at this new optimum, leading to the next LP optimum at the intersection of cutting plane 0 and cutting plane 1. Finally, cutting plane 2 is derived from the information currently available and the resulting LP optimum is the solution to the IP problem.

The procedure illustrated by this diagram may be summarized as follows.

2.2.2. A "Dual" Cutting Plane Procedure

1. Solve the LP problem.
2. If the current LP solution is integer feasible, the IP problem is solved. Otherwise, add a cut that makes the current LP solution infeasible.
3. Reoptimize (with the dual simplex method) to obtain a new LP optimum and return to step 2.

The requirement of step 2 that the cut makes the current LP solution infeasible is, of course, essential or else there would be no change in the LP optimum and there would be nothing to be done at step 3. The dual simplex method is the logical choice for reoptimization, since the LP tableau stays dual feasible when new cuts are added. (The cuts appear as additional variables in the dual problem. If these variables are given the value of 0, the dual solution is unchanged and hence remains feasible.)

As cuts accumulate, a common practice is to discard those that are *nonbinding* at the current LP optimum. A nonbinding cut is one whose slack variable is basic at the current LP optimum and, correspondingly, a

Integer Programming

binding cut is one whose slack variable is nonbasic. The reason for this terminology is that the LP solution occurs geometrically at the intersection of the constraint hyperplanes that correspond to setting the nonbasic variables equal to 0. All other constraints can be discarded without affecting the LP solution. This discarding rule thus helps to keep the size of the problem down. The anticipation is that a cut that is nonbinding at an "intermediate" LP solution is likely to be redundant at the integer optimum, although this is not necessarily so. In any case, nonbinding inequalities are not essential to the current progress of the method, and if it turns out that a worthwhile cut was thereby discarded, another cut to replace it can always be generated at an appropriate point.

We will now show how to derive some simple but fundamental cuts and then illustrate their use by numerical example.

2.2.3. Some Basic Cuts

The cuts we describe here are among the first to have been developed (Gomory, 1958). Others of greater power and complexity (Glover, 1975b; Balas, 1979; Jeroslow, 1990.) have since been proposed, but the ones we consider have the particular merit of being especially easy to generate and understand. In addition, these cuts have sometimes proved to be quite effective in spite of their simplicity. Furthermore, the full potential of their application has by no means been thoroughly plumbed. Many different individual cuts can all be obtained from the same fundamental framework, and no one has yet explored all the possible ways. However, it is well to maintain a healthy skepticism concerning such matters. A number of intuitively based proposals ballyhooed as "improvements" in the past have been notorious flops—if not outright disasters!

2.2.4. Cuts for the Pure IP Problem

We first consider the pure IP problem, in which all variables must be integer valued. (The standard way to assure that a slack variable is integer valued is to require that the starting data consist of integers. For example, the slack variable for the inequality $3x_1 - 2x_2 \leq 7$ must be integer valued because x_1 and x_2 are integer valued.)

Strictly speaking, a cut refers to any linear inequality (or its associated tableau equation) that is implied by the constraints of the IP problem. Thus, a cut need not invariably "cut off" part of the LP feasible region, but it must be *valid* in the sense that it does not exclude any feasible integer solution from the range of acceptable possibilities. We will first show how to generate valid inequalities from given inequalities and then show how to generate cuts directly from tableau equations.

Consider the inequality given by

$$8t_1 - 4t_2 + 3t_3 \leq 11\tfrac{2}{3}$$

where t_1, t_2, and t_3 represent any nonnegative integer variables (e.g., those obtained from the basic or nonbasic variables of any LP tableau). We represent these variables by "t_j" rather than "x_j" because the "x variables" often represent a specific set of initial nonbasic variables. Since the left-hand side of the inequality consists of integer variables and integer coefficients, the value of this left-hand side must be an integer, and hence it is legitimate to write

$$8t_1 - 4t_2 + 3t_3 \leq 11$$

This latter inequality qualifies as a cut, because it is validly implied by the original inequality. Moreover, this cut is *stronger* (i.e., more restrictive) than its source. (Requiring a quantity to be less than or equal to 11 is more restrictive than requiring it to be less than or equal to 11⅔.) In general, dealing with nonnegative variables we will say that one inequality is stronger (more restrictive) than another if it is satisfied by every nonnegative point satisfying the other inequality, but does not admit all of the nonnegative points satisfying this inequality. (That is, the nonnegative solution set of the stronger inequality is a proper subset of that of the second inequality. In pictorial terms, this means that the stronger inequality "cuts off" more of the nonnegative region.)

Now consider the source inequality given by

$$5\tfrac{3}{4}t_1 + 2t_2 - 3\tfrac{1}{2}t_3 \leq 7\tfrac{5}{8}$$

Because the variables are nonnegative, if we decrease the coefficient of any variable (e.g., from a fractional value to the next largest integer below it) we obtain a quantity that is no larger (and possibly smaller) than before. Thus, in particular, we may note that

$$5t_1 \leq 5\tfrac{3}{4}t_1 \quad \text{and} \quad -4t_3 \leq -3\tfrac{1}{2}t_3$$

Suppose the coefficients of t_1 and t_3 are decreased to their next smaller integer values as indicated. Then we may infer from the original inequality that

$$5t_1 + 2t_2 - 4t_3 \leq 7\tfrac{5}{8}$$

This inequality is valid and therefore a cut, but it is weaker (less restrictive) than its source. (In particular, we derived the cut by observing "new left-hand side $\leq$ old left-hand side," and for positive values of t_1 and t_3 this

Integer Programming

inequality holds strictly. Thus it is possible that the cut is satisfied when the source inequality is violated. A simple example is $t_1 = t_2 = 1$ and $t_3 = 0$.)

Of course, there's no particular virtue in obtaining an inequality that is weaker than its source inequality, and our purpose in carrying out this weakening step is to create an "all integer" left-hand side. By this means, it becomes possible to apply the strengthening step previously illustrated. Thus, in our example we finally obtain the cut

$$5t_1 + 2t_2 - 4t_3 \leq 7$$

This cut is neither uniformly stronger nor uniformly weaker than its source inequality (as a result of the combined weakening and strengthening steps). However, the important fact is that it does exclude some of the nonnegative region admitted by the source.

Perhaps surprisingly, a single source inequality can be made to give rise to a wide variety of cuts by the approach demonstrated. For example, consider the inequality

$$8t_1 - 6t_2 - 7t_3 \leq -13$$

The inequality is still valid if we divide it by any positive number, say 3:

$$2\tfrac{2}{3}t_1 - 2t_2 - 2\tfrac{1}{3}t_3 \leq -4\tfrac{1}{3}$$

The foregoing procedure can then be applied to obtain the cut

$$2t_1 - 2t_2 - 3t_3 \leq -5$$

Using algebraic notation, we may summarize the preceding observations to express the source inequality and its cut as follows.

Source inequality: $\quad \sum_{j=1}^{n} a_j t_j \leq a_0$

Cut inequality: $\quad \sum_{j=1}^{n} \left\lfloor \dfrac{a_j}{d} \right\rfloor t_j \leq \left\lfloor \dfrac{a_0}{d} \right\rfloor$

where d is any positive number and $\lfloor . \rfloor$ represents the largest integer $\leq$ the quantity inside (e.g., $\lfloor 2\tfrac{1}{3} \rfloor = 2$, $\lfloor 4 \rfloor = 4$, $\lfloor -5\tfrac{1}{4} \rfloor = -6$).

Adding a slack variable s to change the cut inequality into an equation and writing the result in the condensed tableau format, we obtain

Cut equation: $\quad s = \left\lfloor \dfrac{a_0}{d} \right\rfloor + \sum_{j=1}^{n} \left\lfloor \dfrac{a_j}{d} \right\rfloor (-t_j)$

Significantly, *the slack variable s is integer valued,* because it is defined in terms of integer variables and integer coefficients. This means that the addition of such a cut keeps the IP problem a *pure integer* problem.

In general, the assumptions that support the cut derivation are that the variables are nonnegative and integer valued. However, it is worth noting that if a_j/d is an integer (hence $a_j/d = \lfloor a_j/d \rfloor$), then t_j need not be nonnegative. (Nonnegativity is used only to justify decreasing the coefficient of t_j from a_j/d to $\lfloor a_j/d \rfloor$. If no decrease occurs, nonnegativity is irrelevant.) To derive cuts directly from the tableau equations (explained in more detail below), we can now make the following assertion:
The tableau equation

$$z = a_0 + \sum_{j=1}^{n} a_j(-t_j)$$

in nonnegative integer variables t_j implies the following cuts:

1. *All-integer cut*

$$s = \left\lfloor \frac{a_0}{d} \right\rfloor + \sum_{j=1}^{n} \left\lfloor \frac{a_j}{d} \right\rfloor (-t_j) \quad \text{assuming } z \text{ is nonnegative}$$

but not necessarily integer valued

2. *Fractional cut*

$$s = -f_0 - \sum_{j=1}^{n} f_j(-t_j) \quad \text{assuming } z \text{ is integer valued but not}$$

necessarily nonnegative

where f_j is the "fractional coefficient" $a_j - \lfloor a_j \rfloor$. (In each case, s represents a nonnegative, integer-valued variable.)

The "all-integer" cut is just the cut already derived, noting that $z \geq 0$ immediately implies the previously indicated source inequality. In particular, $z \geq 0$ implies

$$\sum \left(\frac{a_j}{d} \right) t_j \leq \frac{a_0}{d} \quad \text{for } d > 0$$

Then we may replace $\lfloor a_j/d \rfloor$ by $\lfloor a_j/d \rfloor$ since this only decreases (or leaves unchanged) the left side of the preceding inequality. Now, since the left side only takes integer values, we may replace a_0/d by $\lfloor a_0/d \rfloor$. Finally, adding an integer-valued slack variable $s \geq 0$, defined to equal the right side minus the left side of the resulting inequality, gives rise to the all-integer cut as shown above.

The fractional cut is justified simply by selecting a "component inequality" of the tableau equation, selecting $d = 1$ to obtain the all-integer

Integer Programming

cut, then subtracting the tableau equation from the all-integer cut to obtain fractions. Specifically, the tableau equation implies

$$z + \sum_{j=1}^{n} a_j t_j \leq a_0$$

which for $d = 1$ gives the all-integer cut

$$s = \lfloor a_0 \rfloor + 1(-z) + \sum \lfloor a_j \rfloor (-t_j)$$

Subtracting the original tableau equation from this all-integer cut eliminates z and leaves the indicated fractional cut. (Because the z coefficient does not change in deriving the all-integer cut, z does not have to be nonnegative.)

To illustrate the fractional cuts, consider the tableau equation

$$z = 4\frac{5}{8} + 2\frac{3}{8}(-t_1) - 4\frac{1}{8}(-t_2) + 3(-t_3)$$

The "fractional coefficients" $f_j = a_j - \lfloor a_j \rfloor$ are then obtained by the calculations

$$4\frac{5}{8} - 4 = \frac{5}{8}, \quad 2\frac{3}{8} - 2 = \frac{3}{8}, \quad -4\frac{1}{8} - (-5) = \frac{7}{8} \quad \text{and} \quad 3 - 3 = 0$$

thus giving the cut

$$s = -\frac{5}{8} - \frac{3}{8}(-t_1) - \frac{7}{8}(-t_2) - 0(-t_3)$$

(Note that these fractions are exactly the fractions that appear in the source equation, after adding integer amounts to negative coefficients in the source equation to make them positive.)

Similarly, if the tableau equation is given by

$$z = 5\frac{1}{3} - 4\frac{1}{3}(-t_1) - 2(-t_2) + 3\frac{2}{3}(-t_3)$$

the fractional cut is

$$s = -\frac{1}{3} - \frac{2}{3}(-t_1) - 0(-t_2) - \frac{2}{3}(-t_3)$$

From these two examples we observe the important fact that *if the constant term of the tableau equation is not an integer, then the fractional cut is "currently infeasible"*—that is, the constant term of the fractional cut is negative, indicating that this cut is *not satisfied* by the current LP solution. This is what we are looking for, because it means that the fractional cut is capable of cutting off the current LP optimum if one of the variables is not integer feasible (hence has a fractional constant term).

2.2.5. Fractional Cuts in the Tableau: An Example

We now illustrate the use of the fractional cuts in a cutting procedure of the type described earlier. (The all-integer cuts have special uses of their own in other types of cutting procedures, as will be demonstrated subsequently.) Consider the following integer programming problem:

$$\text{Minimize} \quad 14x_1 + 12x_2$$
$$\text{subject to} \quad 35x_1 + 24x_2 \geq 106$$
$$14x_1 + 10x_2 \geq 43$$
$$x_1, x_2 \geq 0 \text{ and integer}$$

We first solve the problem as a linear programming problem by disregarding the integer restriction. Using the condensed tableau representation, we obtain the following initial tableau.

		$-x_1$	$-x_2$
$x_0 =$	0	14	12
$y_1 =$	-106	-35	-24
$y_2 =$	-43	(-14)	-10

This representation arises by maximizing $x_0 = 0 + 14(-x_1) + 12(-x_2)$ and introducing nonnegative slack variables y_1 and y_2 to turn the original inequalities into equations. Hence rewriting the first original inequality by taking the constant term (106) on the opposite side gives $35x_1 + 24x_2 - 106 \geq 0$. Then setting y_1 equal to the left-hand side (assuring $y_1 \geq 0$) and reexpressing the result in terms of the tableau format yields $y_1 = -106 - 35(-x_1) - 24(-x_2)$. Similarly, introducing y_2 leads to the equation that corresponds to the "y_2 row" of the initial tableau. Now, an iteration of the dual simplex method (pivoting on the circled element, -14) yields the following LP optimum:

		$-y_2$	$-x_2$
$x_0 =$	-43	1	2
$y_1 =$	$1\frac{1}{2}$	$-2\frac{1}{2}$	1
$x_1 =$	$3\frac{1}{14}$	$-\frac{1}{14}$	$\frac{5}{7}$

Integer Programming

Remark: To confirm the preceding outcome, the pivot step of the condensed tableau format can be achieved by the following standard "solve and substitute" operation from elementary algebra:

1. Solve for the pivot column variable, x_1 in this case, in the pivot row equation, the last equation in this case. The resulting rearranged form of the equation (with the "solved for" variable on the left and the other terms arranged in the condensed tableau format on the right) becomes the updated pivot row, the last row in the updated tableau in this case.
2. Substitute the expression for the "solved for" variable from step 1 in each of the other tableau equations, thereby eliminating this variable from these equations. Upon collecting terms in the condensed tableau format, each such equation has the proper updated form for the new tableau.

In this case,

$$y_1 = -106 - 35(-x_1) - 24(-x2)$$
$$y_1 = -106 - 35(-3\tfrac{1}{14} + \tfrac{1}{14}(-y_2) - \tfrac{5}{7}(-x_2)) - 24) - x_2)$$
$$y_1 = 1\tfrac{1}{2} - 2\tfrac{1}{2}(-y_2) + 1(-x_2)$$

The solution is not integer feasible. Hence an equation with a fractional constant term may be selected to derive a fractional cut. Selecting the x_1 equation for this role (the y_1 equation is also acceptable), we obtain the cut

$$s_1 = -\tfrac{1}{14} - \tfrac{13}{14}(-y_2) - \tfrac{5}{7}(-x_2)$$

which is then adjoined to the preceding tableau to give

		$-y_2$	$-x_2$	
$x_0 =$	-43	1	2	
$y_1 =$	$1\tfrac{1}{2}$	$-2\tfrac{1}{2}$	1	
$x_1 =$	$3\tfrac{1}{14}$	$-\tfrac{1}{14}$	$\tfrac{5}{7}$	Source Equation
$s_1 =$	$-\tfrac{1}{14}$	$\boxed{-\tfrac{13}{14}}$	$-\tfrac{5}{7}$	Cut

Reoptimizing with the dual method yields the following tableau, in which the selected source equation and its associated cut are indicated as before.

	$-s_1$	$-x_2$	
$x_0 =$	$-43\,1/13$	$1\,1/13$	$1\,3/13$ Source Equation
$y_1 =$	$1\,9/13$	$-2\,9/13$	$2\,12/13$
$x_1 =$	$3/13$	$-1/13$	$10/13$
$y_2 =$	$1/13$	$-1\,1/13$	$10/13$
$s_2 =$	$-12/13$	$-1/13$	$\boxed{-3/13}$ Cut

The choice of the x_0 equation to be the source for the cut in this tableau is perfectly legitimate since x_0 is an integer variable, and the fractional cut does not require the "z" variable of the tableau equation to be nonnegative. The next tableau, obtained after one additional pivot, is not primal feasible. Thus, instead of adjoining a new cut we continue to reoptimize with the dual simplex method.

	$-s_1$	$-s_2$
$x_0 =$	$-48\,1/13$	$2/3$	$5\,1/3$
$y_1 =$	-10	$\boxed{-3\,2/3}$	$12\,2/3$
$x_1 =$	0	$-1/3$	$3\,1/3$
$y_2 =$	-3	$-1\,1/3$	$3\,1/3$
$x_2 =$	4	$1/3$	$-4\,1/3$

The resulting tableau is not primal feasible, and thus a cut is adjoined as shown:

	$-y_1$	$-s_2$
$x_0 =$	$-49\,9/11$	$2/11$	$7\,7/11$
$s_3 =$	$-2/11$	$\boxed{-2/11}$	$-7/11$
$x_1 =$	$10/11$	$-1/11$	$2\,2/11$
$y_2 =$	$7\,7/11$	$-4/11$	$-1\,3/11$
$x_2 =$	$3\,1/11$	$1/11$	$-3\,2/11$

The new cut has not been adjoined to the bottom of the tableau as before, but has instead been "written over" the s_1 equation. This accomplishes the effect of discarding the s_1 cut, which is currently nonbinding. (That is, s_1 is basic at the current LP optimum, and therefore the hyperplane $s_1 = 0$ is not among the hyperplanes $y_1 = 0$, $s_2 = 0$ whose intersection determines this optimum.) The procedure of writing a new cut over an old one keeps the number of rows of the tableau from becoming unmanageably large. However, the procedure may be postponed until the problem reaches a "critical size" on the chance that some of the currently nonbinding cuts may nevertheless become restrictive once again at later iterations.

After one further pivot, we obtain the tableau

		$-s_3$	$-s_2$
$x_0 =$	-50	1	7
$y_1 =$	1	$-5\frac{1}{2}$	$3\frac{1}{2}$
$x_1 =$	1	$-\frac{1}{2}$	$2\frac{1}{2}$
$y_2 =$	1	-2	0
$x_2 =$	3	$\frac{1}{2}$	$-3\frac{1}{2}$

This final tableau is integer optimal and the problem is solved. The current nonbasic variables disclose that the integer optimum occurs at the intersection of the cut hyperplanes $s_3 = 0$ and $s_2 = 0$.

2.3. Chvátal-Gomory Rounding Method

The integer rounding method may be summarized in matrix notation, $Ax \leq b$, in the following three steps.

1. Generate an inequality as a nonnegative combination of other inequalities. Let $u \geq 0$ be a vector of multipliers, one for each existing inequality, and let a_j be the column of A corresponding to the variable x_j.

$$\sum u a_j \leq ub$$

2. Round down the left-hand side.

$$\sum_j \lfloor u a_j \rfloor \leq ub$$

3. Round down the right-hand side. (The left-hand side must be an integer.)

$$\sum_j \lfloor u a_j \rfloor \leq \lfloor ub \rfloor$$

It can be shown that the convex hull of integer solutions can be generated by repeated application of this procedure, sometimes called the *Chvátal-Gomory rounding method* (Chvátal, 1973).

One consequence of generating the convex hull by this procedure is that *every* valid inequality is either a Chvátal-Gomory cut or dominated by one. An inequality is *dominated* by inequality j if combining the polyhedron, P, of the linear programming relaxation with the half-space, J, feasible for inequality j implies inequality i but not vice versa, i.e.,

$P \cap J \subset P \cap I$

Every inequality can be given a ranking based on the iteration of applying the Chvátal-Gomory rounding method that first generates the inequality or generates a dominating inequality. This is referred to as the *Chvátal-Gomory rank*.

2.4. References

Polyhedral combinatorics has generated valid inequalities for many specific problems. Wolsey (1989b) provides a survey of recent results. Balas et al. (1993, 1995) describe an approach called *lift and project*. Other cuts for integer programming have been suggested by Balas and Jeroslow (1980), Dietrich et al. (1993), Glover et al. (1995), Johnson (1989), Martin and Schrage (1985), and Van Roy and Wolsey (1986).

Many cuts have been derived for the knapsack problem and closely related problems. One of the earlier papers was by Padberg (1979). More recent papers include those by Boyd (1991, 1993), Dietrich and Escudero (1989), Park and Park (1994), and Sherali and Lee (1995).

Another area of concentrated research is fixed charge problems, in particular fixed charge network design problems. See, for example, Padberg et al. (1985) and Van Roy and Wolsey (1987). Wolsey (1989a) describes cuts based on submodular functions.

Variable redefinition is a promising approach proposed by Martin (1987) for deriving valid inequalities based on additional variables added to a problem.

Boyd (1994) described cuts based on Fenchel duality.

2.5. Exercises

2.1. For the example illustrated in Figures 5-7 show that the objective function hyperplane can be varied to produce three different integer optima while keeping the same optimal LP vertex.

2.2. It is sometimes suggested to round the LP solution to one of its "nearest-neighbor" integer points to try to obtain an optimal integer solution. Show in the example of Figures 5-7 that none of the nearest-neighbor rounding possibilities (allowing each variable to be rounded up and down, as desired) produces a feasible IP solution, let alone an optimal one. (There is a published example of a problem that has more than a million rounding alternatives, none of which are feasible. Yet the optimal integer solution to this problem can be identified by inspection.)

2.3. One strategy for solving IP problems is to disregard all LP constraints except those that are binding at the LP optimum and to attempt to obtain an integer solution to the relaxed IP that results. In the example of Figures 5–7, show that this strategy would obtain the correct solution for the objective function hyperplane as shown but would obtain wrong solutions for other slopes that maintain the same optimal LP solution.

2.4. More than one fractional cut can be obtained from a single tableau equation as follows. Because the only requirement on the variable is the integer value requirement, it is possible to obtain a fractional cut from the equation

$$hz = ha_0 + \sum ha_j(-t_j)$$

where h is an integer, because then hz is an integer variable. This yields the cut

$$s = -f_{h0} - \sum f_{hj}(-t_j)$$

where f_{hj} is the fractional part of ha_j; that is, $f_{hj} = ha_j - \lfloor ha_j \rfloor$. Show that another way to reach the same conclusion is to select $d = 1/h$ (for h an integer) instead of simply $d = 1$ in deriving the fractional cut from the all-integer cut.

2.5. Consider the tableau equation

$$z = 5\tfrac{3}{8} + 4\tfrac{1}{8}(-t_1) - 2\tfrac{1}{4}(-t_2) + 3\tfrac{1}{2}(-t_3)$$

Generate fractional cuts from this equation by the cut equation of Exercise 2.4, letting h take the integer values 1 through 10 and -1 through -3. Show that the cuts obtained are the same as those obtained from the source equation

$$z^1 = \tfrac{3}{8} + \tfrac{1}{8}(-t_1) + \tfrac{3}{4}(-t_2) + \tfrac{1}{2}(-t_3)$$

i.e., that the integer parts of the source equation coefficients are irrelevant to all of the cuts. Furthermore, on the basis of the cuts generated, explain why there are in fact only eight unique cuts (including the vacuous all-0 cut) that can be obtained from the source equation, no matter what integer value h receives. Finally, state a rule that identifies exactly which one of these eight cuts will be obtained for an arbitrary integer value of h.

2.6. Using the definition of the relative strength of two inequalities indicated earlier, justify the assertion:

The inequality $\sum p_j t_j \leq K$

is stronger than $\sum q_j t_j \leq K$
(under the assumption $t_j \geq 0$) if and only if $p_j \geq q_j$ for all j and $p_j > q_j$ for some j. (Thus, one way to compare inequalities that have constant terms of the same sign is to "normalize" them to the same right-hand side and then apply the foregoing check.)

2.7. Apply the result of Exercise 2.6 to compare the relative strengths of the cuts of Exercise 2.5, using the fact that for $x \geq 0$ the cut equation

$$s = f_{h0} - \sum f_{hj}(-t_j)$$

corresponds to the inequality

$$-\sum f_{hj} \leq -f_{h0}$$

(This disregards the integer requirement on s.) In a similar manner, show that the first cut adjoined to the example problem illustrating the fractional cut method was drastically weaker than other cuts that were available from the same tableau. Finally, graph the inequality corresponding to the adjoined cut (i.e., $-13/14 y_2 - 5/7 x_2 \leq -1/14$, etc.) and compare the result to the graph of two of the stronger inequalities (e.g., $-1/14 y_2 - 2/7 x_2 \leq -13/14$, etc.). How do the graphs disclose the algebraic fact that one cut is stronger than another?

2.8. All of the coefficients of the fractional cuts are nonpositive. However, it is possible to obtain "generalized" fractional cuts that may contain positive coefficients and that therefore provide a dramatic strengthening along the dimension of the positive coefficient. Show, for example, that if the source equation

$$y_1 = 3\tfrac{5}{9} + 6\tfrac{2}{3}(-t_1) + 5\tfrac{1}{9}(-t_2) + 2\tfrac{8}{9}(-t_3)$$

then there exists a value of h slightly larger than 1 that yields the cut

$$s = -\tfrac{5}{9} + \tfrac{1}{3}(-t_1) - \tfrac{1}{9}(-t_2) + \tfrac{1}{9}(-t_3)$$

Graph the two-dimensional cross sections of this cut for the variables t_1, t_2 and the variables t_2, t_3 (i.e., graph the inequalities $1/3 t_1 - 1/9 t_2 \leq -5/9$ and $-1/9 t_2 + 1/9 t_3 \leq -5/9$) and compare the result to the corresponding graphs of the fractional cut obtained for $h = 1$. What do these graphs disclose about the

Integer Programming

relative strengths of the two cuts? (A procedure exists to identify best parameter values for the generalized fractional cuts but has not been tested in practice.)

3. BRANCH AND BOUND PROCEDURES

Branch and bound (B & B) procedures can be viewed in a variety of ways. Some of these are reflected in the names sometimes given to B & B methods—e.g., implicit enumeration, tree search, progressive separation and evaluation, strategic partitioning, and provisional cutting. Regardless of the way it is viewed, branch and bound has two appealing qualities. First, B & B can be applied to the mixed-integer problem and to the pure integer problem in essentially the same way, so that a single method works for both. Second, B & B typically yields a succession of feasible integer solutions (unlike dual cutting approaches, in which a feasible integer solution isn't obtained until the problem is solved), so that when available computer time begins to run out the method can stop and offer the "current best" solution as a candidate for the optimum.

This branch and bound methods can also often be tailored to exploit special problem structures, thereby allowing these structures to be handled with greater efficiency and reduced computer memory (an important feature when solving large-scale problems). In fact, except in very general terms, there is no one B & B method but a whole collection of such methods that happen to share a number of overlapping characteristics.

This tailoring aspect of branch and bound further enables it to be applied directly to many kinds of combinatorial problems without first going through an intermediate step of introducing specific integer variables and linear constraints to yield an IP model formulation. The direct application of methods to problems is enormously important when the use of an intervening IP formulation may drastically increase the problem size or otherwise obscure exploitable problem characteristics. (The direct problem-solving ability is matched in the cutting plane area only by relatively recent developments such as reflected in the "polyhedral annexation" (e.g., Glover, 1975b) and "disjunctive programming" (e.g., Balas, 1979; Jeroslow, 1990) approaches.

Generally speaking, the type of branch and bound method that applies to the standard IP model gives a framework for understanding (and constructing) B & B methods in other settings. A method that handles general IP formulations constitutes a significant tool when one has neither the time nor the resources (which may involve many thousands of dollars of developmental effort) to produce a specially tailored procedure for a specific problem class.

3.1. Branch and Bound for 0-1 Problems (Pure and Mixed)

We will begin by discussing B & B methods for 0-1 integer programs and then show that the ideas apply almost without change to other IP problems. Many of the more effective B & B procedures are based on the use of the simplex methods. These are the types of procedures we will examine first.

The simplex-based methods begin by solving the original problem as a linear program in the hope that the solution will be integer feasible and thus solve the IP problem. (A similar hope—usually misplaced—underlies the use of dual cutting methods.) If an integer solution is not obtained, the next step is to pick one of the integer variables, denoted (say) by x_s, where s is the variable index, and to create two descendants of the original problem, one in which $x_s = 0$ and one in which $x_s = 1$. Since x_s is a 0-1 variable (preferably one that does not receive an integer value in the lP solution), and the descendant problems are exactly the same as the original problem except for the assignment of a specific value to x_s, the solution to one of these two descendants must in fact be the solution to the original problem. Thus, the original problem may be replaced by these descendants, leaving two IP problems to solve instead of one. (However, because each new problem is more restricted than the original and has fewer variables that must be assigned integer values, the expectation is that it may be solved somewhat more readily than its parent.)

The process then repeats, selecting one of the problems that remain to be solved as the current IP problem and treating it exactly the same as the original. (This in turn may create two new problems to replace the one currently under consideration unless, for example, the current problem is "infeasible.") Eventually by this process an integer solution is obtained for one of the current IP problems, whereupon this solution provides a candidate for the optimal solution to the original problem. Keeping track of the current best of these candidates and its associated objective function value provides an additional way to weed out descendants of the original IP problem (perhaps several generations removed) that are unprofitable for exploration. The current best candidate is called the *incumbent*.

The possibilities for this type of approach can be itemized as follows. Let x^* denote the objective function value for the incumbent solution. Then, whenever a current IP maximization problem is solved as an LP problem one of four alternatives arise:

    a. The LP problem has no feasible solution (in which case the current IP problem also has no feasible solution).
    b. The LP problem has an optimal solution with $x_0 \le x^*$ (in which case the current IP optimum must also yield $x_0' \le x^*$ and therefore cannot provide an improved solution).

Integer Programming

c. The optimal solution to the LP problem is both integer feasible and yields $x_0 > x^*$ (in which case the solution is optimal for the current IP problem and provides an improved incumbent solution for the original IP problem, x^* is reset to x_0).
d. None of the foregoing occurs; i.e., the optimal LP solution exists, satisfies $x_0 > x^*$, and is not integer feasible.

In each of the first three cases the current IP problem is disposed of simply by solving the LP problem. An IP problem that is resolved in this way is said to be fathomed. (A problem that is fathomed as a result of (c) yields particularly useful information.) Otherwise, if the problem is not fathomed and winds up in (d), then the problem requires further exploration.

We are now in a position to summarize the steps of the B & B procedure. To start, we set x^* equal to the objective function value of any feasible integer solution that may be known in advance. Otherwise, x^* is set equal to $-M$, for some large number M, so that the first feasible IP solution found will be regarded as an "improved" solution.

A Branch and Bound procedure for pure and mixed 0-1 problems:
0. Initialized: To begin, the list of problems to be solved consists of the original IP problem.
1. Current problem selection. If the problem list is empty, then the solution process is complete and the current best solution is optimal (or if no integer feasible solution was found, then none exists). Otherwise, select (and remove) a problem from the list, and call it the current IP problem.
2. Fathoming check. Solve the LP problem that corresponds to the current IP problem, and determine which of the alternatives holds:
    (a) If the LP problem has no feasible solution, return to step 1.
    (b) If the optimal LP solution yields $x_0 \leq x^*$, return to step 1.
    (c) If the optimal LP solution is integer feasible and satisfies $x_0 \geq x^*$, record this integer solution as the new "current best" solution (and update x^*). Then return to step 1.
    (d) If the optimal LP solution exists, satisfies $x_0 > x^*$, and is not integer feasible, proceed to step 3.
3. Branch creation. Select an integer variable x_s that is not assigned to an integer value in the current IP problem, and create two descendants of the current IP problems by requiring $x_s = 1$ for one descendent and $x_s = 0$ for the other descendant. (Otherwise, the descendants are the same as the current problem). Add these two descendants to the list of problems to be solved and return to step 1.

3.2. Last In, First Out (LIFO)

A useful way to visualize the branch and bound procedure is by means of a tree format, as illustrated in Figure 8. Each node (circle) in Figure 8 represents a current IP problem, and each arc (arrow) represents a branching choice. The number above each node represents the optimal x_0 value for the LP associated with the node.

The variable x_s selected for branching at a particular node and the alternatives $x_s = 0$ and $x_s = 1$ are indicated on the arcs that leave the node. Square nodes identify problems that are fathomed. The symbols following these nodes identify the way in which these problems are fathomed.

Finally, the numbers inside the circles indicate the sequence in which the current IP problems are selected for consideration. The original IP problem, which must always be considered first, is numbered 0. Note that the x_0 value for the LP problem at node 0 is the largest of all x_0 values. In general, the x_0 value for any node is never larger than the x_0 value for its parent node (i.e., its immediate predecessor), because the problem associ-

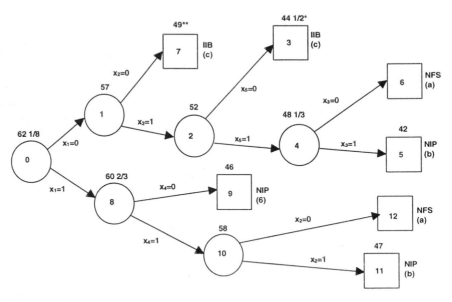

Figure 8 An example tree format for branch and bound (LIFO branch rule) NFS, no feasible solution [fathomed by (a)]; NIP, no improvement possible [fathomed by (b)]; IIB, improved incumbent [fathomed by (c)]; *, value of x^* for first IIB; **, value of x^* for second IIB.

Integer Programming

ated with the given node always has one more restriction (the value assigned to x_s on the arc leading to it) than the problem associated with its parent.

In the tree representation this is conveniently visualized as always selecting one of the two immediate successors of a given node as the next node to examine. However, if the node is fathomed (and hence has no successors), the procedure *backtracks* to the nearest predecessor whose alternative branch has not yet been explored, and selects the next node to be the one reached by this branch. (The branching variables are selected arbitrarily in this example from among the fractional variables. In section 3.4.1 we discuss particular rules for selecting branching variables.)

Thus, the steps of the solution procedure illustrated in Figure 8 can be traced as follows.

Node 0: The original IP problem is solved as a linear program, giving $x_0 = 62\,\tfrac{1}{8}$. No fathoming occurs and x_1 is selected as the branching variable.

Node 1: The branch $x_1 = 0$ is selected first, corresponding to selecting the current IP problem to be the one that adds the restriction $x_1 = 0$ to the original problem. The LP solution to the current IP problem yields $x_0 = 57$ and x_2 is selected as the branch variable.

Node 2: The branch $x_2 = 1$ is selected (from node 1) so that the current IP problem inherits the two assignments $x_1 = 0$, $x_2 = 1$. The LP x_0 value is 52, and x_5 is selected for branching.

Node 3: The choice $x_5 = 0$ (which gives node 3 the inherited restrictions $x_1 = 0$, $x_2 = 1$, $x_5 = 0$) causes the LP solution to be integer feasible, yielding the first candidate for the optimal IP solution. (The solution is recognized as an improved incumbent because $44\,\tfrac{1}{2}$ exceeds the initial x^* value of $-M$.) Thus, node 3 is fathomed by (c), and the method backtracks. The nearest predecessor (most recent ancestor) whose alternative branch has not yet been explored is node 3's immediate predecessor, node 2, and the branch $x_5 = 1$ is now taken.

Node 4: the LP solution value (for the inherited assignment $x_1 = 0$, $x_2 = 1$, $x_5 = 1$) gives $x_0 = 48\,\tfrac{1}{3}$, and x_3 is selected for branching.

Node 5: The choice $x_3 = 1$ yields an LP x_0 value of 42. Since $42 \leq 44\,\tfrac{1}{2}$ (the current x^*), there is no improvement possible along this branch and node 5 is fathomed by (b). (Note that fathoming would occur even if the current LP were integer feasible.) Backtracking identifies node 4 as the nearest predecessor with an unexplored alternative branch, thus leading to node 6 via this branch.

Node 6: The assignment $x_3 = 0$, together with the other restrictions inherited at this node, causes the LP program to have no feasible solution. Thus, the node is fathomed by (a), and the method backtracks. Node 1 is the nearest predecessor whose alternative branch remains to be explored, and this branch leads to node 7.

Node 7: The assignment $x_2 = 0$ (coupled with the inherited assignment $x_1 = 0$) causes the LP solution once again to be integer feasible, with an improved x^* value ($49 > 44\frac{1}{2}$). The improved current best solution is recorded in place of the previous current best, and the method backtracks.

Identification of the steps that lead to the rest of the nodes is left to the reader. It should be noted that the method terminates when a backtracking step occurs and no predecessor of the current node has an unexplored branch.

The LIFO branching rule is relatively popular because it can be implemented by an extremely simple bookkeeping scheme. In fact, this scheme uses only two lists of "numbers" (indexes and values of variables) to keep track of all unexplored branches and to identify which one to examine next.

We can illustrate these lists by indicating their form upon reaching node 3.

Index of selected variable: 1, 2, 5

(at node 3)

Value of variable: 0, 1, 0

The lists indicate that x_1 was selected first with a value of 0, x_2 was selected next with a value of 1, and x_5 was selected third with a value of 0, exactly corresponding to the branches that lead to node 3.

There is one more ingredient to be incorporated into the "index of selected variable" list in order to handle the backtracking step. Specifically, a "star" (i.e., asterisk) is attached to an index in this list if the alternative branch for the variable has already been taken. Then the backtracking step simply looks for the most recent (i.e., rightmost) index that doesn't have a star. This identifies the most recent node whose alternative branch has not been taken and permits the backtracking step to be implemented at once.

For example, node 3 is a fathomed node in Figure 8 and hence leads to backtracking. In the index list, the rightmost unstarred index is 5, indicating that the alternative branch (for $x_5 = 1$) has not been taken. Thus, the procedure takes this alternative branch and attaches a star to the index

Integer Programming

of 5, to indicate that the original branch ($x_5 = 0$) has previously been taken. The lists applicable to node 4 are therefore as follows.

    Index of selected variable: 1, 2, 5*

                                                                    (at node 4)

    Value of variable: 0, 1, 1

For the next step, node 4 is not a fathomed node, and the choice step ($x_3 = 1$) produces the lists

    Index of selected variable: 1, 2, 5*, 3

                                                                      (at node 5)

    Value of variable: 0, 1, 1, 1

Since node 5 is fathomed, we carry out the backtracking step as previously illustrated, identifying 3 as the rightmost unstarred index, to give

    Index of selected variable: 1, 2, 5*, 3*

                                                                    (at node 6)

    Value of variable: 0, 1, 1, 0

Node 6, as it turns out, is also fathomed. Consequently, once again backtracking is required. Now the rightmost unstarred index is 2. This means we pass over the starred indexes (since their alternative branches have been taken) and resume the procedure by taking the alternative branch for x_2. This gives the lists

    Index of selected variable: 1, 2*

                                                                  (at node 7)

    Value of variable: 0, 0

The portions of the lists that were passed over are, of course, dropped, since the branch of the tree now under investigation does not lead to the passed over assignments.

To summarize, the rules for carrying out the LIFO produces can be stated as follows.

The LIFO procedure:
1. Forward step: Add the index of the selected variable and the value assigned to the variable to the bookkeeping lists.
2. Backtracking step: If no unstarred indexes remain, the solution process is completed. Otherwise, identify the rightmost unstarred index, and drop the portions of the lists (on its right) that were passed over to reach this index. Then star this index

and take the alternative branch. (That is, make the alternative value assignment for the associated variable.)

3.2.1. Exercises

3.1. Begin at node 7 in Figure 8 and trace through the remaining nodes of the tree, following the sequence indicated by the numbering of the nodes. Explain the reason for each step.

3.2. Repeat Exercise 3.1 using the index of selected variable and value of variable lists to monitor the forward steps and to identify the appropriate form of the backtracking steps.

3.3. Explain why the absence of unstarred indexes indicates that the solution process is complete when a backtracking step is executed.

3.4. Show that it is possible to generalize the LIFO approach, without requiring more extensive bookkeeping, by the following observation: the arrangement of any sequence of unstarred indexes is "inconsequential" as long as there are no starred indexes within this sequence. Thus, any sequence (or subsequence) of indexes that are all unstarred can be rearranged in any fashion without changing the validity of the procedure (provided the values of variables are rearranged in the same way). Justify this observation by showing that any unbroken string of forward steps can be imagined to have been executed in any order — except that rearrangement can lose the information about the LP x_0 value associated with a given index.

3.5. Show relative to Exercise 3.4 that the LP x_0 value associated with the last position of any rearranged sequence (or subsequence) will not change. Is losing the information about x_0 values of any importance to the method? (That is, is an x_0 value ever used except when it is first obtained?) Would your answer be different if an updated LP tableau was stored for each node of the tree, so that the LP problem for a successor node could be solved at a backtracking step by reoptimizing?

3.3. A Best Bound Rule

The best bound rule, like LIFO, is one of the mainstays of the procedures used to select the current IP problem in branch and bound methods. In fact, virtually all B & B rules used in practice involve some variant or combination of these two rules.

The best bound rule, as its name suggests, always selects the current IP problem (node) to be the one whose LP solution gives the largest (and

Integer Programming

hence most attractive) x_0 value. The "bound" terminology derives from the fact that the LP x_0 value gives an upper bound on the optimal IP x_0 value for the current problem. (This bound is not best in the sense of being the most realistic but only in the sense of being the most inviting—relying on the hope that it somehow approximately resembles the optimal IP x_0 value that can be obtained for the current problem.) We illustrate this rule in Figure 9.

The interpretation of Figure 9 is similar to that of Figure 8, and in fact we suppose that the IP problem for both Figure 8 and Figure 9 is the same. Thus, node 0 represents the original IP problem, whose LP solution gives $x_0 = 62\,\tfrac{1}{8}$ (where, as before, LP x_0 values appear above the nodes). Similarly, the sequence in which the current IP problems are selected for consideration is indicated by the numbers within the nodes (except that now the sequence is changed and some of the fathomed nodes, represented by squares, are unnumbered).

In addition, we have added numbers marked with primes beneath the nodes. The reason for this is that the best bound rule actually incorporates a look-ahead step in its execution. Specifically, whenever two descendants

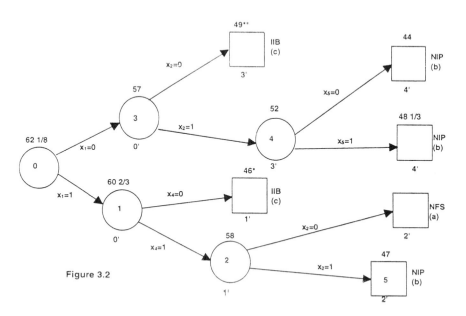

Figure 9 Tree format illustration of the best bound rule. NFS, no feasible solution; NIP, no improvement possible; IIB, improved incumbent; *, value of x^* for first IIB found; II, value of x^* for second IIB found.

of a current problem are created, the best bound immediately solves the IP problem for both of them, in order to obtain the x_0 values that will enable the rule to select the largest of such values. (This is in marked contrast to the LIFO rule illustrated in Figure 8, where only one of the two descendants is examined immediately—by actually selecting its branch—and the other is merely "remembered" so that it can be examined on a subsequent backtracking step.) Thus, the primed number beneath a node identifies the intermediate step at which the associated LP problem was solved. For example, upon first solving the LP problem at node 0, and selecting x_1 as the branching variable, the LP problems are solved for both branches $x_1 = 0$ and $x_1 = 1$ at step 0 (before selecting the next problem to be examined at step 1). This is indicated by the number of 0' that appears beneath each of the immediate successors of node 0. As might be imagined, this look-ahead aspect of the best bound rule changes its character somewhat. The nature of this change will become clear by tracing the steps of the method as they are recorded in Figure 9.

Node 0: Having solved the LP problem for both descendant nodes, the bounds $x_0 = 57$ and $x_0 = 60\frac{2}{3}$ are both available. The best of these is $x_0 = 60\frac{2}{3}$, thereby identifying node 1.

Node 1: The LP problem for this node has already been solved (at step 0') and now the variable x_4 is selected as the branching variable. This leads to solving the LP problems for both branches $x_4 = 0$ and $x_4 = 1$ at step 1'. The LP solution for the $x_4 = 0$ branch turns out to be integer feasible and thus provides the first improved incumbent solution with $x^* = 46$. Note that this solution was found as a by-product of the look-ahead step, without actually choosing this branch as the next to examine. Thus, the node is fathomed at step 1' and does not receive a sequence number that indicates when it is selected for examination. The remaining available x_0 values (bounds) to be considered are $x_0 = 57$ and $x_0 = 58$. Choosing the larger leads to node 2.

Node 2: the variable x_2 is selected for branching, and the LP problems for $x_2 = 0$ and $x_2 = 1$ (together with the assignments inherited at node 2) are solved at step 2'. The branch for $x_2 = 0$ has no feasible solution and thus it is fathomed immediately. The other branch, for $x_2 = 1$, yields an x_0 value of 47. Since this exceeds the current x^* value of 46, the node at this branch is *not* fathomed (although the fact that it is a square node is a giveaway that the node will be fathomed eventually). Again we have two available bounds $x_0 = 57$ and $x_0 = 47$. Selecting the larger leads to node 3.

The remaining steps of the procedure are left as an exercise for the reader.

Integer Programming

3.3.1. Exercises

3.6. Finish tracing the steps of the best bound approach as illustrated in Figure 9. Would it have been possible to fathom node 5 before actually selecting it at step 5? What kind of scanning rule could permit this to be done? Do you think such a rule is worthwhile or a waste of time?

3.7. Show that the general termination rule for the best bound approach is the following: Stop either when no more problems remain to be examined or when the LP x_0 value for the currently selected IP problem does not exceed x^*.

3.8. Compare the steps of the LIFO and best bound approaches for the problem whose branch and bound solution is illustrated by the following tree format.

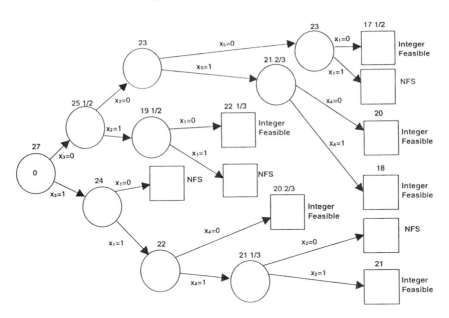

Identify the sequence of the nodes visited by the LIFO rule under the assumption that the $x_s = 0$ branch is always selected before the $x_s = 1$ branch. Also, clearly specify the "prime" (or look-ahead) steps for the best bound method. For both the LIFO and best bound rules, identify the succession of incumbent solutions, and note which integer feasible solutions are bypassed in each case. (Distinguish between integer feasible solutions that are found and bypassed and those that are not found at all.) For each method, identify the nodes of the tree that are not visited at all. Finally, one of the terminal nodes has been assigned an impossible LP x_0 value. Identify this node and explain why its LP x_0 value is impossible.

3.9. What kind of memory requirements apply to the best bound approach? Compare these to the memory requirements of the LIFO method (i.e., the bookkeeping lists of indexes and values).

3.10. A very attractive feature of the LIFO approach, in addition to its bookkeeping simplicity, is the fact that the LP problem for the forward step can always be solved by introducing the constraint $x_s = 0$ or $x_s = 1$ to the currently available LP tableau and then reoptimizing. Since reoptimizing relative to a single added constraint is generally far more efficient than starting over to solve the current LP problem from scratch, this procedure can save a great deal of computational effort. Which steps of the LIFO and best bound procedures can similarly take advantage of reoptimizing? Which steps of these procedures require "starting over" (by introducing the assignments for the current IP problem into the original tableau)? Discuss the trade-offs of data storage and retrieval effort versus the ability to reoptimize if a B & B method stored a "current LP tableau" at every node of the tree. When would a stored tableau be of no further use in the LIFO and best bound procedures?

3.11. Note that the best bound rule shares one kind of choice in common with the LIFO rule–i.e., the choice of which variable to "branch on" at the given node. This type of choice is extremely crucial in practical applications, so that a LIFO approach that makes this choice judiciously may have to examine far fewer problems than a best bound approach that does not make this choice well. However, if the best bound approach and the LIFO approach make this choice in the same way, then the best bound approach will never have to generate more nodes (examine more problems) than the LIFO approach. (This is illustrated by the fact that the tree of Figure 9 is smaller than that of Figure 8.) Justify this statement. Can you apply similar reasoning to show that the statement is also true if any other procedure for selecting the current IP problem is used in place of the LIFO procedure — and in fact that the best bound approach will generate a subset of the nodes generated by the other procedure? Why does the observation not imply

Integer Programming

that the best bound rule will be computationally superior to the LIFO rule?

3.12. Consider the following "LIFO forward, best bound backward" rule: As long as the current node is not fathomed, make a forward step (and take advantage of the ability to reoptimize without having to retrieve an externally stored tableau). But when the current node is fathomed, select the best bound node on the backtracking step. Apply this rule to the example of Exercise 3.8.

3.13. A notable phenomenon in branch and bound is that the LP x_0 values become more and more realistic (in their ability to approximate the optimal x_0 values for the associated IP problem) as one goes deeper into the tree. Thus, an x_0 value for a "shallow" node is apt to overestimate the associated IP optimum by a larger amount than the x_0 value for a "deeper" node. How might the best bound approach be modified to compensate for this discrepancy in realism (or estimating power)? What experiments might be conducted to make this modification more effective? (Devising good modifications is complicated by the fact that the choice of the branching variable can also have a pronounced effect on the realism of the resulting bound. This is an area that has received only a limited amount of attention and undoubtedly deserves more.)

3.4. Branch and Bound Strategies

Integer programming in general, and branch and bound in particular, requires more user control of the solution algorithm than linear programming requires. Naive use of commercial linear programming software, relying on default settings of algorithmic parameters, often causes little performance degradation and seldom has any effect on the ultimate attainment of the optimal solution. In contrast, naive use of branch and bound software frequently leads to noticeable performance degradation and can prevent finding the optimal solution (or even feasible solutions) for an otherwise solvable problem. Most branch and bound software allows a user to select dramatically different solution strategies by adjusting the search parameters. Ideally, the solution strategy should reflect both the model's special structure and the current qualification for solution progress: find a feasible solution, find an improved solution, or prove optimality (or near optimality) of the current solution. Selecting and controlling a good branch and bound code is second in importance only to a tight formulation in the practical solution of integer programming problems. For this reason we devote considerable space to user-controllable solution strategies for branch and bound.

The two most influential strategic choices for branch and bound are the selections of which variable to branch on next and which node to branch on next. Our discussion concentrates primarily on these choices. Other factors that affect branch and bound performance include the type of linear programming algorithm used to bound subproblems, the extent of optimization of subproblems in the tree, and the type of branching allowed at a node. Typically, not all of the options described here will be available within a given software package, and hence software selection plays a major role in limiting a user's branch and bound strategic options. Software surveys (e.g., Saltzman, 1994) are frequently published reflecting the continuing evolution of branch and bound algorithms.

3.4.1. Selecting the Branching Variable

Several reasonable criteria exist for selecting branching variables. The criteria differ principally in how they measure the desirability of a branch: penalties, integer infeasibility, pseudocosts, pseudoshadow costs, and user-specified priorities. These measures share many common elements, and differ largely in the amount of computation required and the quality of the valuations. Section 3.5 describes the penalty approach and Section 3.6 describes the alternative branching criteria, their relative advantages and disadvantages, and some guidelines for selection. Because these criteria are closely related, we explain our first criterion, branching based on penalties, in great detail and the remaining methods in less detail.

The branching criteria are presented in terms of zero-one variables but they easily generalize to general integer variables and specially ordered sets. The following example illustrates the branching process for general integer variables. Suppose the LP relaxation yields $x = 2.4$ where x is a general integer variable. The appropriate branches are $x \leq 2$ and $x \geq 3$. All of the tests and measures below for zero-one variables are based on identifying the outcome of compelling variables to satisfy bounds of 0 and 1 but can just as easily be applied to general integer variables using this branching method. In particular, these measures rest on identifying the "fractional part" of a zero-one variable and apply in the general case simply by identifying the corresponding fractional part of x, which in this example is 0.4. In principle, instead of having two branches for each node the tree could have one branch for each possible value of x. In practice, this usually leads to an unnecessarily large tree, which impairs the search. However, there are exceptions where multiple branches are quite helpful. Application of the branching criteria to specially ordered sets is more involved and is described in Section 4.5.

Integer Programming

3.5. Penalty Analysis and Look-Ahead Strategies

There are several key ways to enhance the effectiveness of the LIFO and best bound approaches and their variants (e.g., the LIFO forward, best bound backward approach of Exercise 3.12). In this section we will illustrate the fundamental ideas underlying some of the more important types of enhancements and then build more fully on these ideas later.

A simple, but often potent, kind of enhancement is the following. At the branching step, one may solve an LP problem for each of the branches $x_s = 0$ and $x_s = 1$ (by reoptimization or otherwise) to provide a "look-ahead" that allows the consequences of different branches—for different choices of branching variables—to be analyzed and compared. Such an approach is called *probing*. Because it may be computationally expensive to carry out a probing step for all potential branching variables, some form of initial screening to isolate more promising variables is relevant. In fact, intelligent screening can embody the same issues that underlie the analysis of information generated by probing, which fall under the classification of *penalty analysis*. A particular form of screening, which in effect is an abbreviated type of probing, is especially useful for illustrating these considerations, because it can be used in place of a complete probing step in very large applications where such a complete probe is costly even for a single variable. This technique, which is sometimes called the *single-pivot look-ahead*, applies the dual simplex method for linear programming for just one iteration upon adjoining the branching constraint.

Rather than carry out a full pivot with the dual method, in this type of abbreviated probe, it is necessary only to update the column of constants, since the "body" of the LP tableau is of no particular relevance. The updated coefficients of the constant column can be generated one at a time, examined to determine the values of the associated variables, and then discarded, so that only a very small amount of additional memory is required to leave the entire LP tableau intact during this operation. Consequently, the tableau does not have to be stored somewhere and subsequently retrieved to perform a similar operation for the alternative branch. The dramatic resulting improvement in efficiency makes the single-pivot look-ahead a great deal faster than complete reoptimization. (Note that in a similar manner, if the entire tableau is not required for subsequent analysis, one can always avoid the work of the last pivot with the dual simplex method simply by updating the constant column and checking for primal feasibility before updating the rest of the tableau.)

Even if the single pivot does not completely reoptimize the problem—which it usually does not—the use of the dual simplex method (and the consequent dual feasibility of each tableau) implies that the updated x_0

value is always less (worse) than or equal to the optimal LP x_0 value. The single-pivot value of x_0 gives a valid bound that can be used in the best bound approach in place of the fully optimized x_0 value. In view of this, to save even more work, some approaches update only this x_0 value to obtain a bound, disregarding the values of the other variables. If this bound does not exceed x^* the associated branch is immediately fathomed by bound, alternative (b) from Section 3.1, just as if a complete reoptimization had occurred. (The same type of observation of course holds for any iteration of the dual simplex method, whether during reoptimization or during a process that starts from scratch. That is, whenever $x_0 \le x^*$ at a dual iteration, the current IP problem is fathomed immediately whether or not LP optimality is achieved.)

The magnitude of change of the LP bound, x_0, achieved by single-pivot look-ahead is frequently referred to as a *penalty* on the original LP bound. "Up" and "down" penalties (p_{Uj}, and p_{Dj}, respectively) are associated with increasing and decreasing the value of the branching variable. By our preceding observations, these penalties have the same uses whether generated by single-pivot look-ahead or by a "full probe" that consists of complete reoptimization. The key difference is that complete reoptimization, as a compensation for taking longer, allows fuller consequences of a branch to be determined, thus in general yielding stronger bounds.

3.5.1. Other Uses of the Single-Pivot Look-Ahead

Due to the speed and abbreviated calculation of the single-pivot look-ahead, the approach can be conveniently incorporated as a screening device for the LIFO method as well as the best bound method (and hence for combinations of the two). This permits the LIFO method, for example, to select the branch that offers the more promising bound of the two available on a forward step (and sometimes results in fathoming one or both of the branches immediately). In fact, in both the LIFO and best bound approaches, the simple "x_0 update" can be used as a guide to selecting the variable x_s to branch on. For example, one can select the variable x_s to be the one for which the bounds for $x_s = 0$ and $x_s = 1$ differ by the greatest amount and then branch in the direction of the more promising bound.

There is a subtle trap to avoid here. At first, it might seem worthwhile just to select the variable x_s to be the one that gives the most promising bounds on both branches, or at least that gives the most promising bound on its preferred branch. To illustrate the potential danger in either of these two types of choices, suppose x_1, x_2, and x_3 are 0-1 variables that have not been assigned specific values in the current IP problem and that bounds for

Integer Programming

x_0 that result for setting each to 0 or 1 are as shown in the following table, where we have assumed the unrestricted LP has $x_0 = 80$:

Alternative values for 0 – 1 variables	Resulting upper bounds for x_0 (e.g., computed by a dual-pivot look-ahead)	Penalties
$x_1 = 0$	79¼	$P_{D1} = ¾$
$x_1 = 1$	77	$P_{U1} = 3$
$x_2 = 0$	70	$P_{D2} = 10$
$x_2 = 1$	73½	$P_{U2} = 6½$
$x_3 = 0$	75	$P_{D3} = 5$
$x_3 = 1$	69	$P_{U3} = 11$

The most promising upper bounds (since we are trying to maximize x_0) are those for x_1, particularly for the branch $x_1 = 0$. Thus, it might seem logical to select x_1 to be the branching variable. However, this is probably the worst possible choice, because the upper bound information available from the branches for x_1 is highly unrealistic in view of the information available from the other branches. To see why this is so, note that at least one of the two upper bounds associated with x_2 must be valid since x_2 must be either 0 or 1. Similar remarks apply to x_3. The consequences of this observation can be demonstrated by the following table.

Alternative values for variables	Respective implied bounds for x_0	Legitimately known bound for x_0 in either case
$x_1 = 0$ or $x_1 = 1$	$x_0 \leq 79¼$ or $x_0 \leq 77$	$x_0 \leq 79¼$
$x_2 = 0$ or $x_2 = 1$	$x_0 \leq 70$ or $x_0 \leq 73½$	$x_0 \leq 73½$
$x_3 = 0$ or $x_3 = 1$	$x_0 \leq 75$ or $x_0 \leq 69$	$x_0 \leq 75$

Most restrictive (hence most accurate) known bound: $x_0 \leq 73½$

In short, this table illustrates that the larger of the two upper bounds on x_0 obtained by setting $x_j = 0$ or $x_j = 1$ is always valid. Consequently, we see that x_1 gives very poor quality information indeed, since even the most pessimistic of its two bound possibilities ($x_0 \leq 77$) still substantially overshoots the known upper bound $x_0 \leq 73\frac{1}{2}$ available from x_2. As a result, branching on x_1 would be a pretty chancy affair, because with such poor information we have no reliable guide to which of the choices $x_1 = 0$ or $x_1 = 1$ is likely to lead to the better solution.

There is another, perhaps even more important reason why x_1 is not likely to be a good variable to branch on. This derives from the fact that x_1 does not have as much influence as the other variables—that is, it does not have as much power to produce a change in the current LP x_0 value (which can be anything not exceeding $79\frac{1}{4}$). In general, the notion of influence is broader, reflecting the power to change the set of optimal solutions to the LP problem. The significance of this is that a variable (or branching alternative) that has little power to change the set of optimal LP solutions will typically result in a reoptimized LP tableau that supplies scarcely more accurate information than the present one. But the ability of a reoptimized tableau to provide more accurate information about the current IP problem—i.e., to model this problem more closely—is the only thing that prevents branch and bound from degenerating into total enumeration. If the same gap exists between the IP problem and LP problem at each stage of assigning a variable a value, then the solution to the LP problem may be unable to resolve the status of the IP problem (i.e., by fathoming) until all of the integer variables have been assigned values by branching. Thus, the concept of an influential variable (or an influential branch for a variable) is exceedingly important.

Admittedly, the single-pivot bound calculation for x_0 gives only a modest clue to the influence of a variable or branch, but by this limited criterion it appears quite likely that x_1 is pretty far down on the scale and therefore should be avoided as a branching variable. Continuing with the example, we are left to decide between x_2 and x_3 as a branching variable, and now the choice is somewhat more difficult. While x_2 gives the most realistic bound (for one of its alternatives), x_3 has a far more dramatic split between the bounds supplied by its two alternatives. If both of these two bounds for x_3 are in error (i.e., overly optimistic) by about the same amount, then their difference accurately reflects the true difference between the alternatives and thus suggests that x_3 is a good branching variable and $x_3 = 0$ is a good branch to select. Further support of x_3 as a branching variable is provided by the fact that the branch $x_3 = 1$ is a bit more influential than either of the x_2 branches. (That is, the branch $x_3 = 1$ restricts x_0 more strongly than the branches for x_2.)

Note that since we are maximizing, smaller bounds are desirable in the preceding table, maximal penalties lead to minimal bounds, and minimal penalties lead to maximal bounds. If we were minimizing instead of maximizing, larger bounds would have been desirable, maximal penalties lead to

Integer Programming

maximal bounds, and minimal penalties lead to minimal bounds. By referring to penalties, we can avoid such confusion and state branch rules (below) generically for either minimization or maximization problems.

The standard branch rules give preference based on one of the following (where θ_j is the value associated with branching on variable i):

1. The maximum difference between the two x_0 bounds, hence between the associated penalties:

   max θ_j

   $\theta_j = |p_{Uj} - p_{Dj}|$

   purpose: increase the chance of choosing the best branch.

2. The value of the most pessimistic bound, hence of the greatest penalty (followed by choosing the other branch),

   max θ_j

   $\theta_j = \max(p_{Uj}, p_{Dj})$

   purpose: rule out the apparent worst branch, effective near an integer solution.

3. The value of most restrictive (realistic) bound, or

   max θ_j

   $\theta_j = \min(p_{Uj}, p_{Dj})$

   purpose: quickly degrade the objective function.

4. The value of the most optimistic bound (which of course leads to disastrous choices).

   min θ_j

   $\theta_j = \min(p_{Uj}, p_{Dj})$

The considerations discussed here constitute an area that has been explored all too briefly in practical applications. Standard branch rules have been used almost exclusively in isolation from the other rules. But the development of a good rule based on assessing the realism of bounds other than the most realistic, combined with evaluation of splits, and making use of a concept of "influence," has not been undertaken.

Penalties can be an effective method for selecting a branching variable, but they have two limitations. First, penalties guarantee a minimal degradation of the objective function associated with a branch, but at the price of a consistent bias toward underestimating actual degradation. At the top of a branch and bound tree, underestimates are especially pronounced and in extreme cases a penalty based branching rule can be close to arbitrary. Second, penalties have generally proved to be more effective on combinatorial optimization problems than they have been on integer programming problems.

3.5.2. Additional Fathoming Possibilities

We have so far skipped over two immediate and conspicuous uses of the information from the example. To illustrate these uses, we augment the preceding table by introducing a row that includes the LP x_0 value before restricting x_1, x_2, or x_3 to 0 or 1.

Alternative values for variables	Respective implied bounds for x_0	Legitimately known bound for x_0
Not restricted (LP solution)	$x_0 \leq 80$ (LP x_0 value)	$x_0 \leq 80$
$x_1 = 0$ or $x_1 = 1$	$x_0 \leq 79\frac{1}{4}$ or $x_0 \leq 77$	$x_0 \leq 79\frac{1}{4}$
$x_2 = 0$ or $x_2 = 1$	$x_0 \leq 70$ or $x_0 \leq 73$	$x_0 \leq 73\frac{1}{2}$
$x_3 = 0$ or $x_3 = 1$	$x_0 \leq 75$ or $x_0 \leq 69$	$x_0 \leq 75$

Most restrictive (hence most accurate) known bound $x_0 \leq 73\frac{1}{2}$

The augmented table immediately suggests that the most restrictive known bound ($73\frac{1}{2}$ in the example) can be used in place of the current LP x_0 value in an attempt to fathom the current IP problem. Thus, in particular, suppose the "current best" solution yields $x^* = 74$. Since $80 > 74$, the LP x_0 value does not fathom the current IP problem. But we know that $x_0 \leq 73\frac{1}{2}$ holds in any case. Furthermore, since $73\frac{1}{2} \leq 74$, the current IP problem can't give an improved incumbent solution, and hence fathoming does in fact occur.

Stated in another way, if the x_0 bounds for both the branches $x_s = 0$ and $x_s = 1$ are less than or equal to x^*, and hence these bounds would lead to fathoming both branches if they were selected, then there are no available alternatives for the current IP problem and hence it is also fathomed. (In the example, the branch $x_2 = 0$ would be fathomed because $70 \leq 74$, and the branch $x_2 = 1$ should be fathomed because $73\frac{1}{2} \leq 74$.)

This point of view leads to the second use of the information from the table. To illustrate this use, let us now suppose that $x^* = 71$. Thus it no longer happens that the current IP problem can be fathomed by $x_0 \leq x^*$, since $73\frac{1}{2} > 71$. However, the branches $x_0 = 0$ and $x_3 = 1$ can be fathomed in this way, because $70 \leq 71$ and $69 \leq 71$. In other words, to obtain an improved current best solution, it is possible to allow either $x_2 = 0$ and $x_3 = 1$. This means that the alternative branches $x_2 = 1$ and $x_3 = 0$ are both compulsory branches. Consequently, they both may be "taken" at once (without regard to the order).

A compulsory branch (or assignment) can be handled very easily by the LIFO bookkeeping scheme described earlier. Since the alternative to the compulsory branch is not available (hence is implicitly examined and discarded), it suffices to attach a "star" to the branching variable (i.e., to its index). The meaning of a starred index, it may be recalled, is exactly that the alternative branch no longer exists. For example, suppose the current lists for the LIFO procedure are:

Integer Programming

Index of variable selected: 6, 8*, 5
Value of variable: 0, 1, 1

From the knowledge that the assignments $x_2 = 1$ and $x_3 = 0$ are compulsory, we obtain the lists

Index of variable selected: 6, 8*, 5, 2*, 3*
Value of variable: 0, 1, 1, 1, 0

3.5.3. The Power of Compulsory Assignments

Compulsory assignments are especially advantageous because they allow the procedure to go deeper into the tree without having to branch in the ordinary sense. Thus, whenever such assignments occur, they should immediately be accompanied by reoptimizing, thereby providing an updated LP tableau that yields stronger (more accurate) information. Handling forced assignments in this way may lead to uncovering additional forced assignments and possibly even to fathoming the current IP problem altogether. For example, upon setting $x_2 = 1$ and $x_3 = 0$ and reoptimizing (in the preceding illustration), it may be that the new LP tableau discloses that x_1 can be compelled to equal a specific value. Then, upon assigning x_1 this value and reoptimizing, it may further be discovered that x_4 likewise has a compulsory assignment, or that the current tableau is integer feasible. Such a "domino effect" of successive compulsory assignments does sometimes happen, and even in the absence of such an effect, the improved information available after reoptimization often leads to better subsequent branching choices.

The chief insights of our foregoing discussion can be summarized as follows.

Summary of main considerations:
1. If $x_0 \le x^*$ at any iteration of the dual simplex method, then the current IP problem can be fathomed without waiting to achieve LP optimally.
2. The single-pivot look-ahead with the dual method can be restricted to identifying only the updated x_0 value, thus requiring a small fraction of the effort of a full pivot. Furthermore, this procedure leaves the tableau unchanged, so that it is possible to effectively identify the updated x_0 values that result for each variable x_j on its two branches $x_j = 0$ and $x_j = 1$ (where x_j is restricted to be a 0-1 variable that is eligible to be the branching variable x_s).
3. The x_0 values for $x_j = 0$ and $x_j = 1$, as x_j ranges over the candidates for x_s, have three important (and interrelated) uses: (1) they can help guide the choice for which variables should be

x_s and which branch should for x_s; (2) the larger of the two x_0 values for $x_j = 0$ and $x_j = 1$ is a valid upper bound on x_0, and any such bound—in particular, the most restrictive—can be used in place of the current LP x_0 value to see if fathoming is possible by condition $x_0 \leq x^*$; (3) if the current IP problem cannot be fathomed by (2), it still may be possible to identify compulsory branches (whose alternative branches are fathomed because their x_0 values are $\leq x^*$).

4. Compulsory branches can be accommodated by the LIFO rule by "starring" the indexes of the branching variables. Such branches should be taken at once, accompanied by reoptimization with the dual simplex method, in order to exploit the advantages of going deeper into the tree.

Finally, we observe that the importance of generating more accurate information about branching alternatives can make it relevant to expend the effort of actually reoptimizing the LP for a subset of alternative branches that look promising. That is, using the single-pivot look-ahead as a screening device, we may then choose a subset of variables that appear to be attractive candidates by the considerations indicated, and follow with a full probe by reoptimizing for both of their branching. It may be useful to include variables whose values are closest to ½, in situations where degeneracy or other aspects of problem structure render the penalties from a single-pivot look-ahead less informative.

The foregoing considerations are vitally important to the efficiency of a branch and bound method and apply by extension to branch and bound procedures for IP problems whose integer variables may not all be 0-1. (Related ideas are also introduced in the following exercises.)

3.5.4. Exercises

3.14. Consider the information contained in the following table.

	Values of x_0 (obtained by dual pivot or reoptimized look-ahead) when	
Variable	$x_j = 0$	$x_j = 1$
x_1	41 ⅗	49
x_2	51 ⅕	53 ⅕
x_5	46 ⅘	43 ⅖
x_8	44	45 ⅖

Integer Programming

Identify the most restrictive upper bound for x_0 in an optimal solution to the current IP problem. Discuss the pros and cons of branching on each of the variables in the table.

3.15. Referring to the data of Exercise 3.14, what is the smallest possible value that could have been obtained for x_0 in an optimal solution to the current LP problem? What is the largest value of x_0 in a optimal solution to the current LP problem? What is the largest value of x_0 that would permit the current IP problem to be fathomed? What information could be obtained if $x_0 = 43\frac{2}{5}$? Show how this information would be handled by the LIFO rule if the current lists for this rule are as follows:

Index of variable selected: 4*, 6, 3
Value of variable: 0, 1, 0

3.16. *Compulsory assignments for nonbasic variables.* Suppose that the x_0 row in the optimal LP tableau for the current IP problem is

$$x_0 = 62\tfrac{1}{2} \quad \begin{array}{cccc} -y_1 & -x_3 & -x_4 & -y_3 \\ 3\tfrac{1}{4} & 7\tfrac{1}{4} & 2\tfrac{1}{2} & 5\tfrac{3}{4} \end{array}$$

where x_3 and x_4 are 0-1 variables. Show that if $x^* = 58$, then $x_3 = 0$ is a compulsory assignment. From this observation, state a rule for identifying compulsory assignments for nonbasic variables. Is this rule valid for any dual feasible tableau? What if the tableau is not dual feasible?

3.17. The compulsory assignment of Exercise 3.16 does not change the optimal LP tableau (except for allowing the x_3 column to be dropped). On the other hand, if x_3 had been basic and had a value other than 0 in the current LP solution (e.g., $x_3 = \frac{1}{2}$), then the assignment $x_3 = 0$ will lead to a changed tableau. What significance might this difference have for deciding which type of compulsory assignment is more important? Can you argue that the relative importance of these two is not always clear-cut? (What if both a basic and a nonbasic variable received compulsory assignments? Could dropping the column of the nonbasic variable lead to a more dramatically changed LP optimum than might otherwise be the case?)

3.18. Suppose in the example of Exercise 3.16 that x_3 and x_4 are not 0-1 variables but are integer variables. Show that $x^* = 58$ still implies $x_0 \le 0$ (hence $x_3 = 0$). Further, show that $x^* = 58$ implies $x_4 = 1$. State a general rule for obtaining upper bounds for nonbasic integer variables. Do you think that upper bounds other than 0 could provide useful information?

3.19. An intuitive kind of branching rule that has been tried in practice is the following: If any of the 0-1 variables are nonbasic, select the one with the largest objective function coefficient (after all compulsory assignments have been made) and branch on this variable by setting it equal to 0. (The logic is that the variable probably ought to be 0 even if the kind of bound information that yields compulsory assignments isn't sharp enough to detect this.) It turns out that this is an extremely bad branching rule! Explain why this might be so in terms of the earlier discussion of influential branches. (After a branch has been taken by this rule, how must the next branch to be taken be evaluated?)

3.20. An extremely important branching concept, intimately related to the notion of influential (and uninfluential) branches, is the following: *don't branch on an alternative that the optimal LP solution will probably select automatically* (e.g., at a later stage). For example, in the branching rule of Exercise 3.24, there is a reasonable chance that a nonbasic variable with a large objective function coefficient will automatically wind up 0 in subsequent solution stages. But branching on such a variable will then unnecessarily require exploring the other alternative. Show how this concept relates to (and is one aspect of) the concept of influence. Can you think of other situations in which this idea might be relevant?

3.6. Alternative Methods for Selection of Branching Variables

Penalties are only one method for selecting branching variables. Depending on the problem type, other methods may be more effective. In this section, we consider branching based on integer infeasibility, pseudocosts, and user-specified priorities. In addition, we describe how these methods can be applied to specially ordered sets.

3.6.1. Integer Infeasibility

An alternative to selecting branching variables based on penalties is to select branching variables based on integer infeasibility (e.g., Forrest, Hirst, and Tomlin, 1974), which we describe in this section. Integer infeasibility does

Integer Programming

not use any information about the objective function but rather concentrates only on integer variables that have fractional values in a LP relaxation with the goal of removing this "infeasibility." For this reason, one would correctly assume that this branching criterion works generally better for finding feasible solutions than criteria that are affected by the objective function. Correspondingly, integer infeasibility can be a very poor criterion when there are many feasible solutions with different objective function values. Feasibility tends to be more important for problems typically described as "combinatorial optimization" problems rather than "integer programming" problems. However, we caution that many problems can reasonably be described either way.

The integer infeasibility of a single integer variable is formally defined to be

$$\min\{f_j, 1 - f_j\}$$

where f_j is the fractional part of x_j. Informally, integer infeasibility is the distance from the closest integer.

Given an integer infeasibility measure, two options arise. Typically the branching variable can be either the variable with the largest integer infeasibility or the variable with the smallest integer infeasibility.

Selecting the variable with the greatest integer infeasibility will generate two branches that are more likely to differ from the current solution than other branching alternatives. The underlying assumption is that the objective function value is more likely to degrade on both branches and the other variables are more likely to change on both branches. If so, such a branch would represent an influential variable. However, the objective function need not change and the influence on integer infeasibility may not be the one desired; i.e., the branching may increase the integer infeasibility of other variables.

Selecting the variable with the smallest integer infeasibility will generate one branch very similar to the current solution and one branch very different from the current solution. The underlying assumption is that the similar branch is where a solution lies and that the different branch will prove uninteresting. The hope for the similar branch is that removing a small amount of infeasibility in one variable will reduce the infeasibility of other variables as well. Typically, the algorithm would further pursue the similar branch in the hope of generating a feasible solution.

Selecting a variable with a large integer infeasibility (i.e., closest to ½) is highly recommended, particularly in cases in which penalty information is weak but still useful. However, some special structure may justify the use of the alternative, as in the context of using branch and bound as a heuristic for finding a solution rather than for proving optimality of a solution. In fact, some codes explicitly avoid branching on variables with small integer infeasibility. Such variables are often called quasi-integer. Similarly, specially ordered sets (described later in this section) with small integer infeasibility are called quasi-satisfied SOS, and these sets are not branched on. In cases where penalty information is completely lacking as a basis for choosing a branching variable, choices may focus on infeasibility close to ⅓ or ¼, as a reasonable trade-off between influence and the ability to discern which branch (from the two alternatives for the chosen variable) is likely to be preferable.

3.6.2. Pseudocosts

Pseudocosts are philosophically similar to penalties; both methods attempt to estimate the degradation in objective function value associated with a particular branch. The difference is that penalties give a lower bound on degradation that consistently underestimates actual degradation, whereas pseudocosts sacrifice the bound (and its additional fathoming uses) to achieve more accurate predictions of actual degradation with less computation.

Pseudocosts estimate the degradation of the objective function per unit of integer infeasibility reduced in a given variable. The pseudocosts associated with branching up and down on a variable, PCU_j and PCL_j respectfully, are not necessarily identical. The alternatives for selecting a branching variable based on pseudocosts are the same as the alternatives based on penalties as discussed earlier. Because penalties are measured in objective degradation and pseudocosts are measured in objective degradation per unit of infeasibility removed we replace p_{Uj} with $PCU_j(1 - f_j)$ and replace p_{Dj} with $PCL_j f_j$ in the evaluations:

(i) The maximum difference between the two x_0 estimates,
Max θ_j
$$\theta_j = |PCU_j(1 - f_j) - PCL_j f_j|$$
purpose: increase the chance of choosing the best branch.

(ii) The value of the most pessimistic estimate (followed by choosing the other branch),
max θ_j
$$\theta_j = \max(PCU_j(1 - j_j), PCL_j f_j)$$
purpose: rule out the apparent worst branch, effective near an integer solution.

(iii) The value of most restrictive (realistic) estimate, or
max θ_j
$$\theta_j = \min(PCU_j(1 - f_j), PCL_j f_j)$$
purpose: quickly degrade the objective function.

Integer Programming

(iv) The value of the most optimistic estimate (which of course leads to disastrous choices).

$$\min \theta_j$$
$$\theta_j = \min(\text{PCU}_j(1 - f_j), \text{PCL}_j f_j)$$

Pseudocosts are typically independent of nodes in the branch and bound tree. The implied assumption is that the proportionality between objective function degradation and infeasibility reduction is constant across the search tree. In principle, pseudocosts could be adjusted based on the current node.

Unlike penalties, there is a wide variety of ways to calculate pseudocosts. Pseudocosts can be calculated empirically based on actual branches:

$$\text{PCU}_j = \frac{\Delta x_0}{(1 - f_j)}$$

$$\text{PCL}_j = \frac{\Delta x_0}{f_j}$$

Since a variable may be branched on many times, this value could be changed as the tree develops to best use the accumulating branching data. In practice, little effort has been made in this direction and pseudocosts tend to be based on observed pseudocosts for either the first branch or the most recent branch involving the variable of interest.

While the empirical method works after branching has taken place, this raises the natural question, "What initial values should be used for pseudocosts?" The easiest initial value is zero. With this value, all branches will appear to be equivalent. As branches are taken, the branched variables will acquire nonzero values and appear to be much more influential and the search will concentrate on branching on these variables first. This should not cause any trouble in a LIFO search but may be counterproductive for a best bound search: the tree will evolve with a largely fixed branching order that is arbitrarily chosen based on zero pseudocosts.

Alternatively, if an incumbent integer solution exists, all pseudocosts can be initialized based on the average change in objective value and integer infeasibility between the LP relaxation and the incumbent solution:

$$\text{PCU}_i, \text{PCL}_i = \frac{|x_0^{\text{LP}} - x_0^*|}{\sum_i \min(f_i^{\text{LP}}, 1 - f_i^{\text{LP}})} \quad \forall i$$

When pseudocosts are initialized to equal values, the branching strategy is equivalent to using integer infeasibility. In fact, if all unit pseudocosts are initialized to a value of one, then the total pseudocost of any branch equals the integer infeasibility resolved by taking that branch. In this sense, pseudocosts are a generalization of integer infeasibility.

A third initialization alternative is to solve a series of dummy subproblems and determine pseudocosts based on the results. For example, one could form separate branches off of the original LP relaxation for each fractional variable, evaluate the implied pseudocosts, and discard the branches. The remaining integer variables (taking on a whole value at the LP relaxation) could lead to similar dummy subproblems if they eventually become fractional.

Theoretically, the alternatives for setting pseudocosts could be explained in two ways. First, users could specify pseudocost values based on knowledge of their problem. Second, the actual degradation of the objective as infeasibility is removed is piecewise linear rather than linear. Hence, a piecewise linear (or other nonlinear function) could be used to evaluate the pseudocost at a particular node.

An apparent improvement on pseudocosts is available in some branch and bound software. Recall that penalties are always a lower bound on objective degradation. Hence, larger pseudocosts may be justified at a particular node, provided that the penalty at that node is sufficiently large:

$$\text{new PCU}_j = \max\left(\text{PCU}_j, \frac{p_{Uj}}{(1 - f_j)}\right)$$

$$\text{new PCL}_j = \max\left(\text{PCL}_j, \frac{p_{Dj}}{f_j}\right)$$

Penalty values that increase pseudocosts represent an opportunity to improve pseudocosts. At the same time, such penalty values imply that the existing pseudocosts are unreliable. The penalties associated with the current node may strengthen some of the pseudocosts but others may not be changed and even the pseudocosts that are changed may return to their previous "unreliable" values at the next node. Furthermore, the original pseudocost values may have been correct in a relative sense, if not an absolute sense, and hence may still lead to good branching choices, while the changed values may no longer be consistent and could lead to worse branching choices.

3.6.3. User-Specified Priorities

One of the historically most effective methods for selecting the branching variable is for the user to directly specify a priority ordering of the variables. The reason this approach is effective is that the user can incorporate model and solution knowledge into the priority order that may not be

Integer Programming

reflected in the mathematical formulation. Three general strategies dominate priority ordering. The first strategy is to rank the variables in terms of their influence on the critical elements of the model. The second strategy, often in conflict with the first, groups together blocks of highly interdependent variables and resolves one block at a time. This strategy is implemented by prioritizing the blocks. Prioritizing by influential variables tends to reduce the objective function quickly, while prioritizing blocks can be an effective component of a heuristic for achieving a reasonable feasible solution. A third strategy is based on the notion that weak LP relaxations are the main difficulty for branch and bound. If branching on particular variables will remove perceived weaknesses in the formulation, then these variables should be branched on first. It is not always clear where the weaknesses in a formulation lie, and rerunning branch and bound with different priority orders may prove worthwhile. Of course, if it is possible to improve the formulation directly rather than "correcting" the formulation with branch and bound, branch and bound could pursue a more useful strategy.

For example, suppose several factories can each be build in several alternative configurations, the configurations are jointly constrained to meet companywide goals, and the selection of configurations represents a large portion of the overall cost. In addition, each factory has less costly minor decisions that are strongly influenced by the choice of factory configuration but that do not interact with the decisions in other factories. In this example, the influential variables are the configurations, and the blocks correspond to the factories. Thus, the first two alternatives are:

1. To branch on configurations first and minor decisions last, or
2. Prioritize the factories and branch on all variables within one factory before another factory.

This particular example affords a third alternative that is not always possible, which is to combine these two alternatives (available here because of the mutual independence of the factories' minor decisions). The priority order for the third alternative is: branch on configurations first, then minor decisions for factory A, minor decisions for factory B, etc.

A modeler can gain further control over the search process by creating additional variables and giving them high priority. We illustrate this principle with two examples.

Suppose several identical, or nearly identical, facilities are to be built and this constitutes the bulk of the cost of an integer program. A problematic behavior frequently arises in such problems, illustrated by a situation in which the LP relaxation minimizes cost with 2.5 facilities and branching on the fractional facility does not change the number of facilities in the LP relaxation; some other facility now becomes fractional. Such branching can create a large tree of nodes with identical, or nearly identical, costs. In some cases, one can prove, for example, that at least three facilities, denoted by x_j, must be built and one would add an appropriate constraint:

$$\sum_j x_j \geq 3$$

However, in general, such inequalities may not be obvious. A general solution is to add a general integer variable, g, that counts the number of facilities

$$g = \sum_j x_j$$

and give g high branching priority. Thus, all descendant relaxations in the example will have either the constraint that $g \geq 3$ or the constraint $g \leq 2$ and this particular problematic behavior is eliminated.

As a second example, consider a modeler's "pivotal" pair of inequalities, and suppose we require that exactly one inequality must be true:

$$\sum_j a_j x_j \geq b_1$$
$$\sum_j a_j x_j \leq b_2$$

where $b_1 > b_2$. Branching on this pair of inequalities can be achieved by adding an integer variable, y:

$$\sum_j a_j x_j \geq b_1 - M_1 y$$
$$\sum_j a_j x_j \leq b_2 + M_2(1 - y)$$

where M_1 and M_2 are constants chosen large enough to relax their respective constraints. We emphasize again that for computational performance it is important to choose these constants no larger than necessary. When $y = 0$, the first original constraint holds and the second original constraint is relaxed. When $y = 1$, the roles are reversed. In isolation, this pair of constraints may not strengthen the LP relaxation. However, our present goal is instead to create the two cases that result from branching on y and the formulation will accomplish this goal. Obviously, this branching strategy relies on the dichotomy of these inequalities being an influential change in the LP relaxation. More generally, the two inequalities need not have the same left-hand side, and the requirement that exactly one inequality be satisfied by the optimal solution can be replaced by the requirement that at least one inequality be satisfied.

3.6.4. Specifying Priorities

The mechanics of determining the branching order is software dependent. In some cases the order in which variables are specified in the formulation automatically becomes the branching order and the user needs to be aware of this "feature" when formulating a model. Preferably, a separate file specifies the order either explicitly or implicitly by attaching weights to each variable. A separate file usually affords the opportunity to specify priorities for only a subset of variables. Remaining variables are assigned a lower priority and their branching is determined using a rule such as those given earlier in this section. In theory, branch and bound needs only a partial order and a rule to choose among variables satisfying the partial order. At present, this level of flexibility has not been reached. Nevertheless, by carefully selecting (possibly nonunique) weights users can gain considerable control over the branching order.

Cost coefficients and reduced cost coefficients may also be used to specify a branching order implicitly. When cost coefficients are used, the priority order is typically to branch on larger coefficients first. In theory this should lead to branching on more influential variables. Alternatively, reduced cost coefficients associated with the initial LP relaxation can be used to determine the branching order. Variables are separated into three categories based on reduced cost: zero, less than zero, greater than zero. All fractional variables will have a reduced cost of zero, and for most problems this group should be branched on before the others. In the remaining categories, variables with reduced costs closer to zero are the most plausible variables to change value and should be branched on next in most cases.

In addition to specifying a branching order, the user may be able to specify the direction of branching (for a given variable) to explore first. This can be a useful feature for finding a feasible solution. For example, if the current branching variable determines whether or not to create a new facility, it is generally better to branch toward creating the facility (which the LP relaxation has already partially created). Typically, the linear program reflects both the need to build a facility and the desire to avoid a large fixed cost. Usually, the cost must be incurred by any feasible integer solution.

Branching orders can be used in two different ways: fixed branching and automatic ordered branching.

Fixed branching branches on variables in the same order throughout the tree, including branching on variables taking whole values. This procedure can lead to creating many unnecessary nodes and branches in the tree, with the danger of causing the tree to become excessively "enumerative." However, it does have two advantages. First, pseudocost estimates are more consistent; two nodes at the same level in the tree have the same variables fixed (but to different values). Second, the bounding process can be implemented more efficiently, because a dual feasible solution for a node is a dual feasible solution for any other node at the same level (or a lower level). Hence, several nodes may be bounded at once and a warm basis can be stored for later use at the same level in the tree. The process of bounding several nodes at once is called a *resource space tour*.

Automatic ordered branching differs from fixed branching only in that it only branches on fractional variables and momentarily will skip over any variable taking a whole value.

Under either fixed branching or automatic ordered branching it is possible to start without any branching order, and instead allow the solver to build a branching order based initially on branches that have been made using other rules. The same process can be used to augment a user specified order covering only a subset of the variables.

3.6.5. Specially Ordered Sets

Specially ordered sets (Beale and Tomlin, 1969) have a dramatic computational advantage over alternative formulations when an SOS can be used. The advantage arises from branching methods, described below, that explicitly use the structure of an SOS. In most cases, it is advantageous to give an SOS set a high branching priority. The reduction in x^* as a result of branching on an SOS tends to be large. (Specially ordered sets are described in Section 1.11.)

Recall from Section 1.12 that the λ-form for modeling a nonlinear function $z(y)$ with a specially ordered set is given by

$$z = \sum_i z_i \lambda_i$$
$$y = \sum_i y_i \lambda_i$$
$$\lambda_i \in SOS2$$

The indicator variable for selecting point i is λ_i, and the values of y and $z(y)$ at point i are reflected in the coefficients y_i and z_i, respectively.

Branching with specially ordered sets requires two decisions to be made:

1. Which set to branch on, and
2. How to branch within the set.

The latter decision is commonly made using a reference row, the z_i coefficients, specified by the user. For any LP relaxation we can calculate an average reference row weight:

$$\bar{z} = \frac{\sum_j z_j \lambda_j}{\sum_j \lambda_j}$$

Integer Programming

The average reference row, in turn, defines λ_k and λ_{k+1}, corresponding to the adjacent reference row values:

$$z_k \leq \bar{z} < z_{k+1}$$

Thus, the reference row defines the "current midpoint" of the specially ordered set, an appropriate place to branch.

For an SOS of type 1, exactly one λ must be nonzero in a feasible solution. Thus, one branch eliminates the variables with reference weights less than the average value, and the other branch eliminates the variables with reference weights greater than the average value:

$$\sum_{j \leq k} \lambda_j = 0$$

$$\sum_{j \geq k+1} \lambda_j = 0$$

For an SOS of type 2, one or two λ must be nonzero in a feasible solution. Thus, one branch eliminates the variables with reference weights less than z_k, and the other branch eliminates the variables with reference weights greater than z_k:

$$\sum_{j \leq k-1} \lambda_j = 0$$

$$\sum_{j \geq k+1} \lambda_j = 0$$

Each summation, for either SOS type, can be viewed as a composite variable and the techniques for a single integer variable can be applied to the composite variable. If the λ_i are nonnegative and additionally constrained to sum to 1, i.e., a convex combination of (y, z) points the first summation can be treated as the fractional part of an integer variable. Otherwise, we need to scale each summation. For SOS of type 1 this yields:

$$\frac{\sum_{j \leq k} \lambda_j}{\sum \lambda_j}, \frac{\sum_{j \geq k+1} \lambda_j}{\sum \lambda_j}$$

The composite variables form a basis for applying penalties, integer infeasibility, and pseudocosts to SOS sets as they are applied to integer variables. The evaluations are, in general, more complicated than the evaluations for a single variable. As with integer variables, these evaluations affect both node selection and which SOS to branch on.

Care should be taken to choose reference row weights meaningfully because of their influence on the branching process.

3.7. Selecting the Branching Node

Selecting the branching node is just as important as selecting the branching variables. As noted in Exercise 3.10, there are essentially two main types of operations to consider: starting over and reoptimizing, which are also called backtracking and immediate strategies, respectively. These types of operations are especially clear-cut for the LIFO method, which starts over on backtracking steps and reoptimizes on forward steps. But the same basic distinction holds for any method—i.e., reoptimization can occur if the next node selected is an immediate descendant of the current node, and starting over must occur otherwise (unless one resorts to recording LP tableaus for other nodes of the tree). Since reoptimizing relative to a single added constraint is generally far more efficient than starting over to solve the current LP problem from scratch, this procedure can save a great deal of computational effort.

3.7.1. Reoptimizing Strategy

The strategic choice made in reoptimizing during a forward step is how many nodes to "bound" (i.e., to generate x_0 bounds for), by solving the associated LP (if this information is not available from earlier probing efforts). There are three alternatives, which rely on an estimate of the attractiveness (discussed in the next section) of the two nodes created by the branching alternatives for the variable under consideration. The first option is to bound the more attractive node. This is essentially a heuristic search and is appropriate when no reasonable solution is known. The second option is to bound both nodes, starting with the least attractive node and finishing with the most attractive. The advantage of this tactic is that the more attractive node is more likely to be branched on next and the node's basis will already be in memory. The third strategy is to bound the most attractive node first and bound the second node contingent on finding the first node now to be less attractive than the second node. This third strategy is an enhanced version of the first strategy. In situations in which the choice of the branching variable is based on a full probe (complete reoptimization), these strategies are superfluous.

3.7.2. Starting Over Strategy

Starting over strategies are both more influential and more complex than reoptimizing strategies; more node choices are available. The measures used to select branching variables are also used in starting over strategies.

Before we describe these new strategies we revisit the best bound and LIFO strategies. The best bound strategy, described earlier, is a pure starting over strategy; it gives no preference to the descendants of the previous nodes. In contrast, LIFO is the easiest strategy to implement because it has

Integer Programming

a fixed starting over strategy that does not require any evaluation. This causes the major disadvantage of LIFO: the search can be very narrow and lead to an extended search of a possibly uninteresting part of the branch and bound tree.

Before discarding LIFO, we recall the related advantage of LIFO; it requires minimal memory to save the branch and bound tree. The limitations of tape storage devices were the original motivation for reducing memory use. Modern computers are far less restricted in memory, but unchecked growth of the branch and bound tree still creates memory problems for other starting over strategies. Furthermore, using minimal tree memory allows the status of the search to be summarized easily should search be halted. Potentially, this information can be used to strengthen a formulation before resuming branch and bound. This would be a challenging proposition with a more general tree representing the status of the search.

3.7.3. Node Evaluation Methods

The main alternatives to LIFO select a node based on some evaluation of the node. The best bound method uses the bound on the node as the evaluation. The following sections describe the other main evaluation methods, involving the use of penalties, integer infeasibility, and pseudocosts. These criteria rely in part on aggregating the same measures used for variable selection across all variables. Thus, the description of node selection criteria builds on the description of branching variable selection criteria.

3.7.4. Penalties

The best bound method can be adapted to use bounds based on penalties rather than just the LP relaxation. Unlike the variable selection methods, there is no ambiguity in which penalty evaluation to use in evaluating a node. We seek the most restrictive (realistic) bound on the node (this was option (iii) in Section 3.5 on penalties for evaluating branching variables):

$$\max \theta_j$$
$$\theta_j = \min(p_{Uj}, p_{Dj})$$

Although this may seem like a large increase in computation compared to the best bound method, in fact each node need only be evaluated once and the resulting penalty can be stored on the node for future comparison and possible selection during backtracking.

3.7.5. Integer Infeasibility

Recall that the integer infeasibility of a single integer variable is

$$\min\{f_j, 1 - f_j\}$$

where f_j is the fractional part of x_j. The integer infeasibility of a solution is the sum of the individual variable infeasibilities:

$$\text{II} = \sum_j \min\{f_j, 1 - f_j\}$$

The integer infeasibility of a solution can be used to estimate the objective function degradation that is required to attain an integer feasible solution from the current solution. One backtracking rule is to choose the node with the smallest integer infeasibility. This is a useful rule when feasible integer solutions are difficult to find and for many combinatorial optimization problems. The primary shortcoming of using integer infeasibility to select a node is that the objective function is ignored.

The best projection (BP) method (e.g., Forrest et al., 1974) attempts to balance integer infeasibility and changes in the objective function value. The premise of the best projection method is that the objective function degradation required to attain an integer feasible solution is proportional to the integer infeasibility. Hence, the method subtracts the expected deterioration in x_0 for the node. The best projection estimate is

$$\text{BP} = x_0 - \lambda p \text{II}$$

where λ is a user-controlled bias, p is the proportionality ratio of objective degradation to integer infeasibility, and II is the sum of infeasibilities. Typically, the user defaults to no bias so $\lambda = 1$. (When $\lambda = 0$ the best projection method becomes the best bound method.) The proportionality constant, p, is usually estimated empirically using the objective function degradation and integer infeasibility improvement, II_0, between the LP relaxation and the incumbent integer solution:

$$p = \frac{x_0 - x_0^*}{\text{II}_0}$$

The best projection method can be biased to emphasize or deemphasize variables that have large integer infeasibility by replacing the sum of infeasibilities, an L_1 norm, with an L_n norm,

$$\text{II} = |\min\{f_j, 1 - f_j\}|^n$$

As n becomes large, individual integer infeasibilities near 0.5 dominate II, and as n approaches 0, fractional variables are of similar importance and II is dominated by the number of fractional variables.

3.7.6. Pseudocosts

To calculate the pseudocost estimate (PCE) for a node we simply subtract the sum of individual variable pseudocost amounts from the bound:

$$\text{PCE} = x_0 - \sum_j \min\{\text{PCL}_j f_j, \text{PCU}_j(1 - f_j)\}$$

Integer Programming

Pseudocost estimates for a node are a generalization of the best projection method: when all pseudocosts equal one, the pseudocost estimate equals the best projection.

3.7.7. Norm Estimate

Norm estimates attempt to scale node estimates by accounting for depth in the tree. Norm estimates can be calculated in several different ways. The following three estimates have been used:

(1) $\quad NE = \dfrac{depth^2}{|PCE - x_0^0|}$

(2) $\quad NE = \dfrac{depth}{|PCE - x_0^0|}$

(3) $\quad NE = \dfrac{depth}{fractional \times |PCE - x_0^0|}$

where *depth* is the depth in the tree, *fractional* is the number of fractional variables, and PCE is the pseudocost estimate. These particular norm estimates may be double counting the effect of depth; in principle the pseudocost estimate is also affected by depth. Replacing PCE with x_0 would remove the apparent double counting. Alternatively, the entire denominator could be replaced by the current bound minus the incumbent solution. For example,

(1') $\quad NE = \dfrac{depth^2}{|x_0 - x_0^*|}$

3.7.8. Percentage Error

The percentage error criterion is designed to generate improved solutions given an existing integer feasible solution. If pseudocost estimates were 100% correct, the best strategy would be to branch on the node with the best estimate. Rather, pseudocost estimates are, by their nature, usually in error. The chances for error are greater when the allowable degradation in x_0 between the current node and the incumbent solution is greatest. Hence, the percentage error attempt to minimize the ratio of the error required for a node to improve on the incumbent, $x_0^* - PCE$, over the allowable degradation in the bound, $x_0 - x_0^*$:

$$PE = \dfrac{x_0^* - PCE}{x_0 - x_0^*}$$

Large negative numbers indicate nodes likely to generate an improved solution. The percentage error criterion is best suited to improving on an existing integer feasible solution rather than finding the first solution or proving the optimality of a solution.

3.7.9. Candidature Rules

Until now, we have presented the node selection strategies in isolation as "pure" strategies, and all too frequently these strategies have been applied in isolation. In this section, we describe strategies that combine the pure strategies, which have historically been called candidature rules. The terminology arises in the following way. One of the pure strategies is used to select the best branching node, but the candidates are restricted to nodes that meet a given threshold according to another criterion. In some implementations, once a node fails to meet a threshold it is removed from the tree, and hence, never becomes a candidate again. This practice frees memory but also forgoes proof of optimality.

Two common candidature rules use the bound provided by the LP relaxation. A cutoff removes nodes within a fixed percentage of the incumbent solution. This is often called an optimality tolerance because, when the search ends, the proof of optimality guarantees only that the incumbent solution is within this percentage of optimality. At first glance, this may appear to be a large restriction. In practice, the sacrifice is quite small, and even small tolerances can enormously improve convergence. This is particularly true when many solutions have identical (or nearly identical) objective function values. (These problems are, in general, quite difficult for branch and bound and reformulating to remove solutions of equivalent value is highly advised.) The second candidature rule allows only nodes whose bound is within a fixed percentage of the cobound, the best bound among unfathomed nodes. As with the norm estimates, this rule can be adapted to account for tree depth. For example, we can use the cobound associated with a given tree depth. The percentage calculations associated with the cutoff and cobound typically are modified to prevent inappropriate behavior when the incumbent solution is near zero. For example, cutting off nodes satisfying the following modified percentage is well behaved when the incumbent is near zero:

$$\frac{|x_0 - x_0^*|}{|x_0^*| + 1} \leq \beta$$

where β is user specified, based on experimentation or experience with similar problems. In principle, the cutoff and cobound candidature rules can be generalized to work with any fitness measure associated with the nodes.

Far more complex search strategies could also be used to search the branch and bound tree. Adaptive strategies that learn from the experience of previous branches may be useful.

Strategies that more gracefully automate the stages of search would be particularly useful for modelers with limited integer programming experience. The following table (adapted from the user's manual for ZOOM/XMP) gives some rough guidelines for selecting a strategy. (Keep in mind that experimentation and adaptation to a particular problem can often do better.) If a given software package does not incorporate the suggested strategy, one of the neighboring strategies should be deployed.

Duality gap	Previous action	Goal	Strategy	Cutoff (%)	Cobound (%)
Unknown or D.g. $\leq 10\%$		Initial solution Proof of optimality	Heuristic search Pseudocost, best bound	Small	Small
$10\% \leq$ D.g. $\leq 25\%$		Improved solution	Percentage error	Small	Small
$25\% \leq$ D.g.	No B & B yet	Feasible solution	Best proj., int. infeasibility	Small	Small
$25\% \leq$ D.g.	Small cutoff	Pare bottom of tree	Best proj., int. infeasibility	Large	Small
$25\% \leq$ D.g.	Large cutoff	Reduce paring	Best proj., int. infeasibility	Reduce	Small
$25\% \leq$ D.g	Small cobound	Feasible, pare tree	Norm estimate	Large	Large
$25\% \leq$ D.g.	Large cobound	Feasible, reduce paring	Norm estimate	Large	Reduce
$50\% \leq$ and bound not improving		Help B & B	Heuristic, reformulate		

Modern metaheuristics (e.g., Kelly and Osman 1995), such as tabu search (e.g., Glover and Laguna 1993), simulated annealing, and genetic algorithms, may also prove useful in searching the tree, particularly when the duality gap is large. (See Chapter 9 in this book.) Glover and Løketangen (1994) describe one attempt at this. Thus far, these techniques have primarily concentrated on building new solutions in combination with heuristics, rather than being married with branch and bound. A few exceptions exist for problems that are characteristically "combinatorial," as where tabu search has produced improved branch and bound methods for graph coloring and maximum stable set problems.

3.8. Other Practical Considerations

Branch selection and node selection are two of the more theoretic choices in branch and bound. However, many other choices can substantially influence the solution time. One choice is the criterion for stopping branch and bound. Another set of choices is the underlying linear program used to bound the nodes. We describe these in this section, but many other choices exist an any particular piece of software. While the default values are usually chosen reasonably, branch and bound is still far from a "black box" and experimentation with the available options often improves performance, sometimes substantially.

3.8.1. Stopping Criteria

First-time users of branch and bound frequently overlook the need for stopping criteria. As a practical matter, many problems will be unsolvable with the initial formulation and search strategy. Gracefully stopping can increase the information collected from an unsuccessful run. In particular, it is potentially valuable to save the cobound, the incumbent solution, and the current tree.

Most branch and bound codes can be stopped when one or more of the following limits is exceeded: CPU time, nodes explored, and cumulative linear programming iterations. Some codes allow the limits to be extended and processing resumed without any loss.

Most codes also allow processing to stop whenever a new incumbent solution (or a specified number of solutions) is found. Because new solutions reduce the duality gap and change the search strategy, it is often worthwhile to restart branch and bound; the existing tree may hinder the search after the strategy changes. If the strategy does not change as a result of finding a new solution, then resuming the search may be the best action.

Finally, most codes will stop when a desired duality gap has been attained. At this point a user can either quit or tighten the bound that expresses the desired gap.

3.8.2. Implementing B & B Methods in the LP Tableau

The natural form of the simplex method (primal or dual) to use in branch and bound applications is the bounded variable simplex method. This is true not only because it avoids enlarging the LP tableau with constraints such as $x_j \leq 1$ but also because the bounded variable method is particularly convenient for carrying out other operations in branch and bound. (This applies also for treating IP problems with general integer variables.)

The two dominant methods for bounding a subproblem are the dual simplex method and parametric optimization. The dual simplex method is generally more efficient for complete optimization of a subproblem. The principal advantage of parametric optimization is that it extends the penalty approach; as the branched variable changes value, both the bound and the penalty change. If the node becomes fathomed during bounding, this fact will be reflected by the penalty before the bound itself, and computation can stop. In contrast, the dual simplex computation must wait for the bound itself to cause fathoming before the computation may stop.

The state of the nodes in the tree varies with software. In principle, nodes may be either solved subproblems, unsolved subproblems, or *prenodes*. Prenodes are subproblems that became undesirable, but not fathomed, during bounding. Hence, optimization is truncated, without fully bounding the node.

3.8.3. References

The following references provide greater detail of branch and bound implementations: Beale (1985a, 1985b), Benichou et al. (1971, 1977), Forrest et al. (1974), Garfinkel (1979), Gauthier and Ribiere (1977), Glover and Tangedahl (1976), Johnson and Powell (1978), and Land and Powell (1979).

Saltzman (1994) provides a summary of currently available commercial software and capabilities. However, this information changes rapidly. Integer programming would seem to be a natural application of parallel computing. Gendron and Crainic (1994) provide a summary of approaches.

Preprocessing and probing approaches use cutting planes and other methods to simplify the search space before using branch and bound. Most of these methods could be applied at any node of the branch and bound tree, and the name *preprocessing* reflects when it is usually done. Nemhauser et al. (1994) describe MINTO, one code emphasizing the use of preprocessing techniques. Crowder et al. (1983) popularized both preprocessing and cutting plane approaches in general by reporting their successful experiments. Additional efforts on preprocessing include Hoffman and Padberg (1985) and Savelsbergh (1994).

4. LAGRANGEAN RELAXATION

Lagrangean relaxation (Everett, 1963) is a useful and established tool of integer programming. Lagrangean relaxation tends to be most useful when a problem can be characterized as "easy, except for some complicating constraints." The relaxation of the complicating constraints leads to a series of easy (relaxed) problems with solution values equal to or better than the optimal solution to the original problem. Generally, these solutions are not feasible solutions to the original problem but do provide an upper bound on solutions to maximization problems. In contrast, feasible solutions to the original problem are no worse than the optimal solution by definition and, hence, provide a lower bound on solutions to maximization problems. Together, relaxations and feasible solutions bracket the range of possible values of the optimal solution. An important use for such a range is to narrow the search for improved solutions by either heuristics or branch and bound algorithms. For minimization problems, the roles are reversed from those mentioned earlier for maximization problems: relaxations define a lower bound, and feasible solutions define an upper bound.

In the following sections we describe (1) relaxations in general, including linear programming relaxations; (2) lagrangean relaxations; (3) the lagrangean dual, i.e., the "best" lagrangean relaxation; (4) iterative methods that converge toward the lagrangean dual, primarily column generation and subgradient optimization; (5) the trade-offs involved in selecting a relaxation; (6) improvements on lagrangean relaxation: surrogate duality and lagrangean decomposition.

Lagrangean relaxation is based on an elegant theory; we present some highlights of the theory in Section 4.3. Lagrangean relaxation is also a powerful computational tool when implemented in a numerically stable way (Section 4.4) on a suitable problem (Section 4.5). We highlight some of the more common pitfalls to implementing lagrangean relaxation in Sections 4.4 and 4.5. Throughout the chapter, we demonstrate these ideas both algebraically and graphically with a two-dimensional example.

4.1. Relaxation In General

The "standard" branch and bound presented in the Section 3 used linear programming *relaxations* to bound each node in the branch and bound tree. By relaxing the integrality restriction, $x \in \{0, 1\}$, and instead allowing a larger range of values, $0 \leq x \leq 1$, the problem becomes a linear programming problem and, hence, easier to solve. The LP problem is called a *relaxation* because it allows additional solutions without disallowing any of the original solutions. Because a relaxation allows additional solutions, it can only improve the objective. Specifically, the objective function value of a relaxation is an upper bound for maximization problems and a lower bound for minimization problems.

Indeed, in common with the LP relaxation, any relaxation can be used to perform three functions:

1. To bound subproblems in branch and bound
2. To provide a worst-case bound on a heuristic solution's deviation from optimality
3. To initialize a perturbation heuristic: seeking a feasible solution "near" the relaxed solution

The importance of problem relaxation derives from three facts:

1. If the relaxed problem has no feasible solution, then neither does the original.
2. The optimum objective function value for the relaxed problem

Integer Programming

provides an "optimistic estimate" and hence a bound for the optimum objective function value of the original problem (an upper bound if maximizing, a lower bound if minimizing).
3. If an optimal solution for the relaxed problem is feasible for the original problem, then it is optimal for the original problem.

Because of these three conditions, a global solution strategy (such as branch and bound) that iteratively uses problem relaxation will succeed at each step in either solving the problem of current interest [as a result of condition (1) or (3)] or in obtaining a bound on the optimum objective function value [as a result of condition (2)]. (In fact, these three conditions are the same three that are at the heart of the "fathoming procedures" of branch and bound.)

To illustrate, consider the 0-1 IP problem

$$\begin{array}{ll} \text{Minimize} & 9x_1 + 5x_2 \\ \text{subject to} & 2x_1 - 3x_2 \leq 3 \\ & -5x_1 - 4x_2 \leq -5 \\ & x_1, x_2 \in \{0, 1\} \end{array}$$

A simple relaxation of this problem is to drop all constraints except the 0-1 restrictions, to give the problem

$$\begin{array}{ll} \text{Minimize} & 9x_1 + 5x_2 \\ \text{subject to} & x_1, x_2 \in \{0, 1\} \end{array}$$

The optimal solution to this relaxed problem is $x_1 = 0$, $x_2 = 0$. The resulting 0 value for the objective function is an optimistic estimate (hence a lower bound) for the optimum value for the original problem. Also, if $x_1 = 0$, $x_2 = 0$ had been feasible for the original problem, then clearly it would have been optimal.

Another relaxation, a bit stronger than the preceding, is to replace the original linear constraints by the constraint $-3x_1 - 7x_2 \leq -2$, which results by adding the original constraints together. Then the problem becomes

$$\begin{array}{ll} \text{Minimize} & 9x_1 + 5x_2 \\ \text{subject to} & -3x_1 - 7x_2 \leq -2 \\ & x_1, x_2 \in \{0, 1\} \end{array}$$

The optimal solution to this problem is $x_1 = 0$, $x_2 = 1$, giving a lower bound of 5 for the original problem. The solution is not feasible for the original problem, and thus no further information is available. If the LP problem for the original problem is taken to be the relaxed problem, i.e.,

Minimize $9x_1 + 5x_2$
subject to $2x_1 - 3x_2 \leq 3$
             $-5x_1 - 4x_2 \leq -5$
             $0 \leq x_1, \quad x_2 \leq 1$

then the optimal LP solution is $x_1 = \frac{1}{5}$, $x_2 = 1$. The objective function value of $6\frac{4}{5}$ provides a stronger bound than before. Again the solution is not feasible for the original (in this case, because x_1 is not an integer).

Finally, consider the relaxation that results by dropping the first linear constraint and keeping only the second.

Minimize $9x_1 + 5x_2$
subject to $-5x_1 - 4x_2 \leq -5$
             $x_1, \quad x_2 \in \{0, 1\}$

The solution to this problem is $x_1 = 1$, $x_2 = 0$. This solution turns out to be feasible for the original problem and hence is optimal.

The ideas illustrated by these examples are very simple but are also very powerful (particularly if a "good relaxation" is identified for the original problem). However, these ideas are not yet complete. More generally, *we allow a relaxed problem to be characterized not only by a less restrictive constraint set but also by a "relaxed" objective function.* A relaxed objective function, broadly speaking, is one that gives an optimistic estimate for the original objective function value. (For example, a linear function that "underestimates" a nonlinear function would qualify as a relaxed objective function in the minimization context, since the linear function always gives an optimistic value for the function it replaces.) More precisely, *any objective function that allows condition (2) to hold qualifies as a relaxed objective function.* (Hence, of course, the original objective function qualifies.)

4.1.1. Change in Condition 3 for a Relaxed Objective Function

If an objective function is "strictly relaxed"—i.e., is *not* the same as the original—then an additional stipulation must be added to condition (3) to yield the following modified statement:

(3′) If an optimal solution for the relaxed problem is feasible for the original problem, *and if in addition this solution gives the same objective function value in both problems*, then the solution is optimal for the original problem.

The italicized portion of (3′) is not included in (3), but otherwise (3) and (3′) are the same. (The italicized phrase is superfluous if the objective functions for the two problems are the same and hence was not required before.)

Integer Programming

We can illustrate a relaxation involving an "underestimating" objective function by reference to the example already discussed. One such relaxation is the following.

$$\text{Minimize} \quad -4x_1 - x_2 + 13$$
$$\text{subject to} \quad x_1, \quad x_2 \in \{0, 1\}$$

The optimal solution to this problem is $x_1 = 1$, $x_2 = 1$, giving an objective function value of $-4 - 1 + 13 = 8$, which provides a lower bound for the true optimum. (The fact that 8 does not exceed the optimum objective function value for the original problem—which we saw earlier to be 9—justifies calling the current objective function a relaxed objective function, regardless of how it was obtained.) In addition, the solution $x_1 = 1$, $x_2 = 1$ is feasible for the original problem. However, the objective function value for this solution in the original problem is 14, and since $14 \neq 8$, condition (3′) does not confirm optimality.

Consider next the relaxation

$$\text{Minimize} \quad -3x_1 + 2x_2 + 10$$
$$\text{subject to} \quad x_1, \quad x_2 \in \{0, 1\}$$

Here the optimal solution is $x_1 = 1$, $x_2 = 0$, yielding an objective function value of $-3 + 10 = 7$. This solution is also feasible for the original problem and gives a value of 9 for the original objective function. Since $9 \neq 7$, condition (3′) is again unable to disclose whether or not this solution is optimal for the original problem. However, as already seen, this solution is optimal. (This shows how it may be possible to "stumble onto optimality" using a weakened objective function without knowing it.)

The basic idea of a relaxation strategy is to identify a relaxed problem that is significantly easier to solve than the original but that is still "strong enough" to have a fair chance of providing an optimal solution to the original, or at least a good bound. (The ability to demonstrate infeasibility by condition (1) is fully as important as achieving optimality.) A crucial determinant of overall efficiency for a global solution method that uses relaxation iteratively lies in balancing the trade-off between the "strength" of the relaxed problems and the effort required to solve them.

4.1.2. Exercises

    4.1. Identify which of the following five problems qualify or fail to qualify as relaxed problems for the preceding example problem. (When the objective function differs from the original, apply the definition of a relaxed objective function to see if it holds.)

Problem 1:
$$\text{Minimize} \quad 9x_1 + 5x_2 + 3$$
$$0 \leq x_1, \quad x_2 \leq 1$$

Problem 2:
$$\text{Minimize} \quad 9x_1 + 5x_2$$
$$\text{subject to} \quad -8x_1 - 11x_2 \leq -7$$
$$x_1, \quad x_2 \in \{0, 1\}$$

(The linear constraint is obtained by adding two times the second constraint to the first.)

Problem 3:
$$\text{Minimize} \quad 9x_1 + 5x_2$$
$$\text{subject to} \quad -9x_1 + 2x_2 \leq -11$$
$$x_1, \quad x_2 \in \{0, 1\}$$

(The linear constraint is obtained by subtracting two times the first constraint from the second.)

Problem 4:
$$\text{Minimize} \quad -3x_1 - 3x_2 + 14$$
$$\text{subject to} \quad x_1, \quad x_2 \in \{0, 1\}$$

Problem 5:
$$\text{Minimize} \quad -x_1 + 3x_2 + 11$$
$$\text{subject to} \quad 2x_1 - 3x_2 \leq 3$$
$$0 \leq x_1, \quad x_2 \leq 1$$

4.2. Explain why each of the conditions (1), (2), and (3) must hold for a relaxed problem that has the same objective function as the original.

4.3. Show that condition (3′) is a direct consequence of condition (2), thereby justifying the "optimality check" of condition (3′) for any weakened objective function.

4.4. Consider the IP problem
$$\text{Minimize} \quad 4x_1 - 3x_2$$
$$\text{subject to} \quad -3x_1 + 5x_2 \leq -3$$
$$4x_1 - 2x_2 \leq 2$$
$$x_1, \quad x_2 \geq 0 \text{ and integer}$$

Consider three relaxations:
1. Drop the first constraint.
2. Drop the second constraint.

3. Drop both constraints and replace them with the sum of the constraints.

Which of these relaxation provides the most useful information and which provides the least?

4.5. Show that the problem

Minimize $\quad -3x_1 \quad +7x_2 \quad +4$
subject to $\quad -x_1 \quad -x_2 \leq -1$
$\quad x_1, \quad x_2 \in \{0, 1\}$

is compatible with the conditions (1), (2), and (3′) relative to the original problem illustrated earlier but is not a relaxed problem. Would there be anything wrong with *defining* a relaxed problem to be one that satisfies these three conditions?

4.2. Lagrangean Relaxation

Our qualitative description of problems suitable for lagrangean relaxation, "easy, except for some complicating constraints," can be restated mathematically. In matrix notation, an integer programming formulation is

IP $\quad \max_{x} \quad cx$
subject to $\quad A_1 x = b_1$
$\quad A_2 x = b_2$
$\quad x \in \{0, 1\}$

where A_1 and A_2 are matrices and b_1, b_2, and x are column vectors. We assume that the second block of constraints are the complicating constraints in this problem. In general, designating the complicating constraints requires good judgment.

Selecting the complicating constraints defines a lagrangean relaxation:

LR $\quad \max_{x} \quad cx + u(b_2 - A_2 x)$
subject to $\quad A_1 x = b_1$
$\quad x \in \{0, 1\}$

The complicating constraints are said to have been *dualized*, and the row vector u has one entry, a *dual price*, for each dualized constraint. The dual prices have an economic interpretation: the dual price associated with a given constraint is the approximate unit value of the resource represented by the constraint. The effect of the dual prices is to penalize resource usage in the objective function. The dual prices are constant within problem LR.

Ideally, the dual prices will be such that the objective function value of LR, z_{LR}, is close to the objective function of IP, z_{IP}. In the next section we discuss the selection of the dual prices in the context of the lagrangean dual.

The derivation of LR from IP is easily extended to handle inequalities. Any inequality can be rewritten as an equality through the use of slack and surplus variables. If all of the complicating constraints are inequalities,

$$A_2 x \leq b_2$$

the transformation to equalities produces

$$A_2 x + I s = b_2$$

where s is a vector of slack variables, $s \geq 0$, and I is the identity matrix. The objective of the relaxed problem becomes

$$\max_{x,s} cx + u(b_2 - A_2 x - Is)$$

or equivalently,

$$\max_{x,s} cx + u(b_2 - A_2 x) - us$$

If any component, i, has $u_i < 0$, then $z_{LR} = \infty$, which is of little interest: the solution of LR reveals nothing about the original problem. Thus, we may restrict $u \geq 0$. With this assumption, $s = 0$ in LR, and the objective becomes

$$\max_{x} cx + u(b_2 - A_2 x)$$

Thus, the objective has the same form for either equalities or inequalities. We merely add the restriction that $u \geq 0$ for $\leq$ inequalities and for similar reasons, $u \leq 0$ for $\geq$ inequalities. By similar reasoning, these inequalities are reversed for a *minimization* problem: $u \geq 0$ for $\geq$ inequalities and $u \leq 0$ for $\leq$ inequalities.

These restrictions are easy to remember: they are consistent with the restrictions on dual variables in linear programming.

A two-dimensional example graphically illustrates lagrangean relaxation:

Maximize $3x - y$

subject to

(1) $\quad x + 4y \leq 12$
(2) $\quad x - 4y \leq 0$
(3) $\quad 2x + 3y \leq 13$
(4) $\quad 2x - 3y \leq 4$

$\qquad x, y \geq 0$, integer

Integer Programming

One lagrangean relaxation is formed by dualizing the last two constraints:

Maximize $3x - y + u_3(13 - 2x - 3y) + u_4(4 - 2x + 3y)$
subject to
(1) $x + 4y \leq 12$
(2) $x - 4y \leq 0$
$x, y \geq 0$, integer

The white points in Figure 10 show the optimal solutions to the LR for selected values of u_3 and u_4, shown in parentheses in the figure. The four points are numbered arbitrarily for reference later in the section. Other values of u_3 and u_4 lead to the same four points shown in the figure, but with different values of z_{LR}. The black points in the figure, along with the two leftmost white points, are the feasible integer solutions to the original problem.

A "natural" strategy for selecting the dual prices is to increase the price associated with violated constraints and decrease the price associated with satisfied constraints. In this sense, lagrangean relaxation is a special case of *penalty* methods for nonlinear programing. In the example, suppose we start with dual prices (0,0). These prices lead to an LR solution $x = 4$, $y = 1$, $z_{LR} = 11$, which violates constraint 4. Thus we increase the dual price on constraint 4, by say 1 unit. The new dual prices are (0,1) and the

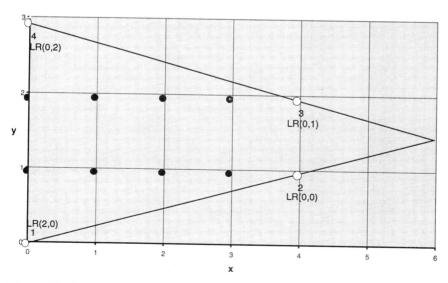

Figure 10 Lagrangean relaxation solutions.

LR solution is $x = 4$, $y = 2$, $z_{LR} = 12$, which violates constraint 3. Thus we increase the dual price on constraint 3 and decrease the dual price on constraint 4, by say 0.5. The new dual prices are (0.5,0.5) and the LR solution is $x = 4$, $y = 1$, $z_{LR} = 12$, which again violates constraint 4. Thus we increase the dual price on constraint 4 and decrease the dual price on constraint 3, by say 0.25. The new dual prices are (0.25,0.75) and the LR solution is $x = 4$, $y = 2$, $z_{LR} = 11.25$, which again violates constraint 3. This process of adjustment can continue for many steps. One difficulty of the natural strategy is choosing how large a step to take in adjusting the dual prices. In Section 4.4 we take up this issue when we discuss a formal version of the natural strategy, subgradient optimization.

The optimal dual prices, derived in Section 4.4, are (0,0.33). With these values, there are two optimal solutions to the lagrangean relaxation: $x = 4$, $y = 1$, and $x = 4$, $y = 2$, with $z_{LR} = 10.67$. A convex combination of these two points (e.g., $x = 4$, $y = 1.33$) violates neither constraint 3 nor 4.

Figure 11 illustrates both lagrangean relaxation and LP relaxation. The solid black points are integer feasible solutions, including the optimal IP solution. The white points represent optimal solutions (or convex combination of optimal solutions) arising from various relaxations:

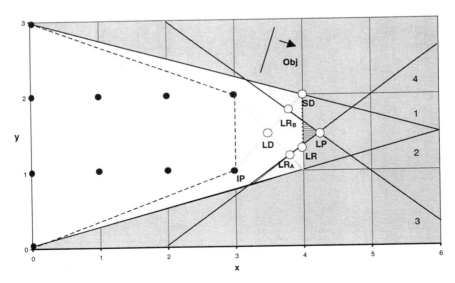

Figure 11 Lagrangean relaxation example.

Integer Programming

LP — the linear programming relaxation
LR — the lagrangean relaxation specified above, dualizing constraints 3 and 4
SD — the surrogate dual resulting from dualizing constraints 3 and 4
LR_A — the lagrangean relaxation resulting from dualizing constraint 4
LR_B — the lagrangean relaxation resulting from dualizing constraint 3
LD — the lagrangean decomposition of LR_A and LR_B

The surrogate dual and the lagrangean decomposition are described in Section 4.6.

Several feasible regions are shown in Figure 11:

1. The convex hull of the integer solutions, the dashed region,
2. The feasible region (also shown in Figure 10) for lagrangean relaxations, the white region combined with the dotted region: constraints 1 and 2, and integrality, the region bounded by the points (0,0), (0,3), (4,2), and (4,1),
3. The feasible region of the LP relaxation, the white region combined with the region shaded with horizontal lines, the region bounded by the points (0,0), (0,3), (3.2,2.2), (4.25,1.5), and (3.2,0.8).

(The intersection of the LR and LP regions is white.)

The objective function is shown in the figure as well. The graph illustrates the relative objective function values of the particular relaxations as well as the integer programming solution:

$$z_{IP} < z_{LD} < z_{LR_B} < z_{SD} < z_{LR_A} < z_{LR} < z_{LP}$$

The more general relationships will be explored as the section continues.

We close this section by proving that z_{LR} is always an upper bound on z_{IP}. Let x^* denote the optimal integer solution. The following two inequalities constitute the proof:

$$cx^* \leq cx^* + u(b - Ax^*) \leq \max_x cx + u(b - Ax)$$

subject to nondualized constraints

The quantity on the left is z_{IP} and the quantity on the right is z_{LR}. The inequality on the left holds because $u(b - Ax^*) \geq 0$. If constraint i is an equality constraint, then $A_i x^* = b_i$ because x^* is feasible. If constraint i is a $\leq$ constraint, then $u \geq 0$ and $A_i x^* \leq b_i$. If constraint i is a $\geq$ constraint, then $u \leq 0$ and $A_i x^* \geq b_i$.

The inequality on the right holds because x^* satisfies the nondualized constraints. A maximum solution value for z_{LR} can only be greater.

Corollary: $z_{IP} = z_{LR}$ if and only if a lagrangean solution x^* satisfies $u(b - Ax^*) = 0$

and hence both of the above inequalities become equalities.

The condition in the corollary is called *complementary slackness*. (This concept appears in several areas of mathematical programming. For example, the optimal solution of an LP satisfies complementary slackness. See Chapters 2 and 5 of this book.) The condition states that each constraint is either unpenalized ($u_i = 0$) or the constraint is satisfied [$(b - Ax^*)_i$]. This condition is seldom satisfied by a lagrangean relaxation; continuous adjustment in u_i may cause discontinuous adjustment in $(b - Ax^*)_i$. Hence, there may not exist a value of u_i that leads to $(b - Ax^*)_i = 0$.

4.2.1. Exercises

4.6. Determine the strongest lower bound that can be obtained for the following example using the lagrangean relaxation approach and relaxing both constraints.

$$\begin{aligned}
\text{Minimize} \quad & 4x_1 + 9x_2 \\
\text{subject to} \quad & 3x_1 + 4x_2 \leq 7 \\
& -5x_1 - 10x_2 \leq -7 \\
& x_1, \quad x_2 \in \{0, 1\}
\end{aligned}$$

Use the natural strategy, described above, for adjusting weights. What intuitive reasons lead you to suspect that the bound you have found is strongest? (What happens if you try to change the weights on either constraint by a small amount?)

4.7. Apply a lagrangean relaxation (using the natural strategy for adjusting weights) to the example problems treated in Exercises 4.1 and 4.4. Identify what you believe to be the strongest bounds in each case, and give intuitive reasons to support your belief.

4.8. Carry out the instructions of Exercise 4.7 for the following problem.

$$\begin{aligned}
\text{Minimize} \quad & -4x_1 + 8x_2 \\
\text{subject to} \quad & -7x_1 + 3x_2 \leq 4 \\
& 2x_1 - 7x_2 \leq -7 \\
& x_1, \quad x_2 \in \{0, 1\}
\end{aligned}$$

4.9. (*"Sharpening" the strategy for adjusting weights.*) What kinds of latitude exist for applying the natural strategy for adjusting weights? How might the changes in weights be made to reflect the amounts by which the constraints are currently over- or undersatisfied? (which weights should undergo the relatively largest changes?) Can you give a "formula" that will tend to make the changes exactly proportional to the over- and undersatisfied amounts?

4.10. Using the ability to detect the optimal solutions for simple example problems by inspection, how can you know in advance for each of the problems treated in Exercises 4.7 and 4.8 that the lagrangean approach will not be able to verify optimality? Identify changed constant terms for the example of Exercise 4.8 that will make it possible to identify a relaxed lagrangean problem that can identify optimality—and determine such a relaxed problem.

4.11. (*A potential pitfall*) Determine a lagrangean relaxation that can identify optimality for the following problem:

$$\begin{array}{ll} \text{Minimize} & 6x_1 + 5x_2 \\ \text{subject to} & -8x_1 + 2x_2 \leq -6 \\ & 2x_1 - 7x_2 \leq -5 \\ & x_1, \quad x_2 \in \{0, 1\} \end{array}$$

What limitation of the lagrangean approach does this disclose? How serious could it be for problems involving many variables? Can you think of a possible way to overcome this limitation for linear constraints? (Hint: If optimality can be verified, the constraints must hold as equalities. Why?)

4.12. If *equations* are taken into the objective function in the lagrangean approach, their weights may have any sign. Explain this by showing what happens if each equation is taken into the objective function as a pair of inequalities.

4.3. The Lagrangean Dual

A natural question arises from the formulation of a lagrangean relaxation: "What values of the dual prices, u, give the best relaxation?" The answer is simple and elegant; the best relaxation for a maximization problem is the one with minimum value and hence the tightest upper bound. This minimization problem is called the *lagrangean dual*. The elegant aspect of the lagrangean dual is that its optimum solution value is equivalent to solving a particular linear programming relaxation. In this section, we explore these ideas more fully.

The lagrangean dual for maximization problems is

LD $\quad z_{LD} = \min_{u} z_{LR}(u)$

Thus, the lagrangean dual selects the lagrangean relaxation with the smallest upper bound on the optimum solution value for the original problem. This is the tightest, and hence most informative, bound among those provided by dualizing particular constraints.

Much of the practical power of lagrangean relaxation stems from a single theorem.

Theorem:

$$z_{LD} = \max_{x} cx$$

subject to

$$x \in \text{conv}(A_1 x = b_1)$$
$$A_2 x = b_2$$
$$0 \le x \le 1$$

where z_{LD} is the objective function value of the lagrangean dual, and conv-$(A_1 x = b_1)$ denotes the convex hull of integer solutions to the "nice" constraints. A simplistic, but graphic, definition of the convex hull of a polyhedron is the polyhedron that would arise from shrinking the polyhedron defined by the original constraints until it becomes the smallest polyhedron that contains all feasible integer solutions. Mathematically, any polyhedron, P, can be defined as a convex combination of extreme points plus extreme rays, indexed by i and j, respectively:

$$P = \bigcup_{x} x = \sum_{i} \lambda_i x^i + \sum_{j} \lambda_j x^j$$

$$\sum_{i} \lambda_i = 1$$

$$\lambda_i \ge 0$$

The linear program defined in the theorem can be seen in the example. (The example generalizes the theorem to general integer variables.) The convex hull of feasible integer solutions to constraints 1 and 2 is the region bounded by the four white points in Figure 10.

The proof of the theorem is constructive; it suggests an iterative method for computing z_{LD}. The proof reformulates the lagrangean dual in three different, but equivalent, ways. Each reformulation is demonstrated with the example before continuing with the proof. Expanding the definition of $LR(u)$ within the definition of LD yields

Integer Programming

$$z_{LD} = \min_{u} \left\{ \begin{array}{l} \max_{x} cx + u(b_2 - A_2 x) \\ \text{s.t. } A_1 x = b_1 \\ x \in \{0, 1\} \end{array} \right\}$$

Replacing the constraints of LR(u) with its convex hull yields

$$z_{LD} = \min_{u} \left\{ \max_{x^i \in \text{conv}(A_1 x = b)} cx^i + u\left(b_2 - A_2 x^i\right) \right\}$$

We pause in our proof to demonstrate this reformulation with the example. The extreme points of the convex hull in this case are the four white points in Figure 10: (0,0), (4,1), (4,2), and (0,3). Thus, one term appears for each of these points in the maximization.

$$z_{LD} = \min_{u} \{ \max\{ 0 + 13u_3 + 4u_4, 11 + 2u_3 - u_4,$$
$$10 - u_3 + 2u_4, -3 + 4u_3 + 13u_4 \} \}$$

The expressions in the maximization above are the lagrangean relaxation objective function values at each of the four points in Figure 10.

Suppose for some u, an extreme ray, j, in the convex hull is such that

$$(c - uA_2)x^j > 0$$

In this case, the maximization will have a value of ∞, and this u will not be minimal. Therefore we can assume

$$(c - uA_2)x^j \leq 0$$

for all extreme rays. The lagrangean dual can be reformulated as a linear program that includes these constraints.

$$\min_{u,w} w$$
subject to
$$w \geq cx^i + u(b_2 - A_2 x^i) \quad \forall \text{ extreme points}$$
$$(c - uA_2)x^j \leq 0 \quad \forall \text{ extreme rays}$$

Rearranging constraints to canonical form yields:

$$\min_{u,w} w$$
subject to
$$w + u(A_2 x^i - b_2) \geq cx^i \quad \forall \text{ extreme points}$$
$$uA_2 x^j \geq cx^j \quad \forall \text{ extreme rays}$$

We pause again in our proof to demonstrate this reformulation in the example.

$$\min \quad w$$

subject to

$$
\begin{aligned}
w - 13u_3 - 4u_4 &\geq 0 \\
w - 2u_3 + 1u_4 &\geq 11 \\
w + 1u_3 - 2u_4 &\geq 10 \\
w - 4u_3 - 13u_4 &\geq -3 \\
u_3, u_4 &\geq 0
\end{aligned}
$$

The solution of this problem is $w = 10.67$, $u_3 = 0$, $u_4 = 0.33$. Hence, $z_{LD} = 10.67$. Constraints 2 and 3, corresponding to points 2 and 3 in Figure 10, are tight at optimality.

The dual of the linear program completes the proof:

$$\max_{\lambda} \sum_i \lambda_i(cx^i) + \sum_j \lambda_j(cx^j)$$

subject to

$$\sum_i \lambda_i(A_2 x^i - b_2) + \sum_j \lambda_j(A_2 x^j) = 0$$

$$\sum_i \lambda_i = 1$$

$$\lambda_i, \lambda_j \geq 0$$

Applying the second constraint to the b_2 term in the first constraint further simplifies the formulation:

$$\max_{\lambda} \sum_i \lambda_i(cx^i) + \sum_j \lambda_j(cx^j)$$

subject to

$$\sum_i \lambda_i(A_2 x^i) + \sum_j \lambda_j(A_2 x^j) = b_2$$

$$\sum_i \lambda_i = 1$$

$$\lambda_i, \lambda_j \geq 0$$

This is the optimization problem in the theorem, restated in terms of extreme points and extreme rays. The first constraint is equivalent to:

$$A_2 x = b_2$$

and the second constraint guarantees the solution belongs to conv$(A_1 x = b_1)$. This completes the proof.

We demonstrate the last formulation with our example. [The coeffi-

Integer Programming

cients associated with $\lambda_1 (x = 0, y = 0)$ evaluate to 0, except for the last constraint, and, hence, appear to be missing from the formulation.]

$$
\begin{aligned}
\text{Maximize} \quad & 11\lambda_2 + 10\lambda_3 - 3\lambda_4 \\
\text{subject to} \quad & \\
& 11\lambda_2 + 14\lambda_3 + 9\lambda_4 \leq 13 \\
& 5\lambda_2 + 2\lambda_3 - 9\lambda_4 \leq 4 \\
\lambda_1 + & \lambda_2 + \lambda_3 + \lambda_4 = 1 \\
\lambda_1, & \lambda_2, \lambda_3, \lambda_4 \geq 0
\end{aligned}
$$

Each column corresponds to a point in Figure 10. The solution to this linear program is $\lambda_1 = 0$, $\lambda_2 = 0.67$, $\lambda_3 = 0.33$, $\lambda_4 = 0$. Again, $z_{LD} = 10.67$. The solution is a convex combination of points 2 and 3. This solution corresponds to the white point marked LR in Figure 11.

Readers familiar with Dantzig-Wolfe decomposition may recognize that the proof also establishes that the lagrangean dual is equivalent to Dantzig-Wolfe decomposition applied to the original problem: each lagrangean relaxation is a *subproblem*, and the last linear program is the *master problem*.

4.4. Iterative Methods

Two iterative methods historically have dominated the calculation of z_{LD}: *column generation* (e.g., Shapiro, 1993; Ribeiro and Soumis, 1994; Barnhart et al., 1995) and *subgradient optimization* (Held et al., 1974). Both methods alternate between solving lagrangean relaxations and adjusting the dual prices, u. Column generation adjusts the dual prices by solving a *revised master problem*, related to the Dantzig-Wolfe master problem from the previous section. Subgradient optimization adjusts the dual prices by determining a direction, a *subgradient*, that decreases z_{LD}.

Both methods need some initial dual prices to start the iterative process. The simplest method is to initialize the dual prices to 0. The economic interpretation is that the dualized constraints do not influence the solution. Should this prove false, the solution to the lagrangean relaxation will violate these constraints and either method will respond by changing the dual prices to penalize this violation. Another method is to solve the linear programming relaxation of the original problem and use the dual prices it generates for the dualized constraints.

4.4.1. Column Generation

Column generation alternates between solving lagrangean relaxations and solving the revised master problem. In this section we describe the implementation details of column generation: the alternating iterations, initializing the master problem, separable subproblems, and the relative advantages and disadvantages of the method.

The revised master problem contains a subset of columns in the Dantzig-Wolfe master problem. Each solution of a lagrangean subproblem generates a solution that becomes an additional column in the revised master problem. Each solution of the master problem leads to revised dual prices that change the objective function of the next lagrangean relaxation. The solution process alternates between these iterations until the method converges to a solution or meets some other stopping criteria.

Initially, the revised master problem has only the column generated by the first lagrangean relaxation, and the problem may not be feasible. One way to make the problem feasible is to add at least one solution that is feasible for the original problem, perhaps generated by a heuristic. Another way to make the problem feasible is to add artificial variables, s^+ and s^-, to the formulation and penalize these variables in the objective, with a penalty vector p.

$$\max_\lambda \sum_i \lambda_i(cx^i) + \sum_j \lambda_j(cx^j) + p(s^+ + s^-)$$

subject to

$$\sum_i \lambda_i(A_2 x^i) + \sum_j \lambda_j(A_2 x^j) + s^+ - s^- = b_2$$

$$\sum_i \lambda_i = 1$$

$$\lambda_i, \lambda_j \geq 0$$

The penalties should be chosen high enough that the artificial variables do not appear in the solution that the iterations converge to. At the same time, the penalties should be chosen reasonably small to prevent distorting the solutions generated by the lagrangean relaxations.

We illustrate column generation with our example.

Iteration 1. We initialize the lagrangean relaxation with $u_3 = 0$ and $u_4 = 0$. The lagrangean relaxation is

Maximize $3x - y$

subject to

(1) $x + 4y \leq 12$
(2) $x - 4y \leq 0$
     $x, y \geq 0$, integer

The solution is $x = 4$, $y = 1$. This is point 2 in Figure 10. This solution translates into a column of the revised master problem. The

Integer Programming

objective function value and the left-hand side of the dualized constraints associated with this solution become

$$3x - y = 11$$
(3) $2x + 3y = 11$
(4) $2x - 3y = 5$

The revised master problem is initialized with this "left-hand side solution" as the column associated with the variable λ_2, defined in the master problem above. We complete the formulation by adding artificial variables, s^-, to assure the existence of a feasible solution. We set the penalties on the artificial variables at a value of 2, following the rough guidelines stated earlier, that we want the penalties to be "large enough" to drive the artificial variables to 0 if possible, but no larger than necessary. The last constraint requires the solution to be a convex combination of the extreme points of constraints (1) and (2) generated thus far.

Maximize $11\lambda_2 - 2s_3^- - 2s_4^-$
subject to

$$11\lambda_2 - s_3^- \leq 13$$
$$5\lambda_2 - s_4^- \leq 4$$
$$\lambda_2 = 1$$
$$\lambda_i \geq 0$$

The solution is $\lambda_2 = 1$, $s^- = (0, 1)$, $u = (0, 2)$, $z = 9$.

Iteration 2. The lagrangean relaxation is

Maximize $-x + 5y + 8$
subject to
(1) $x + 4y \leq 12$
(2) $x - 4y \leq 0$
    $x, y \geq 0$, integer

The solution is $x = 0$, $y = 3$. We add this column to the revised master problem and resolve.

Maximize $11\lambda_2 - 3\lambda_4 - 2s_3^- - 2s_4^-$
subject to

$$11\lambda_2 + 9\lambda_4 - s_3^- \leq 13$$
$$5\lambda_2 - 9\lambda_4 - s_4^- \leq 4$$
$$\lambda_2 + \lambda_4 = 1$$
$$\lambda_i \geq 0$$

The solution is $\lambda_2 = 0.929$, $\lambda_4 = 0.071$, $s^- = (0,0)$, $u = (0,1)$, $z = 10$.

Iteration 3. The lagrangean relaxation is

Maximize $\quad x + 2y + 4$

subject to

(1) $\quad x + 4y \le 12$
(2) $\quad x - 4y \le 0$
$\quad\quad x, y \ge 0$, integer

The solution is $x = 4$, $y = 2$. We add this column to the revised master problem and resolve.

Maximize $\quad 11\lambda_2 + 10\lambda_3 - 3\lambda_4 - 2s_3^- - 2s_4^-$
subject to

$$\begin{aligned} 11\lambda_2 + 14\lambda_3 + 9\lambda_4 - s_3^- &\le 13 \\ 5\lambda_2 + 2\lambda_3 - 9\lambda_4 \quad\quad - s_4^- &\le 4 \\ \lambda_2 + \lambda_3 + \lambda_4 &= 1 \\ \lambda_1 \quad\quad &\ge 0 \end{aligned}$$

The solution is $\lambda_2 = 0.67$, $\lambda_3 = 0.33$, $\lambda_4 = 0$, $s^- = (0, 0)$, $u = (0, 0.67)$, $z = 10.67$.

Iteration 4. The lagrangean relaxation is

Maximize $\quad 2.33x + 1.33$

subject to

(1) $\quad x + 4y \le 12$
(2) $\quad x - 4y \le 0$
$\quad\quad x, y \ge 0$, integer

There are alternative optimal solutions, either $x = 4$, $y = 1$ or $x = 4$, $y = 2$. Both columns are already in the master problem. Therefore, column generation has converged. Observe two things: first, these two solutions are the active solutions in the previous revised master problem, and second, the column associated with λ_1 was never generated because it was not necessary.

In addition to producing optimal dual prices, column generation produces a primal solution. However, the primal solution will frequently have fractional values for integer variables. In our example, the primal solution

Integer Programming

to the third revised master problem of $\lambda_2 = 0.67$, $\lambda_3 = 0.33$, $\lambda_4 = 0$, corresponds to a solution in the original primal variables.

$$\begin{bmatrix} x \\ y \end{bmatrix} = \begin{bmatrix} 4 \\ 1 \end{bmatrix} \lambda_2 + \begin{bmatrix} 4 \\ 2 \end{bmatrix} \lambda_3 + \begin{bmatrix} 0 \\ 3 \end{bmatrix} \lambda_4 = \begin{bmatrix} 4 \\ 1.33 \end{bmatrix}$$

This fractional solution can be the starting point for a branch and bound algorithm. In this case, we could branch on $y : y \leq 1$ or $y \geq 2$. These constraints can be passed on to the lagrangean relaxations. The revised master problem must be updated to exclude columns corresponding to solutions that violate the constraints. Alternatively, one could solve the revised master problem from scratch for each node of the branch and bound tree; lagrangean relaxation would regenerate any columns that should not have been discarded. This alternative would increase solution time, but it may also reduce the programming requirements.

4.4.2. Separable Subproblems

A variant of the column generation presented thus far frequently arises from lagrangean relaxations with separable subproblems. Specifically, the subproblems can be solved independently, usually with significantly greater efficiency. (Additional gains can be achieved by solving the problems with parallel computing.)

We suggest an example, the *generalized assignment problem* (e.g., Laguna et al., 1995):

Maximize $\quad \sum_{k,l} c_{k,l} x_{k,l}$

subject to

$\sum_{l} a_{k,l} x_{k,l} \leq b_k \quad \forall k$

$\sum_{k} x_{k,l} = 1 \quad \forall l$

$x_{k,l} \in \{0, 1\}$

One application is assigning jobs, indexed by l, to machines, indexed by k. In this case, the parameters can be interpreted as:

$c_{k,l}$—the value of assigning job l to machine k
$a_{k,l}$—the time to process job l on machine k
b_k—the time available on machine k

The first constraint is the machine capacity constraint, and the second constraint requires each job to be assigned to exactly one machine.

Dualizing the second constraint of the generalized assignment problem creates the lagrangean relaxation

Maximize $\sum_{k,l} c_{k,l} x_{k,l} + \sum_{l} u_l \left(1 - \sum_{k} x_{k,l}\right)$

subject to

$\sum_{l} a_{k,l} x_{k,l} \leq b_k \quad \forall k$

$x_{k,l} \in \{0, 1\}$

This problem separates into one subproblem for each k. Rearranging terms and omitting the constant in the objective, Σu_l, leads to subproblem(k):

Maximize $\sum_{l} (c_{k,l} - u_l) x_{k,l}$

subject to

$\sum_{l} a_{k,l} x_{k,l} \leq b_k$

$x_{k,l} \in \{0, 1\}$

This is a simple knapsack problem and can be solved efficiently. [In the terminology of complexity theory, the knapsack problem can be solved in pseudopolynomial time. See, for example, Garey and Johnson (1979) or Boyd (1991).]

The separability exploited above can be further exploited by modifying the revised master problem. The revision is to create one column for each subproblem solution and one convexity constraint for each subproblem. For our example, let $\theta^{i,k,l}$ and $\theta^{j,k,l}$ equal 1 if $x_{k,l} = 1$ in extreme point i or extreme ray j, respectively. The master problem is

$\max_{\lambda} \sum_{k} \left(\sum_{i} \lambda_{i,k} (cx^{i,k}) + \sum_{j} \lambda_{j,k} (cx^{j,k}) \right)$

subject to

$\sum_{k} \left(\sum_{i} \lambda_{i,k} (\theta^{i,k,l}) + \sum_{j} \lambda_{j,k} (\theta^{j,k,l}) \right) = b_2$

$\sum_{i} \lambda_{i,k} = 1 \quad \forall k$

$\lambda_i, \lambda_j \geq 0$

Generalizing, the master problem for any lagrangean relaxation that separates into subproblems, indexed by k, is

$\max_{\lambda} \sum_{k} \left(\sum_{i} \lambda_{i,k} (cx^{i,k}) + \sum_{j} \lambda_{j,k} (cx^{j,k}) \right)$

subject to

Integer Programming

$$\sum_k \left(\sum_i \lambda_{i,k}(A_2 x^{i,k}) + \sum_j \lambda_{j,k}(A_2 x^{j,k}) \right) = b_2$$

$$\sum_i \lambda_i = 1 \quad \forall k$$

$$\lambda_i, \lambda_j \geq 0$$

The advantage of this formulation is that it reduces the number of columns that exist in the master problem. In the original master problem the extreme points correspond to a cross-product of the extreme points of the subproblems. For example, suppose there are three subproblems and each has 10 extreme points. The original master problem would have $10 \times 10 \times 10 = 1000$ extreme points, whereas the modified master problem has only $10 + 10 + 10 = 30$ extreme points.

4.4.3. Lagrangean Relaxation in Branch and Bound

Lagrangean relaxation can be the bounding method in branch and bound. With either column generation or subgradient optimization the lagrangean dual need *not* be solved to optimality to obtain a worthwhile bound. Stopping short of optimality can substantially reduce computation time with little degradation in the bound.

The choice of methods for optimizing the lagrangean dual affects the branch and bound algorithm. In the case of column generation the following pitfall must be avoided: The columns cannot be used as the branching variables. The problem with branching on the columns, λ, is that the branch $\lambda = 0$ is ineffectual: the column will be regenerated by the next relaxed problem if the column has a nonzero value in the master problem. Instead, one must use the branch on the original variables (or any pair of disjoint cuts). These branches are added to the relaxed problem.

4.4.4. Advantages and Disadvantages

The main advantage of the column generation approach is that it retains all of the information generated by the lagrangean relaxations in the revised master problem. Thus, the dual prices generated at any iteration may reflect both the most recent lagrangean relaxation and relaxations from previous iterations.

Two criticisms have been made of the column generation approach historically. First, if the lagrangean dual has the same bound as the linear programming relaxation, the iterative process may be quite slow in comparison. Second, extensive programming is required to implement the procedure. This criticism has become dated with the advent of improved algebraic modeling languages: the master problem, subproblems(s), and iteration control are typically easy to describe in an algebraic modeling language.

4.4.5. Subgradient Optimization

Subgradient optimization is an alternative to column generation for solving the lagrangean dual. Unlike column generation, subgradient optimization neglects previous lagrangean relaxations and uses only the current relaxation to adjust the dual prices.

In this section, we first will show that the lagrangean dual is a convex optimization problem that enables the use of nonlinear optimization methods. More specifically, the lagrangean dual is piecewise linear and continuous. Although many nonlinear methods use gradients to optimize over smooth surfaces, the lagrangean dual need not have a smooth surface. The gradient generalizes to a set of subgradients that can take the place of a gradient at a nonsmooth point on the surface.

Next, we suggest a step size for adjusting current dual prices in the subgradient direction. Although the suggested step size is a straightforward formula and converges in theory, practical success requires that several common pitfalls be avoided. We address pitfalls that are important to be aware of.

Finally, we discuss the relative merits of subgradient optimization. The primary advantage of subgradient optimization is that the calculations of a given iteration are straightforward and fast. The primary disadvantage is that even though the lagrangean dual may be converging, the solutions in the lagrangean relaxations may oscillate wildly without ever generating a solution that is close to satisfying the original constraints.

4.4.6. A Piecewise Linear Optimization Problem

Subgradient optimization depends on the fact that the lagrangean dual is convex. The graph of a function of one variable conveys the general notion of convexity: for any two points on the curve, a straight line connecting these points will lie above the function. For example the dashed line between u^1 and u^2 (or any other two points on the graph of the function) in Figure 12 shows that this piecewise linear function is convex.

This informal definition of convexity can also be used for a function of several variables. First, we introduce notation: let u be a convex combination of two points, u^1 and u^2,

$$u = \lambda u^1 + (1 - \lambda)u^2$$
$$0 \leq \lambda \leq 1$$

Definition: A function, f, is *convex* if for any two points, u^1 and u^2, and any convex combination, u, of these points, $f(u)$ is no greater than a convex combination of the original function values, $f(u^1)$ and $f(u^2)$,

Integer Programming

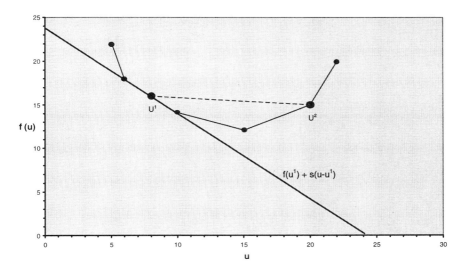

Figure 12 A convex function.

$$f(u) \leq \lambda f(u^1) + (1 - \lambda) f(u^2)$$

For lagrangean relaxation, the function $f(u)$ is the objective function value of the relaxation at u, $z_{LR}(u)$.

We are now prepared to state the main theorem of this section.

Theorem (Geoffrion, 1974): The lagrangean dual is convex.

Proof: Recall this previous stated formulation of the lagrangean dual:

$$z_{LD} = \min_{u} \left\{ \max_{x^i \in conv(A_1 x = b)} cx^i + u(b_2 - A_2 x^i) \right\}$$

Let $f(u)$ denote the linear expression, indexed by i, that attains the maximum value for a given u.

$$f(u) = cx^i + u(b_2 - A_2 x^i)$$

Linearity allows $f(u)$ to be expanded in terms of u^1 and u^2.

$$f(u) = (\lambda + (1 - \lambda))cx^i + (\lambda u^1 + (1 - \lambda)u^2)(b_2 - A_2 x^i)$$

Rearranging terms yields

$$f(u) = \lambda(cx^j + u^1(b_2 - A_2x^j)) + (1 - \lambda)(cx^j + u^2(b_2 - A_2x^j))$$

The last expression is a combination of terms with the properties

$$cx^i + u^1(b_2 - A_2x^i) \le f(u^1)$$
$$cx^i + u^2(b_2 - A_2x^i) \le f(u^2)$$

With a linear combination of these inequalities,

$$f(u) \le \lambda f(u^1) + (1 - \lambda)f(u^2)$$

the proof is completed. □

We can verify the theorem for two particular points in the example, such as

$$u^1 = \begin{bmatrix} 0 \\ 0 \end{bmatrix}, \quad u^2 = \begin{bmatrix} 0 \\ 2 \end{bmatrix}$$

Suppose, $\lambda = 0.5$. This defines

$$u = \lambda u^1 + (1 - \lambda)u^2 = \begin{bmatrix} 0 \\ 1 \end{bmatrix}$$

Given these assumptions,

$$f(u) = \max\{0 + 13u_3 + 4u_4, 11 + 2u_3 - 1u_4, 10 - 1u_3 + 2u_4,$$
$$-3 + 4u_3 + 13u_4\}$$
$$= \max\{4, 10, 12, 10\} = 12$$

and similarly, $f(u^1) = 11$ and $f(u^2) = 23$. Evaluating the expressions in the proof leads to

$$f(u) = 12 = \frac{1}{2}(10 - 1 \times 0 + 2 \times 0)$$
$$+ \frac{1}{2}(10 - 1 \times 0 + 2 \times 2)$$
$$= \frac{1}{2}10 + \frac{1}{2}14 \le \frac{1}{2}(11 + 2 \times 0 - 1 \times 0)$$
$$+ \frac{1}{2}(-3 + 4 \times 0 + 13 \times 2)$$
$$= \frac{1}{2}11 + \frac{1}{2}23 = 17$$

Convex optimization problems, such as the lagrangean dual, can be solved by local improvement methods. If the lagrangean dual were smooth, i.e., continuous and with a continuous gradient, a dual solution could be improved by moving in the direction of the gradient.

Integer Programming

However, the lagrangean dual's gradient is not continuous, because the function is piecewise linear. The piecewise segments correspond to extreme points of $\text{conv}(A_1 x = b_1)$. The function *is* smooth along a given segment: the gradient is constant. In our example, the gradient is

$$\left[\frac{\partial f}{\partial u_3}, \frac{\partial f}{\partial u_4} \right]$$

The most recently used formulation of the lagrangean dual reveals the gradient. The expression achieving the maximum value for a set of dual prices corresponds to the segment: the coefficients on the dual variables comprise the gradient. When $u = (0, 1)$, the expression is $10 - 1u_3 + 2u_4$, and the gradient is $[-1\ 2]$. In the general case, the gradient is $b_2 - A_2 x^i$, where i is associated with the maximizing expression of $f(u)$. Changing u in the direction of the gradient is a "proportional change" strategy: the changes in u_i are proportional to the violation or slack of constraint i.

4.4.7. Subgradients

When two or more segments meet, the lagrangean relaxation has alternative optimal solutions for a given set of dual prices, and the foregoing procedure for calculating the gradient is no longer well defined; alternative "gradients" exist. In fact, these alternatives are called *subgradients*, and any convex combination of these subgradients is also a subgradient. We generalize this informal definition to a formal definition of a subgradient for an arbitrary convex function.

Definition: A subgradient, s, at u^* satisfies

$$f(u^*) + s(u - u^*) \leq f(u) \quad \forall u$$

If $f(u)$ is concave, then the subgradient is defined by

$$f(u^*) + s(u - u^*) \geq f(u) \quad \forall u$$

Definition: The set of all subgradients is called a *subdifferential*.

Theorem (Geoffrion, 1974): The expression, $b_2 - A_2 x^i$, suggested earlier is a subgradient for the lagrangean dual if x^i is the unique solution to LR(u).

Before proving the theorem, we can verify that it holds in Figure 12. Observe that $f(u)$ is never below the solid line extending from the point u^1. Relabeling u^1 as u^*, this line corresponds to the expression on the left side of the subgradient definition. The inequality becomes strict, $f(u)$ rises above the line, as the segments change; the solution to the lagrangean relaxation changes as the curve changes segments.

Proof:

$$f(u^*) + (u - u^*)(b_2 - A_2 x^i)$$
$$= cx^i + u^*(b_2 - A_2 x^i) + (u - u^*)(b_2 - A_2 x^i)$$
$$= cx^i + u(b_2 - A_2 x^i) \leq \max_{x_i \in \text{conv}(A_1 x = b)} cx^i + u(b_2 - A_2 x^i)$$
$$= f(u) \quad \square$$

The subgradients for the example can be shown graphically. Figure 13 is divided into four regions, separated by bold lines, one for each lagrangean relaxation solution, x^i. The subgradient for each region, indicated by arrows, is determined by the associated solution. Where two or more regions meet, a set of subgradients exists. The figure also shows contours of constant objective function value: $z = 11, 12, 13$, and 20. The subgradients are shown as arrows orthogonal to the contours.

4.4.8. Step Size

Having found a direction of improvement, a subgradient, the remaining question is, "What step size should be taken in this direction?" We derive a particular formula (Held et al., 1974) that has been used heavily in practice. The derivation motivates the formula and helps illuminate the main pitfalls of implementing subgradient optimization.

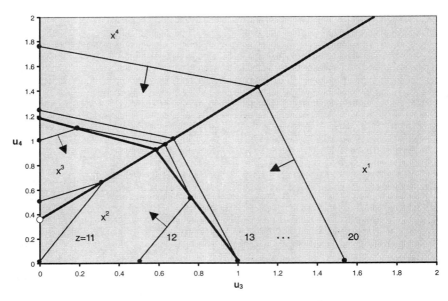

Figure 13 Lagrangean relaxation contours and subgradients.

First we normalize the subgradient:

$$\frac{1}{\|A_2 x^i - b_s\|^2} (A_2 x^i - b_2)$$

A step of this magnitude generates one unit of objective function improvement, assuming the subgradient remains unchanged as we change u. (In general, the subgradient will change, and, hence, this step is normalized only in a local sense.)

Next, assuming the direction is toward the optimal u, the ideal step size would be large enough to achieve the optimal u. Thus, the step should be scaled by the difference between the value of $z_{LR}(u)$ at step k, z^k_{LR}, and the estimated value of z_{LD}, z^*:

$$\frac{z^k_{LR} - z^*}{\|A_2 x^i - b_2\|^2} A_2 x^i - b_2$$

Finally, to compensate for the aforementioned assumptions and estimations, we add a parameter, $\pi^k > 0$, to adjust during subgradient optimization. Initially, $\pi^k = 2$. As subgradient optimization progresses π^k is systematically reduced to attain convergence. Adding π^k yields

$$\frac{\pi^k (z^k_{LR} - z^*)}{\|A_2 x^i - b_2\|^2} A_2 x^i - b_2$$

Denoting the scalar step by t^k, we now can write the formula for updating u. Subgradient optimization dual price update:

$$u^{k+1} = u^k + t^k (A_2 x^i - b_2)$$

$$t^k = \frac{\pi^k (z^k_{LR} - z^*)}{\|A_2 x^i - b_2\|^2}$$

These formulas are appropriate for equality constraints, but the formulas need to be adapted for inequalities. In particular, suppose a redundant constraint is added to the formulation. In this case, the dual price should be 0 and the constraint will have slack.

$$b_2 - A_2 x^i > 0$$

The previous update formula will force the dual price to a negative value. However, the dual price is restricted to be nonnegative. Thus the update formula for an inequality is

$$u^{k+1} = \max\{0, u^k + t^k (A_2 x^i - b_2)\}$$

Furthermore, the addition of redundant constraints should not affect the other constraints. If a constraint, c, has slack and its dual price is 0, the dual price will be unaffected by any step. The constraint is redundant and should not affect the other constraints, and therefore constraint c should not be included in

$$\|A_2 x^i - b_2\|^2$$

We demonstrate the iterations of subgradient optimization with the example. For the example, we have used the optimal integer solution to estimate $z^* = 8$.

Iter.	u_3	u_4	x	y	z_{LR}	$(A_2 x - b_2)_3$	$(A_2 x - b_2)_4$	$\|A_2 x - b_2\|^2$	π	t
1	0	0	4	1	11	−2	1	1	2	6
2	0	6	0	3	75	−4	−13	169	2	0.79
3	0	0	4	1	11	−2	1	1	1	3
4	0	3	0	3	36	−4	−13	169	1	0.40
5	0	0	4	1	11	−2	1	1	0.5	1.50
6	0	1.5	0	3	16.5	−4	−13	169	0.5	0.20
7	0	0	4	1	11	−2	1	1	0.25	0.75
8	0	0.75	4	2	11.5	1	−2	5	0.25	0.18
9	0.18	0.39	4	1	10.97	−2	1	5	0.25	0.15
10	0	0.54	4	2	11.08	1	−2	5	0.25	0.16
11	0.16	0.22	4	1	11.1	−2	1	5	0.25	0.16
12	0	0.38	4	2	10.76	1	−2	5	0.25	0.14
13	0.14	0.1	4	1	11.18	−2	1	5	0.25	0.16
14	0	0.26	4	1	10.74	−2	1	1	0.13	0.36
15	0	0.62	4	2	11.24	1	−2	5	0.13	0.08
16	0.08	0.46	4	2	10.88	1	−2	5	0.13	0.07
17	0.15	0.32	4	1	10.98	−2	1	5	0.13	0.08
18	0	0.40	4	2	10.8	1	−2	5	0.13	0.07
19	0.07	0.26	4	1	10.88	−2	1	5	0.13	0.07
	0	0.33	4	1	10.68					

4.4.9. Practical Implementation

Thus far, we have ignored implementation details such as choice of parameters in favor of illustrating the basic ideas. Indeed, the parameter settings and their effect on the rate of convergence matter less on simple problems than they do in typical applications of lagrangean relaxation. Specifically, in the above example, the reduction of π was ad hoc; when the method didn't seem to be converging, π was reduced. In addition, a larger value of z^* would have given faster convergence, with less adjustment of π. Also, the example is well scaled and no scaling was done. Finally, we terminated the iterations somewhat arbitrarily; the objective seemed to be improving

Integer Programming

slowly. The following paragraphs suggest more careful ways of controlling subgradient optimization.

Generally, π^k is initialized at 2 and $\pi^{k+1} = 0.5\pi^k$ if the best objective function value

$$z_{LR}^k = \min_{k' \leq k} z_{LR}(u^{k'})$$

has not improved in some fixed number of iterations. The number of iterations depends on the problem size and tuning this parameter requires experimentation. Typical values are in the range 5–30 iterations. (The ad hoc tuning in the example above averaged 3.3 iterations per π^k value.) If π^k is too large, convergence may be slow. If π^k is too small, subgradient optimization may converge to a non-optimal solution.

Ideally, z^* will exactly equal z_{LD}. In practice, z_{LD} is unknown and must be estimated. Frequently, z^* is estimated using the best known feasible solution. By definition, z^* certainly will not be too large, $z^* \leq z_{LD}$. However, this value may be too small as it was in the example. If z^* is too small, subgradient optimization will converge more slowly. Although reducing π^k could compensate for this, π^k is not adjusted every iteration; an accurate choice of z^* will more smoothly reduce the step size. This is evident in the example: $(z_{LR} - z^*) = 2.68$ in the final iteration when it would be 0.1 if we accurately estimated $z^* = 10.67$. If convergence is slow, it is appropriate to increase z^*. If z^* is too large, subgradient optimization may converge to a nonoptimal solution. If this is suspected, it is appropriate to decrease z^*. If z^* is adjusted, either up or down, then π^k should be reset to 2, to restart convergence.

The scaling of the dualized constraints can strongly affect the steps. For example, suppose we multiplied constraint 3 by 10:

$$20x + 30y \leq 130$$

The objective of the lagrangean relaxation becomes:

$$\max 3x - y + u_3(130 - 20x - 30y) + u_4(4 - 2x + 3y)$$

The function, $z_{LR}(u)$ becomes:

$$z_{LR}(u) = \max\{0 + 130u_3 + 4u_4,\ 11 + 20u_3 - 1u_4,\ 10 - 10u_3 + 2u_4,\ -3 + 40u_3 + 13u_4\}.$$

The first 12 iterations of subgradient optimization for the rescaled problem yields the following table.

Iter.	u_3	u_4	x	y	z_{LR}	$(A_2x - b_2)_3$	$(A_2x - b_2)_4$	$\|A_2x - b_2\|^2$	π	t
1	0	0	4	1	11	−20	1	1	2	6
2	0	6	0	3	75	−40	−13	169	2	0.79
3	0	0	4	1	11	−20	1	1	1	3
4	0	3	0	3	36	−40	−13	169	1	0.40
5	0	0	4	1	11	−20	1	1	0.5	1.50
6	0	1.5	0	3	16.5	−40	−13	169	0.5	0.20
7	0	0	4	1	11	−20	1	1	0.25	0.75
8	0	0.75	4	2	11.5	10	−2	104	0.25	0.00841
9	0.084	0.73	0	0	13.87	−130	−4	16916	0.25	0.0000868
10	0.073	0.73	0	0	12.41	−130	−4	16916	0.25	0.0000651
11	0.065	0.73	4	1	11.57	−20	1	401	0.25	0.00223
12	0.019	0.73	4	2	11.27	10	−2	104	0.25	0.00786

Notice that the first seven iterations are unaffected because the third constraint is satisfied during these iterations. The remaining iterations change dramatically because of the effect of the scaling on the step size, t. During these iterations, u_4 is nearly unchanged. The large coefficients cause the subgradient to emphasize improvement with respect to u_3. The result is that we have two levels of convergence: first converging with respect to u_3 and then with respect to u_4. However, while converging at the second level, subgradient optimization can fall back to the first level, effectively restarting and leading to very slow convergence. In the foregoing example, we started at the second level (improving u_4), reduced π for the second level (to 0.25), and then fell back to the first level (improving u_3), where π was far too small.

There are two schools of thought about scaling. One approach is to achieve "natural" scaling. This qualitative notion includes consistent units, few small or large coefficients, and ease of readability. The alternative approach is to use a generic formula to normalize the constraints. One example of a generic formula is to scale the right-hand sides to equal one. Another example is to treat each row as a vector and scale the vectors to a length of one. Our experience has been better with natural scaling, even though the concept is not well defined. The example used thus far has natural scaling: the coefficients are simple and of similar magnitude. If the nonzero right-hand sides were scaled to 1 or the rows were treated as vectors, the coefficients would be far less reasonable.

The ideal termination condition is

$$z_{LR}(u) = z_{IP}$$

This condition can be attained only when

Integer Programming

$$z_{LD} = z_{IP}$$

Otherwise, even the optimal dual prices will not cause subgradient optimization to terminate. In the example, the optimal dual prices are $u_3 = 0$, and $u_4 = 0.33$. With these dual prices, the solution $x = 4$, $y = 1$ and the solution $x = 4$, $y = 2$ are both optimal in the lagrangean relaxation. Either solution will also lead to changing the dual prices.

Failing an optimal integer solution, the only reasonable termination condition is when progress slows sufficiently. The π parameter is a natural measure of slowing progress: each time the objective function fails to improve, π is reduced. A good termination condition is that π is sufficiently small, for example, $\pi \leq 0.005$.

The suggested update formula has been proved to converge to the lagrangean dual when

$$\sum_{k=0}^{\infty} t^k \to \infty, \qquad t^k \to 0 \text{ as } k \to \infty$$

In practice, the convergence may be unacceptably slow. Typically, the improvement in the best bound generated thus far is rapid during early iterations but slows in later iterations. Alternative step size formulas to the one discussed here do exist. However, these formulas do not have a proof of convergence.

4.4.10. Branch and Bound

Adapting branch and bound to use subgradient optimization to generate the bound requires some modification. Instead of fractional variables that require branching, there are constraints, i, that lead to a duality gap:

$$DG = \sum_{i} DG_i$$

$$DG_i = |u_i(b_2 - A_2x)_i| > 0$$

One strategy is to branch on variables in the constraint with the largest DG_i value and thereby reduce the duality gap.

4.4.11. Advantages and Disadvantages

The primary advantage of subgradient optimization over column generation is that the individual iterations are faster than solving a linear program. If the numbers of iterations in both methods are comparable, subgradient optimization should be more efficient.

The disadvantages of subgradient optimization stem chiefly from failure to retain previous primal solutions. First, the primal solutions may oscillate between dramatically different solutions and thereby give little indication of the nature of an optimal solution. This is a handicap for initiating a heuristic or branch and bound based on the dual solution. Second, the lagrangean dual may never generate a good integer solution, even if $z_{LD} = z_{LP}$. Consider the following trivial example:

Maximize x

subject to

$x \le 3$

x integer and $0 \le x \le 10$

Dualizing the constraint creates the lagrangean relaxation:

Maximize $x + u(3 - x)$

subject to

x integer and $0 \le x \le 10$

The two extreme point solutions are $x = 10$ and $x = 0$. Neither is a useful solution to the original problem.

In contrast, a column generation model would find the optimal integer solution by solving:

Maximize $10\lambda_1$

subject to

$$10\lambda_1 \le 3$$
$$\lambda_1 + \lambda_2 = 1$$
$$0 \le \lambda_1, \lambda_2 \le 1$$

where λ_1 is the weight for the extreme point, $x = 10$, and λ_2 is the weight for the extreme point, $x = 0$. The solution to this linear program, $\lambda_1 = 0.3$, $\lambda_2 = 0.7$, maps into the integer solution, $x = 3$.

Subgradient optimization is often replaced by specialized methods that adjust dual prices using the structure of a given problem. These methods are called *multiplier adjustment methods*. Typically, exploiting special structure leads to a much faster algorithm. A related approach is to solve the dual of a linear program by the same kind of adjustments. Such methods are often referred to as *dual ascent* methods.

4.4.12. Exercises

 4.13. In the following problem, start with $u_1^* = u_2^* = 0$, and use the "proportional change" strategy to adjust the weights (Exercise 4.9), until an iteration is reached that provides a local (and hence global) optimum for the dual.

Integer Programming

$$\begin{array}{rl} \text{Minimize} & 10x_1 + 7x_2 \\ & 3x_1 - 2x_2 \le 2 \\ & -5x_1 + 3x_2 \le -3 \\ & x_1, \quad x_2 \in \{0, 1\} \end{array}$$

In solving this problem, identify the subgradient at each step, and explain the criterion you selected for the step size. What argument did you use to verify local optimality? [In practice, ascent procedures often stop when the step size or the change in $z_{LR}(u)$ on successive iterations becomes "negligible."]

4.14. The restriction $u \ge 0$ implies that it is not always possible to move in the direction of the subgradient because some parameter may then become negative. Can you argue intuitively on the basis of two- and three-dimensional examples (functions of one and two variables) that this will not lead to difficulties?

4.15. Consider the following argument in the context of Exercise 4.14. If some parameter is "blocked" at 0, then the problem temporarily reduces to one with fewer parameters (for which the ascent method is "valid"). Furthermore, if a local optimum is attained for this reduced problem, and the subgradient does not indicate that the blocked variable should be increased, this optimum must be global. What elaboration—if any—do you think this argument may need? Express this argument appropriately for the case when several variables may be blocked at 0 (simultaneously or alternately).

4.16. Show by reference to the following problem that the absence of an upper limit on u can lead to the case where $z_{LR}(u)$ has no finite maximum for $u \ge 0$. What interpretation does this have for the original problem? (Could you also support this interpretation without reference to the example?)

$$\begin{array}{rl} \text{Minimize} & 3x_1 + 5x_2 \\ & -5x_1 + 2x_2 \le -6 \\ & x_1, \quad x_2 \in \{0, 1\} \end{array}$$

4.17. (*Multiple subgradients at u^*: A potential hitch*) When more than one subgradient exists at a point u^*, it is entirely possible that one (or many) of them will *not* point into the improving region. Thus an ascent method that finds only such "misdirecting" subgradients may find it impossible to locate a true improving direction. Illustrate this with a two-dimensional drawing involving two subgradients d_1^* and d_2^*, such that the intersection of the "improving regions" $ud_1^* \geq u^*d_1^*$ and $ud_2^* \geq u^*d_2^*$ is an acute angle cone in the plane. (The point u^* is the vertex of this cone; i.e., it lies on the intersection of the hyperplanes $ud_1^* = u^*d_1^*$ and $ud_2^* = u^*d_2^*$. Why?) Since all improving values of u must lie in both half-spaces, they must lie in this cone. But show that the direction vectors d_1^* and d_2^* (orthogonal to the hyperplanes $ud_1^* = u^*d_1^*$ and $ud_2^* = u^*d_2^*$) do not point to the cone from the point u^* and hence will not lead to improved parameter values.

4.18. (*Overcoming the hitch*) In the graphical example for Exercise 4.17, identify a collection of half-spaces, all of whose hyperplanes pass through u^*, such that every point in the "improving region" of the cone lies within each of these half-spaces. Precisely characterize the full set of half-spaces with this property (relative to the example). Explain why each of these half-spaces (i.e., its associated d^*) qualifies as a subgradient. Finally, show that the direction vectors d^* for some of these half-spaces do point into the cone and thereby provide "reliable" subgradients that can be used by an ascent method. (The algebraic analog of these observations is discussed in the next two exercises.)

4.19. The set of half-spaces identified in Exercise 4.18 is the set of all nonnegative linear combinations excluding the 0 combination of the two half-spaces that define the cone. Demonstrate this by showing that the hyperplane corresponding to each such convex combination must pass u^* and that this hyperplane rotates from one of the "extreme" hyperplanes to the other as the parameters of the combination change their relative values. Further, show algebraically that each half-space thus obtained must contain all points in the cone (i.e., all points that satisfy both $ud_1^* \geq u^*d_1^*$ and $ud_2^* \geq u^*d_2^*$). Finally, show algebraically that no other half-space whose hyperplane goes through u^* (i.e., no other nonzero linear combination of the extreme half-spaces) can contain all points of the cone.

4.20. From the example of Exercises 4.18 and 4.19, show algebraically that the direction vector for the convex combination ($\frac{1}{2}$, $\frac{1}{2}$) — i.e., $\frac{1}{2}d_1^* + \frac{1}{2}d_2^*$ — must point into the cone from u^*. (Specifically, show that the point $u = u^* + hd$, for $d = \frac{1}{2}d_1^* + \frac{1}{2}d_2^*$, lies in the cone for all $h \geq 0$.) In general, it can be proved that one can always find a convex combination of half-spaces that define a polyhedral convex cone whose direction vector "points into" the cone. Describe the significance of this result if the lagrangean problem

$$\text{Minimize}_{x \in X} \quad f(x) + u^*g(x)$$

has more than one optimal solution x^*. [Note, it has been established that the subgradient $g(x^*)$, as x^* ranges over all optimal solutions to the lagrangean, are sufficient to define the cone of the "improving region."]

4.5. Selecting a Relaxation

Integer programming problems frequently offer several alternative lagrangean relaxations — alternative sets of constraints to relax. Frequently, it is not clear which constraints to relax. The main issues are the speed of solving the relaxations and the bound achieved by the lagrangean dual. The choice of a relaxation has connections to the choice of a formulation and polyhedral theory. Central to this discussion is the integrality property, which characterizes certain constraint sets.

4.5.1. Integrality Property

A constraint set has the *integrality property* if the linear program formed by adding any linear objective function is guaranteed to have an optimal integer solution. For example, all networks have the integrality property: the optimal flows in a network have integer values for any linear objective. (More precisely, every extreme point solution yields integer values for all flows.) The following theorem about the integrality property is influential in choosing a lagrangean relaxation.

Theorem (Geoffrion, 1974): If a lagrangean relaxation has the integrality property, then $z_{LD} = z_{LP}$.

Proof. Recall that

$$z_{LD} = \max_x cx$$
subject to
$x \in \text{conv}(A_1 x = b_1)$ and
$A_2 x = b_2$
$0 \le x \le 1$

$$z_{LP} = \max_x cx$$
subject to
$A_1 x = b_1$
$A_2 x = b_2$
$0 \le x \le 1$

If the lagrangean relaxation has the integrality property then $A_1 x = b_1$ has an integer solution for any objective function. In other words:

$$x \in \text{conv}(A_1 x = b_1) \Leftrightarrow x \in A_1 x = b_1$$

This completes the proof. □

4.5.2. Trade-offs

The reasons for using lagrangean relaxation instead of a linear programming relaxation are (1) to find an upper bound more quickly, or (2) to find a smaller upper bound, or both ideally, if rarely. If a lagrangean relaxation has the integrality property, the second reason is not satisfied and the first reason must justify lagrangean relaxation. This usually means that multiplier adjustment methods or subgradient optimization will be used to solve the lagrangean dual efficiently. If a lagrangean relaxation does not have the integrality property, the second reason is satisfied, but the concerns are that (1) each lagrangean will be too difficult to solve quickly, and (2) the improvement in the bound may be too small to justify the extra time to solve the relaxations.

4.5.3. Attributes of Typical Applications

Recall the generalized assignment problem with jobs, indexed by l, assigned to machines, indexed by k, described during column generation.

$$\text{Maximize} \quad \sum_{k,l} c_{k,l} x_{k,l}$$

subject to

$$\sum_{l} a_{k,l} x_{k,l} \leq b_k \quad \forall k$$

$$\sum_{k} x_{k,l} = 1 \quad \forall l$$

$$x_{k,l} \in \{0, 1\}$$

This problem is far more typical of lagrangean relaxation than the example we have been using. First, there are competing relaxations: either the first or the second constraint block may reasonably be relaxed. Second, either relaxation leads to separable subproblems that can be solved efficiently. Third, the original problem is sufficiently difficult for most instances to prevent solution by branch and bound using a linear programming relaxation. (In defense of the example, it is sufficiently small to demonstrate the theorems and algorithms, in some cases graphically.)

Dualizing the first constraint block of machine capacity constraints creates the lagrangean relaxation

$$\text{Maximize} \quad \sum_{k,l} c_{k,l} x_{k,l} + \sum_{k} u_k \left(b_k - \sum_{l} a_{k,l} x_{k,l} \right)$$

subject to

$$\sum_{k} x_{k,l} = 1 \quad \forall l$$

$$x_{k,l} \in \{0, 1\}$$

This problem separates into one subproblem for each job, indexed by l. Rearranging terms and omitting the constant in the objective, $\Sigma\ u_k b_k$, leads to subproblem(l):

$$\text{Maximize } \sum_k (c_{k,l} - u_k a_{k,l}) x_{k,l}$$

subject to

$$\sum_k x_{k,l} = 1$$

$$x_{k,l} \in \{0, 1\}$$

The solution to this problem is obvious: the $x_{k,l}$ with the largest coefficient equals 1 and all others equal 0. Hence, the lagrangean relaxation has the integrality property and therefore, $z_{LD} = z_{LP}$.

As we have already seen, dualizing the second constraint block (requiring each job to be assigned to exactly one machine) leads to separable subproblems: knapsack problems for each machine. This relaxation has $z_{LD} < z_{LP}$. Furthermore, each knapsack problem can be solved efficiently.

Separability of subproblems arises naturally in many models. Typically, the subproblems are nearly identical, differing only in time or location. The dualized constraints are those that reflect the connection of adjacent time periods or interactions among locations respectively.

4.5.4. Mixed Master Problems

If a lagrangean relaxation has separable subproblems and an integer solution to a set of subproblems guarantees that the integrality condition will be satisfied for the remaining subproblems, the following mixed approach to the master problem may be appropriate. Columns are generated as before for the subproblems without the integrality property. For the subproblems with the integrality property, the original variables are used in the master problem instead of extreme points. The master problem coefficients for these variables in the relaxed constraints are the original coefficients. The "convex combination" constraints for these subproblems in the master problem are unnecessary and are omitted. Instead, the master problem contains the original constraints from these subproblems. The advantage of this approach is that the master problem may be smaller if the subproblem is stated in terms of the original variables and constraints instead of the extreme points and the convex combination constraint.

One example in which a mixed approach to the master problem can be an improvement is problems with setup constraints:

$$y_{d,t} - x_{d,t} + x_{d,t-1} \geq 0 \quad \forall d,t$$

where $y_{d,t}$ is the setup variable for period t for device d, typically with an associated setup cost, $x_{d,t}$ is an on-off variable for the device, and both types of variables are binary variables. The constraint forces a setup in any period of transition in which the device switches from off to on. This constraint might be relaxed to separate the problem into different subproblems for each time period. The subproblems may contain additional constraints restricting the combined operation of the devices. The setup variables are unconstrained as a result of the relaxation.

A "pure" master problem would generate a subproblem for each of these variables. The subproblem has two extreme points: $y_{k,t} = 0$ and $y_{d,t} = 0$ and $y_{d,t} = 1$. The master problem would have one column for each extreme point and a convex combination constraint.

Instead, a "mixed" master problem could use the variable $y_{d,t}$ directly. The master problem constraints would be

$$y_{d,t} - \sum_i I_{d,t,i}\lambda_{t,i} + \sum_i I_{d,t-1,i}\lambda_{t-1,i} \geq 0 \quad \forall d,t$$

where the λ variables correspond to the extreme points of each time period and $I_{k,t,i}$ is the indicator of the value of $x_{d,t}$ at extreme point i of the subproblem. In this case, the formulation is clearly a better linear program: fewer constraints and variables. Furthermore, no effort is expended in solving a subproblem for the setup variables. In general, the mixed approach may be suitable when the integrality property holds, the number of constraints in the subproblem is small, and the number of extreme points exceeds the number of original variables.

4.5.5. Connections to Formulation

Lagrangean relaxation is closely related to the choice of formulation in three ways.

First, only the constraints in the formulation may be dualized. If a given formulation is difficult to solve and doesn't have an effective lagrangean relaxation, alternative formulations should be considered.

Second, constraints that would be redundant in the original problem may strengthen a lagrangean relaxation. In the example, the constraint

$$2x - 1.5y \leq 6.25$$

would be redundant in both the integer program and the linear programming relaxation: it is dominated by a combination of constraints 3 and 4. However, this constraint is not redundant in the lagrangean relaxation; constraints 3 and 4 have been dualized. The addition of this constraint

Integer Programming

improves z_{LR} from 10.67 to 9.6. Graphically, the solution moves from the point marked LR to the point marked RC in Figure 14. The change is caused by reducing the feasible region (eliminating the slashed region) of the lagrangean relaxation: the point (4,1) is no longer feasible.

Third, lagrangean relaxation is equivalent to replacing the nondualized constraints with the constraints defining their convex hull and solving the linear programming relaxation of this problem. If the lagrangean relaxations are easy to solve, these constraints may be easy to find. If so, it may be more efficient to solve a single linear program with these constraints than to solve a series of lagrangean relaxations without the constraints. If the convex hull has a large number of constraints, the most efficient option may be to add constraints only if they change the linear programming relaxation. Algorithms for detecting these *violated* constraints are called *fractional cutting plane algorithms*. Gomory's cuts (described in Section 2) were the first fractional cutting plane algorithms. One could interpret this addition of constraints as a form of improved formulation. In the example, lagrangean relaxation can be replaced by adding the constraint

$x \leq 4$

to the linear programming relaxation.

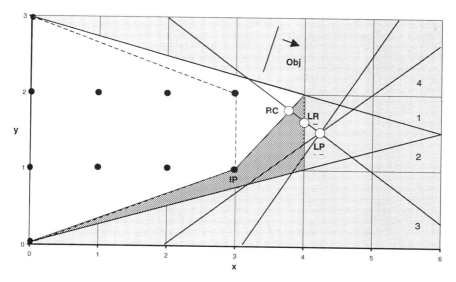

Figure 14 Lagrangean relaxation solutions.

4.5.6. Exercises

4.21. (*Selecting the constraints to discard.*) Consider the problem

$$
\begin{array}{rlrlrlrl}
\text{Minimize} & 3x_1 & +7x_2 & +5x_3 & +8x_4 & \\
& 9x_1 & +4x_2 & +8x_3 & +10x_4 & \leq 15 \\
& 11x_1 & +7x_2 & +9x_3 & +3x_4 & \leq 12 \\
& x_1 & +x_2 & & & = 1 \\
& & & +x_3 & +x_4 & = 1 \\
& x_1, & x_2, & x_3, & x_4 & \geq 0 \text{ and integer}
\end{array}
$$

Suggest all reasonable lagrangean relaxations for this problem and indicate the advantage of each relaxation. Can you predict how the bounds produced by these relaxations will compare to z_{IP} and z_{LP}?

4.6. Improvements

The bound produced by the lagrangean dual is dominated by two closely related techniques: *surrogate duality* (Glover, 1975a) and *lagrangean decomposition* (Glover and Klingman, 1988; Guignard and Kim, 1987a, 1987b). Surrogate duality generates an additional constraint for the relaxation by combining the dualized constraints. Lagrangean decomposition is a special case of lagrangean relaxation that combines the benefits of alternative lagrangean relaxations by creating duplicate variables for each relaxation.

4.6.1. Surrogate Duality

Surrogate relaxation, like lagrangean relaxation, relaxes a set of constraints. Instead of using dual multipliers to generate a term in the objective function, surrogate relaxation uses the multipliers to add a single constraint to the relaxed problem. Using the same notation as in lagrangean relaxation and again relaxing the second block of constraints lead to a surrogate relaxation.

$$
\text{SR} \quad
\begin{array}{ll}
\max_{x} & cx \\
\text{subject to} & uA_2 x = ub_2 \\
& A_1 x = b_1 \\
& x \in \{0, 1\}
\end{array}
$$

Let $z_{SR}(u)$ denote the solution to the surrogate relaxation. The advantage of surrogate relaxation is in the following theorem.

Integer Programming

Theorem (Greenberg and Pierskalla, 1970): $z_{IP} \leq z_{SR}(u) \leq z_{LR}(u)$. In words, surrogate relaxation provides a bound on the optimal integer solution no worse than that of lagrangean relaxation.

Proof. The left-hand inequality holds because any solution to the integer program is feasible in the surrogate relaxation, and both problems have the same objective function. The right-hand inequality holds because the surrogate constraint can be derived from optimal weights for the lagrangean (since then the lagrangean is a direct relaxation of the surrogate problem), and the choice of an optimal surrogate constraint can only strengthen the inequality. □

The surrogate dual is defined to be

$$z_{SD} = \min_u z_{SR}(u)$$

As with the lagrangean dual, the surrogate dual evaluates the best relaxation.

We apply surrogate duality to the example, with $u_3 = 0.25$ and $u_4 = 0.75$. (In general, the combination of inequalities into a single inequality can be a convex combination: the dual prices sum to one.)

Maximize $3x - y$

subject to

$$2x - 1.5y \leq 6.25$$
$$(1) \quad x + 4y \leq 12$$
$$(2) \quad x - 4y \leq 0$$
$$x, y \geq 0, \text{ integer}$$

The optimal solution is $x = 4$, $y = 2$, $z_{SR} = 10$. This is also the optimal solution of the surrogate dual, $z_{SD} = 10$. This is the point labeled SD in Figure 11.

The structure of the surrogate dual is quite different from the structure of the lagrangean dual. Instead of being a continuous piecewise linear function of u, the function is discontinuous and piecewise constant. The constant regions correspond to integer solutions of the relaxed problem. In the example, dual values of $u_3 = 0.3$ and $u_4 = 0.7$ yield the same surrogate dual solution with a different surrogate constraint.

$$2x - 1.2y \leq 6.7$$

Optimization of the surrogate dual has received less attention than that of the lagrangean dual. An effective method for finding an optimal solution to the surrogate dual exists when only two constraints are relaxed. This method can be applied to the general case of multiple relaxed constraints by subdividing the relaxed constraints into two groups. The multipliers of each group are used to produce two "surrogate source" constraints, to which the method can then be applied.

Both the column generation and subgradient optimization methods for the lagrangean dual may be used. Of these, subgradient optimization would be more appropriate: the master problem converges to the lagrangean dual, not the surrogate dual. Surrogate duality theory shows that the subgradient is defined differently than for the lagrangean. However, this difference has not to our knowledge been exploited.

This theory also applies to combined surrogate-lagrangean relaxations. In a combined relaxation both a surrogate constraint and a penalty term in the objective are created, typically using different multipliers. Combined relaxations are stronger than either the surrogate or the lagrangean taken separately. In the example we intentionally chose the surrogate constraint to equal the redundant constraint of the previous section. Optimization with the redundant constraint, in this case, is equivalent to combining a surrogate constraint with lagrangean relaxation. The combination improves the bound over pure surrogate duality from 10 to 9.6. Limited experimentation suggests the difference between the combined relaxation and the surrogate relaxation is small.

In addition, the device of generating an additional surrogate constraint is, of course, stronger still. The key is generally to seek different surrogate constraints that represent different subsets of constraints with particular structures.

The primary disadvantage of the surrogate dual is that the extra constraint destroys properties such as the integrality property that often motivate relaxation in the first place. It also can remove convenient structural features, as when lagrangean relaxation leads to separable problems. The additional constraint means that the problem is no longer separable.

Yet there are significant applications in which the surrogate dual is relevant. One is where the surrogate subproblems consist of networks with a single additional (surrogate) constraint, because effective procedures have been designed for such problems. Another is where multiple surrogate constraints are generated to capture different types of information from the original constraints, and these surrogate constraints are used to generate tests for compelling zero-one variables to receive particular values. Surrogate constraints can also be used to generate source inequalities for generating cutting planes, again applying methods previously discussed.

Integer Programming

4.6.2. Lagrangean Decomposition

Lagrangean decomposition (Glover and Klingman, 1988; Guignard and Kim, 1987a, 1987b) is appropriate when a problem has two or more competing relaxations. The method converts the problem to a new problem and then applies lagrangean relaxation. The lagrangean relaxation will generate a bound that equals or improves each of the competing relaxations. The steps are as follows:

1. Create a separate copy of each variable for each relaxation.
2. Add constraints equating the copies.
3. Relax these constraints.
4. Solve each relaxation.
5. Update the dual prices based on the difference between the copies.

The lagrangean dual of this problem is no greater than any of the individual relaxations. In fact, lagrangean decomposition is appropriate when it leads to a value strictly lower than each of the individual relaxations.

Using notation similar to that used for lagrangean relaxation, the original problem can be represented as follows:

$$
\begin{aligned}
\text{IP} \quad \max_{x} \quad & cx \\
\text{subject to} \quad & A_1 x = b_1 \\
& A_2 x = b_2 \\
& A_3 x = b_3 \\
& A_4 x = b_4 \\
& x \in \{0, 1\}
\end{aligned}
$$

Relaxation A is defined by relaxing blocks 2 and 3 and relaxation B is defined by relaxing blocks 2 and 4. Let x_A and x_B denote the vector of variables for relaxation A and relaxation B, respectively, and rewrite the original problem:

$$
\begin{aligned}
\text{IP} \quad \max_{x} \quad & cx_A \\
\text{subject to} \quad & A_1 x_A = b_1 \\
& A_1 x_B = b_1 \\
& A_2 x_A = b_2 \\
& A_3 x_B = b_3 \\
& A_4 x_A = b_4 \\
& x_A - x_B = 0 \\
& x \in \{0, 1\}
\end{aligned}
$$

Block 2 has been grouped with x_A above. This is arbitrary and does not affect the final solution because this block is relaxed in both relaxations. Similarly, the original objective, c, is arbitrarily grouped with x_A above.

The relaxations are

$$\text{LR}_A \quad \max_x \quad cx_A + u_2(b_2 - A_2 x_A) - u_5 x_A$$
$$\text{subject to} \quad A_1 x_A = b_1$$
$$A_4 x_4 = b_4$$
$$x_A \in \{0, 1\}$$

and

$$\text{LR}_B \quad \max_x \quad u_5 x_B$$
$$\text{subject to} \quad A_1 x_B = b_1$$
$$A_3 x_B = b_3$$
$$x_B \in \{0, 1\}$$

The same integer program used to illustrate lagrangean relaxation also illustrates lagrangean decomposition. In relaxation A, constraint 3 (block 3) is relaxed. In relaxation B, constraint 4 (block 4) is relaxed. (With these two relaxations, there are no constraints like block 2.) In general, block 1 or 2 may not exist.

The integer program with the copied variables is given below.

IP Maximize $3x_A - y_A$
   subject to
   (1a) $x_A + 4y_A \leq 12$
   (1b) $x_B + 4y_B \leq 12$
   (2a) $x_A - 4y_A \leq 0$
   (2b) $x_B - 4y_B \leq 0$
   (3) $2x_B + 3y_B \leq 13$
   (4) $2x_A - 3y_A \leq 4$
   (5) $x_A - x_B = 0$
   (6) $y_A - y_B = 0$
   $x_A, y_A \geq 0$, integer
   $x_B, y_B \geq 0$, integer

Relaxing constraints 5 and 6 yields:

Integer Programming

LR_A Maximize $3x_A - y_A - u_5 x_A - u_6 y_A$
subject to
(1) $x_A + 4y_A \leq 12$
(2) $x_A - 4y_A \leq 0$
(4) $2x_A - 3y_A \leq 4$
$x_A, y_A \geq 0$, integer

and

LR_B Maximize $u_5 x_B + u_6 y_B$
subject to
(1) $x_B + 4y_B \leq 12$
(2) $x_B - 4y_B \leq 0$
(3) $2x_B + 3y_B \leq 13$
$x_B, y_B \geq 0$, integer

Figure 15 illustrates the feasible regions for LR_A (marked by forward slashes) and LR_B (marked by backward slashes). The intersection of these regions is the feasible region for lagrangean decomposition (marked by a crosshatch pattern). Both relaxations A and B, individually, are stronger than the original lagrangean relaxation: fewer constraints are relaxed: $z_A =$

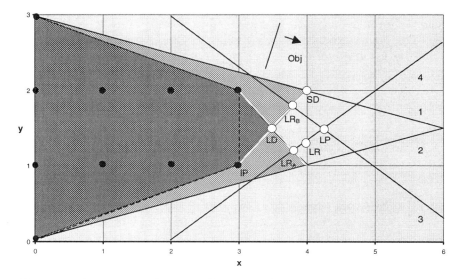

Figure 15 Lagrangean decomposition example.

9.6 and $z_B = 10.2$ versus $z_{LR} = 10.66$. Figure 15 illustrates the solutions with points marked LR_A and LR_B. Combined, lagrangean decomposition is even stronger, as we shall show.

As before, the lagrangean dual may be solved by either column generation or subgradient optimization. If column generation is used, the separable subproblem version, described earlier, can be chosen. The relationship between the lagrangean dual and the convex hull is specialized for lagrangean decomposition in the following theorem.

Theorem:

$$z_{LD} = \max_x cx$$

subject to

$$x \in \text{conv}(A_1 x = b_1, A_4 x = b_4)$$
$$x \in \text{conv}(A_1 x = b_1, A_3 x = b_3)$$
$$A_2 x = b_2$$
$$0 \leq x \leq 1$$

Proof: The two convex hulls represent the separable subproblems and together comprise the convex hull of the lagrangean relaxation. □

One consequence of the theorem is that if the integrality property holds for one of the relaxations, that relaxation may be omitted from the decomposition. If there are only two competing relaxations, omitting one leads to simple lagrangean relaxation.

The theorem implies that we can calculate the lagrangean dual by generating the master problem. The master problem for the example is:

Maximize $\quad 10\lambda_2 + 8\lambda_3 - 3\lambda_4$
subject to

$$4\lambda_2 + 3\lambda_3 \qquad\qquad -3\gamma_2 - 4\gamma_3 \qquad = 0$$
$$2\lambda_2 + \lambda_3 \quad 3\lambda_4 \quad -2\gamma_2 - \gamma_3 - 3\gamma_4 = 0$$
$$\lambda_1 + \lambda_2 + \lambda_3 + \lambda_4 \qquad\qquad\qquad = 1$$
$$\gamma_1 + \gamma_2 + \gamma_3 + \gamma_4 = 1$$
$$\lambda_i, \quad \gamma_i \quad \geq 0$$

The first and second constraints require each convex combination to have the same x and y values, respectively. The third and fourth constraints require a convex combination of the solutions to each subproblem to be used. The coefficients for each column are the x and y values from the subproblem solution that generates the column. (The coefficients of the B subproblem are multiplied by -1.) If the example had a block of type 2,

Integer Programming

the master problem would have additional constraints. These constraints have the same form as in lagrangean relaxation.

$$\sum_i \lambda_i (A_2 x^i) + \sum_j \lambda_j (A_2 x^j) = b_2$$

The solution to the master problem (omitting unused columns) is $\lambda_2 = 0.5$, $\lambda_3 = 0.5, \gamma_2 = 0.5$, $\gamma_3 = 0.5$. The objective is $z_{LD} = 9$ and is labeled LD in Figure 11. The corresponding solution in the original variables is $x = 3.5$, $y = 1.5$. While the bound from lagrangean decomposition exceeds the optimal integer solution, $z_{IP} = 8$, it does improve on both individual relaxations in isolation, $z_A = 9.6$, $z_B = 10.2$.

4.7. References

Several good surveys of lagrangean relaxation exist: Beasley (1993), Fisher (1981, 1985), Geoffrion (1974), and Shapiro (1979b). Geoffrion (1976a) suggests that lagrangean relaxation has value as a tool for model validation.

Held et al. (1974) provide more detail on subgradient optimization.

Surrogate duality is explored in more detail in Glover (1975a), Greenberg and Pierskalla (1970), Karwan and Rardin (1979), and Rardin and Unger (1976). Glover et al. (1979) describe alternative relaxations.

Lagrangean decomposition was developed separately by Glover and Klingman (1988) and Guignard and Kim (1987a, 1987b).

5. APPLICATIONS

A comprehensive summary of integer programming applications is beyond our scope, but we do suggest a select guide of integer programming applications in the areas of capacity expansion, facility location, logistics, assembly line operation, production planning, lot sizing, process planning, and scheduling. In general, the selected applications have been solved by the methods described in this chapter. These references are a cross section of surveys, classic papers, applications, and recent research.

Luss (1982) modeled the fixed charge aspects of capacity expansion. Hiller and Shapiro (1986) incorporated learning effects. Sahinidis and Grossmann (1992) applied reformulation techniques to expansion of chemical processes. Kuby et al. (1995) considered expansion in the context of a limited-capacity transportation network. Weintraub et al. (1994) studied expansion as part of forest planning.

Love et al. (1988) provide an overview of facilities location in general and integer programming models in particular. Polyhedral theory has been applied to this area. Both Cornuejols and Thizy (1982) and Leung and Magnanti (1989) defined facets of the simple plant location problem. Additional valid inequalities and facets have been derived by Aardal et al. (1995). Many challenging problems exist in telecommunications network expansion. Magnanti et al. (1995) provide one example of applying polyhedral theory in this field.

Many areas of logistics require integer programming. Arntzen et al. (1995) studied total supply chain management. Purchasing is a related and frequently overlooked area. Two efforts in this area are described by Katz et al. (1994) and Bender et al. (1985).

Distribution has a long and continuing history. Geoffrion and Graves (1974) applied Benders decomposition to a distribution problem. Geoffrion (1976a) also wrote a less formal description. More recently, Robinson et al. (1993) solved a distribution problem. Vehicle routing by column generation methods is a promising area. For example, see Shapiro (1993) and Ribeiro and Soumis (1994).

Balancing assembly lines minimizes worker idle time and continues to be a problem of interest. Baybars (1986) presents a survey of exact methods. Deckro (1989) considered problems of simultaneously determining the cycle time of the assembly line and the number of workstations. Graves and Lamar (1983) considered a little studied problem with *nonidentical* candidate stations and hence an *unbalanced* workload. They solved the problem by branch and bound with a twist. Only one set of variables had the integrality condition relaxed. Thus each node of the branch and bound tree was an IP rather than an LP and generated a stronger bound with additional computation time.

Diaby (1993) describes an efficient algorithm for zero-one minimax problems. Assembly line balancing problems can be formulated in this way. Other potential applications are Benders decomposition and facility location.

The *just in time* approach to assembly line management uses kanban (cards) to manage the flow of material. Bitran and Chang (1987) considered the optimal number of kanban at each station.

Hax and Meal (1975) created hierarchical production planning, a formal method for approaching the different levels of detail involved in planing. Graves (1982) suggested a lagrangean relaxation approach, consistent with the hierarchical approach.

A considerable amount of integer programming research has taken place and continues in lot sizing. Both Shapiro (1993) and Baker (1993) provide overviews of select research. Research has considered several versions of the problem: capacitated vs. uncapacitated, single-item vs. multi-item, and various material flows. Pochet (1988) described valid inequalities for the capacitated single-item problem.

Lasdon and Terjung (1971) and Thizy and Van Wassenhove (1985) applied lagrangean relaxation to the capacitated multi-item problem. Barany et al. (1984) and Leung et al. (1989) described valid inequalities for this problem. Eppen and Martin (1987) applied the idea of variable redefinition to generate valid inequalities. These inequalities were extended by Jeroslow (1990). Trigeiro et al. (1989) added the notions of setup time and setup cost to the single-product problem, and they solved the problem with lagrangean relaxation and dynamic programming.

Afentakis et al. (1984) considered the uncapacitated problem in which the product has an assembly structure. The approach included lagrangean relaxation, subgradient optimization, branch and bound, and heuristics. Afentakis and Gavish (1986) extended this result to a general product structure.

Two other papers are related to the lot sizing problem. Speranza and Ukovich (1994) considered transportation vs. inventory costs on a single link. Magnanti and Vachani (1990) presented valid inequalities for the problem of production scheduling with changeover cost.

Planning in the process industries typically requires nonlinear programming or successive linear programming (e.g., Baker and Lasdon, 1985). In some cases nonconvexity requires the use of integer programming, as described in Shapiro (1993).

We divide scheduling applications of integer programming into three groups: machine scheduling, personnel scheduling, and project scheduling. A general reference on scheduling is Conway et al. (1967). Blazewicz et al. (1991) provide a survey of mathematical programming formulations for scheduling.

Potts and Van Wassenhove (1985) applied lagrangean relaxation, multiplier adjustment, and branch and bound to the single machine scheduling problem with the objective of minimizing total weighted tardiness. Applegate and Cook (1991) combined new valid inequalities with previously derived inequalities to solve job shop scheduling problems. In addition they used branch and bound and heuristics.

Many personnel scheduling problems may be solved as linear programs, but some require integer programming. Bechtold and Jacobs (1990) considered a problem with flexible break assignments. Franz and Miller (1993) focused on resolving the infeasibility inherent in many personnel scheduling problems.

Airlines have crew and plane scheduling problems and have been the subject of considerable integer programming efforts. One of the earlier papers on crew planning was by Marsten et al. (1979). More recently, Barnhart et al. (1995) applied column generation to the problem, and they reference other work on the problem. Another interesting airline problem is the aircraft loading problem; Ng (1992) suggests an integer goal programming approach.

Project scheduling is normally a network flow problem. However, the addition of resource constraints leads to an integer programming problem. Christofides et al. (1987) approached this problem with four separate relaxations for use in branch and bound, including a linear programming relaxation and a lagrangean relaxation.

Project selection also requires integer programming. Hall et al. (1992) describe one application.

In general, any LP can become an MIP when additional constraints are imposed. A good example of this is adding resource constraints to the assignment problem, which has application in such diverse areas as personnel scheduling, job shop loading, and facility location. Mazzolla and Neebe (1986) suggest a branch and bound approach using subgradient optimization to bound a lagrangean relaxation.

6. CONCLUSIONS

As emphasized in the discussions and illustrations of the preceding sections, practical applications of pure and mixed-integer programming abound. From an engineering perspective, such applications center about determining a best design or a most effective utilization of resources, as commonly encountered in problems in such areas as mechanical, chemical, geophysical, architectural, industrial, and computer engineering. Additional recent applications are emerging in biochemical areas such as biomolecular design. From an economic perspective, the issues of accommodating appropriate planning horizons and of heeding the cost and revenue implications of alternative "solutions" are embedded in the fabric of intelligent integer programming models.

The technology for solving integer programming problems has come a long way during the past two decades, spurred by advances in computer technology but no less importantly by many theoretical and practical discoveries that have led to developing improved methods. State-of-the-art software from commercial vendors is greatly superior to what it was even one or two years ago (although not all vendors offer software of comparable quality), and promises to continue to improve. Many practical applications can be solved to optimality with a sound formulation and algorithm. Specifically, the practitioner must select and control algorithms based on a solid understanding of the algorithm rather than treating the algorithm as a black box (as is the custom in linear programming, for example). In addition, such as understanding will help practitioners recognize hard problems and realize when heuristics must be used.

Still, we emphasize that a long way remains to go for achieving the goal of routinely obtaining optimal solutions to practical IP and MIP problems in acceptable amounts of time. Because of the inherent difficulty of these problems, we may always expect to encounter instances that resist the assaults of our best solution procedures. One of the promising avenues for responding to this situation is to marry *exact* algorithms (that guarantee optimal solutions, although perhaps in galactic time) with *heuristic* procedures (which are more flexible, but without such guarantees). Theoretically, under many conditions it is just as hard to find a good "near-optimal" solution as it is to find an optimal solution. However, experience suggests that the universe of practical problems may be sufficiently benign to allow considerable strides toward the goal of solving problems *approximately*. The integration of exact methods with heuristic methods may well work to the benefit of each type of approach.

Ultimately, it must be remembered that solution methods are not the whole story in the integer programing domain. The form of the models selected is absolutely critical—not only for the ultimate "solvability" of a problem but also for capturing its essential features in a useful way, potentially leading to improved insights. The intimate links between modeling and problem solving, as we have undertaken to illustrate in the preceding sections, are the core to achieving the most successful applications of integer programming.

REFERENCES

Aardal, K., Pochet, Y., Wolsey, L. A., "Capacitated Facility Location: Valid Inequalities and Facets," *Mathematics of Operation Research*, Vol. 20, No. 3, pp. 562–582, 1995.

Adams, W. P., Sherali, H. D., "Linearization Strategies for a Class of Zero-One Mixed Integer Programming Problems," *Operations Research*, Vol. 38, No. 2, pp. 217–226, 1990.

Adams, W. P., Sherali, H. D., "Mixed-Integer Bilinear Programming Problems," *Mathematical Programming*, Vol. 59, No. 3, pp. 279–306, 1993.

Afentakis, P., Gavish, B., "Optimal Lot-Sizing Algorithms for Complex Product Structures," *Operations Research*, Vol. 34, No. 2, pp. 237–249, 1986.

Afentakis, P., Gavish, B., Karmarkar, U., "Computationally Efficient Optimal Solutions to the Lot-Sizing Problem in Multistage Assembly Systems," *Management Science*, Vol. 30, No. 2, pp. 222–239, 1984.

Applegate, D., Cook, W., "A Computational Study of the Job-Shop Scheduling Problem," *ORSA Journal on Computing*, Vol. 3, No. 2, pp. 149–156, 1991.

Arntzen, B. C., Brown, G. G., Harrison, T. P., Trafton, L. L., "Global Supply Chain Management at Digital Equipment Corporation," *Interfaces*, Vol. 25, No. 1, pp. 69–93, 1995.

Baker, K. R., "Requirements Planning," in *Handbooks in Operations Research and Management Science*, 4, *Logistics of Production and Inventory*, ed. S. C.

Graves, A. H. G. Rinnooy Kan, and P. H. Zipkin, pp. 571–627. North-Holland, Amsterdam, 1993.

Baker, T. E., Lasdon, L. S., "Successive Linear Programming at Exxon," *Management Science*, Vol. 31, No. 3, pp. 264–274, 1985.

Balas, E., "Disjunctive Programming," *Annals of Discrete Mathematics*, Vol. 5, pp. 3–51, 1979.

Balas, E., Jeroslow, R. G., "Strengthening Cuts for Mixed Integer Programs," *European Journal of Operations Research*, Vol. 4, pp. 224–334, 1980.

Balas, E., Ceria, S., Cornuejols, G., "A Lift-and-Project Cutting Plane Algorithm for Mixed 0-1 Programs," *Mathematical Programming*, Vol. 58, pp. 295–324, 1993.

Balas, E., Ceria, S., Cornuejols, G., "Mixed 0-1 Programming by Lift-and-Project in a Branch-and-Cut Framework," Carnagie Mellon University, Pittsburgh, PA, 1995.

Balinski, M. L., Spielberg K., "Methods for Integer Programming: Algebraic, Combinatorial and Enumerative," in *Progress in Operations Research, Relationship Between Operations Research and the Computer*, Vol. III, ed. J. S. Aranofsky, 195–292. Wiley, New York, 1969.

Barany, I., Van Roy, T. J., Wolsey, L. A., "Strong Formulations for Multi-item Capacitated Lot Sizing," *Management Science*, Vol. 30, No. 10, pp. 1255–1261, 1984.

Barnhart, C., Johnson, E. L., Nemhauser, G. L., Sigismondi, G., Vance, P., "Formulating a Mixed Integer Programming Problem to Improve Solvability," *Operations Research*, Vol. 41, No. 6, pp. 1013–1019, 1993.

Barnhart, C., Hatay, L., Johnson, E. L., "Deadhead Selection for the Long-Haul Crew Pairing Problem," *Operations Research*, Vol. 43, No. 3, pp. 491–499, 1995.

Baybars, I., "A survey of exact algorithms for the simple assembly line balancing problem," *Management Science*, Vol. 32, No. 8, pp. 909–932, 1986.

Beale, E. M. L., "Integer Programming," in *Computational Mathematical Programming*, ed. K. Schittkowski, 1–24. Springer-Verlag, Berlin, 1985a.

Beale, E. M. L., "The Evolution of Mathematical Programming," *Journal of the Operational Research Society*, Vol. 36, pp. 357–366, 1985b.

Beale, E. M. L., Tomlin, J. A., "Special Facilities in a General Mathematical Programming System for Non-convex Problems Using Ordered Sets of Variables," in *Proceedings of the 5th International Conference on Operations Research*, ed. J. Lawrence. Tavistock, London. 1969.

Beasley, J. E., "Lagrangean Relaxation," in *Modern Heuristic Techniques for Combinatorial Problems*, ed. C. R. Reeves, 243–298. Blackwell Scientific Publications, Oxford, 1993.

Bechtold, S. E., Jacobs, L. W., "Implicit Modeling of Flexible Break Assignments in Optimal Shift Scheduling," *Management Science*, Vol. 36, No. 1, pp. 1339–1351, 1990.

Bender, P. S., Brown, R. W., Isaac, M. H., Shapiro, J. F., "Improving Purchasing Productivity at IBM with a Normative Decision Support System," *Interfaces*, Vol. 15, No. 3, pp. 106–115, 1985.

Benichou, M., Gauthier, J. M., Girodet, P., Hentges, G., Ribiere, G., Vincent, O., "Experiments in Mixed-Integer Programming," *Mathematical Programming*, Vol. 1, pp. 76–94, 1971.

Benichou, M., Gauthier, J. M., Hentges, G., Ribiere, G., "The Efficient Solution of Large-Scale Linear Programming Problems Some Algorithmic Techniques and Computational Results," *Mathematical Programming*, Vol. 13, pp. 280–322, 1977.

Bitran, G. R., Chang, L., "A Mathematical Programming Approach to a Deterministic Kanban System," *Management Science*, Vol. 33, No. 4, pp. 427–441, 1987.

Blazewicz, J., Dror, M., Weglarz, J., "Mathematical Programming Formulations for Machine Scheduling: A Survey," *European Journal of Operational Research*, Vol. 51, pp. 283–300, 1991.

Boyd, E. A., "A Pseudopolynomial Network Flow Formulation for Exact Knapsack Separation," TR91-04, Rice University, Houston, 1991.

Boyd, E. A., "Polyhedral Results for the Precedence-Constrained Knapsack Problem," *Discrete Applied Mathematics*, Vol. 41, pp. 185–201, 1993.

Boyd, E. A., "Fenchel Cutting Planes for Integer Programs," *Operations Research*, Vol. 42, No. 1, January-February, pp. 53–64, 1994.

Bradley, S., Hax, A., Magnanti, T., *Applied Mathematical Programming*. Addison-Wesley, Reading, MA, 1977.

Christofides, N., Alvarez-Valdez, R., Tamarit, J. M., "Project Scheduling with Resource Constraints: A Branch and Bound Approach," *European Journal of Operational Research*, Vol. 29, pp. 262–273, 1987.

Chvátal, V., "Edmonds Polytopes and a Hierarchy of Combinatorial Problems," *Discrete Mathematics*, Vol. 4, pp. 305–337, 1973.

Conway, R. W., Maxwell, W. L., Miller, L. W., *The Theory of Scheduling*. Addison-Wesley, Reading, MA, 1967.

Cornuejols, G., Thizy, J. M., "Some Facets of the Simple Plant Location Polytope," *Mathematical Programming*, Vol. 23, pp. 50–74, 1982.

Crowder, H., Johnson, E. L., Padberg, M. W., "Solving Large Scale Zero-One Programming Problems," *Operations Research*, Vol. 31, pp. 803–834, 1983.

Deckro, R. F., "Balancing Cycle Time and Workstations," *IIE Transactions*, Vol. 21, No. 2, pp. 106–111, 1989.

Diaby, M., "Implicit Enumeration for the Pure Integer 0/1 Minimax Programming Problem, *Operations Research*, Vol. 41, No. 6, pp. 1172–1176, 1993.

Dietrich, B. L., Escudero, L. F., "New Procedures for Preprocessing 0-1 Models with Knapsack-like Constraints and Conjunctive and/or Disjunctive Variable Upper Bounds," INFOR 29, 305–317, 1989.

Dietrich, B. L., Escudero, L. F., Change, F., "Efficient Reformulation for 0-1 Programs, Methods and Computational Results," *Discrete Applied Mathematics*, Vol. 42, pp. 147–175, 1993.

Eppen, G. D., Martin, R. K., "Solving Multi-Item Capacitated Lot-Sizing Problems Using Variable Redefinition," *Operations Research*, Vol. 35, No. 6, pp. 832–848, 1987.

Everett, H., "Generalized Lagrange Multiplier Method for Solving Problems of

Optimum Allocation of Resources," *Operations Research*, Vol. 11, pp. 399–417, 1963.

Fisher, M. L., "The Lagrangian Relaxation Method for Solving Integer Programming Problems," *Management Science*, Vol. 27, pp. 1–18, 1981.

Fisher, M. L., "An Applications Oriented Guide to Lagrangian Relaxation," *Interfaces*, Vol. 15, No. 2, pp. 10–21, 1985.

Forrest, J. J. H., Hirst, J. P. H., Tomlin, J. A., "Practical Solution of Large Mixed Integer Programming Problems with UMPIRE," *Management Science*, Vol. 20, pp. 736–773, 1974.

Franz, L. S., Miller, J. L., "Scheduling Medical Residents to Rotations: Solving the Large-Scale Multiperiod Staff Assignment Problem," *Operations Research*, Vol. 41, No. 2, March-April, pp. 269–279, 1993.

Garey, M., Johnson, D., *Computers and Intractability*. Freeman, San Francisco, 1979.

Garfinkel, R. S., "Branch and Bound Methods for Integer Programming," in *Combinatorial Optimization*, ed. N. Christofides, pp. 1–20, 1979.

Gauthier, J. M., Ribiere, G., "Experiments in Mixed-Integer Linear Programming Using Pseudo-Costs," *Mathematical Programming*, Vol. 12, pp. 26–47, 1977.

Gendron, B., Crainic, T. G., "Parallel Branch-and-Bound Algorithms: Survey and Synthesis," *Operations Research*, Vol. 42, No. 6, November-December, pp. 1042–1066, 1994.

Geoffrion, A. M., "Lagrangean Relaxation for Integer Programming," *Mathematical Programming Study*, Vol. 2, pp. 82–114, 1974.

Geoffrion, A. M., "Better Distribution Planning with Computer Models," *Harvard Business Review*, Vol. 54, pp. 92–99, July-August, 1976a.

Geoffrion, A. M., "The Purpose of Mathematical Programming Is Insight, Not Numbers," *Interfaces*, Vol. 7, No. 1, pp. 81–92, 1976b.

Geoffrion, A. M., Graves, G. W., "Multicommodity Distribution System Design by Benders Decomposition," *Management Science*, Vol. 20, No. 5, pp. 822–844, 1974.

Glover, F., "Surrogate Constraint Duality in Mathematical Programming," *Operations Research*, Vol. 23, pp. 434–451, 1975a.

Glover, F., "Polyhedral Annexation in Mixed Integer and Combinatorial Programming," *Mathematical Programming*, Vol. 8, pp. 161–188, 1975b.

Glover, F., "Integer Programming and Combinatorics," in *Handbook of Operations Research*, Vol. 1, eds. J. J. Moder and S. E. Elmaghraby, pp. 120–146. Van Nostand Reinhold, New York, 1978.

Glover, F., Klingman, D., "Layering Strategies for Creating Exploitable Structure in Linear and Integer Programs," *Mathematical Programming*, Vol. 40, pp. 165–181, 1988.

Glover, F., Laguna, M., "Tabu Search," in *Modern Heuristics Techniques for Combinatorial Problems,*" ed. C. R. Reeves, pp. 70–150. Blackwell Scientific Publications, Oxford, 1993.

Glover, F., Løkketangen, A., "Probabilistic Tabu Search for Zero-One Mixed Integer Programming Problems," School of Business, University of Colorado, Boulder, 1994.

Glover, F., Sommer, D. C., "Pitfalls of Rounding in Discrete Management Decision Problems," *Decision Science*, Vol. 6, pp. 211-220, 1975.

Glover, F., Tangedahl, L., "Dynamic Strategies for Branch and Bound," *International Journal of Management Science*, Vol. 4, No. 5, pp. 571-576, 1976.

Glover, F., Karney, D., Klingman, D., "A Study of Alternative Relaxation Approaches for a Manpower Planning Problem," in *Quantitative Planning and Control*, eds. Y. Ijiri and A. B. Winston, pp. 141-164. Academic Press, New York, 1979.

Glover, F., Klingman, D., Phillips, N. V., "A Network-Related Nuclear Power Plan Model with an Intelligent Branch-and-Bound Solution Approach," *Annals of Operations Research*, Vol. 21, pp. 317-332, 1989.

Glover, F., Sherali, H. D., Lee, Y., "Generating Cuts from Surrogate Constraint Analysis for Zero-One and Multiple Choice Programming," presented at INFORMS National Meeting, Los Angeles, April 23-26, 1995.

Gomory, R. E., "Outline of an Algorithm for Integer Solutions to Linear Programs," *Bulletin of the American Mathematical Society*, Vol. 64, pp. 275-278, 1958.

Gomory, R. E., "Solving Linear Programming Problems in Integers," in *Combinatorial Analysis*, eds. R. E. Bellman and M. Hall, Jr., pp. 211-216. American Mathematical Society, 1969.

Gomory, R. E., "An Algorithm for Integer Solutions to Linear Programs," in *Recent Advances in Mathematical Programming*, eds. R. Graves and P. Wolfe, pp. 269-362. McGraw-Hill, New York, 1963a.

Gomory, R. E., "An All-Integer Programming Algorithm," in *Industrial Scheduling*, eds. J. R. Muth and G. L. Tompson, pp. 193-206. Prentice-Hall, Englewood Cliffs, NJ, 1963b.

Gomory, R. E., "Properties of a Class of Integer Polyhedra," in *Integer and Nonlinear Programming*, ed. J. Abadie, pp. 353-365, North Holland, Amsterdam, 1970.

Graves, S. C., "Using Lagrangean Techniques to Solve Hierarchical Production Planning Problems," *Management Science*, Vol. 28, No. 3, pp. 260-275, 1982.

Graves, S. C., Lamar, B. W., "An Integer Programming Procedure for Assembly System Design Problems," *Operations Research*, Vol. 31, No. 3, pp. 522-545, 1983.

Greenberg, H. J., Pierskalla, W. P., "Surrogate Mathematical Programs," *Operations Research*, Vol. 18, pp. 924-939, 1970.

Guignard, M., Kim, S., "Lagrangean Decomposition for Integer Programming: Theory and Applications," *Operations Research*, Vol. 21, pp. 307-323, 1987a.

Guignard, M., Kim, S., "Lagrangean Decomposition: A Model Yielding Stronger Lagrangean Bounds," *Mathematical Programming*, Vol. 39, pp. 215-228, 1987b.

Hall, N. G., Hershey, J. C., Kessler, L. G., Stotts, R. C., "A Model for Making Project Funding Decisions at the National Cancer Institute," *Operations Research*, Vol. 40, No. 6, pp. 1040-1052, 1992.

Hax, A. C., Meal, H. C., "Hierarchical Integration of Production Planning and

Scheduling," in *Studies in Management Sciences*, Vol. 1, *Logistics*, ed. M. A. Geisler, pp. 53-69. North Holland, Amsterdam Elsevier, New York, 1975.

Held, M., Wolfe, P., Crowder, H. P., "Validation of Subgradient Optimization," *Mathematical Programming*, Vol. 6, pp. 62-88, 1974.

Hiller, R. S., Shapiro, J. F., "Optimal Capacity Expansion Planning When There Are Learning Effects," *Management Science*, Vol. 32, No. 9, pp. 1153-1163, 1986.

Hoffman, K., Padberg, M., "LP-Based Combinatorial Problem Solving," in *Computational Mathematical Programming*, ed. K. Schittkowski, pp. 55-124. Springer-Verlag, Berlin, 1985.

Hoogeveen, J. A., Oosterhout, H., Van de Velde, S. L., "New Lower and Upper Bounds for Scheduling Around a Small Common Due Date," *Operations Research*, Vol. 42, No. 1, pp. 102-110, 1994.

Jeroslow, R. G., "Two Mixed Integer Programming Formulations Arising in Manufacturing Management," *Discrete Applied Mathematics*, Vol. 26, pp. 137-157, 1990.

Johnson, E. L., "Modeling and Strong Linear Programs for Mixed Integer Programming," in *Algorithms and Model Formulations in Mathematical Programming*, ed. S. W. Wallace, Springer-Verlag, New York, 1989.

Johnson, E. L., Powell, S., "Integer Programming Codes," in *Design and Implementation of Optimization Software*, ed. H. Greenberg, pp. 225-240. Sijthoff & Noordhoff, Alphen aan den Rijn, The Netherlands, 1978.

Karp, R. M., Papadimitriou, C. H., "On Linear Characterizations of Combinatorial Optimization Problems," *SIAM Journal on Computing*, Vol. 11, pp. 620-632, 1982.

Karwan, M. H., Rardin, R. L., "Some Relationships Between Lagrangean and Surrogate Duality in Integer Linear Programming, *Mathematical Programming*, Vol. 17, pp. 320-334, 1979.

Katz, P., Sadrian, A., Tendick, P., "Telephone Companies Analyze Price Quotations with Bellcore's PDSS Software," *Interfaces*, Vol. 24, No. 1, pp. 50-63, 1994.

Kelly, J. P., Osman, I., eds, *Metaheuristics for Combinatorial Optimization*, Kluwer, Boston; 1995.

Kuby, M., Qingqi, S., Watanatada, T., "Planning China's Coal and Electricity Delivery System," *Interfaces*, Vol. 25, No. 1, pp. 41-68, 1995.

Laguna, M., Kelly, J. P., Gonzalez-Velarde, J. L., Glover, F., "Tabu Search for the Multilevel Generalized Assignment Problem," *European Journal of Operational Research*, Vol. 82, pp. 176-189, 1995.

Land, A., Powell, S., "Computer Codes for Problems of Integer Programming," *Annals of Discrete Mathematics*, Vol. 5, pp. 221-269, 1979.

Lasdon, L. S., Terjung, R. C., "An Efficient Algorithm for Multi-item Scheduling," *Operations Research*, Vol. 19, pp. 946-969, 1971.

Leung, J. M. Y., Magnanti, T. L., "Valid Inequalities and Facets of the Capacitated Plant Location Problem," *Mathematical Programming*, Vol. 44, pp. 271-291, 1989.

Leung, J. M. Y., Magnanti, T. L., Vachani, R., "Facets and Algorithms for Capacitated Lot Sizing," *Mathematical Programming*, Vol. 45, pp. 331-360, 1989.

Love, R. F., Morris, J. G., Wesolowsky, G. O., *Facilities Location: Models and Methods*. Elsevier Science Publishing, New York, 1988.

Luss, H., "Operations Research and Capacity Expansion Problems: A Survey," *Operations Research*, Vol. 30, No. 5, pp. 907–947, 1982.

Magnanti, T. L., Vachani, R., "A Strong Cutting Plane Algorithm for Production Scheduling with Changeover Costs," *Operations Research*, Vol. 38, pp. 456–473, 1990.

Magnanti, T. L., Mirchandani, P., Vachani, R., "Modeling and Solving the Two-Facility Capacitated Network Loading Problem," *Operations Research*, Vol. 43, No. 1, pp. 142–157, 1995.

Marsten, R. E., Morin, T. L., "A Hybrid Approach to Discrete Mathematical Programming," *Mathematical Programming*, Vol. 14, No. 1, pp. 21–40, 1978.

Marsten, R. E., Muller, M. R., Killion, C. L., "Crew Planning at Flying Tiger: A Successful Application of Integer Programming," *Management Science*, Vol. 25, No. 12, pp. 1175–1183, 1979.

Martello, S., Toth, P., *Knapsack Problems: Algorithms and Computer Implementations*, Wiley, New York, 1990.

Martin, R. K., "Generating Alternative Mixed-Integer Programming Models Using Variable Redefinition," *Operations Research*, Vol. 35, No. 6, pp. 820–831, 1987.

Martin, R. K., Schrage, L., "Subset Coefficient Reduction for 0/1 Mixed-Integer Programming," *Operations Research*, Vol. 33, pp. 505–526, 1985.

Mazzola, J. B., Neebe, A. W., "Resource-Constrained Assignment Scheduling," *Operations Research*, Vol. 34, No. 4, pp. 560–572, 1986.

Murty, K. G., *Linear and Combinatorial Programming*. Wiley, New York, 1976.

Nemhauser, G. L., "The Age of Optimization: Solving Large-Scale Real-World Problems," *Operations Research*, Vol. 4, No. 1, pp. 5–13, 1994.

Nemhauser, G. L., Wolsey, L. A., *Integer and Combinatorial Optimization*. Wiley, New York, 1988.

Nemhauser, G. L., Savelsbergh, M. W. P., Sigismondi, G. C., "MINTO: A Mixed INTeger Optimizer," *Operations Research Letters*, Vol. 13, pp. 47–58, 1994.

Ng, K. Y. K., "A Multicriteria Optimization Approach to Aircraft Loading," *Operations Research*, Vol. 40, No. 6, November-December, pp. 1200–1205, 1992.

Padberg, M. W., "Covering, Packing, and Knapsack Problems," in *Discrete Optimization I*, eds. P. L. Hammer et al. *Annals of Discrete Mathematics*, Vol. 4, pp. 265–287, 1979.

Padberg, M., Rinaldi, G., "A Branch-and-Cut Algorithm for the Resolution of Large-Scale Symmetric Traveling Salesman Problems," *SIAM Review*, Vol. 33, pp. 60–100, 1991.

Padberg, M. W., Van Roy, T. J., Wolsey, L. A., "Valid Linear Inequalities for Fixed Charge Problems," *Operations Research*, Vol. 33, No. 4, pp. 842–861, 1985.

Papadimitriou, C. H., Steiglitz, K., *Combinatorial Optimization: Algorithms and Complexity*. Prentice-Hall, Englewood Cliffs, NJ, 1982.

Park, K., Park, S., "Lifting Cover Inequalities for the Precedence-Constrained

Knapsack Problem," Korea Advanced Institute of Science and Technology, 1994.

Parker, G., Rardin, R., *Discrete Optimization*. Academic Press, New York, 1988.

Pochet, Y., "Valid Inequalities and Separation for Capacitated Economic Lot Sizing," *Operations Research Letters*, Vol. 7, No. 3, pp. 109-115, 1988.

Potts, C. N., Van Wassenhove, L. N., "A Branch and Bound Algorithm for the Total Weighted Tardiness Problem," *Operations Research*, Vol. 33, No. 2, pp. 363-377, 1985.

Rardin, R. L., Unger, V. E., "Surrogate Constraints and the Strength of Bounds in 0-1 Mixed Integer Programming," *Operations Research*, Vol. 24, pp. 1169-1175, 1976.

Ribeiro, C. C., Soumis, F., "A Column Generation Approach to the Multiple-Depot Vehicle Scheduling Problem," *Operations Research*, Vol. 42, No. 1, pp. 41-52, 1994.

Robinson, E. P., Jr., Gao, L. L., Muggenborg, S. D., "Designing an Integrated Distribution System at Dowbrands, Inc., *Interfaces*, Vol. 23, No. 3, May-June, pp. 107-117, 1993.

Rosander, R. R., "Multiple Pricing and Suboptimization in Dual Linear Programing Algorithms," *Mathematical Programming Study*, Vol. 4, pp. 108-117, 1975.

Sahinidis, N. V., Grossmann, I. E., "Reformulation of the Multiperiod MILP Model for Capacity Expansion of Chemical Process," *Operations Research*, Vol. 40, Suppl. 1, pp. S127-S144, 1992.

Salkin, H. M., Mathur, K., *Foundations of Integer Programming*, Elsevier, New York, 1989.

Saltzman, M. J., "Survey: Mixed Integer Programming," *OR/MS Today*, April, pp. 42-45, 1994.

Savelsbergh, M. W. P., "Preprocessing and Probing Techniques for Mixed Integer Programming Problems," *ORSA Journal on Computing*, Vol. 6, No. 4, pp. 445-454, 1994.

Schrijver, A., *Theory of Linear and Integer Programming*. Wiley, New York, 1986.

Shapiro, J. F., *Mathematical Programming, Structures and Algorithms*. Wiley, New York, 1979a.

Shapiro, J. F., "A Survey of Lagrangean Techniques for Discrete Optimization," in *Discrete Optimization II*, eds. P. L. Hammer, E. L. Johnson, B. H. Korte, *Annals of Discrete Mathematics*, Vol. 5, pp. 113-138, 1979b.

Shapiro, J. F., "Mathematical Programming Models for Production Planning and Scheduling," in *Handbooks in Operations Research and Management Science*, Vol. 4, *Logistics of Production and Inventory*, eds. S. C. Graves, A. H. G. Rinnooy Kan, P. H. Zipkin, pp. 371-443. North-Holland, Amsterdam, 1993.

Sherali, H. D., Lee, Y., "Sequential and Simultaneous Liftings of Minimal Cover Inequalities for GUB Constrained Knapsack Polytopes," *SIAM Journal on Discrete Mathematics*, Vol. 8, pp. 133-153, 1995.

Speranza, M. G., Ukovich, W., "Minimizing Transportation and Inventory Costs

for Several Products on a Single Link," *Operations Research*, Vol. 42, No. 5, pp. 879-894, 1994.
Subramanian, R., Scheff, R. P., Jr., Quillinan, J. D., Wiper, D. S., Marsten, R. E., "Coldstart: Fleet Assignment at Delta Air Lines," *Interfaces*, Vol. 24, No. 1, pp. 104-120, 1994.
Thizy, J. M., Van Wassenhove, L. N., "Lagrangian Relaxation for the Multi-Item Capacitated Lot-Sizing Problem," *IIE Transactions*, Vol. 17, No. 4, pp. 308-313, 1985.
Trigeiro, W. W., Thomas, L. J., McClain, J. O., "Capacitated Lot Sizing with Setup Times, *Management Science*, Vol. 35, No. 3, pp. 353-366, 1989.
Van Roy, T. J., Wolsey, L. A., "Valid Inequalities for Mixed 0-1 Programs," *Discrete Applied Mathematics*, Vol. 14, pp. 199-213, 1986.
Van Roy, T. J., Wolsey, L. A., "Solving Mixed Integer Programming Problems Using Automatic Reformation," *Operations Research*, Vol. 35, No. 1, pp. 45-57, 1987.
Weintraub, A., Jones, G., Magendzo, A., Meacham, M., Kirby, M., "A Heuristic System to Solve Mixed Integer Forest Planning Models," *Operations Research*, Vol. 42, No. 6, pp. 1010-1024, 1994.
Williams, H. P. *Model Building in Mathematical Programming*. Wiley, New York, 1993.
Wolsey, L. A., "Submodularity and Valid Inequalities in Capacitated Fixed Charge Networks," *Operations Research Letters*, Vol. 8, pp. 119-124, 1989a.
Wolsey, L. A., "Strong Formulations for Mixed Integer Programming: A Survey," *Mathematical Programming*, Vol. 45, pp. 173-191, 1989b.
Zionts, S. C., *Linear and Integer Programming*, Prentice-Hall, Englewood Cliffs, NJ, 1974.

4

Graph Theory and Networks

James R. Evans
University of Cincinnati, Cincinnati, Ohio

1. INTRODUCTION

Graph theory and network optimization are fields of mathematics and operations research that have wide practical application, not only in industrial engineering but also in diverse areas such as psychology, management, marketing, and transportation planning. In this part of the chapter we define some elementary concepts necessary to understand the models and algorithms that follow.

A *graph* G is a set X of points called *vertices* and a set E of pairs of vertices called *edges*. The graph G is denoted by (X, E). An example of a graph is shown in Figure 1. Here, $X = \{1,2,3,4,5\}$ and $E = \{(1,2),(1,3),(2,4),(3,4),(3,5),(4,5)\}$. Whenever E consists of unordered pairs of vertices, we have an *undirected graph*. Thus, in Figure 1, the edge $(1,2)$ could have been written as $(2,1)$. If the edges correspond to ordered pairs of vertices, we have a *directed graph*. In a directed graph, an edge (x, y) is represented by an arrow directed from vertex x to vertex y. Figure 2 shows a directed graph corresponding to the set E if the vertices are assumed to be ordered. We usually call vertices and edges in directed graphs *nodes* and *arcs*, respectively. A *network* is a graph with one or more numbers associated with each edge or arc. These numbers might represent distances, costs, reliabilities, or other relevant parameters.

Graph and network optimization is an important area of operations research and industrial engineering for several reasons. Graphs and net-

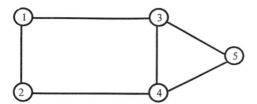

Figure 1 An undirected graph.

works can be used to model many diverse applications such as transportation systems, communication systems, vehicle routing problems, production planning, and cash flow analysis. They are accepted more readily by nontechnical people because they are often related to the physical system, which enhances model understanding. Finally, graphs and networks have special properties that allow solution of much larger problems than can be handled by any other optimization technique and are much more computationally efficient than other types of solution procedures.

1.1. Some Basic Concepts and Definitions

The field of graphs and networks is rich in terminology, which often differs from one author to another. To maintain some consistency, we present some fundamental concepts and definitions that will be used later.

A vertex and an edge are said to be *incident to one another* if the vertex is an endpoint of the edge. Thus, in Figure 1, edges (1,2) and (1,3) are incident to vertex 1. Two edges are said to be *adjacent* if they are both incident to the same vertex. In Figure 1, edges (1,2) and (1,3) are adjacent.

Suppose that $x, y, u \ldots v, z$ is any sequence of vertices. A *path* is any sequence of edges $(x,y), (y,u) \ldots (v,z)$ that connects the vertices. A *cycle* is a path whose initial and terminal vertices are identical. In Figure 1, the edges (1,2), (2,4), (4,3), (3,1) form a cycle.

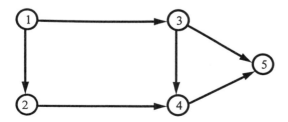

Figure 2 A directed graph.

Graph Theory and Networks

A graph is *connected* if there is a path joining every pair of distinct vertices in the graph. If a graph is not connected, it consists of two or more *connected components*. (A single node is considered a connected component.)

A graph is called a *tree* if it is connected and has no cycles. Figure 3 shows two examples of trees taken from the graph G in Figure 1. In Figure 3b, we see that all vertices of the original graph are included in the tree; such a tree is called a *spanning tree* of G. An edge of a spanning tree is called a *branch*; an edge in the graph that is not in the spanning tree is called a *chord*. A set of unconnected trees is called a *forest*.

Trees have some special properties. Note that a tree with one edge contains two vertices and a tree with two edges contains three vertices. In general, a tree with $m - 1$ edges contains m vertices. Since a tree is connected, there is at least one path between any pair of vertices. Suppose there was more than one. These paths would create a cycle. But a tree cannot have a cycle; therefore, there must be a unique path between any two vertices in a tree.

A *flow* is a way of moving objects through a network. Flows might represent a pumping of oil through a network of pipelines, the transportation of products from a factory to a warehouse, or the transmission of messages in a computer network. Flow networks have one or more nodes at which flows originate; these are called *source nodes*. Similarly, the nodes at which flows terminate are called *sink nodes*. Figure 4 shows an example of a flow network. Node 1 is a source node with a supply of 10; nodes 2 and 4 are sink nodes with demands of 5. The number next to each arc is the capacity of the arc. Capacities are simple upper bounds of the flow variables. If $f(x,y)$ is the amount of flow from node x to node y and $u(x,y)$ is the capacity of arc (x,y), we express the capacity with the constraint

$$f(x, y) \leq u(x, y)$$

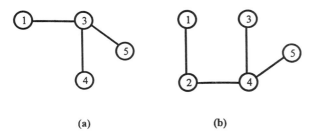

Figure 3 Examples of trees.

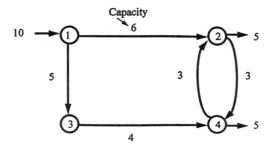

Figure 4 A flow network.

The fundamental equation governing flows in networks is known as *conservation of flow*. This means that at any node

Flow out − flow in = 0

Supplies at source nodes are considered as "flows in," while demands at sink nodes are treated as "flows out." For the network in Figure 4, the conservation of flow equations are

$$
\begin{aligned}
\text{node 1:} &\quad f(1,2) + f(1,3) - 10 = 0 \\
\text{node 2:} &\quad f(2,4) + 5 - f(1,2) - f(4,2) = 0 \\
\text{node 3:} &\quad f(3,4) - f(1,3) = 0 \\
\text{node 4:} &\quad f(4,2) + 5 - f(2,4) - f(3,4) = 0
\end{aligned}
$$

If we rewrite these equations, putting all constants on the right-hand side, we obtain

$$
\begin{aligned}
f(1,2) + f(1,3) &= 10 \\
f(2,4) - f(1,2) - f(4,2) &= -5 \\
f(3,4) - f(1,3) &= 0 \\
f(4,2) - f(2,4) - f(3,4) &= -5
\end{aligned}
$$

Written in this form, the conservation of flow equations have the following properties:

1. Each variable $f(x,y)$ appears in two equations, once with a $+1$ coefficient and once with a -1 coefficient.
2. A positive right-hand side value corresponds to a supply at a source; a negative value corresponds to a demand at a sink.

Graph Theory and Networks

Property 1 defines the *node-arc incidence matrix of the network*. A node-arc incidence matrix is one whose rows correspond to nodes and whose columns correspond to arcs. Arc (i,j) has a $+1$ in row i and a -1 in row j. The node-arc incidence matrix for the network in Figure 4 is

Node \ Arc	(1,2)	(1,3)	(2,4)	(3,4)	(4,2)
1	1	1			
2	-1		1		-1
3		-1		1	
4			-1	-1	1

Note that this is simply the coefficient matrix of the conservation of flow equations.

A flow is called *feasible* if it satisfies conservation of flow equations and any applicable capacity restrictions. Because the conservation of flow equations and capacity constraints are linear, most network flow problems can be expressed as linear programs. For the network in Figure 4, the objective function might be to minimize the cost of transporting the flow from the source to the sinks.

1.2. Graph and Network Optimization Problems

Many different types of problems can be expressed as optimization problems on graphs and networks. Space permits us only to discuss a few of them. In this chapter we describe models and optimization algorithms for four types of problems:

1. Spanning trees
2. Shortest paths
3. Minimum-cost flows
4. Arc routing

Spanning tree problems involve finding a minimum-cost spanning tree on a graph. A typical example is the optimal design of communication or pipeline systems. For example, a cable TV company might face the problem of connecting a set of homes or neighborhoods using the minimum amount of coaxial cable. If we represent the homes or neighborhoods as vertices and the possible connections among them as edges of a graph, the optimal solution would be a minimum-cost spanning tree.

Shortest-path problems involve finding a path from one vertex to another whose distance is a minimum. Such problems are commonly encountered in transportation, communication, and production applications.

For example, trucks that enter a state must file a route plan with the state government. They are not allowed to travel on roads with prohibitive weight restrictions. On the other hand, the trucker wants to reach his or her destination in the shortest possible time. We may construct a graph corresponding to the highway network and delete the edges on which the truck is prohibited. By knowing the mileage corresponding to each edge and the legal speed limit, we can compute the travel time on each edge. The shortest path from point of entry to the final destination would provide the quickest route.

Minimum-cost flow problems constitute one of the most useful classes of network optimization problems. In a minimum-cost flow problem, we have a directed network in which each arc has a cost per unit of flow and possibly arc capacities. Each node has a net supply or demand (which may be zero). The objective is to find a feasible flow that minimizes the total cost. Many problems in distribution, production scheduling, and service scheduling can be modeled as minimum-cost flow problems.

Arc routing problems involve finding paths or cycles that traverse a set of edges or arcs of a network. Many practical applications fall into this category. These include postal delivery, garbage collection, electric meter reading, and snow and ice removal. For example, a problem in postal delivery is to minimize the time or distance required to deliver to all homes in a residential neighborhood. This problem can be conceptualized by finding a cycle on a graph that crosses all the edges at least once in minimum total distance (starting at and returning to the post office). This problem is called the *Chinese postman problem* and is discussed in this chapter.

In the following sections we address each of these classes of problems. For each, we first present mathematical programming formulations and then present graph and network-based algorithms for solving the problem.

2. SPANNING TREE PROBLEMS

2.1. Mathematical Programming Approaches

One example of a minimum spanning tree problem is designing electrical connections on a printed circuit board. Figure 5 shows a layout of possible connections that do not interfere with other elements of the circuit board. The numbers by each edge represent the length of the connection — our measure of cost. We seek to connect all nodes of the network at minimum total cost.

Because each edge is either in the tree or not, we can think of this problem in terms of logical, zero-one variables. Therefore, we define $x_i = 1$ if edge i is in the tree and 0 otherwise.

Graph Theory and Networks

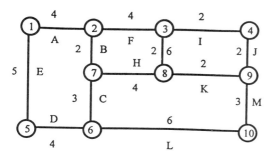

Figure 5 Circuit board connections.

Can we model this problem as a linear or integer program? First, notice that for any tree, there must be at least one branch connected to each node. This tells us that any solution must at least satisfy $\Sigma x_i \geq 1$, where the summation includes all edges connected to a particular node. For example, at node 1 we must have

$$x_A + x_E \geq 1$$

At node 7 we need

$$x_B + x_C + x_H \geq 1$$

and so forth.

Also, since a spanning tree contains exactly one less branch than the number of nodes, we must have exactly nine branches in the tree, or

$$x_A + x_B + x_C + x_D + x_E + x_F + x_G + x_H + x_I + x_J + x_K + x_L + x_M = 9$$

The objective is to minimize the cost of the edges included in the tree, or

$$\min 4x_A + 2x_B + 3x_C + 4x_D + 5x_E + 4x_F + 2x_G + 4x_H + 2x_I + 2x_J + 2x_K + 6x_L + 3x_M$$

The complete linear program is

Minimize $4x_A + 2x_B + 3x_C + 4x_D + 5x_E + 4x_F + 2x_G + 4x_H + 2x_I + 2x_J + 2x_K + 6x_L + 3x_M$

subject to

$$x_A + x_E \geq 1$$
$$x_A + x_B + x_F \geq 1$$

$$x_F + x_G + x_I \geq 1$$
$$x_I + x_J \geq 1$$
$$x_D + x_E \geq 1$$
$$x_C + x_D + x_L \geq 1$$
$$x_B + x_C + x_H \geq 1$$
$$x_H + x_G + x_K \geq 1$$
$$x_J + x_K + x_M \geq 1$$
$$x_L + x_M \geq 1$$
$$x_A + x_B + x_C + x_D + x_E + x_F + x_G + x_H + x_I + x_J + x_K + x_L + x_M = 9$$
$$x_i \geq 0$$

whose solution is

$$x_A = 1$$
$$x_B = 4$$
$$x_D = 1$$
$$x_G = 1$$
$$x_I = 1$$
$$x_M = 1$$

This solution clearly does not form a spanning tree; in fact, one of the variables has a value greater than one. If we enforce the 0-1 restriction on the variables and solve the associated integer program, we obtain the following solution:

$$x_A = 1$$
$$x_B = 1$$
$$x_C = 1$$
$$x_D = 1$$
$$x_G = 1$$
$$x_I = 1$$
$$x_J = 1$$
$$x_K = 1$$
$$x_M = 1$$

On close examination, you see that this solution does not provide a spanning tree either. Although all variables are 0 or 1, the variables corresponding to branches I, J, K, and G form a cycle that is disconnected from the rest of the tree. If we add a constraint that eliminates this cycle:

$$x_I + x_J + x_K + x_G \leq 3$$

we get the following solution:

$$x_A = 1$$
$$x_B = 1$$
$$x_C = 1$$
$$x_D = 1$$
$$x_H = 1$$
$$x_I = 1$$
$$x_J = 1$$
$$x_K = 1$$
$$x_M = 1$$

This solution is indeed a spanning tree and is the optimal solution.

Unfortunately, there was no guarantee that adding this constraint would result in a feasible solution. We might have ended up with a cycle at some other place. To ensure that no cycles would be formed, we would have to write out a large number of these constraints. There actually is a way of modeling this problem as a linear program, but the theory is too complicated to describe here (Edmonds, 1971). Despite the difficulty in modeling the minimum spanning tree problem as a mathematical program, in the next section we present incredibly simple algorithms to solve it.

2.2. Minimum Spanning Tree Algorithms

We present two approaches for finding minimal spanning trees on a network. We are given a network with edge weights $w(x,y)$ and seek a spanning tree with minimum total weight. To motivate the approach, let us consider the simple network shown in Figure 6.

In this example, a spanning tree consists of any two edges. Since this network is a simple cycle, we must *remove* some edge to create a tree. Clearly, the edge to remove is the one with the largest weight, or edge (1,3). Thus, the minimum-weight spanning tree must consist of edges (1,2) and (2,3).

We can extend this idea by observing that whenever we add some edge to a tree that creates a cycle, we must remove the edge having the largest weight in order to have the minimum-weight tree among the edges that we

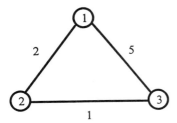

Figure 6 A simple network.

have considered. We can select edges in an arbitrary order and make a decision on whether or not each edge should remain in the optimal spanning tree. Essentially, we are solving a series of optimization problems.

Let us order the edges in G arbitrarily, calling them $e_1, e_2, \ldots, e_n$. Let the sets $S_1, S_2, \ldots, S_n$ be defined as

$$S_1 = \{e_1\}$$
$$S_2 = \{e_1, e_2\}$$
$$\vdots$$
$$S_n = \{e_1, e_2, \ldots e_n\}$$

Let T_i be the minimum spanning forest over the set of edges in S_i for $i = 1, 2, \ldots, n$. We easily see that $T_1 = S_1$ and that $T_2 = S_2$. Now consider S_i, for some $i > 2$. $T_{i-1} \cup e_i$ must either be a forest itself or contain a cycle. If $T_{i-1} \cup e_i$ is a forest, then $T_i = T_{i-1} \cup e_i$ must be the minimum-weight spanning forest over the set of edges in S_i. If $T_{i-1} \cup e_i$ creates a cycle, we can find the minimum-weight spanning forest over S_i simply by deleting the largest-weight edge in the cycle. By induction, we see that T_n must be a minimum-weight spanning tree in G.

This argument provides a simple algorithm for finding minimum spanning trees:

1. Select each edge arbitrarily and add it to the current forest.
2. If a cycle is formed, delete the edge in the cycle having the largest weight.
3. Continue until all edges have been examined.

Applying this procedure to the network in Figure 7 results in the following:

Graph Theory and Networks

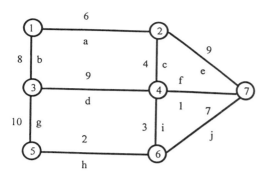

Figure 7 Spanning tree example.

Edge order	Action	Current forest
a		a
b		a,b
c		a,b,c
d	cycle; delete d	a,b,c
e		a,b,c,e
f	cycle; delete e	a,b,c,f
g		a,b,c,f,g
h		a,b,c,f,g,h
i	cycle; delete g	a,b,c,f,h,i
j	cycle; delete j	a,b,c,f,h,i

If we select the edges in order of *nondecreasing weight*, then whenever a cycle is created, the edge that we delete must be the last edge considered. This simplifies the computations because we can stop as soon as a spanning tree is formed because all remaining edges must have weights at least as large as those currently in the tree. This variation of the algorithm is called *Kruskal's algorithm* (Kruskal, 1956). In the example in Figure 7 we would have the following:

Edge order	Action
f	In tree
h	In tree
i	In tree
c	In tree
a	In tree
j	Out of tree
b	In tree – stop. Spanning tree formed.

An alternative algorithm for the minimum spanning tree problem was developed by Prim (1957). It can be described as follows:

1. Begin with any vertex. Select the edge of least weight that connects this vertex with another.
2. At any intermediate iteration we have a subtree (a tree, but not spanning). Select the edge of least weight that connects some vertex in the subtree to a vertex not in the subtree.
3. If the current tree is spanning, stop. Otherwise return to step 2.

To illustrate Prim's algorithm, let us start at node 1 in Figure 7. The smallest-weight edge incident to node 1 is edge a. The current subtree spans vertices 1 and 2. The set of edges connecting these vertices to those not in the subtree is $\{b,c,e\}$. We select the edge having the smallest weight, c. The new subtree spans the set of vertices $\{1,2,4\}$. Continuing, we have

Candidate edges	Least weighted edge	Vertices in subtree
$\{b,d,e,f\}$	f	$\{1,2,4,7\}$
$\{b,d,i,j\}$	i	$\{1,2,4,7,6\}$
$\{b,d,g,h\}$	h	$\{1,2,4,7,6,5\}$
$\{b,d,g\}$	b	$\{1,2,4,7,6,5,3\}$

Try these algorithms on the example in Figure 5 to verify the solution we obtained through linear programming.

For a graph with m vertices and n edges, Kruskal's algorithm may take up to n iterations. Each iteration requires the determination of the smallest edge available; this can be done in $O(\log n)$ time using appropriate data structures. Hence, the overall complexity of the algorithm is $O(n \log n)$. Prim's algorithm takes $m - 1$ iterations to complete because one edge is added at each iteration. Finding the next edge to add can be done in $O(m)$ time. Thus, the overall complexity of this algorithm is $O(m^2)$.

3. SHORTEST-PATH PROBLEMS

3.1. Linear Programming Formulation

We may model a shortest-path problem on a network $G = (X,E)$ between two nodes s and t as an ordinary network flow problem. Let $a(x,y)$ be the length of arc (x,y). The length of each arc in the network is the cost associated with a unit of flow along that arc. We treat node s as a source with a supply of 1 and node t as a sink with demand of 1. Because there is only one unit of supply and demand, it must travel along some path from s to t. By minimizing the total cost of this flow, we will have solved the shortest-path problem.

Stated formally, the formulation is

$$\text{Minimize} \sum_{(x,y) \in E} a(x,y) f(x,y)$$

subject to

$$\sum_y f(s,y) - \sum_y f(y,s) = 1$$

$$\sum_y f(x,y) - \sum_y f(y,x) = 0 \quad \text{for } x \neq s,t$$

$$\sum_y f(t,y) - \sum_y f(y,t) = -1$$

$$f(x,y) \geq 0 \quad \text{for all } (x,y)$$

Although the linear programming formulation is straightforward, more efficient solution procedures exist. In the next section we discuss one of these.

3.2. A Shortest-Path Algorithm

Many different algorithms exist for finding shortest paths in graphs and network. In this section we present *Dijkstra's algorithm* (Dijkstra, 1959). Dijkstra's algorithm will not only find the shortest path from one node to another but also find the shortest paths from one node to *all* others in a graph with nonnegative arc lengths. Most other algorithms are variations of fundamental concepts embodied in Dijkstra's algorithm.

The basic idea underlying the approach is simple. Suppose we know the k vertices that are closest in total length to vertex s and also a shortest path from s to each of these vertices. Label vertex s and these k vertices with their shortest distance from s. Then the $(k+1)$st closest vertex to s is found as follows. For each unlabeled vertex y, construct k distinct paths from s to y by joining the shortest path from s to x with the arc (x,y) for all labeled vertices x. Select the shortest of these k paths and let it tentatively be the shortest path from s to y.

Which labeled vertex is the $(k+1)$st closest vertex to s? It must be the unlabeled vertex with the shortest tentative path length from s as calculated above. This follows because the shortest path from s to the $(k+1)$st closest vertex to s must use only labeled vertices as its intermediate vertex (assuming that all arc lengths are nonnegative).

Therefore, if the k closest vertices to s are known, the $(k+1)$st can be determined as we described. Starting with $k = 0$, this process can be repeated until the shortest path from s to all nodes is found. We can now formally state the Dijkstra shortest-path algorithm.

Dijkstra Shortest-Path Algorithm:

Step 1. Initially, all arcs and nodes are *unlabeled*. Assign a number $d(x)$ to each node x to denote the tentative length of the shortest path from s to x that uses only labeled nodes as intermediate nodes. Initially, set $d(s) = 0$, $d(x) = a(s,x)$, and $p(x) = s$ for all $x \neq s$. [If there is no edge from s to x, then $d(x) = \infty$ and $p(x) = 0$.] The function $p(x)$ is called the *predecessor function* and tells the preceding node on the shortest path to x.

Step 2. Let y be the unlabeled node having the smallest value for $d(x)$. Label node y. If all nodes are labeled, stop — the shortest paths from s to all nodes have been found. If $d(x) = \infty$ for all unlabeled nodes x, no path exists from s to any unlabeled node and we may stop.

Step 3. For each unlabeled node x, redefine $d(x)$ as follows:

$$d(x) = \min\{d(x), d(y) + a(y,x)\}$$

This can be performed efficiently by considering only the nodes adjacent to node y. If the $d(x)$ changed, replace the current value of $p(x)$ with y. Return to step 2. At termination, $d(x)$ represents the length of the shortest path from s to x.

The predecessor function defines a unique path from each node back to s. The edges along these paths form a tree rooted at s, called the shortest-path tree. The unique path from s to any other node x contained in any shortest-path tree is a shortest path from s to x.

Let us perform the Dijkstra shortest-path algorithm to find a shortest path from node s to all others in the graph in Figure 8.

Step 1. Initially, node s is labeled with $d(s) = 0$. Set

$d(1) = a(s,1) = 4:$ $p(1) = s$
$d(2) = a(s,2) = 7;$ $p(2) = s$
$d(3) = a(s,3) = 3;$ $p(3) = s$

Set $d(x) = \infty$ and $p(x) = 0$ for all other nodes.

Step 2. Select the unlabeled node with the smallest distance value. This is node 3.

Step 3. Recompute distances for the unlabeled nodes adjacent to node 3 as follows:

Graph Theory and Networks

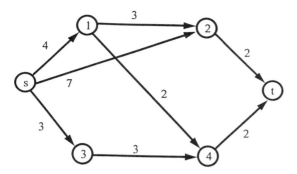

Figure 8 Shortest-path example.

$$d(4) = \min\{d(4), d(3) + a(3,4)\} = \min\{\infty, 3 + 3\} = 6$$
$$p(4) = 3$$

Step 2. The unlabeled node with the smallest distance value is node 1.
Step 3. Recompute distances for nodes 2 and 4:

$$d(2) = \min\{d(2), d(1) + a(1,2)\} = \min\{7, 4 + 3\} = 7$$
$$d(4) = \min\{d(4), d(1) + a(1,4)\} = \min\{6, 4 + 2\} = 6$$

In these cases, we have a tie, indicating that alternate optimal solutions exist. If we wish, we could define an alternate precedence function identification to indicate this.

Step 2. The minimum distance on the unlabeled nodes is $d(4) = 6$. Let $y = 4$.
Step 3.

$$d(t) = \min\{d(t), d(4) + a(4, t)\} = \min\{\infty, 6 + 2\} = 8$$
$$p(t) = 4$$

Step 2. Select node 2.
Step 3.

$$d(t) = \min\{d(t), d(2) + a(2,t)\} = \min\{8, 7 + 2\} = 8$$

At this point, all nodes have been labeled. Also, arc $(4,t)$, which determined $d(t)$, is labeled. The shortest-path tree consists of the arcs $(s,3)$, $(s,1)$, $(3,4)$, $(s,2)$, and $(4,t)$ (see Figure 9).

The algorithm terminates with the following predecessor function:

$$p(1) = s$$
$$p(2) = s$$

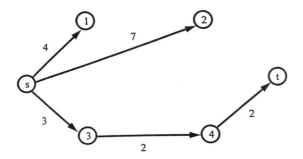

Figure 9 Shortest-path tree.

$p(3) = s$
$p(4) = 3$
$p(t) = 4$

To trace the final path from node t, for example, we need only apply this function recursively from node t. Thus, $p(t) = 4$; $p(4) = 3$; $p(3) = s$ represents the shortest path.

4. MINIMUM-COST FLOW PROBLEMS

4.1. Linear Programming Formulation

The minimum-cost flow problem on a network $G = (X,E)$ can be formulated as a linear program:

$$\text{Minimize} \sum_{(x,y) \in E} c(x,y) f(x,y) \tag{1}$$

subject to

$$\sum_y f(x,y) - \sum_y f(y,x) = b(x) \quad \text{for all } x \in X \tag{2}$$

$$f(x,y) \geq 0 \tag{3}$$

Here, $c(x,y)$ = unit cost of flow on arc (x,y), $f(x,y)$ = flow arc (x,y), and $b(x)$ = net supply at node x. In the most general case we may have upper and/or nonzero lower bounds on the variables. However, these can be removed through simple transformations (see Evans and Minieka, 1992, pp. 130–133), so we will assume that the problem is in the form of Eqs. (1)–(3).

4.2. The Network Simplex Method

The simplex algorithm can be applied to problem (1)–(3) in a very specialized fashion that leads itself to fast and efficient computation. We call this specialization the *network simplex algorithm*. Computer implementations of the network simplex algorithm are hundreds of times faster than general-purpose linear programming codes.

Observe that the left-hand side of Eq. (2) is the node-arc incidence matrix A of the underlying network. That is, each variable $f(x,y)$ in (2) has a $+1$ in the row corresponding to node x and a -1 in the row corresponding to node y. If we add all the constraints of (2) we obtain $0 = 0$; therefore, the rows of this matrix are linearly dependent. It is easy to prove the following:

Theorem: The rank of a node-arc incidence matrix is equal to the number of rows minus one.

Proof: If m is the number of rows, the rank of A must be less than m. We need only show that there exists an $(m-1)$ by $(m-1)$ submatrix that is nonsingular. Let T be any spanning tree in the network. We know that T consists of m nodes and $m-1$ arcs that do not form a cycle. Let A_T be the submatrix of A associated with the arcs in T. Since T is a tree, there must be at least one node x having only one incident arc. Therefore, the corresponding row in A_T becomes

$$\begin{bmatrix} \pm 1 & 0 \\ q & A_{T'} \end{bmatrix}$$

The matrix $A_{T'}$ corresponds to the tree obtained by removing node x and its incident arc. Since T' is also a tree, it must have at least one node with only one incident arc. Permute the rows and columns of $A_{T'}$ in the same manner as A_T so that the nonzero entry corresponding to that node is in the upper left corner. If this procedure is repeated $m-1$ times and then we delete the last row, the resulting $(m-1)$ by $(m-1)$ matrix will be lower triangular with nonzero diagonal elements. Therefore, it must be nonsingular. □

We may transform A into one with full row rank by adding a column consisting of all zeros and a single one. By convention, we will position the 1 in the first row (node 1). We may think of this column as "half an arc," one whose head is not incident to any node. This additional arc may be regarded as an artificial variable in the linear programming (LP) formulation. We call this artificial arc the *root arc* and its incident node the *root node*. Any spanning tree in the network plus this artificial arc (called a *rooted spanning tree*) will therefore correspond to a basis. For any feasible solution, this artificial variable must be basic and the flow must be zero (why?). Thus, we have shown that a *basis for the minimum-cost network flow problem (1)–(3) is characterized as a rooted spanning tree in the network.*

Now all we need is to show how to perform the steps of the simplex algorithm using this network representation of the basis rather than a matrix representation. Let us review the steps in the simplex algorithm:

Step 1. Begin with a basic feasible solution.

Step 2. Compute the complementary dual solution, that is, a dual solution that satisfies complementary slackness. The dual linear program is

$$\text{Maximize} \sum_{x \in X} b(x) w(x)$$

subject to

$$w(x) - w(y) \leq c(x,y) \quad \text{for all } (x,y) \in E$$

$w(x)$ unrestricted in sign

Complementary slackness conditions imply that

If $f(x,y)$ is basic, then $w(x) + w(y) - c(x,y) = 0$

If $w(x) - w(y) < c(x,y)$, then $f(x,y) = 0$

Step 3. Price out the nonbasic variables by computing the reduced costs $\bar{c}(x,y) = c(x,y) - w(x) + w(y)$. If $\bar{c}(x,y) \geq 0$, stop; the current basis is optimal. Otherwise select a nonbasic variable with $\bar{c}(x,y) < 0$ to enter the basis.

Step 4. Determine the basic variable to leave the basis and pivot, bringing the new variable into the basis and updating the values of all other basic variables. Return to step 2.

Let us assume that we have an initial basic feasible solution in the form of a rooted spanning tree on the network. If the network has n nodes, there are n dual variables $w(x)$. However, only $n - 1$ flows $f(x,y)$ are basic. Thus, we have $n - 1$ complementary slackness conditions in n variables. We can arbitrarily set some $w(x)$ to any value. It is customary to set the dual variable corresponding to the root node to zero. Once this is done, the remaining dual variables can be uniquely determined as follows. Select any node x for which the dual variable $w(x)$ is known. Consider the set of arcs (x,y) such that $w(y)$ has not yet been determined. Each of these arcs corresponds to a complementary slackness condition. Since $w(x)$ is known, we can solve uniquely for $w(y)$ as

$$w(y) = w(x) - c(x,y)$$

We continue in this fashion until all dual values have been assigned. Since there is a unique path between any pair of nodes in a spanning tree, all remaining dual variables are uniquely determined from the root. Notice that all computations are simple additions and subtractions.

We previously saw that the reduced costs are easily computed as

$$\bar{c}(x,y) = c(x,y) - w(x) + w(y)$$

This computation also consists solely of additions and subtractions.

Since the entering variable is not currently basic, its arc is not part of the basis spanning tree. Any arc not in a spanning tree creates a unique cycle when added to the tree. To preserve conservation of flow at all nodes, the flows along every arc in this cycle must reflect any increase in the flow of the entering arc. Consider traversing the cycle in the direction of the entering arc. As the entering arc flow is increased, we must also increase the flow on any arc in the cycle that is oriented in the same direction and decrease the flow on any arc that is oriented in the opposite direction (see Figure 10). Since the flow on arcs being decreased will eventually hit zero and they must all change by the same amount, the increase in the entering arc flow is limited by the minimum of the arc flows being decreased. The arc on which this occurs will leave the basis [or, if there is a tie, one will leave the basis and the others will remain basic at a zero (degenerate) level].

Thus, we have a simple network interpretation of the familiar minimum ratio rule of linear programming. Pivoting simply consists of adjusting the flows on the arcs of the cycle by the limiting value to preserve conservation of flow. Again, we point out that all computations are additions and subtractions. Thus, the entire simplex algorithm can be implemented on a network using elementary arithmetic operations. [In practice, sophisticated data structures are used to streamline the operations. The reader is referred to Bradley et al. (1977) for an excellent example of how this is done.]

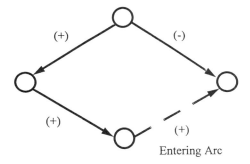

Entering Arc

Figure 10 Adjusting flows in a cycle.

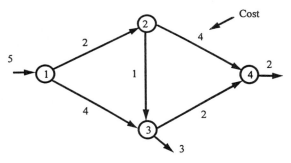

Figure 11 A minimum-cost network example.

We will illustrate the network simplex method using the example in Figure 11. Figure 12 shows an initial basic feasible solution and its complementary dual solution. We price out the nonbasic arcs as follows:

$$c(2,3) = 1 - (-2) - 4 = -1$$
$$c(2,4) = 4 - (-2) - 6 = 0$$

Since $c(2,3) < 0$, we bring arc (2,3) into the basis. Adding this arc to the current basis spanning tree creates the following cycle:

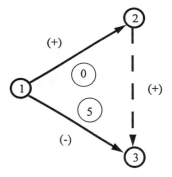

Traversing this cycle in the direction of arc (2,3), we find that we must decrease the flow on arc (1,3) and increase the flow on arc (1,2). The minimum flow on decreasing arcs is 5; therefore we adjust all flows by this amount. The new basis and complementary dual solution are shown in Figure 13. Pricing out the nonbasic arcs, we have

$$c(1,3) = 4 - 0 - 3 = 1 > 0$$
$$c(2,4) = 4 - (-2) - 5 = 1 > 0$$

Therefore, the current solution is optimal.

Graph Theory and Networks

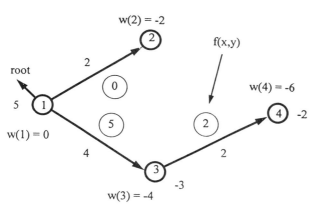

Figure 12 Initial basic feasible solution.

5. ARC ROUTING PROBLEMS

Many practical routing problems involve finding paths or cycles that traverse a set of arcs in a graph. In general, we call such problems arc routing problems. Arc routing problems have many practical applications. For example, routing street sweepers, snowplows, police patrol cars, electric line inspectors, or automated guided vehicles in a factory involves determining minimum-cost paths that cover edges of the graph that models the situation.

Finding a minimum-cost cycle in an undirected graph that crosses every edge at least once is called the *Chinese postman problem* (*CPP*). The

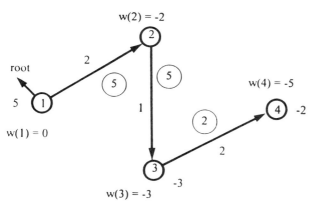

Figure 13 Optimal basic feasible solution.

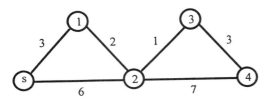

Figure 14 An Euler graph.

CPP was first defined by a Chinese mathematician, Kwan Mei-Ko, who sought to optimize postal carrier routes (Mei-Ko, 1962). The CPP can be generalized to graphs in which all edges are directed (the directed postman problem) or in which only some of the edges are directed (the mixed postman problem). We shall only consider the classic CPP in this chapter.

5.1. Euler Tours

The mathematician Leonhard Euler showed that it is impossible to find a cycle in a graph that includes every edge *exactly once* unless all vertices have an even number of incident edges. We call any cycle in a graph that crosses each edge exactly once an *Euler tour*. Any graph that possesses an Euler tour is called an *Euler graph*.

Figure 14 shows an example of an Euler graph. There are several different Euler tours in the graph starting from vertex s. For example, each of the following four routes is an Euler tour:

Route 1: $(s,1), (1,2), (2,3), (3,4), (4,2), (2,s)$
Route 2: $(s,1), (1,2), (2,4), (4,3), (3,2), (2,s)$
Route 3: $(s,2), (2,3), (3,4), (4,2), (2,1), (1,s)$
Route 4: $(s,2), (2,4), (4,3), (3,2), (2,1), (1,s)$

Each of these four routes traverses each edge exactly once; thus, the total length of each route is $3 + 2 + 1 + 3 + 7 + 6 = 22$.

If a graph possesses an Euler tour, then the Chinese postman problem is solved; we need only find some Euler tour. If a graph has no Euler tour, then at least one edge must be crossed more than once. In a vehicle routing context, we call this *deadheading*, because the vehicle is not performing any productive work. In Figure 15, for example, there is no way for the postman to traverse edge (2,3) only once. No Euler tour exists. A shortest tour that traverses each edge at least once is $(s,1), (1,2), (2,3), (3,4), (4,5), (5,3), (3,2), (2,s)$. Can you find another optimal solution? The total length of the tour is the length of all the edges $(3 + 2 + 5 + 1 + 3 + 7 + 6)$ plus the length of deadheading (5), or 32. It is easy to see that because the sum of all

Graph Theory and Networks

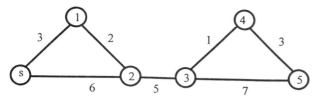

Figure 15 A graph with no Euler tour.

edge lengths is constant, the Chinese postman problem can be interpreted as minimizing the amount of deadheading necessary in the graph.

5.1.1. Constructing Euler Tours

Let us suppose that we have an Euler graph $G = (X,E)$. How can we construct an Euler tour in G? A simple algorithm for finding an Euler tour proceeds as follows (Edmonds and Johnson, 1973):

> *Step 1.* Begin at any vertex s and construct a cycle C. This can be done by traversing any edge (s,x) incident to vertex s and marking this edge "used." Next, traverse any unused edge incident to vertex x. Repeat this process of traversing unused edges until you return to vertex s. (This process must return to vertex s because every vertex has an even number of incident edges and every visit to a vertex leaves an even number of unused edges incident to that vertex. Hence, every time a vertex is entered, there is an unused edge for departing from that vertex.)
>
> *Step 2.* If C contains all the edges of G, stop. If not, then the subgraph G' in which all edges of C are removed must be Euler because each vertex of C must have an even number of incident edges. Since G is connected, there must be at least one vertex v in G' in common with C.
>
> *Step 3.* Starting at v, construct a cycle in G', say C'.
>
> *Step 4.* Splice together the cycles C and C', calling the combined cycle C. Return to step 2.

We illustrate this method using the graph in Figure 16. If we begin at vertex 1, we might construct the cycle C consisting of edges a, f, h, and i. Let v be vertex 2. The remaining graph has a unique cycle C' consisting of edges b, c, d, g, and e. We splice the cycles together as follows. Begin at the initial vertex of C and proceed around the cycle until the common vertex v is reached. At this point, traverse C' until you return to vertex v. Then continue from v to the initial vertex along the remaining path in C. The

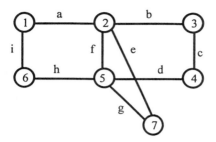

Figure 16 Euler tour example.

combined cycle in the example above would consist of edges a, b, c, d, g, e, f, h, and i, which is an Euler tour in G.

The basic ideas in this algorithm were formalized by Edmonds and Johnson (1973). Their implementation maintains a list of vertices $L_x(1)$, $L_x(2), \ldots, L_x(k), \ldots$ to visit next when vertex x is reached the kth time.

> *Step 1.* Select any vertex w. Let $v = w$ and $k_x = 0$ for all vertices x (k_x is an index denoting the number of times that node x is reached). All edges are labeled "unused."
>
> *Step 2.* Randomly select an unused edge incident to vertex v. Mark this edge "used." Let y be the vertex at the other end of the edge. Set $k_y = k_y + 1$ and $L_y(k_y) = v$. If vertex y has any incident unused edges, go to step 3. Otherwise vertex y must be v_0. In this case, go to step 4.
>
> *Step 3.* Set $v = y$ and return to step 2.
>
> *Step 4.* Let v_0 be any vertex that has at least one used edge and one unused edge incident to it. Set $v = v_0$ and return to step 2. If no such vertex exists, go to step 5.
>
> *Step 5.* To construct the tour, start at the original vertex w and proceed to vertex $L_w(k_w)$. The first time vertex x is reached, leave it by going to vertex $L_x(k_x)$. Set $k_x = k_x - 1$ and continue, each time going from vertex x to vertex $L_x(k_x)$.

We will illustrate this algorithm using Figure 16.

Step 1. $w = 1 = v$. $k_x = 0$ for $x = 1, \ldots, 7$.
Step 2. Select edge a; $y = 2$. $k_2 = 1$; $L_2(1) = 1$. Mark edge a used.
Step 3. $v = 2$.
Step 2. Select edge b; $y = 3$; $k_3 = 1$; $L_3(1) = 2$. Mark edge b used.
Step 3. $v = 3$.

Step 2. Select edge c; $y = 4$; $k_4 = 1$; $L_4(1) = 3$. Mark edge c used.
Step 3. $v = 4$.
Step 2. Select edge d; $y = 5$; $k_5 = 1$; $L_5(1) = 4$. Mark edge d used.
Step 3. $v = 5$.
Step 2. Select edge h; $y = 6$; $k_6 = 1$; $L_6(1) = 5$. Mark edge h used.
Step 3. $v = 1$.
Step 2. Select edge i; $y = 1$; $k_1 = 1$; $L_1(1) = 6$. Mark edge i used. $y = v_0$. Go to step 4.
Step 4. Set $v_0 = 2 = v$. Return to step 2.
Step 2. Select edge e; $y = 7$; $k_7 = 1$; $L_7(1) = 2$. Mark edge e used.
Step 3. $v = 7$.
Step 2. Select edge g; $y = 5$; $k_5 = 2$; $L_5(2) = 7$. Mark edge g used.
Step 3. $v = 5$.
Step 2. Select edge f; $y = 2$; $k_2 = 2$; $L_2(2) = 5$. Mark edge f used. Go to step 4 and determine that no new vertex exists. Go to step 5.
Step 5. Begin at $w = 1$. Go to $L_1(1) = 6$; set $k_1 = 0$. From node 6, go to $L_6(1) = 5$ and set $k_6 = 0$. From node 5, go to $L_5(2) = 7$ and set $k_5 = 1$. At this point, observe that we are switching to the second cycle found during the course of the algorithm. We then proceed to vertices, 2, 5, 4, 3, 2, and 1, in order. Notice that the Euler tour we construct is in the reverse order from the way in which it was generated.

5.2. The Chinese Postman Problem

This section describes how to solve the CPP on an undirected graph $G = (X,E)$. We have seen that if G is an Euler graph, any Euler tour solves the postman problem and no deadheading is necessary. If G is not an Euler graph, we seek to minimize the amount of deadheading that is required. Let $a(i,j)$ be the length of edge (i,j) in G.

In any postman route, the number of times the postman enters a vertex equals the number of times the postman leaves that vertex. Consequently, if vertex x does not have even degree, then at least one edge incident to vertex x must be repeated by the postman.

Let $n(i,j)$ denote the number of times that edge (i,j) is repeated by the postman. Thus, edge (i,j) is traversed $n(i,j) + 1$ times by the postman. Construct a new graph $G^* = (X,E^*)$ that contains $n(i,j) + 1$ copies of each edge (i,j) in graph G. Clearly, an Euler tour of graph G^* corresponds to a postman route in graph G.

We wish to select nonnegative integer values for the $f(i,j)$ variables so that

1. All nodes in G^* have an even number of incident edges.
2. $\Sigma\, a(i,j)n(i,j)$, the total length of repeated edges, is minimized.

If vertex x is an odd-degree vertex in graph G, an odd number of edges incident to vertex x must be repeated by the postman, so that in graph G^* vertex x has even degree. Similarly, if vertex x is an even-degree vertex in graph G, an even number of edges (zero is an even number) incident to vertex x must be repeated by the postman, so that in graph G^* vertex x has even degree.

If we trace out as far as possible a path of repeated edges starting from an odd-degree vertex, this path must necessarily end at another odd-degree vertex. Thus, the repeated edges form paths whose initial and terminal vertices are odd-degree vertices. Of course, any such path may contain an even-degree vertex as one of its intermediate vertices. Consequently, the postman must decide (1) which odd-degree vertices will be joined together by a path of repeated edges and (2) the precise composition of each such path.

One method of solving the problem is to arbitrarily join the odd-degree vertices by paths of repeated edges and use the following theorem that was proved by Mei-Ko:

Theorem: A feasible solution to the Chinese postman problem is optimal if and only if

(i) no more than one duplicate edge is added to any original edge and

(ii) the length of the added edges in any cycle does not exceed one-half the length of the cycle.

This result provides a method of successively improving feasible solutions until an optimal solution is found — we need only check the two conditions (i) and (ii). The only problem is that the number of cycles that must be checked in condition (ii) grows exponentially in the size of the graph. Therefore, this algorithm cannot be performed efficiently. However, an efficient algorithm exists for the CPP (Edmonds and Johnson, 1973).

We can determine a shortest path between each pair of odd-degree vertices in graph G and determine which pairs of odd-degree vertices are to be joined by a path of repeated edges as follows. Construct a graph $G' = (X', E')$ whose vertex set consists of all odd-degree vertices in G and whose edge set contains an edge joining each pair of vertices. Let the length of each edge equal the length of a shortest path between the corresponding two vertices in graph G. Since graph G' has an even number of vertices, k, and each pair of vertices in G' is joined by an edge, we can *match* (associate together) $k/2$ pairs of vertices. If we can find a minimum-length matching of vertices, their corresponding shortest paths will define the edges that

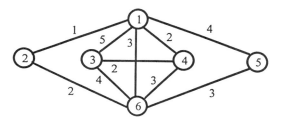

Figure 17 Postman problem.

should be repeated by the postman. Efficient algorithms for minimum-length matching exist, but we will not discuss them here. (Small problems can be solved by inspection.) The interested reader should consult Evans and Minieka (1992).

Let us find an optimal postman route for the undirected graph in Figure 17. Notice that vertices 1, 3, 4, and 6 have odd degree. Form the graph G' shown in Figure 18. The vertices of G' are the odd-degree vertices 1, 3, 4, and 6 of graph G. The shortest distances are shown on the edges (you can verify these using Dijkstra's algorithm).

We can find a minimum-length matching of all the edges in G' by enumeration. Three matchings are possible:

Matching	Weight
(1,3), (4,6)	4 + 3 = 7
(1,4), (3,6)	2 + 4 = 6
(1,6), (3,4)	3 + 2 = 5

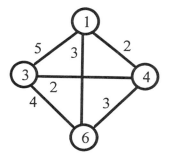

Figure 18 Graph G'.

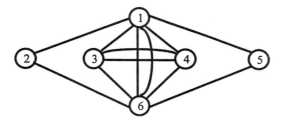

Figure 19 Graph G^*.

Consequently, the minimum-length matching is (1,6), (3,4) and the postman should repeat the shortest path from 1 to 6, which is edge (1,6), and should repeat the shortest path from 3 to 4, which is edge (3,4). Figure 19 shows graph G^*, in which edges (1,6) and (3,4) have each been duplicated once. All vertices in G^* have an even number of incident edges, and an optimal postman route for the original graph in Figure 17 corresponds to an Euler tour of graph G^* in Figure 19. The technique described in Section 5.1 for Euler graphs can be applied to graph G^*. An optimal route is (1,2), (2,6), (6,5), (5,1), (1,3), (3,6), (6,4), (4,3), (3,4), (4,1), (1,6), (6,1), which traverses each edge in G^* exactly once and traverses each edge in G at least once. Only edges (1,6) and (3,4) are repeated in graph G. The total length of this route is 34 units, which is 5 units more than the sum of the edge lengths.

Beginning with Section 6, we present some applications of the concepts and models described in this chapter to industrial engineering.

6. FACILITY LAYOUT

Graph theory has useful applications in facility layout (Foulds and Tran, 1986). We can always represent any layout as a planar graph as shown in Figure 20. We associate a vertex with each department in the layout and draw an edge between two vertices if the departments are adjacent. The more difficult problem is to find an appropriate graph in order to design a layout.

Suppose that we wish to lay out m facilities, for example, different departments in a library. A traffic study might be conducted, resulting in the construction of a matrix of specifying the number of trips made between each pair of facilities. This matrix is as follows:

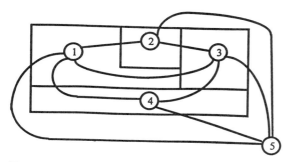

Figure 20 A department layout and associated graph.

		1	2	3	4	5
1.	Entrance	—				
2.	Catalogue	200	—			
3.	Photocopy	4	77	—		
4.	Journals	80	125		—	
5.	New books	32	42	19	26	—

We would like to locate the departments so as to maximize the sum of the values of adjacent pairs of departments. We may represent each department by a vertex of a complete graph. The number of trips between each pair of departments will be assigned as a "weight" of the corresponding edge of the graph (Figure 21). The solution is to find a subgraph having no edges crossing each other that contains all the vertices and has maximum

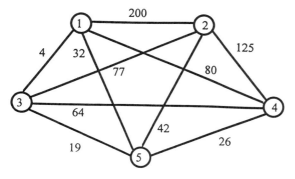

Figure 21 Complete graph for facility layout example.

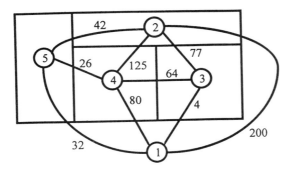

Figure 22 Feasible solution to facility layout problem.

total edge weight. Optimal algorithms for this problem are somewhat complicated and, therefore, are not discussed here. However, good solutions can often be found visually by trial and error. Figure 22 shows an example of a feasible solution. Can you find a better solution? Of course, the graph theoretic solution is only the first step. The actual layout must be designed to reflect the appropriate size of the departments and realistic architectural configurations. Nevertheless, graphs provide a convenient way of modeling such problems.

7. CELLULAR MANUFACTURING

Cellular manufacturing is an approach to designing manufacturing systems that groups parts into *part families* and machines into *manufacturing cells* for the production of part families. Cellular manufacturing has advantages over traditional process layouts (functional groupings of like machines) by reducing material handling and hence reducing waiting and throughput times (Flynn and Jacobs, 1986).

An important problem in designing manufacturing cells is machine grouping. Typically we are given a machine-part incidence matrix A for which $a_{ij} = 1$ if part j must be processed by machine i and 0 otherwise. Denote by M_i the set of parts that require processing by machine i. For any two machines, i and k, define a distance measure $d(i,k)$ as follows:

$$d(i,k) = \frac{M_i \oplus M_k}{M_i \cup M_k}$$

The numerator is the symmetric difference (exclusive OR) of the sets M_i and M_k, namely the number of parts requiring processing *only* on machine i or machine k but not both. The denominator is the number of parts

Graph Theory and Networks

Machines   Parts

	1	2	3	4	5	6	7	8	9	10	11	12	13
1		1	1				1	1	1			1	1
2		1	1				1	1			1	1	
3	1					1							
4				1	1				1				
5				1			1	1	1			1	1
6					1					1	1		
7				1						1			
8				1						1			
9						1							

Figure 23 Machine-part incidence matrix.

requiring processing on either machine or both. The distance function is a measure of the relative dissimilarity of two different machines with respect to their parts processing requirements. A value of zero indicates that the two machines have the same set of parts to process, and a value of one means that they process no parts in common.

We construct a complete graph in which each vertex represents a machine. Each edge (i,k) is assigned the weight $d(i,k)$. The minimum-weight spanning tree in this graph minimizes the total dissimilarity that includes all machines. If we wish to group the machines into K cells, we delete the $K - 1$ largest edges of the spanning tree. The result will be a forest of K trees. The machines corresponding to the vertices in each tree provide the machine groupings.

To illustrate this approach, a machine-part incidence matrix is shown in Figure 23. To illustrate the calculation $d(i,k)$, consider $i = 1$ and $k = 2$. $M_1 = \{2,3,7,8,9,12,13\}$ and $M_2 = \{2,7,8,11,12\}$. Thus $M_1 \oplus M_2 = 4$, and $M_1 \cup M_2 = 8$. Therefore $d(1,2) = 4/8 = 0.5$. The complete matrix of $d(i,k)$ values is shown in Figure 24.

	1	2	3	4	5	6	7	8	9
1	–	0.50	1.00	0.89	0.14	1.00	1.00	1.00	1.00
2		–	1.00	1.00	0.62	1.00	1.00	1.00	1.00
3			–	1.00	1.00	1.00	1.00	1.00	0.50
4				–	0.87	0.67	0.75	0.75	1.00
5					–	1.00	1.00	1.00	1.00
6						–	1.00	1.00	1.00
7							–	0.00	1.00
8								–	1.00

Figure 24 $d(i,k)$ values.

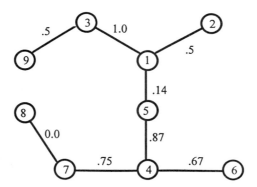

Figure 25 Minimum spanning tree for cellular manufacturing example.

The minimum spanning tree on the complete graph using these distances is shown in Figure 25. Thus, if we wished to form three machine cells, we would remove the three largest edges from the tree. The resulting connected components would consist of the vertices {1,2,5}, {3,9}, and {4,6,7,8}. Parts 2,3,7,8,9,11,12, and 13 would be assigned to the first cell; parts 1 and 6 to the second cell; and parts 4,5, and 10 to the third cell.

8. PRODUCTION LOT SIZING

A common problem encountered in production planning is to determine a schedule of production runs when the product's demand varies deterministically over time (Johnson and Montgomery, 1974). We are given a set of demands $D_1, D_2, \ldots, D_n$ over an n-period planning horizon. At each period j, we incur a fixed set up cost A_j if we produce any positive amount as well as a unit production cost C_j. Any amount in excess of the demand for that period is held in inventory until the next period, incurring a per-unit holding cost of H_j.

This problem can be modeled as a mixed-integer linear program. Let Q_j be the amount produced in period j and I_j be the amount held over in inventory to period $j + 1$. Y_j is a 0-1 variable that is equal to 1 if we produce in period j and equal to 0 otherwise. The model is

Minimize $\sum(C_j Q_j + H_j I_j + A_j Y_j)$
$Q_j + I_{j-1} - I_j = D_j \quad \text{for } j = 1, \ldots, n$
$Q_j \leq M Y_j$
$Q_j \geq 0, \quad Y_j = 0,1$

Graph Theory and Networks

where M is a very large positive number.

It can be shown that an optimal solution will have the property that $Q_j = D_j + D_{j+1} + \cdots + D_j$ for some $k \geq j$. That is, production in any period must be a sum of the demands for some set of future periods. Under these conditions, the problem can be reformulated as a shortest-path problem on a directed graph. Let W_{jk} be the total cost associated with producing $D_j + D_{j+1} + \cdots + D_k$ units in period j. Then,

$$W_{jk} = A_j + C_j Q_j + \sum H_j(I_j + \cdots + I_{k+1})$$

where

$$I_j = D_{j+1} + \cdots + D_k$$
$$I_{j+1} = D_{j+2} + \cdots + D_k$$
$$I_{k-1} = D_k$$

The network for a four-period problem is shown in Figure 26. Any directed path from node 1 to node 5 represents a sequence of production decisions. The minimum-cost path represents the optimal solution.

9. EMPLOYEE SCHEDULING

In many industries, the demand for services varies considerably over time (Bennington, 1974). For example, a mail-order company needs customer service telephone operators on a 24-hour basis, 365 days per year. However, the number of operators needed may vary by the hour, day of the week, and even the month of the year. It would be too expensive to maintain a full-time work force at a staffing level to meet peak period demands during the holiday ordering season. As a result, many companies hire part-time employees. Some costs are associated with recruiting and training these

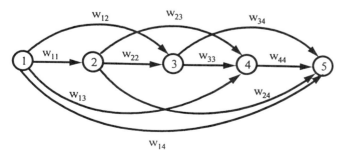

Figure 26 A four-period production lot-sizing network.

employees. Thus, a firm would be interested in planning a staffing schedule that minimizes these expenses.

Define x_{ij} to be the nonnegative number of employees hired at the start of month i and retained through month $j - 1$. Let c_{ij} be the total employment-related cost for this strategy, and let r_k be the minimum number of employees required in month k. A feasible schedule must satisfy the following constraints:

$$\sum_{j=2}^{n} x_{1j} \geq r_1$$

$$\sum_{i=1}^{k} \sum_{j=k+1}^{n} x_{ij} \geq r_k, \quad \text{for } k = 2, \ldots, n - 2$$

$$\sum_{i=1}^{n-1} x_{in} \geq r_{n-1}$$

The objective function would be:

$$\text{Minimize} \sum_{i=1}^{n-1} \sum_{j=i+1}^{n} c_{ij} x_{ij}$$

To illustrate this, suppose that the requirements for a 3-month period are 15, 10, and 20 employees. The requirements constraints are

$$x_{12} + x_{13} + x_{14} \geq 15$$
$$x_{13} + x_{14} + x_{23} + x_{24} \geq 10$$
$$x_{14} + x_{24} + x_{34} \geq 20$$

At first glance, these three constraints bear no apparent relationships to network flows. However, with a little algebra, we may transform these constraints to four conservation of flow equations. First add a nonnegative surplus variable s_k, $k = 1,2,3$, to convert them to equalities. Then subtract the first equation from the second, the second from the third, and multiply the third equation by -1. This results in the following.

$$
\begin{aligned}
x_{12} + x_{13} + x_{14} & & & - s_1 & & = 15 \\
-x_{12} & & + x_{23} + x_{24} & & + s_1 - s_2 & = -5 \\
& - x_{13} & - x_{23} & + x_{34} & + s_2 - s_3 & = 10 \\
& & - x_{14} & - x_{24} - x_{34} & + s_3 & = -20
\end{aligned}
$$

Since each column has exactly one $+1$ and one -1, these equations represent the node-arc incidence matrix of a network flow problem. If we associate with each equation a node, we have the network shown in Figure 27. Notice that the right-hand sides specify the net supply at each node. The

Graph Theory and Networks

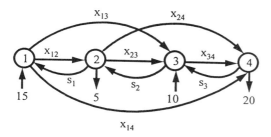

Figure 27 Network model for employment scheduling problem.

solution to this minimum-cost flow problem provides the optimal employment schedule.

10. SNOW AND ICE CONTROL

An important generalization of the Chinese postman problem arises in snow and ice control on streets and highways. For example, a state Department of Transportation (DOT) district operates a fleet of salt trucks and snow plows to remove snow and ice from the roads in its jurisdiction. The DOT seeks a set of routes that meet time or capacity restrictions. Time restrictions arise in plowing, as the agency would like to have all roads plowed within 2 or 3 hours from the time of dispatch. Knowing the time it takes to traverse each arc of the road network, we seek a set of routes that cover all arcs yet remain within the time restrictions.

Capacity restrictions arise in salt spreading, because trucks can carry only a maximum amount of salt, say 10 tons. The salt spreading rate is usually fixed, say at 200 lb per lane-mile. Therefore, the demand on each road segment can be computed as the distance of the road segment times the salt spreading rate. The agency seeks a set of routes that covers all arcs of the network but meets the capacity constraint. These problems are called *capacitated Chinese postman problems* (CCPP) (Golden and Wong, 1981). The CCPP is much more difficult to solve than the ordinary Chinese postman problem. We must resort to heuristic solutions to address these problems. A typical heuristic is to begin at the depot and make a decision of which arc to travel on next whenever a node is reached using some logical rule such as "stay as close to the depot as possible" or "go further away from the depot as long as the truck has at least half its capacity." Whenever a truck reaches the end of its route, we can find the shortest path back to the depot. A software package called *SnowMaster*, written by the author, has been used by many Midwestern government agencies to help plan their snow removal fleets and route the vehicles.

11. FINAL REMARKS

Graphs and networks can be used to model many practical situations. They provide decision makers with intuition about the structure of systems and generally are quite efficient to solve. In this chapter we discussed basic concepts of networks and graphs, spanning tree problems, shortest paths, minimum-cost flows, and arc routing. We also illustrated applications in facility layout, cellular manufacturing, production lot sizing, employee scheduling, and snow and ice control that are of interest to industrial engineers. However, we have barely scratched the surface of both applications and techniques. Hundreds of articles and books exist that describe other algorithms and applications of graphs and networks. We encourage you to learn more about this fascinating subject and discover new applications.

REFERENCES

Bennington, G. E., "Applying Network Analysis," *Ind. Eng.*, Vol. 6, No. 1, pp. 17–25, 1974.

Bradley, G. H., Brown, G. G., Graves, G. W., "Design and Implementation of Large Scale Primal Transshipment Algorithms," *Management Science*, Vol. 24, pp. 1–34, 1977.

Dijkstra, E. W., "A note on Two Problems in Connexion with Graphs," *Numer. Math.*, Vol. 1, pp. 269–271, 1959.

Edmonds, J., "Matroids and the Greedy Algorithm," *Math Prog.*, Vol. 1, pp. 127–136, 1971.

Edmonds, J., Johnson, E. L., "Matching, Euler Tours, and the Chinese Postman," *Math. Prog.*, Vol. 6, pp. 88–124, 1973.

Evans, J. R., Minieka, E., *Optimization Algorithms for Networks and Graphs*, 2nd ed. Marcel Dekker, New York, 1992.

Flynn, B. B., Jacobs, F. R., "A Simulation Comparison of Group Technology with Traditional Job Shop Manufacturing," *Int. J. Prod. Res.*, Vol. 24, pp. 1171–1192, 1986.

Foulds, L. R., Tran, H. V., "Library Layout via Graph Theory," *Comput. Ind. Eng.*, Vol. 10, pp. 245–252, 1986.

Golden, B., Wong, R. T., "Capacitated Arc Routing Problems," *Networks*, Vol. 11, pp. 305–315, 1981.

Johnson, L. A., Montgomery, D. C. *Operations Research in Production Planning, Scheduling, and Inventory Control*. Wiley, New York, 1974.

Kruskal, J. B., "On the Shortest Spanning Subtree of a Graph and the Traveling Salesman Problem," *Proc. AMS*, Vol. 7, pp. 48–50, 1956.

Mei-Ko, K., "Graphic Programming Using Odd or Even Points," *Chinese Math.*, Vol. 1, pp. 273–277, 1962.

Prim, R. C., "Shortest Connection Networks and Some Generalizations," *Bell Syst. Tech. J.*, Vol. 36, pp. 1389–1401, 1957.

5

Dynamic Programming

Eric V. Denardo
Yale University, New Haven, Connecticut

1. INTRODUCTION

A *sequential decision process* is an activity in which a sequence of actions is taken toward some goal. Human affairs abound with sequential decision processes. Surprisingly, when sequential decision processes are analyzed, certain themes recur over and over. For this reason, the collection of tools that are used in the analysis of sequential decision processes has been given the name *dynamic programming*. Dynamic programming describes the analysis of decision-making problems that unfold over time.

The names "linear programming" and "dynamic programming" are similar, but this similarity is misleading. Linear programming concerns a particular model, that in which a linear objective is maximized subject to finitely many linear constraints. Dynamic programming concerns a family of models, those of sequential decision processes.

The sequential decisions processes for which dynamic programming has proved to be useful fall into two large classes—those in which uncertainty plays an incidental role and those in which uncertainty is crucial. In the first class, for which uncertainty can be ignored, dynamic programming handles the sorts of nonconvexities that can wreak havoc on linear programming and its generalizations. These include setup costs and other economies of scale. For the second class, in which uncertainty cannot be ignored, dynamic programming is the principal model.

This chapter surveys dynamic programming from the perspective of industrial and systems engineering. The writer has chosen an inductive style. Rather than presenting general ideas and then applying them to examples, the writer presents a series of examples and abstracts general ideas from them. These examples touch issues of interest to industrial engineers — project management, production planning, and capacity expansion, to name three. The aims of this chapter are to show what dynamic programming is, when it is useful, and how to make use of it.

The key to dynamic programming is an object with the odd name "functional equation." A functional equation is a particular type of recursion. But what is a recursion? A section of this chapter is devoted to that question. Recursions are vital to dynamic programming, and their uses outside dynamic programming are legion.

Section 2 focuses on recursions. It sets the stage for what follows. Section 3 presents the simplest types of problem for which dynamic programming is appropriate. These are shortest- and longest-path problems in networks. Section 4 describes the basic ideas of dynamic programming, which are states, functional equations, and the principle of optimality. It uses the prior examples to illustrate these ideas.

Sections 2 through 4 prepare the reader for three different avenues of study. To develop the art of formulating optimization problems as dynamic programs, read Section 5. To see how dynamic programming is used in deterministic settings, read Sections 6 through 8. To see how dynamic programming is used in decision making under uncertainty, read Sections 9 through 11. Section 12 summarizes the chapter and discusses the limitations of dynamic programming. Section 13 provides a guide to the literature on which this chapter is based.

2. RECURSIONS

A *recursion* is a set of one or more equations in which the unknown quantity or quantities appear just to the left of the equal sign and also on the right-hand side. The functional equation of dynamic programming is a particular type of recursion.

Recursions are so basic to dynamic programming that they are discussed first, before dynamic programming is introduced. Recursions play an important role not merely in dynamic programming, but in nearly every branch of mathematical analysis. For that reason, time invested studying recursions pays rich rewards.

Through a series of examples, this section introduces recursions and develops the reader's facility with them. Each example is solved. Before reading each solution, the reader is urged to pause and try to construct a recursion. This section culminates with tips on how to discover recursions.

Dynamic Programming

These tips are designed to help the reader find the recursions of dynamic programming.

The first example poses a question whose answer is well known. The novelty lies in the method that will be used to get that answer, which is obtained from a recursion.

Example 1: Find the sum of the infinite series $(1 + x + x^2 + x^3 + \cdots)$.

For a value of x for which the series $(1 + x + x^2 + x^3 + \cdots)$ converges, denote its sum as $f(x)$. Observe that

$$\begin{aligned} f(x) &= (1 + x + x^2 + x^3 + \cdots) \\ &= 1 + (x + x^2 + \cdots) \\ &= 1 + x(1 + x + \cdots) \\ &= 1 + xf(x) \end{aligned} \quad (1)$$

Equation (1) is a particularly simple recursion. The quantity $f(x)$ appears just to the left of the equals sign and also on the right-hand side. To solve this recursion for $f(x)$, rewrite it as $(1 - x)f(x) = 1$, and conclude that $f(x) = 1/(1 - x)$.

Recursions like (1) can be used to sum other series. Consider:

$$\begin{aligned} g(x) &= (1 + 2x + 3x^2 + 4x^3 + \cdots) \\ &= (1 + x + x^2 + \cdots) + (0 + x + 2x^2 + 3x^3 + \cdots) \\ &= \frac{1}{1-x} + x(1 + 2x + 3x^2 + \cdots) \\ &= \frac{1}{1-x} + xg(x) \end{aligned}$$

because we had learned that $(1 + x + x^2 + \cdots) = 1/(1 - x)$. This recursion for $g(x)$ is easy to solve: $g(x) = 1/(1 - x)^2$.

Equally simple recursions convert repeating decimals to fractions. Consider this example: $x = .126126126 \cdots = .126 + .000126126 \cdots = .126 + (.001)(.126126 \cdots) = .126 + .001x$. Solving this recursion gives $x = 126/999 = 14/111$.

Example 2: Consider a set S that contains n elements. Each element of S can be painted red or blue. There are 2^n different ways to paint these elements because there are 2 ways to paint each of them. In how many different ways can these elements be painted so that exactly k of them are painted red?

A recursion will be constructed that answers the question for all possible values of n and k. For values of n and k such that $n \geq k \geq 0$, denote by $f(n, k)$ the number of different ways to paint k of n balls red. In the case $k = n$, all balls must be painted red, so

$$f(n, n) = 1, \quad \text{each } n \qquad (2.1)$$

In the case $k = 0$, all balls must be painted blue, so

$$f(n, 0) = 1, \quad \text{each } n \qquad (2.2)$$

The remaining case is that in which $n > k > 0$. For this case, we pick any particular element of S and paint it red or blue. If we paint it red, we must paint $k - 1$ of the remaining $n - 1$ balls red, and that can be done in $f(n - 1, k - 1)$ ways. Alternatively, if we paint this particular ball blue, we must paint k of the remaining $n - 1$ balls red, and that can be done in $f(n - 1, k)$ different ways. In other words,

$$f(n, k) = f(n - 1, k - 1) + f(n - 1, k), \quad \text{for } n > k > 0 \qquad (2.3)$$

Once $f(n - 1, k)$ is known for each k, the recursion in (2) specifies $f(n, k)$ for each k. Executing this recursion forms *Pascal's triangle*, the first seven rows of which are evident in Table 1.

Readers who are familiar with "n choose k" may have realized that

$$f(n, k) = \binom{n}{k} = \frac{n!}{k!(n-k)!}$$

This is a succinct way to specify n choose k. When doing mathematical manipulations, it is very handy. But the recursion in (2) provides a simple way to *calculate* n choose k.

Table 1 Table of f(n, k) for n ≤ 6

Row n	Column k						
	0	1	2	3	4	5	6
0	1						
1	1	1					
2	1	2	1				
3	1	3	3	1			
4	1	4	6	4	1		
5	1	5	10	10	5	1	
6	1	6	15	20	15	6	1

Dynamic Programming

Example 3: In the game of craps, the "shooter" rolls two fair dice. If the first roll shows a 7 or an 11 (as the sum of the pips on the two dice), the shooter wins. If the first roll shows a 2, 3, or 12, the shooter loses. If the first roll shows any other number, that number is the shooter's "point." The shooter must roll again and again until he or she rolls the point or a 7. The shooter wins if the point occurs before a 7. The shooter loses if a 7 occurs before the point.

Suppose the first roll shows a 5. Given 5 as the point, compute two quantities — the probability $p(5)$ that the shooter will win and the expectation $t(5)$ of the number of additional rolls that will occur until the shooter wins or loses.

The probability of rolling a 5 equals 4/36. The probability of rolling a 7 equals 6/36. Probabilities sum to 1, so the probability of rolling other numbers than 5 or 7 equals 26/36. A "probability tree" that depicts this situation is presented as Figure 1.

By conditioning the probability of winning on the outcome of the first roll, one gets the recursion

$$p(5) = (4/36)(1) + (6/36)(0) + (26/36)p(5) \qquad (3.1)$$

Solving this recursion gives $p(5) = 4/10 = 0.4$. Similarly, by conditioning the time until play ends on the outcome of the first roll, one gets the recursion

$$t(5) = 1 + (26/36)t(5) \qquad (3.2)$$

Solving that recursion gives $t(5) = 36/10 = 3.6$.

Example 4: A prairie dog is in room A of its modest three-room burrow depicted in Figure 2. Being hungry, the prairie dog is eager to leave the burrow. This particular prairie dog has lost its sense of direction. It is equally likely to leave each room by any of its connecting tunnels, independent of how it entered that room. Shown within the tunnels are their travel times, measured in minutes. What is the expectation of the time it takes the prairie dog to emerge from its burrow?

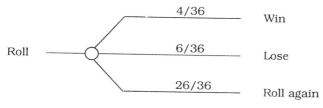

Figure 1 A probability tree for Example 3.

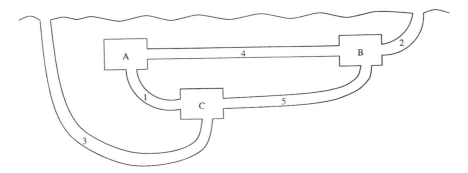

Figure 2 A three-room burrow, with travel times in minutes.

The upcoming recursion consists of three equations, one per room. For each room i, let $f(i)$ denote the expectation of the number of minutes from now to the moment at which the prairie dog emerges from the burrow given that it has just entered room i. Example 4 asks for a recursion for $f(A)$.

The prairie dog begins in room A. It is equally likely to leave room A by either of its adjacent tunnels. The expectation of the time that it will spend in its first tunnel equals $(1/2)[4 + 1]$. The expectation of the time that it will spend traveling after emerging from its first tunnel equals $(1/2)[f(B) + f(C)]$. The expectation of the sum equals the sum of the expectations, so

$$f(A) = (1/2)[4 + 1] + (1/2)[f(B) + f(C)] \tag{4.A}$$

Similarly,

$$f(B) = (1/3)[4 + 5 + 2] + (1/3)[0 + f(A) + f(C)] \tag{4.B}$$
$$f(C) = (1/3)[1 + 5 + 3] + (1/3)[0 + f(A) + f(B)] \tag{4.C}$$

This is a system of three linear equations in three unknowns. It can be solved by elimination of variables. Its solution is

$$f(A) = 15, \quad f(B) = 51/4, \quad f(C) = 49/4$$

This prairie dog problem is a whimsical example of an important calculation in Markov chains. Consider a finite set S whose elements are called states, with random transitions from state to state. For each pair i and j of states, let P_{ij} denote the probability that the next transition will occur to state j, given that state i is now observed. Similarly, for each pair i and j of states, let t_{ij} denote the time that it will take for transition to occur,

Dynamic Programming

given that the transition will occur from state i to state j. Probabilities are nonnegative, and they sum to 1. That is,

$$P_{ij} \geq 0, \quad \text{for each } i \in S \text{ and each } j \in S$$
$$\sum_{j \in S} P_{ij} = 1, \quad \text{for each } i \in S$$

Let us suppose that state i has just been observed. The P_{ij}'s for each j are the transition probabilities from state i to the various states. These probabilities sum to 1. Also, the expectation of the time that it will take for transition to occur equals $\sum_{j \in S} P_{ij} t_{ij}$.

For the prairie dog illustration in Figure 2, the Markov chain has four states, which we can label A, B, C, and O (short for out). This Markov chain has $P_{AB} = 1/2$ and $t_{AB} = 4$. It has $P_{BA} = 1/3$ and $t_{BA} = 4$. And so forth.

Consider a state k to which transition is certain to occur, sooner or later. Let $f(i)$ denote the expectation of the time until the first transition occurs to state k given that the Markov chain is now in state i. The expectation of the sum equals the sum of the expectations:

$$f(i) = \sum_{j \in S} P_{ij} t_{ij} + \sum_{j \in S - k} P_{ij} f(j)$$

where $S - k$ denotes the set containing all elements in S other than k. With n as the number of states in S, the recursion is a system of $n - 1$ linear equations in $n - 1$ variables. Please check that (4) is exactly this sort of recursion.

Example 5: Each member of a large homogeneous population has k offspring with probability p_k, where $1 = p_0 + p_1 + p_2 + \cdots$. This population is so large that the offspring of different individuals are independent of each other. Write a recursion for the probability ρ that all descendants of a particular individual die out. Solve it for these data: $p_0 = .2$, $p_1 = .6$, and $p_2 = p_3 = .1$.

Aiming to develop a recursion, consider the case in which an individual has exactly two offspring. This case occurs with probability p_2. Due to the assumed independence, the probability that all descendants of both offspring die out equals $\rho \cdot \rho$. Thus, the probability that an individual has two offspring and that all of their descendants die out equals $p_2 \rho^2$. The pattern is clear. By conditioning on the number of first-generation offspring, one gets

$$\rho = p_0 \rho^0 + p_1 \rho^1 + p_2 \rho^2 + p_3 \rho^3 + \cdots \qquad (5)$$

Equation (5) can have many solutions. Check that this equation is satisfied by setting $\rho = 1$. It can be shown that the probability ρ that all descendants of a single individual die out is the smallest positive solution to Eq. (5).

For the data provided

$$\rho = .2\rho^0 + .6\rho^1 + .1\rho^2 + .1\rho^3$$

Recognize this as a cubic equation. Setting $\rho = 1$ satisfies it. Rearrange this equation and then factor out the term $(\rho - 1)$ as follows:

$$0 = .1\rho^3 + .1\rho^2 - .4\rho + .2$$
$$= (\rho - 1)(.1\rho^2 + .2\rho - .2)$$

The quadratic equation in brackets has two solutions, $(\sqrt{2} - 1)$ and $(-\sqrt{2} - 1)$. As mentioned earlier, the probability ρ that all descendants of a single individual die out is the smallest of the positive solutions to Eq. (5), so $\rho = \sqrt{2} - 1 \approx 0.414$.

These examples show that recursions have a great many uses. Recursions help one sum a series, count, compute the probability of a complicated event, compute various expectations, and avoid numerical difficulties.

Exactly why is a recursion useful? A recursion breaks a complicated calculation into one or more simpler calculations. To illustrate this point, reconsider Example 3. The probability $p(5)$ of rolling a 5 before a 7 could be computed directly; this probability equals the sum of the infinite series $(4/36) + (26/36)(4/36) + (26/36)^2(4/36) + \cdots$. Similarly, the expectation $t(5)$ of the number of rolls equals the sum of the infinite series, $1(10/36) + 2(26/36)(10/36) + 3(26/36)^2(10/36) + \cdots$. It is possible to write these series and to sum them, but it is easier to write the recursions and solve them. To reinforce this point, reconsider Example 4. The expectation $f(A)$ of the time until the prairie dog leaves the burrow could be computed directly by accounting for each of the paths that lead from room A to the outside. It is far simpler to write a recursion and solve it.

But how can one dream up a recursion? There is no sure-fire answer, but here is one guidepost. It is the idea of a state:

> *Property of states:* A recursion of one equation per state. Transition occurs from state to state. Each state contains enough information about what has happened in the past to specify the law that governs the transition to the next state.

Examples 3 and 4 illustrate this property of states. The probability $p(5)$ of rolling a 5 before a 7 has nothing to do with the prior rolls. All that matters is that play has not ended, as is evident in Figure 1. Similarly, for Example 4, the expectation $f(B)$ of the time that remains until the prairie dog is above ground is independent of what happened before it reached room B.

Dynamic Programming

This property of states speaks of past and future. Inherent in this property is a sense of time. But recursions arise in cases in which no sense of time is present. Example 2 asks one to compute the number of different subsets of k elements that can be taken from a given set S of n elements. Here is a further clue to devising a recursion:

If a sense of time is lacking, introduce one, and see what states arise. When introducing a sense of time, aim for one that makes the recursion simple, not complicated.

For Example 1, think of the sum $(1 + x + x^2 + \cdots)$ as the first number plus the sum of the rest: $(1 + x + x^2 + x^3 + \cdots) = 1 + (x + x^2 + x^3 + \cdots) = 1 + x(1 + x + x^2 + \cdots)$, which leads directly to the recursion $f(x) = 1 + xf(x)$. For Example 2, look at the elements of S in any sequence. Looking at the first element results in a recursion.

When introducing a sense of time, one may get a recursion that is easy to solve if — and only if — one is clever about it. Consider:

Example 6: A group of n cars have formed a line. These n cars joined the line in random order, each sequence being equally likely. Each car has its own "natural" speed, and each car's speed is different. The line starts to move as a unit, and it proceeds along a narrow road, with no possibility of passing. The first car travels at its natural speed. Each of the other cars travels at the slowest of the natural speeds of the cars that are ahead of it. Thus, the cars form one or more clumps, each clump being a group of cars that travel at the same speed. Use a recursion to compute the expectation $f(n)$ of the number of clumps that form.

It is evident that $f(1) = 1$. To find a recursion for $f(n)$, we can add the cars to the line in any sequence that we wish, provided that we insert each car in random position. Is there a sequence that leads to a simple recursion for $f(n)$?

Yes. It is fastest last. With probability of $1/n$, the fastest car is first in line, in which case it forms a clump of its own. Otherwise, the fastest car joins the clump into which it is inserted. In other words,

$$f(n) = \frac{1}{n}[1 + f(n-1)] + \frac{n-1}{n}f(n-1) \quad \text{for } n > 1$$

$$= \frac{1}{n} + f(n-1)$$

$$= \frac{1}{n} + \frac{1}{n-1} + \cdots + \frac{1}{2} + 1 \quad (6)$$

the last by repeated substitution.

To see that fastest last is an adroit sequence, we see what would occur if we inserted cars in random position, but slowest last. If the slowest car finds i cars in front of it, the number of clumps that form equals $1 + f(i)$, with $f(0) = 0$. The slowest car is inserted in a random position, so it finds i cars in front of it with probability $1/n$. Thus,

$$f(n) = 1 + 1/n[f(1) + f(2) + \cdots + f(n-1)] \quad \text{for } n > 1$$

This recursion is correct. Its solution is $f(n) = 1 + 1/2 + \cdots + 1/n$. But this solution is much easier to find from (6).

Incidentally, the series $(1 + 1/2 + \cdots + 1/n)$ that appears in (6) is famous; it's the *harmonic series*: As n increases, the quantity $[1 + 1/2 + \cdots + 1/n - \log_e(n)]$ decreases monotonically to the number $0.5778 \ldots$, which is known as *Euler's constant*.

As mentioned earlier, the recursions that arise in dynamic programming have a special name. They are called functional equations. This section has prepared the reader for a study of functional equations.

3. THE SIMPLEST DYNAMIC PROGRAM

The simplest dynamic program is the computation of a shortest path in a network that is directed and acyclic. This section presents and solves a few shortest-path problems. The next section uses these examples to reveal some of the common themes of dynamic programming. The terms we use to describe directed networks are introduced in Chapter 4 of this book. We illustrate them here but do not repeat them.

Figure 3 depicts a directed network that has nine nodes, each of which is depicted as a circle with a number inside. As the arrow suggests, this network contains directed arc (1, 2). It does not contain arc (2, 1), although it could. This network contains several paths, including (2, 4, 7). It contains no cycle (path that starts and ends at the same node), so it is acyclic. Each arc (i, j) in this network has a length has an associated length t_{ij}, which is

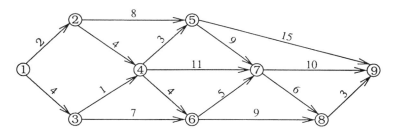

Figure 3 An acyclic network, with arc lengths.

Dynamic Programming

displayed next to it; e.g., $t_{34} = 1$ and $t_{47} = 11$. In the present discussion, the length of a path is defined as the sum of the lengths of the arcs in it; e.g., path (1, 3, 4) has length $5 = 4 + 1$. A shortest-path problem is the problem of finding a shortest path from one designated node to another designated node, e.g., from node 1 to node 9. As mentioned earlier, the shortest-path problem is the simplest dynamic program.

Example 7: Sandy cycles to work. He wishes to use the route that gets from his home to his office as quickly as possible. Sandy has accumulated the data in Figure 3. Node 1 represents his home, node 9 represents his office, the other nodes represent road intersections, each arc represents a road, and each arrow points in the direction in which Sandy might wish to travel the road that it represents. Adjacent to each road is its travel time in minutes. Which is the quickest route from Sandy's home to his office, and how long does it take?

No fancy theory is needed to solve Sandy's problem. A few glances at Figure 3 will convince the reader that the shortest path from node 1 to node 9 is (1, 3, 4, 6, 8, 9) and that the length of this path is 21. Sandy's problem has been solved by inspection. It will be solved again, to introduce the reader to dynamic programing.

Sandy's routing problem is now posed as the solution to a recursion that consists of nine equations, one per node. With $f(9) = 0$, define $f(i)$ for each $i \neq 9$ by

$f(i)$ = the length of the shortest path from node i to node 9

A recursion is now developed for $f(i)$. Please recall that t_{ij} denotes the length of arc (i, j); e.g., $t_{57} = 0$. Interpret

$$t_{ij} + f(j)$$

as the length of the shortest path from node i to node 9 whose first arc is (i, j). The quantity $f(i)$ is the length of the shortest path from node i to node 9, so

$$f(i) \leq t_{ij} + f(j)$$

This inequality hods for each arc (i, j) in Figure 3.

The shortest path from node i to node 9 has some arc (i, j) as its initial arc, and its remaining arcs form a path whose length equals $f(j)$. Thus, the displayed inequality holds as an equation for at least one j. In other words,

$$f(i) = \min_{j} \{t_{ij} + f(j)\} \quad \text{for } i \neq 9, \quad \text{with } f(9) = 0 \tag{7}$$

where it is understood that the minimum is taken over all j such that (i, j) is an arc. In dynamic programming, recursions like (7) are known as functional equations.

In Figure 3, each arc (i, j) has $i < j$. Since $f(9) = 0$, the recursion in (7) can be solved in decreasing i, first for $i = 8$, then for $i = 7$, and so forth, ending with $i = 1$. Beginners are urged to solve this recursion themselves and compare their result with what follows.

$$f(8) = 3 + f(9) = 3 + 0 = 3$$

$$f(7) = \min\begin{cases} 6 + f(8) \\ 10 + f(9) \end{cases} = \min\begin{cases} 6 + 3 \\ 10 + 0 \end{cases} = 9$$

$$f(6) = \min\begin{cases} 5 + f(7) \\ 9 + f(8) \end{cases} = \min\begin{cases} 5 + 9 \\ 9 + 3 \end{cases} = 12$$

$$f(5) = \min\begin{cases} 9 + f(7) \\ 15 + f(9) \end{cases} = \min\begin{cases} 9 + 9 \\ 15 + 0 \end{cases} = 15$$

$$f(4) = \min\begin{cases} 3 + f(5) \\ 4 + f(6) \\ 11 + f(7) \end{cases} = \min\begin{cases} 3 + 15 \\ 4 + 12 \\ 11 + 9 \end{cases} = 16$$

$$f(3) = \min\begin{cases} 1 + f(4) \\ 7 + f(6) \end{cases} = \min\begin{cases} 1 + 16 \\ 7 + 12 \end{cases} = 17$$

$$f(2) = \min\begin{cases} 4 + f(4) \\ 8 + f(5) \end{cases} = \min\begin{cases} 4 + 16 \\ 8 + 15 \end{cases} = 20$$

$$f(1) = \min\begin{cases} 2 + f(2) \\ 4 + f(3) \end{cases} = \min\begin{cases} 2 + 20 \\ 4 + 17 \end{cases} = 21$$

Figure 4 shows what this calculation has accomplished. For each node $i \neq 9$, Figure 4 includes one arc (i, j) for which j attains the minimum in Eq. (7). Figure 4 records $f(i)$ above node i. The arcs in Figure 4 include the

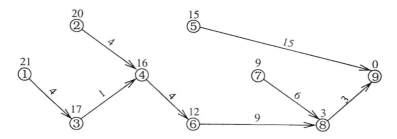

Figure 4 Shortest paths to node 9.

Dynamic Programming

shortest path from each node to node 9. Solving Eq. (7) recursively has produced the "tree" of shortest paths from each node to node 9.

Many writers call Eq. (7) a *backward* recursion because it starts at the end (node from which no arc emanates) and works backward toward the beginning. By contrast, the *forward* recursion is as follows: With $F(1) = 0$, let $F(j)$ denote the length of the shortest path from node 1 to node j. An argument just like the above produces the recursion

$$F(j) = \min_i \{F(i) + t_{ij}\} \quad \text{for } j \neq 1, \text{ with } F(1) = 0$$

where the minimization occurs over all i such that (i, j) is an arc. This forward recursion can be solved in ascending j. Beginners are urged to do so. Solving this recursion produces the tree of shortest paths from node 1 to all others. Of course, this recursion gives $F(9) = 21$.

For Example 7, the length of a path was defined to be the sum of the lengths of its arcs. It can be convenient to define the length of a path as the longest of the lengths of its arcs. Consider:

Example 8: This warm morning, Sandy has an important meeting. He is anxious to look his best when he arrives at work. Now, the length of each arc in Figure 3 is the maximum slope in percent that he must climb on the road it depicts; e.g., arc (1, 2) depicts a road that has a maximum grade of 4% when going from node 1 to node 2. Today, Sandy wishes to select a route that minimizes the largest grade that he must climb. What route is that?

For Example 8, it is natural to define the *length* of a path as the largest of the lengths of its arcs. With that definition of path length, Sandy seeks the shortest path from home to office. With $f(9) = 0$, let $f(i)$ denote the length of the shortest path from i to node 9. The shortest path whose first arc is (i, j) has length (maximum grade) equal to the larger of t_{ij} and f_j, which is written succinctly as $\max\{t_{ij}, f(j)\}$. A familiar argument shows that $f(i) \leq \max\{t_{ij}, f(j)\}$ for each j such that (i, j) is an arc, moreover, that this inequality is satisfied as an equation for some j. In short,

$$f(i) = \min_j \{\max\{t_{ij}, f(j)\}\} \quad \text{for } i \neq 9, \text{ with } f(9) = 0 \quad (8)$$

Equation (8) is a "backward" recursion. It can be solved in decreasing i. Solving it computes a tree of shortest paths to node 9. One such tree is presented in Figure 5. The reader is asked to find a slightly different tree of shortest paths to node 9.

Examples 7 and 8 are shortest-path problems in directed acyclic networks. Both have been solved by recursions that will soon be called functional equations.

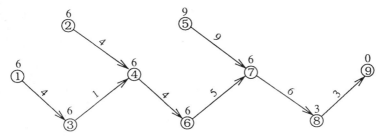

Figure 5 A tree of shortest paths to node 9, with the length of each path defined as the largest of the lengths of its arcs.

Is it perverse to seek the longest path through a directed acyclic network? Decidedly not. Consider:

Example 9 (Critical Path Problem): Sandy is about to undertake a project that consists of six tasks, which he has labeled 1 through 6. Table 2 specifies the time that it takes to complete each task, in weeks. Table 2 also specifies the predecessors of each task, namely the tasks that must be completed before it begins. For instance, task 5 takes 6 weeks, and work on task 5 cannot begin until tasks 3 and 4 are complete. How quickly can Sandy finish the project, and which are the tasks that he cannot delay?

For a reason that will soon be evident, the tasks that Sandy cannot delay form a "critical path." Example 9 is depicted as the directed network in Figure 6. Each task becomes a node in that network. That network has two extra nodes, which are nodes S (for start) and E (for finish). Directed arc (i, j) means that task j cannot begin until task i has been completed. The time that it takes to complete task i is displayed below node i.

For each node i in Figure 6, let t_i denote the number of weeks that it takes to complete the task that node i depicts. Thus, $t_6 = 2$ and $t_E = 0$. Call t_i the *length* of node i. Define the *length* of each path as the sum of the lengths of its nodes. For instance, the path (3, 6, E) has length $10 = 8 + 2 + 0$. Each path corresponds to a sequence of tasks that must be done in the sequence in which they are enumerated. And the length of each path is the

Table 2 Data for Example 9

Task	1	2	3	4	5	6
Completion time	3	7	8	3	6	2
Predecessor(s)	—	—	1	1,2	3,4	3,4

Dynamic Programming

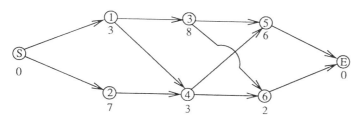

Figure 6 A network representation of Example 9.

time to complete the sequence of tasks that the path represents. In particular, the length of the longest path from node S to node E is the shortest project completion time. A task is said to be *critical* if that task is part of a longest path from node S to node E. The problem of finding the critical tasks is called the *critical path* problem.

For each $i \neq E$, let $f(i)$ denote the length of the longest path from node i to node E. This leads directly to a recursion:

$$f(i) = t_i + \max_j \{f(j)\} \quad \text{for } i \neq E, \text{ with } f(E) = 0 \qquad (9)$$

where the maximization occurs over all j such that (i, j) is an arc. Solving this recursion shows that the project can be completed in 17 weeks. The longest (or critical) path is (S, 1, 3, 5, E). Tasks 1, 3, and 5 are critical; delaying any of them would increase the project completion time. Task 2 can be delayed by 1 week, but not longer. Task 6 can be delayed by 4 weeks but not longer.

Critical path calculations are not mere classroom exercises. They have become a routine part of the management of large-scale projects. The *New York Times* reported on their use in the repair of the World Trade Center after the 1993 bombing of its underground parking garage, for instance.

The preceding three examples illustrate a property that is true in general: For any acyclic network, a tree of shortest (or longest) paths is easily found from a recursion. This is true whether or not the arc lengths are nonnegative, whether or not the nodes are numbered so that each arc (i, j) has $i < j$, and whether or not the network is planar.

Shortest-path methods also apply to networks that have cycles, provided that no cycle has negative length. For cyclic networks with no cycle whose length is negative, trees of shortest paths are characterized by recursions much like the above, but these recursions cannot be solved quite as easily.

If, on the other hand, a directed network has a cycle whose length is negative, the "shortest" path problem is ill-posed. A path can include any number of repetitions of the negative cycle, so the "shortest" paths have minus infinity as their lengths.

This section barely skims a vast literature on shortest- and longest-path problems. For a modern survey, see Ahuja et al. (1993) or Denardo (1982). As presented, the critical path problem (Example 9) does not allow "crashing," i.e., deployment of resources to shorten the times needed to complete critical activities. For an adaptation that allows crashing and connects the critical path problem to nonlinear programming, see Denardo et al. (1994).

4. LANGUAGE OF DYNAMIC PROGRAMMING

Good ideas have deep roots, but Richard Bellman (1952, 1957) is rightly credited with creating the field of dynamic programming and coining the language that describes it, including the terms *functional equation*, *policy*, and *principle of optimality*. This section strives to explain the language he created. An optimization problem is said to be a *dynamic program* if it evolves from state to state, as follows:

1. The decision maker observes the current state. If that state has no actions, decision making ceases. If that state has one or more actions, the decision maker must select one of these actions.
2. Selecting an action earns a reward (or incurs a cost) and causes transition to a new state, whereupon (1) is repeated.
3. The reward (or cost) and the law of motion can depend on the current state and the current action but not on prior states or actions.

Let us interpret Example 7 as a dynamic program. Each node is a state. Suppose state i is observed, with $i \neq 9$. The decision maker must select an action j whose arc (i, j) points away from node i. Selecting action j costs t_{ij} and causes transition to state j, whereupon the decision-making cycle repeats, provided that $j \neq 9$. Decision making ceases when state 9 is observed because state 9 has no actions (arcs pointing away from it). The cost t_{ij} and the state j to which transition occurs depend on the current state i and the action j, but they do not depend on anything that happened before state i was reached.

The shortest- and longest-path problems in the prior section are dynamic programs. The definition of dynamic program specifies neither a particular law of motion from state to state nor a specific way in which the rewards (or costs) are aggregated. Dynamic programs can differ greatly from the shortest- and longest-path problems that have been presented.

For each state i in a dynamic program, let $f(i)$ denote the largest total return (or the least total cost) that the decision maker can earn (or incur) if

Dynamic Programming

the dynamic program begins with the observation of state i and continues until the observation of a state that has no action.

A dynamic program's *functional equation* is a recursion that links the $f(i)$'s as follows:

1. This recursion consists of one equation per state. The solution to this recursion specifies $f(i)$ for each state i.
2. The equation for state i has $f(i)$ just to the left of its equals sign and can have f values for other states on its right-hand side. Just to the right of each equals sign appears a maximization or minimization operation, such as

   "$\max_k$" or "$\min_k$"

   where maximization or minimization occurs over all actions k that can be taken while at state i.

The recursion in (7) is now interpreted as a functional equation. This recursion consists of one equation per state. Its solution specifies $f(i)$ for each i. The equation for $f(i)$ has $f(i)$ to the left of the equals sign and has min j to the right, the minimum being taken over each action j that can be taken if state i is observed.

In a dynamic program, a *policy* is a rule that specifies for each state any one of the actions that the decision maker might take when that state is observed. For the recursion in Eq. (7), state 8 has one possible action (arc pointing away from it), state 4 has three possible actions, and six other states have two actions apiece. That formulation has 192 policies because $192 = 2^6 \times 3$.

The optimization problem for state i is to pick a policy whose total reward is $f(i)$. A policy can be optimal for one state but not for another. To be optimal for a particular state, must a policy be less than optimal for another? Not in dynamic programming. A theme that unifies dynamic programming is the

Principle of optimality. There exists a policy that is optimal for every state.

For functional equation (7), Figure 4 depicts a policy that is optimal, simultaneously, for every state. Using that policy produces the shortest path from each node i to node 9. Similarly, for functional equation (8), Figure 5 depicts a policy that is optimal for every state.

The traditional form of the principle of optimality is now presented This version concerns deviations from the optimal policy. In each of these deviations, the first action is changed, but all other actions are held as they were.

Principle of optimality (traditional version). An optimal policy has the property that whatever the initial state and action are, the remaining actions must constitute an optimal policy with regard to the state resulting from the first transition.

The traditional version is virtually identical to the functional equation.

Dynamic programming is the process of modeling an optimization problem as a dynamic program and solving it. In other words, dynamic programming consists of constructing functional equations and solving them.

Is the "principle" of optimality a fundamental truth? Not quite. Examples 7 through 9 indicate that the principle of optimality holds for three definitions of path length — as the sum of the lengths of its arcs, as the longest of the lengths of its arcs, and as the sum of the lengths of its nodes. The principle of optimality fails for some definitions of path length. Consider:

Example 10 (Trouble): For the network in Figure 3, define the length of each path as the sum of its arc lengths, with alternating signs; for instance, path (i, j, k, m) has length $t_{ij} - t_{jk} + t_{km}$. For the data in Figure 3, the length of path $(1, 3, 6, 8, 9)$ is $4 - 7 + 9 - 3 = 3$.

For Example 10, we shall see that the principle of optimality is violated, specifically, that there is no tree of shortest paths from each node to node 9. Let $f(i)$ denote the length of the shortest path from node i to node 9. Let us calculate $f(7)$ and $f(6)$. The shortest path from node 7 to node 9 is $(7, 8, 9)$, and $f(7) = 3 = 6 - 3$. The shortest path from node 6 to node 9 is $(6, 7, 9)$, and $f(6) = -5 = 5 - 10$. The shortest path from node 6 to node 9 moves first to node 7 and then *avoids* the shortest path from node 7 to node 9. Thus, there is no tree of shortest paths. Similarly, since $-5 = f(6) \neq 2 = \min_j\{t_{6j} - f(j)\}$, the $f(i)$'s do not satisfy a functional equation.

The reader may have guessed that the trouble in Example 10 lies with the equals signs. All dynamic programs for which the principle of optimality is known to hold satisfy the

Monotonicity condition: Increasing an f value on the right-hand side of the recursion for any state i cannot decrease $f(i)$.

Notice that the recursions in Eqs. (7)–(9) do satisfy the monotonicity condition. For Example 10, the recursion for state i would be $f(i) = \min_j\{t_{ij} - f(j)\}$. Increasing $f(j)$ can decrease $f(i)$, so the monotonicity condition is violated. Mitten (1964, 1974) recognized the importance of monotonicity in

Dynamic Programming

dynamic programming. See also Shapley (1953), Blackwell (1965), and Denardo (1965, 1967).

5. FORMULATING DYNAMIC PROGRAMS

The set of states in a dynamic program is called its *state space*. Formulating an optimization problem as a dynamic program can be tricky. There is art in it. The art lies in constructing the state space. This guidepost helps:

Each state is a *regeneration point* — states summarize what came before in enough detail to evaluate current actions and to enable transitions to occur from state to state.

To build a dynamic program, find the regeneration points.

This section contains three optimization problems. Each can be formulated as a dynamic program. These formulations are not easy. Each presents a new difficulty. After reading each example, pause and try to formulate it as a dynamic program. Seek a set of states (regeneration points) and a recursion that links them. Then read the "solution" provided and compare it with your own.

5.1. Minus Signs

In Example 10, the minus signs violated the monotonicity property. That example could not be formulated as a dynamic program whose state space is the set $\{1, 2, \ldots, 9\}$ of nodes. Example 10 can be formulated as a dynamic program, but with a different state space. Consider:

Example 10 (Minus Signs): In Figure 3, use dynamic programming to find the shortest path from node 1 to node 9, where the length of each path is the sum of its arc lengths with alternating signs; e.g., path (1, 2, 3) has length $t_{12} - t_{23}$.

The key is to replicate each node. Create a network whose nodes are $\{1, \hat{1}, 2, \hat{2}, \ldots, 9, \hat{9}\}$. Paths in the expanded network alternate between unhatted and hatted nodes, as follows. For each arc (i, j) in Figure 3, create two arcs, an arc $(i, \hat{j})$ whose length equals t_{ij} and an arc $(\hat{i}, j)$ whose length equals $-t_{ij}$. In the expanded network, states 9 and $\hat{9}$ have no decisions. The expanded network has a tree of shortest paths, each of which ends at node 9 or at node $\hat{9}$. These shortest-path lengths satisfy the functional equation:

$$f(i) = \min_j \{t_{ij} + f(\hat{j})\} \quad \text{for } i \neq 9 \text{ with } f(9) = 0$$
$$f(\hat{i}) = \min_j \{-t_{ij} + f(j)\} \quad \text{for } \hat{i} \neq \hat{9} \text{ with } f(\hat{9}) = 0 \quad (10)$$

This functional equation satisfies the monotonicity principle because the coefficients of the f values on its right-hand side are positive, not negative. One can solve this functional equation recursively, in decreasing i and $\hat{\imath}$. Doing so would verify that the shortest path from node 1 to either "ending" node is $(1, \hat{2}, 4, \hat{5}, 9)$, and $f(1) = -14$.

The key to Example 10 was to replicate the node set. This example may seem contrived. But it and Eq. (10) do suggest, correctly, that dynamic programming applies to games of alternating play, where "states" i and $\hat{\imath}$ describe the status of the game, including the player whose move it is.

5.2. Grouping Jobs into Labor Grades

The next formulation problem describes a trade off faced by factory managers.

Example 11 (forming Labor Grades) (Denardo et al., 1982): A company has N unionized jobs, which are numbered 1 through N, by increasing degree of skill and wage. The natural wage for job i is w_i dollars per month, with $w_1 \leq w_2 \leq \cdots \leq w_N$. Job i requires n_i workers. Of the workers who perform job i, a fraction p_i leave the company each month. Replacements for departed workers must be found and trained.

The union contract allows the company to group the jobs into labor grades as it chooses, subject to three rules. First, a labor grade must consist of consecutive jobs, e.g., of jobs i through j with $i \leq j$. Second, all workers in a labor grade must receive the highest of the natural wages of the jobs in that labor grade. Third, each vacancy must be filled by promotion from the next lower labor grade. Consequently, if an employee in the highest paid of n labor grades quits, each of $n - 1$ employees is promoted to the next higher paid labor grade, and a new employee is hired into the lowest-paid labor grade.

The cost of training an employee to fill a job in a labor grade that consists of jobs i through j is given as t_{ij}. Show how to compute the number n of labor grades and the set of jobs in each grade that minimize the sum of the payroll and training costs.

Example 11 involves a trade-off between wage costs and training costs. Many labor grades means lots of training. Few labor grades means extra wage expense. A balance must be struck. The balance partitions the set $\{1, \ldots, N\}$ of the first N integers in the least expensive way.

One conceivable recursion would partition the set $\{1, \ldots, j\}$ of the first j jobs as cheaply as possible. For it, the natural action is to form a labor grade that consists of jobs i through j, where $i \leq j$. That action causes transition to the problem of partitioning jobs $\{1, \ldots, i - 1\}$ as cheaply as possible. Does this work? It does if the training and wage costs can be accounted for. With $f(0) = 0$, set

Dynamic Programming

$f(j)$ = the least-cost partition of jobs 1 through j into labor grades, including the training costs for all promotions and hiring into jobs 1 through j

The first sum in Eq. (11), below, accounts for the monthly wage cost for a labor grade that consists of jobs i through j. The second sum accounts for all training within that labor grade, including training due to departures from jobs $j + 1$ through N.

$$f(j) = \underset{i=1,\ldots,j}{\text{minimum}} \{f(i - 1) + w_j[\sum_{k=i}^{j} n_k] + t_{ij}[\sum_{k=i}^{N} n_k p_k]\} \quad (11)$$

Thus, the singleton j suffices as a state. State j signifies the partition of jobs 1 through j into labor grades.

This formulation problem has recursions other than Eq. (11). For example, there is a recursion in which state i signifies the partition of jobs i through N into labor grades. With $F(N + 1) = 0$, this formulation gives

$$F(i) = \underset{j=1,\ldots,N}{\text{minimum}} \{w_j[\sum_{k=i}^{j} n_k] + t_{ij}[\sum_{k=i}^{N} n_k p_k] + F(j + 1)\}$$

Curiously, the number of labor grades into which these jobs are partitioned has not been part of the definition of a state. It is possible to take the pair (n, j) as a state, this denoting the partition of jobs 1 through j into n labor grades. With that definition of a state, the functional equation would be exactly like (11), but with $f(n, j)$ and $f(n - 1, i - 1)$ replacing $f(j)$ and $f(i - 1)$, respectively. Equation (11) makes do with a single *state variable*, j. It is correct to use the pair (n, j) of state variables, but n is superfluous. Extra state variables make for extra work, of course.

5.3. Devising an Aggregate Plan

The next formulation problem involves planning for capacity over an extended period of time and on an aggregated basis. An *aggregate plan* focuses on the total production, which has been aggregated over the types of products that the company produces. The aggregate planning model in Example 12 schedules employment to meet a cyclic pattern of demand.

Example 12 (an Aggregate Plan): It is January 1. The personnel manager of Drossel & Meyer, the toy makers, wishes to set a program for hiring and training workers for the current year. The aggregate demand for trained workers is cyclic, as indicated below.

Period i	Jan/Feb 1	Mar/Apr 2	May/June 3	July/Aug 4	Sept/Oct 5	Nov/Dec 6
Number R_i of trained workers required	100	105	130	110	140	120

Drossel-Meyer never makes toys in anticipation of the demand. Drossel-Meyer's personnel policy is kindly; fire no one. Each new worker undergoes a 2-month apprenticeship under the guidance of a trained worker. During that period, the apprentice and the trainer make toys at half of the normal rate for one trained worker. Thus, training an apprentice consumes 2 months of the apprentice's time and 1 month of the time of a trained worker. Experience indicates that 8% of the trainees leave at the end of their 2-month training periods. Similarly, 8% of the trained workers quit at the end of each 2-month period.

At this moment, January 1, Drossel-Meyer has 123 trained workers in its employ. Aiming to minimize cost, Drossel-Meyer wishes to hire as few workers as possible and to hire them as late as possible. How many should they hire, and when? When answering this question, do not worry about next year. Find:

$f(i)$ = the smallest number of trained workers that must be employed at that start of period i in order to meet the demand during periods i through 6

$t(i)$ = the number of apprentices to hire at the start of period i

Hint: Account for two cases, that in which $f(i)$ equals the requirement in period i and that in which $f(i)$ exceeds that requirement.

With R_i as the requirement for trained workers during period i,

$$f(i) \geq R_i, \quad \text{for } i = 1, \ldots, 6$$

If $f(i)$ equals R_i, then no workers can be trained during period i, so $t(i) = 0$.

Discussion focuses on the case in which $f(i)$ exceeds R_i. Since $f(i)$ is the smallest number of workers needed to satisfy the demand during periods i through 6, the excess supply, $[f(i) - R_i]$, of trained workers must be used to train apprentices. In a 2-month period, a trained worker can teach two apprentices, so the number $t(i)$ of trainees in period i satisfies

$$t(i) = 2[f(i) - R_i] \quad \text{if } f(i) > R_i$$

Of the $f(i)$ trained workers and $t(i)$ trainees during period i, 8% will leave, and the remaining 92% will be just enough be to meet the demand during the remaining periods;

$$f(i + 1) = .92[f(i) + t(i)] = .92[3f(i) - 2R_i] \quad \text{if } f(i) > R_i$$

Solving this equation for $f(i)$ gives the bottom line of the functional equation that follows. With $f(7) = 0$.

$$f(i) = \max \begin{cases} R_i \\ \frac{2}{3}R_i + \frac{1}{3}\frac{f(i+1)}{.92} \end{cases} \quad \text{for } i = 1, \ldots, 6 \qquad (12)$$

Combining $t(i) = 2[f(i) - R_i]$ with the above gives:

$$t(i) = \max \begin{cases} 0 \\ \frac{2}{3}\left[\frac{f(i+1)}{.92} - R_i\right] \end{cases} \quad \text{for } i = 1, \ldots, 6$$

Solving these recursions produces the aggregate plan in Table 3.

The aggregate plan in Table 3 calls for training fractional numbers of workers. One issue is how to convert these fractions into a hiring plan. There are other issues, however, whose importance swamps roundoff. Three of these issues are now discussed.

Does the plan in Table 3 keep enough people on the payroll at the end of the year to build for next year's peak? Yes. Peak employment is 140 workers at the start of period 5. Of that number, $140(.92)^2 = 118.5$ remain next January, which exceeds the start-of-year requirement of 109.3.

What should Drossel-Meyer do this year, which starts with 123 trained workers, not 109.3? To be "with" the plan in Table 3 at beginning of March, they should train a number $t(1)$ in January such that $.92[123 + t(1)] = 117.7$. Thus, Drossel-Meyer needs to hire $t(1) = 4.93$ trainees now, not the 18.6 indicated in Table 3.

The aggregate plan in Table 3 spends the entire 8-month period from January through August building capacity for the pre-Christmas peak, which occurs in period 5. Given the uncertain demand for toys, would it be

Table 3 An Aggregate Plan for Drossel & Meyer

i	1	2	3	4	5	6
R_i	100.0	105.0	130.0	110.0	140.0	120.0
$t(i)$	18.6	25.4	3.3	28.1	0	0
$f(i)$	109.3	117.7	131.6	124.1	140	120

prudent to build extra capacity by training a few more than 5 workers now? If so, how many?

That's a "newsboy problem" whose solution is offered with a mere sketch of an explanation. To avoid abstraction, three assumptions are made. First, assume that the demand D_i for workers in period i is a random variable whose expectation equals R_i, as given in Table 3, and whose variance equals 180. Second, assume that the demands in different periods are independent of each other. Thus the total demand S_5 during periods 1 through 5 is a random variable whose mean equals 585 (because $R_1 + \cdots + R_5 = 585$) and whose variance equals 900 (because variances of independent random variables add and $900 = 5 \times 180$). Thus, the standard deviation of S_5 equals 30 (because $30 = \sqrt{900}$). Third, assume that S_5 has the normal distribution, approximately. Then the extra number x of employees to train in January should satisfy the *critical fractile* equation.

$$\text{Prob}\{Z \le \frac{x}{30}\} = \frac{(.92)^4 C_u}{C_o + (.92)^4 C_u}$$

where Z is a (standard) normal random variable whose mean equals zero and whose variance equals one, the "underage cost" C_u is the loss of profit due to having one worker too few during period 5, and the "overage cost" C_o equals the wage cost of hiring one employee now and paying that person for the year or until he or she leaves, whichever comes first. The factor $(.92)^4 C_u$ appears in the critical fractile equation because only the fraction $(.92)^4$ of the extra worker hired on January 1 remains at the start of period 5.

Example 11 is a rudimentary aggregate planning model. This example allows hiring but not firing. It does not allow for making toys before they are needed. When coupled with a "newsboy" calculation, it does treat the uncertainty in the demand for toys.

Typically, an aggregate plan is reviewed periodically, on a "rolling" basis, e.g., every 2 months. When demand is cyclic, it is wise to plan for an interval of time that ends with a "valley" in demand, e.g., on January 1. The capacity that remains at the beginning of the valley is likely to suffice as a starting point for the next cycle.

Aggregate plans are sometimes set by coupling linear programming with a method that deals with the major sources of uncertainty. A linear program that solves Drossel-Meyer's aggregate planning problem is presented below:

Minimize $\{I_1 + I_2 + \cdots + I_6\}$ subject to the constraints

$$.92(I_k + x_k) = I_{k+1}, \quad \text{for } k = 1, \ldots, 6$$
$$x_k \le 2(I_k + R_k), \quad \text{for } k = 1, \ldots, 6$$
$$I_k \ge 0, x_k \ge 0, \quad \text{for } k = 1, \ldots, 6$$

Dynamic Programming

In this linear program, the decision variable I_k equals the number of trained workers on hand at the start of period k, and x_k equals the number hired and trained during period k. We have provided a newsboy scheme for dealing with the uncertainty in this model; for another, see Chapter 8.

6. THE DYNAMIC LOT SIZE MODEL

The dynamic lot size model addresses an important planning problem. Production quantities of one product must be scheduled for each of several periods. The demand for the product in each period is assumed to be known. Each unit of demand must be satisfied, either by producing that unit during the period when it is needed or by producing the unit in an earlier period and storing it until it is needed. There are economies of scale in the production facility; e.g., doubling a production quantity can increase the production cost by a factor that is below 2. This model gets its name from the fact that the production quantities (lot sizes) can vary with the period in response to fluctuating (dynamic) demands.

The economies of scale preclude direct application of the methods of linear and convex nonlinear programming, since we shall be minimizing a concave function. But this model is easy to solve by dynamic programming. For a planning interval of n periods, computing a least-cost production plan requires a number of computer operations that is proportional to n^2 or less. A dynamic lot size model is presented as:

Example 13 (Production Planning with Known Demand and Economies of Scale) (Wagner and Whitin, 1958): Production of a single product must be scheduled for n periods, which are numbered 1 through n. For each j, the demand for the product that occurs during period j is the known nonnegative number d_j of units. Each unit of demand must be satisfied, either by producing that unit in the period when the demand for it occurs or by producing it in some earlier period and storing it until needed. For convenient exposition, it is assumed that the d_j's are integer valued, that the quantity produced during each period must be integer valued, that the initial inventory of the product equals zero, and that the final inventory must equal zero.

The goal is to satisfy the demands for the n periods at minimum total cost. The cost of producing x units during period j is specified by the function $p_j(x)$. The cost of holding I units in inventory at the *start* of period j is specified by the function $h_j(I)$. For convenient exposition, it is assumed that producing nothing costs nothing and that holding no inventory costs nothing, so that $0 = p_j(0) = h_j(0)$ for each j. For each period j, the production and holding cost functions are assumed to satisfy

$$p_j(x + 1) - p_j(x) \le p_j(x) - p_j(x - 1), \quad \text{for } x = 1, 2, \ldots \quad (13)$$
$$h_j(I + 1) - h_j(I) \le h_j(I) - h_j(I - 1), \quad \text{for } I = 1, 2, \ldots \quad (14)$$

Expression (13) states that the incremental cost, $p_j(x + 1) - p_j(x)$, of producing unit $x + 1$ during period j does not exceed the incremental cost of producing unit x during that period. Expression (14) states that the incremental cost of holding item $I + 1$ in inventory at the start of period j does not exceed the incremental cost of holding item I. It is assumed that not all of these inequalities hold as equations. (In the case of linear costs, this planning problem is easy.)

Figure 7 exhibits two cost functions that satisfy (13). The dashed lines in Figure 7 indicate that these functions become concave when linear interpolation is used to extend their domains to the nonnegative numbers. Expressions (13) and (14) identify the case of concave costs, equivalently, of *economies of scale*. The function at the left in Figure 7 models a common situation, namely a production cost that is the sum of a fixed charge for producing any units plus a constant charge per unit.

A functional equation (recursion) will be developed for Example 13. We could obtain a forward recursion that plans for the first i periods or a backward recursion that plans for the final i periods. The backward recursion will prove to be more convenient. Aiming for it, we number the periods backward. Period 1 occurs last. Period n occurs first. By period j, it is meant that j periods remain until the moment at which planning ends. In particular, d_1 is the demand in the final period.

The dynamic lot sizing problem is presented below as Program 1. The decision variables in Program 1 are, for each j, the number x_j of units to produce during period j and the number I_j of units of inventory on hand at the *start* of period j.

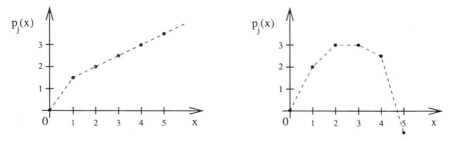

Figure 7 Cost functions that exhibit economy of scale.

Dynamic Programming

Program 1: Compute $f(n) = \text{Minimum}_{x,t} \sum_{j=1}^{n} \{p_j(x_j) + h_j(I_j)\}$, subject to

$I_n = 0$ and $I_0 = 0$

$I_j + x_j - d_j = I_{j-1}$, for $j = 1, \ldots, n$

$x_j \geq 0$, x_j integer valued, for $j = 1, \ldots, n$

$I_j \geq 0$, I_j integer valued, for $j = 1, \ldots, n-1$

The objective of Program 1 measures the total cost over the n periods for which production is planned. The first constraint represents the boundary conditions; the inventory position I_n at the start of the earliest period equals zero, and the inventory position I_0 at the end of the final period equals zero. The second group of constraints are of a "conservation" sort; the inventory I_{j-1} on hand at the end of period j equals the inventory I_j on hand at the start of period j plus the production x_j during period j less the demand d_j during period j. The remaining constraints keep the decision variables nonnegative and integer valued. The optimal value of Program 1 has been labeled $f(n)$ in anticipation of a recursion on n.

Figure 8 interprets Program 1 as a network flow problem. for $j \geq 1$, the conservation-of-flow constraint for node j is that the flow in, which equals $x_j + I_j$, equals the flow out, which equals $d_j + I_{j-1}$. A fixed flow of d_j emerges from node j. A fixed flow of $(d_1 + \cdots + d_n)$ enters node 0. Each of the variable flows in Figure 8 has a cost; e.g., the cost of the flow x_j equals $p_j(x_j)$, and the cost of the flow I_j equals $h_j(I_j)$.

To relate Figure 8 to standard network flow formulations, let the x_j's

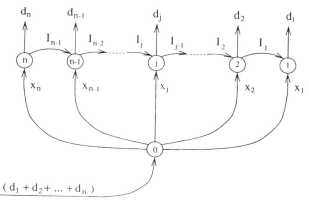

Figure 8 A network flow formulation of Program 1.

and the I_j's take any nonnegative values, and use linear interpolation to define $p_j(x_j)$ and $h_j(I_j)$ for every nonnegative real number x_j and I_j. As the dashed lines in Figure 7 indicate, linear interpolation makes the cost functions concave. Thus, the network flow problem in Figure 8 minimizes a concave function subject to linear constraints and nonnegative variables. It can be shown that the global minimum occurs at an extreme point of the set of feasible solutions. But every extreme point can be a local minimum. For that reason, direct application of the methods of linear and convex programming fail to solve Program 1.

The key to solving Program 1 by dynamic programming is found in:

Proposition 1: At least one optimal solution to Program 1 has

$$I_k x_k = 0, \quad \text{for } k = 1, 2, \ldots, n$$

Remark: The proofs can be omitted with no loss of continuity. The reader may wish to skip or skim them, at least on first reading.

Proof: Program 1 has finitely many feasible solutions, so it has at least one optimal solution. Aiming for a proof by contradiction, we assume that every optimal solution has $I_j x_j > 0$ for at least one j. Among the optimal solutions, pick one that has $I_k x_k = 0$ for $k = j + 1$ through n with j minimized. Since $I_n = 0$, we have $j < n$. Since the theorem is assumed to be false, we have $j \geq 1$. Necessarily, I_j and x_j are positive.

Pick as s the smallest integer such that $s > j$ and $x_s > 0$. Such an s must exist because inventory I_j remains from production during some prior period. By construction, $I_s = 0$. Let y denote the smaller of I_j and x_j. Perturb the optimal solution by adding y to x_s, by adding y to I_k for each k such that $s > k > j$, and by subtracting y from x_j. This perturbed solution remains feasible. Effectively, this perturbed solution ships y units around the "loop" $(0, s, s - 1, \ldots, j, 0)$. Next, perturb the optimal solution by subtracting y from x_s, by subtracting y from I_k for each k such that $s > k \geq j$, and by adding y to x_j. This perturbed solution remains feasible because $I_s = 0$ and $x_s \geq I_k \geq I_j \geq y$ for all k having $s > k \geq j$. This perturbed solution ships $-y$ units around the loop $(0, s, s - 1, \ldots, j, 0)$.

The optimal solution is the average of these two feasible solutions. Since the cost functions are concave, the cost of the optimal solution is at least as large the average of the costs of these two feasible solutions. Thus, all three solutions are optimal. One of the perturbed solutions reduces to zero the smaller of I_j and x_j. That perturbed solution satisfies $I_k x_k = 0$ for all $k \geq j$. This contradicts the minimality of j, which proves Proposition 1. □

The condition $I_k x_k = 0$ states that x_k cannot be positive unless I_k equals zero. In other words, production need not occur in any period k

whose starting inventory I_k is not zero. For a feasible solution to Program 1, call period k a *regeneration point* if its start-of-period inventory equals zero, that is, if $I_k = 0$. The initial and final conditions mean that periods n and 0 are regeneration points. Proposition 1 states:

At lease one optimal solution to Program 1 causes production to occur only at regeneration points.

These regeneration points will be the states in a dynamic program. In order to build this dynamic program, it is necessary to introduce some notation. Let us designate by D_i the total demand in the final i periods. With $D_0 = 0$,

$$D_i = d_1 + d_2 + \cdots + d_i \quad \text{for } i \geq 1$$

Imagine that periods i and $j < i$ are regeneration points, and imagine that no period between them is a regeneration point. Then the quantity x_i produced in period i must equal the total demand in periods i through $j + 1$, so $x_i = D_i - D_j$. This leads us to define $c(i, j)$ by

$c(i, j)$ = the cost incurred during periods i through $j + 1$
 if $I_i = 0$ and if $x_i = (D_i - D_j)$

To specify $c(i, j)$, first note that $p_i(D_i - D_j)$ equals the cost of producing $(D_i - D_j)$ units during period i. For each period k such that $i > k > j$, the number of units that remain in inventory at the start of period k equals $(D_k - D_j)$, and the cost of holding that inventory equals $h_k(D_k - D_j)$. Thus,

$$c(i, j) = p_i(D_i - D_j) + \sum_{k=j+1}^{i-1} h_k(D_k - D_j)$$

Program 1 interprets $f(n)$ as the least cost for all n periods of the plans that satisfy the demands, with beginning inventory $I_n = 0$. Similarly, $f(i)$ is the least cost for the final i periods of the plans that satisfy the demands during those periods, with beginning inventory $I_i = 0$. A functional equation for $f(i)$ is presented in:

Proposition 2: With $f(0) = 0$, one has, for $i = 1, \cdots, n$,

$$f(i) = \text{minimum}\{c(i, j) + f(j) : j = 0, \ldots, i - 1\} \tag{15}$$

Proof: Equation (15) follows directly from Proposition 1; e.g., $f(n)$ is satisfied by a plan that sets $x_n = D_n - D_j$ for some j, and that plan costs $c(n, j) + f(j)$. □

To compute $f(i)$ from functional equation (15), one must know the values $f(0), f(1), \ldots, f(i - 1)$ that might appear on its right-hand side.

For that reason, $f(i)$ can be computed in ascending i, starting with $f(1)$ and ending with $f(n)$.

To help construct a plan whose cost equals $f(n)$, one can record optimizers while solving functional equation (15). Specifically, while computing $f(i)$, record as $m(i)$ a value of j that minimizes the right-hand side of (15). Let us break any possible ties by defining $m(i)$ as the smallest such value:

$$m(i) = \text{minimum}\{j: f(i) = c(i, j) + f(j)\}, \quad \text{for } i \leq n \quad (16)$$

These $m(i)$'s let one *backtrack* a production plan whose cost equals $f(n)$. This is accomplished by the

Routine to backtrack an optimal production plan:
Set $i = n$.
While $i \geq 1$ do
  begin
    Set $j = m(i)$ and $x_i = (D_i - D_j)$
    Set $x_k = 0$ for $i > k > j$
    Set $i \leftarrow j$.
  end

A solution to the dynamic lot size model has now been presented. We have seen that $f(n)$ is the cost of the least expensive plan that satisfies demand for the n periods. We have seen that $f(n)$ can be found from recursion (15). We have seen that an optimal plan can be constructed by backtracking optimizers to (15).

The dynamic lot size model has been solved by a backward recursion. The reader is encouraged to develop the comparable forward recursion, as well as its backtracking routine.

The *work* needed by an algorithm equals the number of computer operations (memory accesses, additions, multiplications, comparisons, etc.) that its execution requires. The work that is needed to solve functional equation (15) is the subject of:

Proposition 3: The work needed to find $f(n)$ by solving (15) in ascending i is proportional to n^2.

Proof: Let us first build a table of D_0 through D_n from the recursion $D_i = d_i + D_{i-1}$ with $D_0 = 0$. This takes work proportional to n. Let us next build a "holding cost" table $H(i, j)$ from $H(j, j) = 0$ and $H(i - 1, j) = H(i, j) + h_i(D_i - D_j)$, which takes work proportional to n^2. Let us next compute a table of values of $c(i, j)$ from $c(i, j) = p_i(D_i - D_j) + H(i, j)$, which takes work proportional to n^2. Finally, we solve (15) in ascending i, which also takes work proportional to n^2. □

Dynamic Programming

To interpret the dynamic lot size model as a shortest-path problem, we build a network whose nodes are the integers 0 through n. For each pair (i, j) having $n \geq i > j \geq 0$, this network has arc (i, j) whose length equals $c(i, j)$. Proposition 2 shows that $f(n)$ equals the length of the shortest path in this network from node n to node 0.

The number of arcs in this network equals $(1 + 2 + \cdots + n) = n(n + 1)/2 \approx n^2/2$. For that reason, the work bound of n^2 in Proposition 3 is not surprising.

An important theme in dynamic programming is to use the structure of a particular sequential decision process to compute its optimal policy with as little effort as is possible. The dynamic lot size model illustrates that theme. We have seen that its optimal solution satisfies $I_k x_k = 0$ for each k, which lets us compute an optimal policy for an n-period planning problem with work proportional to n^2. For an important special case, that work bound can be reduced to n; see Federgruen and Tzur (1991), Wagelmans et al. (1992), and/or Aggarwal and Park (1993).

One limitation of the dynamic lot size model lies in its extreme simplicity. This model treats only one type of item. It treats only one stage of production. It excludes any restriction of production capacity. The essential features of this model can be preserved, however, in more elaborate contexts that allow multiple items, stages of production, and capacity constraints. For entry into the literature on those subjects, the student is referred to Graves (1982), Luss (1986), and Roundy (1993).

A second limitation of the dynamic lot size model lies in its assumption that the demands are known. It is more typical that accurate information about costs and demands is available for a *data horizon*, T, that is well short of the problem's natural *planning interval*, n. The dynamic lot size model is robust enough to enable calculation of the error bound (worst possible error) that can occur due to lack of information in periods beyond the data horizon. Several authors have provided conditions under which the error bound equals zero. This error bound has been computed by Lee and Denardo (1986) and, more efficiently, by Chen and Lee (1995).

7. CAPACITY EXPANSION WITH KNOWN REQUIREMENTS

The section studies the problem of timing increases in capacity in the presence of requirements that are known and are growing with an economy of scale in the cost of adding capacity. This model generalizes the dynamic lot size model. It is presented as Example 14.

Example 14 (Capacity Planning with Known Requirements and Economies of Scale) (Zangwill, 1966; Manne and Veinott, 1967): Capacity must be set for n periods, which are numbered 1 through n in their natural order. Period 1 occurs first. Period $j + 1$ follows period j. Let the datum D_j denote the total capacity that is required by the *start* of period j. For convenient exposition, assume that no capacity exists at the start of period 1 and that none is required then, so that $D_1 = 0$.

Capacity can be built during many period, but exactly D_{n+1} units of capacity must be built by the end of period n. For each period j, if insufficient capacity exists at the beginning the period, the deficit must be rented for that period. Similarly, if excess capacity exists at the beginning of period j, the excess can be rented out for that period.

The object is to add the D_{n+1} units of capacity in a way that minimizes the total cost incurred over the n-period planning interval. The cost of adding x units of capacity during period j is specified by the function $p_j(x)$. The cost of renting y units of capacity from others during period j is specified by the function $r_j(y)$. Similarly, the revenue earned from renting out z units of capacity to others during period j is specified by the function $R_j(z)$.

For each j, the (cost) functions $p_j(x)$, $r_j(y)$, and $R_j(z)$ of x, y, and z are assumed to be concave. This models economies of scale. The cost functions are assumed to satisfy $r_j(y) \geq R_j(y)$ for each y, which precludes arbitrage in the rental market. For convenient exposition, it is also assumed that $0 = p_j(0) = r_j(0) = R_j(0)$ for each j.

In the capacity expansion model, period 1 occurs first (not last) because we shall develop a forward recursion. To see that the capacity expansion model encompasses the dynamic lot size model, imagine that the cost $r_j(x)$ of renting capacity from others is so large that it never pays to rent from others. In that case, one builds (produces) in anticipation of the demand, exactly as in the prior section.

Let the decision variable X_j denote the total amount of capacity that has been build by the start of period j. Notice that $X_{j+1} = X_j + x_j$, where x_j equals the amount of capacity that is built during period j. It has been assumed that $X_1 = 0$ and that $X_{n+1} = D_{n+1}$. The problem of adding capacity in the most economical way appears below as Program 2.

Program 2: Minimize $\displaystyle\operatorname*{}_{x,y,z} \left[\sum_{j=1}^{n}\{p_j(x_j) + r_j(y_j) - R_j(z_j)\}\right]$, subject to

$X_1 = 0$ and $X_{n+1} = D_{n+1}$
$X_{j+1} = X_j + x_j$, for $j = 1, \ldots, n$
$x_j \geq 0$, for $j = 1, \ldots, n$
$D_j - X_j = y_j - z_j$, for $j = 2, \ldots, n$
$y_j \geq 0, z_j \geq 0$, for $j = 1, \ldots, n$

The objective of Program 2 measures the net cost incurred over the n

Dynamic Programming

periods for which capacity expansion is planned. The first constraint represents the boundary conditions; the initial capacity X_1 equals 0, and final capacity X_{n+1} must equal the final requirement D_{n+1}. The next constraints relate total capacity to incremental capacity and require each increment x_j to be nonnegative. The final set of constraints equate the "excess" $D_j - X_j$ of demand over supply at the start of period j to the difference $y_j - z_j$ between capacity y_j rented from others and capacity z_j rented to others during that period.

Program 2 will look more familiar after the variables X_1 through X_n have been eliminated from it. Preparing to do so, define the datum d_j as the increase in the requirement for capacity that occurs during period j:

$$d_j = D_{j+1} - D_j$$

The constraints in Program 2 give

$$\begin{aligned} y_{j+1} - z_{j+1} &= D_{j+1} - X_{j+1} \\ &= (D_j + d_j) - (X_j + x_j) \\ &= (D_j - X_j) + d_j - x_j \\ &= (y_j - z_j) + d_j - x_j \end{aligned}$$

This lets Program 2 be recast as:

Program 2: $\displaystyle \text{Minimize}_{x,y,z} \left[\sum_{j=1}^{n} \{p_j(x_j) + r_j(y_j) - R_j(z_j)\} \right]$, subject to

$$\begin{aligned} y_{j+1} - z_{j+1} &= y_j + d_j - z_j - x_j, \quad \text{for } j = 1, \ldots, n \\ y_1 &= z_1 = 0 \\ y_{n+1} &= z_{n+1} = 0 \end{aligned}$$

$$x_j \geq 0, \, y_j \geq 0, \, z_j \geq 0 \quad \text{for } j = 1, \ldots, n$$

Figure 9 depicts this version of Program 2 as a network flow problem. Its flow conservation constraint for node j is that flow in, which equals $y_{j+1} + z_j + x_j$, equals flow out, which equals $y_j + d_j + z_{j+1}$. A fixed flow of d_j emerges from node j. A fixed flow of $(d_1 + \cdots + d_n)$ enters node 0. Each of the variable flows in Figure 9 has a cost; the cost of the flow x_j equals $p_j(x_j)$, the cost of the flow y_j equals $r_j(y_j)$, and the cost of the flow z_j equals $-R_j(z_j)$.

Figure 9 interprets Program 2 as a network flow problem with concave costs. Because arbitrage in the rental market is assumed to be unprofitable, its global minimum occurs at an extreme point.

An optimal solution to Program 2 will be found from a functional equation, much like that of the dynamic lot size model. Period i is now called a *regeneration point* if $y_i = z_i = 0$. If period i is a regeneration

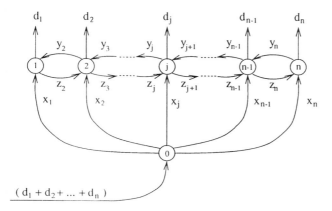

Figure 9 A network flow formulation of Program 2.

point, then the total capacity X_i built by the start of period i equals the total requirement D_i at that moment in time. Periods 1 and $n + 1$ are required to be regeneration points. Other periods may be regeneration points. Periods i and $j > i$ are now called *consecutive* regeneration points if they are regeneration points and if no period between them is a regeneration point. The key to Program 2 is found in:

Proposition 4: At least one optimal solution to Program 2 has these properties: $y_j z_j = 0$ for each j and, if i and $j > i$ are consecutive regeneration points, then $x_k = (D_j - D_i)$ for some k such that $i \leq k < j$.

Sketch of Proof: Consider a feasible solution to Program 2. In this feasible solution, an arc is now said to be "active" if the flow on that arc is nonzero. As in the proof of Proposition 1, a flow whose active arcs contain a "loop" is not an extreme point because it can be perturbed by shipping ϵ and $-\epsilon$ around the loop. A flow in which y_j and z_j are positive contains the loop $(j - 1, j, j - 1)$, hence cannot be an extreme point. Suppose i and $j > i$ are consecutive regeneration points. A flow that has x_p and x_q positive for periods p and q such that $i \leq p < q < j$ has a loop $(0, p, \ldots, q, 0)$, hence cannot be an extreme point.

Program 2 minimizes a concave function. Its objective is bounded because arbitrage is unprofitable. Thus, an extreme point is optimal for Program 2. That extreme point must have $y_j z_j = 0$ for each j. If i and $j > i$ are consecutive regeneration points, it must have $x_k = D_j - D_i$ for some k having $i \leq k < j$. □

Dynamic Programming

Proposition 4 states:

An optimal solution to Program 2 can be found from among the feasible solutions in which capacity is increased only once between consecutive regeneration points.

This property leads, by a familiar route, to a functional equation. It remains to account for the cost that occurs between transitions. Define $c(i, j, k)$ for each i, j, and k that satisfy $1 \leq i \leq k < j \leq n + 1$ by

$c(i, j, k)$ = the cost incurred during periods i through $j - 1$
if i and j are consecutive regeneration points
and if $x_k = (D_j - D_i)$

The cost $c(i, j, k)$ includes the cost $p_k(D_j - D_i)$ of building $(D_j - D_i)$ units of capacity during period k. For each period p such that $i < p \leq k$, the deficit in capacity at the start of period p equals $(D_p - D_i)$, which costs $r_p(D_p - D_i)$. For each period q such that $k < q < j$, the surplus in capacity at the start of period q equals $(D_j - D_q)$, which earns $R_q(D_j - D_q)$. Therefore,

$$c(i, j, k) = p_k(D_j - D_i) + \sum_{p=i+1}^{k} r_p(D_p - D_i) - \sum_{q=k+1}^{j-1} R_q(D_j - D_q)$$

If periods i and j are consecutive regeneration points, then capacity $(D_j - D_i)$ is added in the period k that solves

$$c(i, j) = \underset{k=i,\ldots,j-1}{\text{minimum}} \{c(i, j, k)\} \tag{17}$$

Let $f(j)$ denote the least cost incurred during periods 1 through $j - 1$ with periods 1 and j as regeneration points. Hence, $f(n + 1)$ is the optimal value of Program 2. A forward recursion for $f(j)$ is given by

Proposition 5: With $f(1) = 0$,

$$f(j) = \underset{i=1,\ldots,j-1}{\text{minimum}} \{f(i) + c(i, j)\}, \quad \text{for } j = 2, \ldots, n + 1 \tag{18}$$

Proof: Immediate from Proposition 4. □

Functional equation (18) solves the capacity expansion model. To compute $f(j)$ from this functional equation, one must know the values $f(1)$, $\ldots$, $f(j - 1)$ that might appear on its right-hand side; $f(j)$ can be computed in ascending j. It is not difficult to see that $f(n)$ can be found from (17) and (18) with work proportional to n^3, including the work needed to build tables of $c(i, j, k)$ and $c(i, j)$. If the sole nonlinearities are setup costs for increasing capacity, this work bound can be reduced to $n^2 \log n$.

If, in addition, there is no arbitrage opportunity in inventory, the work bound reduces to n^2.

The use of capacity expansion models is surveyed by Luss (1982), where it is observed that the natural planning interval can be a decade or longer and that the optimal plan can be sensitive to the discount factor. With discounting, interpret $p_i(x)$ as the present value at the beginning of the planning period of the building x units of capacity during period i, for instance. Sensitivity to the discount factor motivates planning on a rolling basis, with a bound on the error, as in Chen and Lee (1995).

8. DECISION TREES

Uncertainty about the future lies at the heart of many decision-making situations. A company that decides to purchase a piece of equipment has imprecise knowledge of its useful lifetime, its reliability, and its maintenance costs. A store that purchases merchandise has inexact knowledge of what its customers will wish to buy. An airline that sets a flight schedule does not know what demands will occur. A medical patient elects to undergo a course of treatment without knowing its outcome.

Although the future is unknown, it may obey a simple probability law. To illustrate, we consider the demand for a consumer product, say, for a particular brand of cereal or a particular type of automobile. Typically, the demand for such a product has these four properties:

1. The number of potential customers is large.
2. No single customer accounts for a significant fraction of the total demand.
3. Different customers make their purchase decisions more or less independently of each other.
4. There is some randomness in the time between successive purchases by any given customer.

Different customers do tend to make independent decisions: If interest rates rise, many customers may delay purchases of new cars; their decisions depend on the interest rate, not on each other's decisions.

For a demand process that satisfies these four conditions, there is this surprisingly little-known *superposition theorem* (see Çinlar, 1972): The total demand D over any given period of time is a Poisson random variable, which means

$$\Pr\{D = k\} = e^{-\lambda}\left(\frac{\lambda^k}{k!}\right), \quad \text{for } k = 0, 1, 2, \ldots$$

where $\lambda = ED$. Thus, to describe this sort of demand, one need only estimate its mean, λ, over the time period in question.

Dynamic Programming

Similarly, the uncertainty in a course of medical treatment can be expressed in terms of probabilities, e.g., the probability that chemotherapy will eradicate a particular type of cancer in a certain class of patient. In this context, the probabilities may model our lack of information, rather than inherent randomness.

There is a well-developed theory of decision making in the presence of an uncertain future, and dynamic programming lies at the heart of it. This section begins with an example of decision making under uncertainty. It ends with a more general discussion. The example is taken from systems engineering.

Example 15: Imagine that you are the Florida state official who is responsible for controlling crop damage from pests. A localized infestation of medflies has just been detected. It threatens the citrus region, where it would cause tens of billions of dollars in damage.

The only surefire way to get rid of the medfly is to burn the area in which it has been detected. Burning costs $60 million if done now. The area of infestation is likely to double in 2 months. Burning would cost $84 million if delayed 1 month and $120 million if delayed 2 months.

Spraying with malathion might eliminate the infestation. The infected area can be sprayed now, at a cost of $3 million. If spraying starts within a few days, it will be learned in 1 month whether or not the medfly has been wiped out. The probability that spraying will eradicate them is 1/3. If spraying fails, it can be repeated 1 month from now at a cost of $4 million, with the same probability of success.

If spraying is undertaken now, a backup option is available, due to the fact that the female medfly mates but once. Should the female mate with a sterile male, no larvae are produced. A gigantic crop of sterile males can be incubated in a laboratory for 1 month, sterilized by x-ray, and then released if they are needed. The incubation cost is $6 million. The sterile males can be released or destroyed at no additional cost. If released, the probability that they eradicate what's left after the failure of an application of spray is 1/2.

The governor just phoned you. Mindful of what happened to the career of ex-Governor Ed "Moonbeam" Brown of California, she has given you 24 hours to present her with a plan for dealing with the pest. Organize your presentation. (The medfly example was developed by Ludo Van der Heyden and this writer for use at Yale's School of Management.)

This problem is ripe for analysis using a decision tree. A *decision tree* consists of choice nodes, chance nodes, and arcs. A *choice node* is a situation in which an action must be selected. A *chance node* is a point at which an element of uncertainty is resolved. In a decision tree, the choice nodes are drawn as squares with numbers inside. The chance nodes are drawn as circles. An *arc* (line segment) is drawn between two nodes if transition can occur from one to the other. By convention, time moves from left to right. Thus, a decision tree "grows" sideways.

In the pest control problem what actions are available today? There are three sensible choices: burn now, spray now, and spray now while incubating a crop of sterilized males. There are some dumb choices too; e.g., do nothing for 1 month, then burn. The sensible choices form the tree in Figure 10. Displayed in this tree are a cost on each arc that represents an expense and a probability on each arc that emerges (rightward) from a chance node.

The choice nodes in Figure 10 are numbered 1 through 5, and these nodes have been numbered so that each transition occurs to a higher-numbered node. In the language of dynamic programming, each choice node is a state, and each arc pointing rightward from a state is an action.

A decision tree indicates what courses of action might be taken. The tree specifies each action's cost and transition probabilities. A decision tree helps the participants in the decision reach a common understanding of the alternatives and their risks. For this reason, you might even present the decision tree in Figure 10 to the governor!

A decision-making situation is said to be *strategic* if it can have dire consequences that jeopardize the life, health, or well-being of the individual

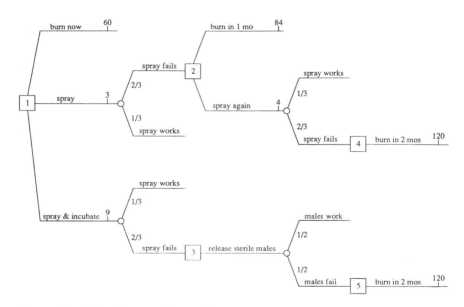

Figure 10 A decision tree for medfly eradication.

Dynamic Programming

or organization. By contrast, a decision-making situation is said to be *operational* if the risks it entails are routine.

Virtually no decision is purely operational. For the medfly example, the worst financial outcome is a $129 million dollar loss. This is a routine expense for the state of Florida; it does not threaten the state with bankruptcy. On the other hand, the governor's career could be wrecked by a really bad policy, e.g., do nothing for 2 months and then burn. As governor, she faces an operational decision that has strategic overtones.

On a worst-case basis, the best plan is to burn now. This costs $60 million, but not more. For operational decisions, the prudent manager would select a plan whose expectation of cost is smallest, rather than one whose worst-case cost is smallest.

What strategy minimizes the expectation of the cost of eradicating the medfly? A functional equation will answer this question. For each state (choice node) i, define $f(i)$ by

$f(i)$ = the lowest attainable value of the expectation of the cost incurred from the moment choice node i is observed until decision making ceases

It is evident in Figure 10 that:

$$f(5) = 120$$
$$f(4) = 120$$
$$f(3) = \tfrac{1}{2} 0 + \tfrac{1}{2} f(5) = 60$$

The expectation of the sum equals the sum of the expectations. In particular, the expected total cost equals the cost that is incurred now plus the expectation of the cost that is incurred in the future. This leads to

$$f(2) = \min \begin{cases} 84 \\ 4 + \tfrac{1}{3} 0 + \tfrac{2}{3} f(4) \end{cases} = \min \begin{cases} 84 \\ 84 \end{cases} = 84$$

$$f(1) = \min \begin{cases} 60 \\ 3 + \tfrac{2}{3} f(2) + \tfrac{1}{3} 0 \\ 9 + \tfrac{1}{3} 0 + \tfrac{2}{3} f(3) \end{cases} = \min \begin{cases} 60 \\ 59 \\ 49 \end{cases} = 49$$

Thus, the least value of the expected cost is $49 million. Figure 11 reproduces Figure 10 but records the computations that we have just made. Each node i is annotated with $f(i)$. Each action that does not achieve $f(i)$ is "cut" by a pair of line segments, with the cost of taking that action and proceeding optimally recorded nearby.

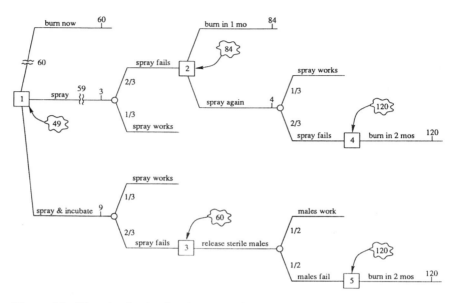

Figure 11 The plan having least expected cost.

The optimizers in Figure 11 describe a plan whose expected cost is $49 million. That plan is this: Spray while incubating sterile males. If the spray works, discard the males and stop. If the spray fails, release the sterile males. If the sterile males work, stop. If they fail, burn.

Describe this plan to the governor. Note that this plan has 1 chance in 3 of requiring burning 2 months from now. Account for this plan's expected cost of $49 million; 49 equals $9 + (1/3)120$. Show her that it is cheaper than the next-best plan by $10 million (because $10 = 59 - 49$). Point out that on a worst-case basis the cheapest thing to do is to burn now; that costs $60 million, but not more.

Like any model, the decision tree in Figure 10 may not capture all of the salient information. The cost of burning measures its direct impact on the state treasury; it omits the trauma of evacuation, as well as all costs not borne by the state. Similarly, the cost of spraying does not measure its impact on the environment. Spraying destroys many creatures other than the medfly. There are political costs as well; for instance, the press might scorch the governor if she opts to burn without trying a less aggressive strategy first.

After presenting the tree itself, guide the governor toward a discussion of the information that the tree might omit. Try to incorporate the informa-

Dynamic Programming

tion that concerns her into the tree. Help her weigh it systematically when deciding what course of action to take.

A recommended plan must be *robust*: It must perform well (if not best) under ranges of the estimates that are broad enough to be realistic. To the extent that the data in a decision tree are estimates, a sensitivity analysis is needed to determine the robustness of the recommended plan. For the medfly eradication plan, you might wish to do a sensitivity analysis on the probability that the spray will fail and on the probability that sterile males will fail.

It can happen that no plan is robust. In that event, there is no rational basis for recommending one.

The medfly example illustrates a new type of functional equation. It is

$$f(i) = \minimum_{k} \{R_i^k + \sum_j P_{ij}^k f(j)\}, \quad \text{for each } i \tag{19}$$

Interpret $f(i)$ as the least expected cost starting in state i. The data in (19) have these interpretations:

k is an action that can be selected if state i is observed.

R_i^k is the cost incurred (possibly negative) if state i is observed and action k is selected.

P_{ij}^k is the probability of transition to state j given that state i is observed and action k is selected.

From Figure 10 or 11, we see that the data for the medfly example include

$$R_1^{spray} = 3, \quad P_{12}^{spray} = \frac{2}{3}$$

Equation (19) is a recursion. The policy whose expected cost is smallest can be found by *rolling back*, that is, by starting with the states that occur last and computing $f(i)$ recursively, in decreasing time. While doing so, record for each i a decision $k = \delta(i)$ that attains the minimum. The resulting policy δ is optimal. Specifically, if one starts in *any* state i and uses this policy δ, one incurs expected cost of $f(i)$.

If the nonzero transition probabilities are from lower-numbered to higher-numbered states, then (19) can be solved by starting with the highest-numbered state and ending with the lowest-numbered state, exactly as was done in the medfly example.

In a profit-maximizing variation of (19), "maximum" replaces "minimum," and R_i^k is the income earned (possibly negative) if state i is observed and action k is selected.

Invariably, the rewards or costs (the R_i^k's) and the transition probabilities (the P_{ij}^k's) can depend on the state i that is observed and on the action k that is selected, but not on any prior states or actions.

9. DECISION ANALYSIS

The prior section modeled an operational decision. A decision tree laid out the alterative courses of action. A plan having least expected cost was found by rolling back the tree.

Let us ask ourselves: Is it sensible to minimize expected cost, or equivalently, to maximize expectation profit? Yes, provided that two conditions are met. The consequences of each action must be measured in monetary terms. And the amounts must be modest. If important consequences are nonmonetary, minimizing expected cost misses the point. If the monetary consequences can include enormous gains or losses, minimizing expected cost may fail to express the decision maker's preferences. To illustrate the latter point, imagine that you were offered two options: receive $10,000 for certain or a 50-50 gamble on winning $0 or $22,000. The gamble has higher expectation. But would you choose it?

This section introduces *decision analysis*, which is a theory of rational decision making in the presence of uncertainty. Decision analysis shows how an individual who accepts certain postulates of "rational behavior" reaches a decision that is consistent with his or her beliefs.

Decision analysis applies to all decision-making situations that one decision maker might face while acting alone. It applies when the consequences of one's actions can include large monetary gains and losses. It applies when important consequences are nonmonetary. It applies to operational and to strategic situations.

Decision analysis generalizes the approach taken in the preceding section. As before, a decision tree is built. Now, each endpoint is assigned a utility (or disutility). Then the tree is rolled back to find the policy whose expected utility is largest (or whose expected disutility is smallest).

An example has been chosen to introduce decision analysis. That example is a medical decision—whether or not a pregnant woman should undergo a procedure that is known as amniocentesis. Every pregnancy, if carried to term, could result in the birth of a child with Down's syndrome. That child is likely to be severely retarded, to require continuous care, and to die at a young age. A test known as *amniocentesis* produces information about whether a pregnant woman's fetus has Down's syndrome. A needle is thrust through the wall of her uterus, and a sample of her amniotic fluid is taken. The sample is analyzed.

Being invasive, amniocentesis increases the probability of a miscarriage (spontaneous abortion). That risk also increases with the mother's

Dynamic Programming

age. On the other hand, not doing amniocentesis risks the birth of a child with Down's syndrome. That risk also increases with the mother's age.

The decision of whether or not to undergo amniocentesis will serve two purposes. First, it will be used to show how a person who accepts certain postulates of rational behavior would reach a decision. Then it will be used to describe and interpret those postulates.

Like all tests, amniocentesis is imperfect. The *sensitivity* of the amniocentesis test is .99, which means that 99% of the fetuses that have Down's syndrome will test positive. The *specificity* of this test is .995, which means that 99.5% of the fetuses that do not have Down's syndrome will test negative.

To describe the sensitivity as the conditional probability $P(+|D)$, some nomenclature is introduced. Let D denote the event that a specific fetus has Down's syndrome, and let $\overline{D}$ denote the event that this fetus does not have Down's syndrome. Let $+$ denote the event that the test outcome is positive, and let $-$ denote the event that this outcome is negative. The sensitivity of .99 and specificity of .995 are stated precisely as

$$P(+|D) = .99, \quad P(-|\overline{D}) = .995$$

where, for instance, $(+|D)$ denotes the event in which a test outcome is positive given that it is applied to a fetus having Down's syndrome.

As mentioned earlier, the risks and benefits of amniocentesis vary from woman to woman. To focus on the decision faced by a specific individual, we consider:

Example 16: Ms. C is a 36-year-old industrial engineer who is married to a 38-year-old research psychologist. They have one child, a 6-year-old son who is healthy and bright. Ms. C has had no prior miscarriages. She has been pregnant for 15 weeks. It had taken 10 months for her to become pregnant. For Ms. C and for women like her, the relevant data are:

.005 = the probability that her fetus has Down's syndrome
.032 = the probability of spontaneous abortion
.037 = the probability of spontaneous abortion if amniocentesis is done now

Should Ms. C undergo amniocentesis? Who should decide?

Let it be emphasized: There is no correct answer to either question. Some women would make this decision themselves. Other women would involve the father, their doctor, or their priest. Some women would elect amniocentesis. Others would not. This section shows how a woman who accepts the postulates of rational behavior would reach her decision. This section also presents those postulates and critiques them.

Let us first see how good a test amniocentesis is. If Ms. C undergoes amniocentesis, the probability $P(+)$ that the test outcome will be positive is given by

$$P(+) = P(+, D) + P(+, \overline{D})$$
$$= P(+|D)P(D) + P(+|\overline{D})P(\overline{D})$$
$$= (.99)(.005) + (.005)(.995) = .01$$

Necessarily, the probability $P(-)$ of a negative test outcome is given by $P(-) = .99$.

Imagine that Ms. C undergoes amniocentesis and that the test outcome is positive. The conditional probability $P(D|+)$ that her fetus has Down's syndrome is given by

$$P(D|+) = \frac{P(D, +)}{P(+)} = \frac{P(+|D)P(D)}{.01} = \frac{(.99)(.005)}{.01} = .5$$

The equation $P(D|+) = .5$ has this meaning: If Ms. C has the test and learns that the test result is positive, there is a 50-50 chance that she is carrying a *healthy* fetus. Thus, a positive test is far from definitive.

On the other hand, the conditional probability $P(D|-)$ that her fetus is affected given a negative test outcome is

$$P(D|-) = \frac{P(D, -)}{P(-)} = \frac{P(-|D)P(D)}{.99} = \frac{(.01)(.005)}{.99} = .00005$$

In other words, a negative test outcome means that there is 1 chance in 20,000 that her fetus has Down's syndrome. A negative test outcome gives excellent assurance that her fetus is unaffected.

Of the people in Ms. C's risk group, only half of those who have positive test outcomes carry fetuses with Down's syndromes. And amniocentesis causes an abortion with probability of 1/200 because .005 = .037 − .032. These calculations suggest that amniocentesis is ineffective. Yet, Pauker and Pauker (1987), who have used decision analysis to counsel hundreds of couples on whether or not to undergo amniocentesis, report that a substantial majority did elect to do so.

9.1. Personal Beliefs and Preferences

The first step in a decision analysis is to elicit one's personal beliefs and preferences. What are the personal beliefs of Ms. C and her spouse? They feel that the amniocentesis decision is hers alone to make. She fears giving birth to a child with Down's syndrome. It would strain her marriage. The child might require full-time care, in which case one parent would need to abandon his career. With a single wage earner and high expenses, the family would be in financial duress.

Ms. C is loathe to abort a healthy fetus, but she is equally loathe to the risk of aborting it spontaneously and as a result of amniocentesis. For Ms. C, there are four key outcomes, and she ranks them from best to worst, as listed in Table 4. The "indifference probability" and "disutility" columns of this table will be explained shortly.

Let it be emphasized that Table 4 reports one woman's ranking of outcomes. Some women rank abortion last; it would be irrational for them to undergo amniocentesis.

In a decision analysis, disutilities play the role of costs. Ms. C has assigned her most preferred outcome a disutility of 0 and her least preferred outcome a disutility of 100. The numbers 0 and 100 are arbitrary; all that is needed is that the disutility (cost) of the worst outcome exceeds that of the best.

The indifference probability and disutility of each outcome are found from a break-even analysis like that in Figure 12.

The theory of rational behavior postulates that for the situation in Figure 12 and any like it each individual has his or her own *indifference probability* $p(k)$; for values $p > p(k)$, the individual prefers the outcome. For values $p < p(k)$, the individual prefers the gamble. Furthermore, if $p = p(k)$, the individual is indifferent between the two choices.

Figure 13 represents Ms. C's break-even analysis for the outcome that she ranks third, which is to abort a healthy fetus.

The break-even analysis in Figure 13 offers Ms. C a choice between outcome 3 (abort a healthy fetus) and a "gamble" that gets her the outcome she ranks worst with probability p and gets her the outcome she ranks best with probability $1 - p$. Consider the extreme cases: If $p = 1$, she prefers the abortion. Why? If $p = 0$, she prefers to give birth. Why?

Table 4 Ms. C's Ranking of Outcomes, Number 1 Being Most Preferred, with the Indifference Probability and Disutility of Each

Outcome	Rank k	Indifference probability $p(k)$	Disutility $d(k)$
Give birth to a healthy child	1	0.0	0
Abort a Down's syndrome fetus	2	.05	5
Abort a healthy fetus	3	.3	30
Give birth to a child with Down's syndrome	4	1.0	100

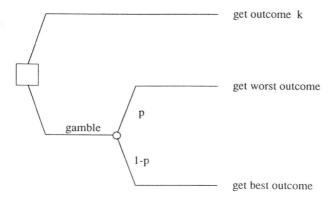

Figure 12 A break-even analysis for the indifference probability $p(k)$.

Ms. C's indifference probability $p(3)$ can be estimated by the process of *bracketing*: "Ms. C, when $p = 1$, the gamble gets you a baby with Down's syndrome, your least preferred outcome. Thus, when $p = 1$, you prefer the abortion. What about the $p = .8$?" The probability is lowered until Ms. C's answer switches. At that point, her indifference probability has been bracketed. Further questions can reduce the interval of ambiguity in p.

For Ms. C, it is supposed that $p(3) = .3$. In other words, it is imagined that she is indifferent between aborting a healthy fetus and a gamble that gives her the outcome she ranks worst (giving birth to a Down's syn-

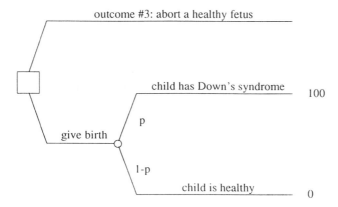

Figure 13 Ms. C's break-even analysis for outcome #3.

Dynamic Programming

drome child) with probability of .3 and the outcome she ranks best (giving birth to a healthy baby) with probability of .7. If the probability of giving birth to a Down's syndrome child exceeds .3, she prefers the abortion. If the probability of a Down's syndrome child is below .3, she prefers to take her chances on the gamble.

Figure 13 reveals a difficulty of decision analysis. Specifying an indifference probability can require an individual to make a choice that is hypothetical. A woman may choose between aborting the fetus in her womb or carrying *that fetus* to term. She cannot face the choice in Figure 13.

Ms. C must also specify her indifference probability $p(2)$ for the outcome that she ranks second, which is to abort a Down's syndrome fetus. Let us suppose that $p(2) = .05$. Thus, she prefers to abort a Down's syndrome fetus to giving birth if the chance of getting a baby that has Down's syndrome is below 1/20.

Table 4 reports that $p(1) = 0$ and that $p(4) = 1$. Note that for each outcome k, Ms. C is indifferent between outcome k and the gamble that gets her worst outcome with probability $p(k)$ and her best outcome with probability $[1 - p(k)]$.

Table 4 also reports a disutility $d(k)$ for each outcome k, with $d(1) = 0$ and $d(4) = 100$. Disutilities are assigned using this rule:

Choices between which the decision maker is indifferent have the same expected disutility.

Ms. C is indifferent between these choices: outcome k and a gamble that gets outcome 4 with probability $p(k)$ and outcome 1 with probability $[1 - p(k)]$. These choices have the same expected disutility:

$$d(k) = p(k)d(4) + [1 - p(k)]d(1)$$
$$= p(k)100 + [1 - p(k)]0 \qquad (20)$$
$$= 100p(k), \quad \text{each } k$$

In particular, $d(2) = 100p(2) = 5$ and $d(3) = 100p(3) = 30$. The relationship between $d(k)$ and $p(k)$ takes the simple form $d(k) = 100p(k)$ because the best and worst outcomes have been assigned disutilities of 0 and 100, respectively. The main result (theorem) of the theory of rational behavior is:

Each decision maker should act so as to minimize the expectation of her (his) disutility.

Ms. C's decision has been cast as the problem of minimizing expected disutility (cost). Let us pause to recast it as the problem of maximizing expected utility (profit). To express Ms. C's preferences in terms of utilities, we assign her most preferred outcome a utility of 100 and her least preferred outcome a utility of 0. (The numbers 100 and 0 are arbitrary; all that is needed is that the utility of the best outcome exceed that of the worst.) Let $u(k)$ denote the utility of outcome k, so $u(1) = 100$ and $u(4) = 0$. Choices between which the decision maker is indifferent have the same expected utility:

$$u(2) = p(2)u(4) + [1 - p(2)]u(1)$$
$$= .05(0) + .95(100) = 95$$
$$u(3) = p(3)u(4) + [1 - p(3)]u(1)$$
$$= .3(0) + .7(100) = 70$$

Notice that $u(k) + d(k) = 100$. Why is that? It makes no difference whether Ms. C maximizes expected utility or minimizes expected disutility; the same plan is best. Why is that?

9.2. A Decision Tree

Figure 14 is Ms. C's decision tree. It models her decision—whether or not to undergo amniocentesis. This tree omits foolish choices; e.g., do amniocentesis and then ignore the test outcome.

Each arc to the right of a chance node has its probability recorded below it. These probabilities have the same meaning as in earlier decision trees. Each arc's probability is conditional; it is the probability of crossing the arc if one is at the chance node to its left. The arcs that point rightward from a chance node have probabilities that sum to 1. Why is that? Each of the probabilities in this tree was calculated earlier in this section. For instance, the probability of .00005 at the bottom of the tree equals the probability that a birth results in a child with Down's syndrome given a negative test outcome; $.00005 = P(D|-)$.

Each endpoint (final outcome) in this tree has been assigned one of four disutilities, $d(1)$ through $d(4)$. (In this tree, it eases numerical computation to work with disutilities, rather than utilities.)

By rolling back Ms. C's tree, one can find the plan whose expected disutility is smallest. The expected disutilities of two plans are reported in the tree. The preferred plan for Ms. C is to undergo amniocentesis and to have an abortion if the test result is positive.

Can meaning be given to the numbers 1.32 and 1.44 in Figure 14? Yes. Let us recall from Eq. (20) that each outcome k has disutility $d(k) = 100p(k)$. This lets us interpret .0132 and .0144 as indifference probabilities. Specifically, Ms. C is indifferent between letting her pregnancy proceed normally and this gamble—give birth to a Down's syndrome child (the outcome she ranks worst) with probability .0144 and give birth to a healthy

Dynamic Programming

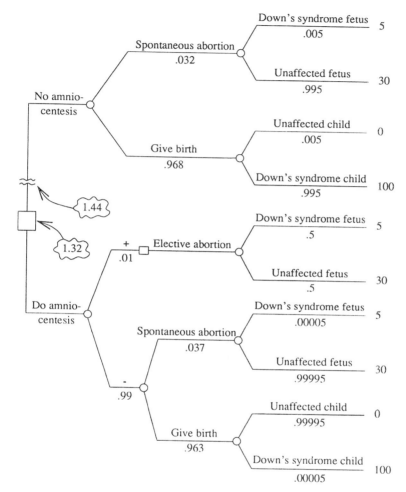

Figure 14 Ms. C's decision tree, minimizing expected disutility.

child (the outcome she ranks best) with probability [1 − .0144]. Similarly, she is indifferent between having amniocentesis followed by an abortion if the test outcome is positive and this gamble—give birth to a Down's syndrome child with probability .0132 and give birth to a healthy child with probability [1 − .0132].

Utility and disutility are code words for indifference probability. For each outcome k, there is a probability $p(k)$ such that the decision maker is indifferent between outcome k and the gamble in Figure 12. That gamble gets the decision maker's worst outcome with probability $p(k)$ and best outcome with probability $[1 - p(k)]$. As a consequence, each plan of action has an indifference probability in the same gamble, and:

> Each decision maker should pick a plan whose indifference probability of getting her (his) worst outcome is smallest.

To compute Ms. C's minimum indifference probability from Figure 13, divide each utility by 100 because $p(k) = d(k)/100$. The same plan is optimal, and the lowest indifference probability equals .0132.

9.3. Robustness

The reader may be skeptical that any person can state his or her indifference probabilities with precision. This writer certainly is. In this writer's opinion, the use of decision analysis *must* be accompanied by a sensitivity analysis that demonstrates the robustness of recommended plan.

For Ms. C, is the decision to undergo the amniocentesis a robust one? A break-even analysis will answer this question. In Figure 14, let us replace 5 and 30, respectively, by $d(2)$ and $d(3)$. Let us find the values of $d(2)$ and $d(3)$ for which the two actions, do amniocentesis and don't, have the same expected disutility. A bit of arithmetic (that is omitted) boils that calculation down to the "indifference" line

$$.00484d(2) + .00979d(3) = .479$$

Since $d(k) = 100p(k)$ for each k, and equivalent indifference line is

$$.484p(2) + .979p(3) = .479$$

One point on this line is $p(2) = 0$ and $p(3) = 0.49$. Another point is $p(2) = 0.99$ and $p(3) = 0$. A third is $p(2) = p(3) = 0.33$. Thus, Ms. C chooses amniocentesis if her indifference probabilities $p(2)$ and $p(3)$ lie within the "do amnio" region of Figure 15.

The probabilities $p(2) = .05$ and $p(3) = .3$ lie comfortably within Ms. C's "do amnio" region. For her, that decision is robust.

9.4. Realistic Choices and Consistency Checks

Specifying $p(2)$ and $p(3)$ required Ms. C to choose between alternatives that could not occur at the same time, e.g., between aborting a healthy fetus and carrying a different fetus to term. "Hypothetical" decisions like that can be difficult to make with conviction.

Figure 16 presents Ms. C with a choice that is more realistic, less hypothetical. This choice is to abort or deliver a fetus that has Down's

Dynamic Programming

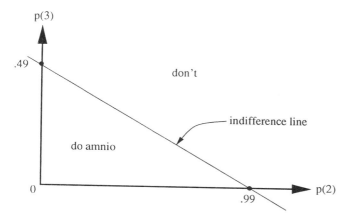

Figure 15 Whether or not to undergo amniocentesis in terms of the indifference probability $p(3)$ for aborting a Down's syndrome fetus and the indifference probability $p(2)$ for aborting a healthy fetus.

syndrome with probability p. If p equals 0, Ms. C prefers to deliver. Why? If p equals 1, she prefers to abort.

Let p denote the probability for which Ms. C is indifferent between the choices in Figure 16. The choices between which she is indifferent have the same expected disutility, so

$$p100 = pd(2) + (1 - p)d(3)$$

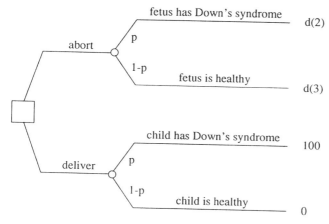

Figure 16 Ms. C's decision to abort.

Since $d(k) = 100p(k)$, the preceding equation relates p to her indifference probabilities, $p(2)$ and $p(3)$. Specifically,

$$p = pp(2) + (1-p)p(3)$$
$$= p(3) + p[p(2) - p(3)]$$
$$= \frac{p(3)}{1 + p(3) - p(2)}$$

Ms. C has $p(2) = .05$ and $p(3) = .3$, so her indifference probability p for Figure 16 should satisfy $p = .3/1.25 = .24$. If p differs significantly from .24, she should revise her estimate of $p(2)$ and/or of $p(3)$.

When eliciting indifference probabilities, try to compare alternatives that can occur simultaneously. An indifference analysis like that in Figure 16 can improve one's estimates of the indifference probabilities, as well as one's confidence in them.

9.5. Postulates of Rational Behavior

This presentation of rational decision making has been vague, to put it kindly. The postulates (axioms) that describe rational behavior have not yet been articulated. Now they are. These postulates will seem familiar; they have been used to build and analyze Ms. C's decision tree.

These axioms are stated with respect to the final outcomes in a decision tree. After each axiom is stated, the amniocentesis decision is used to interpret it. The next section critiques these hypotheses.

The modern theory of rational decision making is aptly attributed to John Von Neumann and Oskar Morgenstern (1947). Credit is due to earlier work by Frank P. Ramsey (1931) and to later work by Leonard J. Savage (1954) and others. Of the many extant sets of axioms under which it is optimal to maximize expected utility, we set forth and critique one.

Axiom 1 (Complete and Transitive Ordering): The decision maker is able to rank the final outcomes in the decision tree from best to worst, with ties allowed.

Table 4 illustrates this axiom. Ms. C assigns the highest rank to the birth of a healthy baby. She assigns the next highest rank to the abortion of a Down's syndrome baby, the third highest rank the abortion of a healthy baby, and the lowest rank to the birth of a Down's syndrome baby.

Axiom 1 is said to represent an *ordinal* preference. An ordinal preference expresses choices between alternatives but fails to describe the extent to which one alternative is preferred over another.

The next three axioms concern preferences over "lotteries." A *lottery* is any probability distribution over the final outcomes of a decision tree. A

Dynamic Programming

lottery over n final outcomes could assign each of them the probability $1/n$. A lottery can assign different probabilities to different outcomes. A degenerate lottery assigns a single outcome the probability 1 and assigns the others a probability of zero.

Axiom 2 (Monotonicity): Suppose that the decision maker prefers outcome A to outcome B. Then, between two lotteries that have A and B as their only outcomes, the decision maker prefers the lottery with the higher probability of obtaining outcome A.

To illustrate, suppose Ms. C is offered a pair of lotteries whose outcomes are giving birth to a healthy baby (her most preferred outcome) and to a Down's syndrome baby. Axiom 2 states that she prefers the lottery that assigns the higher probability to giving birth to a healthy baby.

Axiom 3 (Continuity): Suppose the decision maker prefers final outcome A to B and B to C. Then there exists a probability p that satisfies $0 < p < 1$ such that the decision maker is indifferent between receiving final outcome B and a lottery that assigns probability p to outcome C and $1 - p$ to outcome A.

Taken together, Axioms 1–3 state that it is possible to carry out the indifference analyses in Figures 12 and 13. True or false: Axioms 1–3 guarantee that the indifference probability is unique?

Before presenting the fourth and final axiom, we introduce some notation. With L and L' as any lotteries, the notation $pL + (1 - p)L'$ designates the two-stage lottery whose first stage is to toss a coin that is "heads" with probability p; if that coin is heads, one gets lottery L; if not, one gets lottery L'.

Axiom 4 (Substitution): Suppose the decision maker is indifferent between lotteries L_1 and L_2. Then, for any lottery L and any probability p, the decision maker is indifferent between the lotteries $pL + (1 - p)L_1$ and $pL + (1 - p)L_2$.

To illustrate, consider a lottery X that has three or more final outcomes, with outcome A preferred to B and with B preferred to C. The substitution and continuity axioms show that the decision maker is indifferent between lottery X and a lottery X' where X' has the same outcomes as X, except that X' omits outcome B. What does repetition of this argument achieve?

A decision maker is now said to be *rational* with respect to a decision tree if the decision maker's preferences satisfy Axioms 1–4. The principal theorem of decision analysis is that a rational decision maker can assign a utility or disutility to each final outcome and select the policy whose expected utility is largest, equivalently, whose expected disutility is smallest.

Axioms 1–4 enable use of a break-even analysis to assign a utility or disutility to each outcome. These break-even analyses make it clear that utility and disutility are shorthand for indifference probability.

9.6. Critique of These Postulates

These axioms may seem plausible, but situations have been found in which thoughtful people violate them. Consider:

(*Incomparable outcomes*): There are three outcomes — you get $50, you get $10, and you get shot dead.

Getting shot is so onerous as to question the "continuity" axiom. Is there a probability p that is strictly between 0 and 1 such that you are indifferent between winning $10 and a lottery that gets you shot dead with probability p and gets you $50 with probability $1 - p$?

Several different types of situations have been found in which thoughtful individuals behave in ways that violate these postulates. These include the paradoxes of Allais (1953), Ellsberg (1961), and Kahneman and Tversky (e.g., 1979). In this writer's view, the main drawback to decision analysis lies not in these paradoxes but in a practical consideration. In many settings, the decision maker cannot specify indifference probabilities precisely enough to achieve robust conclusions.

Medical decision making is one setting in which this theory seems to be particularly useful. In medical decisions, the outcomes are relatively few in number, they are sharply defined, their indifference probabilities can be assessed, and break-even lines like that in Figure 15 can demonstrate robustness.

The great strength of decision analysis is that it quantifies the decision maker's preferences. Each outcome is described by a single figure of merit, its indifference probability. The decision maker is indifferent between receiving outcome k and a "standard lottery" that gives the worst outcome with probability $p(k)$ and the best outcome with probability $[1 - p(k)]$.

9.7. Summary

Let us recapitulate. Decision analysis shows how a person whose preferences satisfy certain axioms should reach a decision. That person should:

Rank the outcomes in a decision tree from best to worst, with ties allowed.
For each outcome other than the best and the worst, assign an indif-

Dynamic Programming

ference probability p, with $0 < p < 1$, such that the decision maker is indifferent between getting that outcome and a lottery that gets the worst-ranked outcome with probability p and the best-ranked outcome with probability $1 - p$.

Assign higher utility to the best-ranked outcome than to the worst-ranked outcome. To each outcome other than those two, assign utility so that choices between which the decision maker is indifferent have the same expected utilities.

Roll the tree back to select a plan (policy) whose expected utility is largest.

Any decision that is based on utility maximization should include a sensitivity analysis on the indifference probabilities that demonstrates robustness. One should also check that one's preferences do satisfy Axioms 1–4.

Even the tree is personal. Ms. C's tree is simpler than it might be because she is equally loathe to spontaneous and planned abortions, with and without amniocentesis. To reveal another way in which her tree is simplified, suppose Ms. C elects an abortion. Should she be told whether or not the fetus was healthy? Perhaps. But the decision to tell her affects her decision tree, as her regret, in the sense of Bell (1982), for electing to abort a fetus that proved to be healthy must be measured and incorporated in her tree, which becomes more elaborate than Figure 14.

10. THE MARKOV DECISION MODEL

The medfly example illustrates a model that is central to decision making in the presence of uncertainty. This model dovetails Markov chains with decision making. It is presented as:

Example 17 (Markov Decision Model): This model's states form the set S. Each state i in S has a nonempty set A_i of actions. Each time a state i is observed, an action k in A_i must be selected. Observing state i and selecting action k has these implications:

R_i^k = the expectation of the income earned between now and the moment at which the next state is observed, given that state i has just been observed and given that action k in A_i has just been selected

P_{ij}^k = the probability that state j will be observed next, given that state i has just been observed and action k has just been selected

In this model, the reward (the R_i^k's) need not be positive. Some or all of them can be negative or zero. The transition probabilities satisfy:

$$P_{ij}^k \geq 0 \quad \text{for each } i, j \in S \text{ and each } k \in A_i$$

$$\sum_{j \in S} P_{ij}^k \leq 1 \quad \text{for each } i \in S \text{ and each } k \in A_i$$

The transition probabilities must be nonnegative, of course, and the sum over j of P_{ij}^k cannot exceed 1 because transition either occurs to some future state or to no state at all.

In this model, transition to no state at all marks the end of the decision-making process. For each state i and action k in A_i, define the *stopping probability* s_i^k by

$$s_i^k = 1 - \sum_{j \in A_i} P_{ij}^k \tag{21}$$

Interpret s_i^k as the probability that decision making stops with the observation of state i and selection of action k. Particular state-action pairs can have $s_i^k = 0$. Possibly, every state-action pair has $s_i^k = 0$, in which case decision making continues forever.

The Markov decision model has been presented as a profit maximization problem. To recast it as a cost minimization problem, interpret each R_i^k as a cost.

The medfly example (Figure 10) illustrates a Markov decision model whose objective is to minimize expected cost. That example has five states: $S = \{1, 2, 3, 4, 5\}$. Figure 10 indicates that state 1 has three actions: $A_1 = \{b, s, s\&i\}$ and that the costs of these three actions are $R_1^b = 60$, $R_1^s = 5$, and $R_1^{s\&i} = 9$. Figure 10 also indicates that the "spray now" action has these transition probabilities: $P_{12}^s = 2/3$ and $P_{1j}^s = 0$ for each $j \neq 2$.

A Markov decision model is said to be *finite* if it has finitely many states and actions. This section surveys finite Markov decision models. Markov decision models that have infinitely many states or actions lie beyond the scope of this chapter.

A *policy* is a rule δ that specifies for each state i an action $\delta(i)$ in A_i. The *policy space* Δ is the set of all such policies. For the decision tree in Figure 10, let us count the policies. State 1 has three actions, state 2 has two actions, and the other three states have one action each. This example has six policies because $6 = 3 \times 2 \times 1 \times 1 \times 1$. Even a small model can have a great many policies. An example that has 100 states and 2 actions per state has 2^{100} policies.

Each policy δ has a *return function* v^δ, with this interpretation:

$v^\delta(i) =$ the expectation of the total income earned if decision making starts at state i and if for each state j action $\delta(j)$ is selected whenever state j is observed

Dynamic Programming

A recursion for $v^\delta(i)$ is presented and is then interpreted:

$$v^\delta(i) = R_i^{\delta(i)} + \sum_{j \in S} P_{ij}^{\delta(i)} v^\delta(j), \quad \text{each } i \in S \tag{22}$$

The expectation of the sum equals the sum of the expectations. Interpret $R_i^{\delta(i)}$ as the expectation of the income earned *prior to the first transition* if decision making starts at state i and if policy δ is used. Thus, it suffices to see that the "+" sign in (22) is followed by the expectation of the income earned after the first transition if policy δ is used. Interpret $v^\delta(j)$ as the expectation of the income earned after transition occurs to state j if policy δ is used then and thereafter. That transition occurs with probability of $P_{ij}^{\delta(i)}$, so the + sign is followed by the desired expectation.

For this maximization problem, the *optimal return function* f is defined by

$$f(i) = \text{maximum}\{v^\delta(i) : \delta \in \Delta\}, \quad \text{each } i \in S \tag{23}$$

Interpret $f(i)$ as the largest attainable value of expected income if state i is observed initially. Policy δ is said to be *optimal for state i* if $v^\delta(i) = f(i)$. Policy δ is said to be *optimal* if $v^\delta(i) = f(i)$ for each i.

Familiar questions assert themselves. Does there exist an optimal policy? If so, how can it be computed? Does the optimal return function f satisfy the functional equation given below as (24)? If so, is f the unique solution to this equation?

$$f(i) = \underset{k \in A_i}{\text{maximum}}\{R_i^k + \sum_{j \in S} P_{ij}^k f(j)\}, \quad \text{each } i \in S \tag{24}$$

These questions are easy to answer—in the affirmative—for the type of Markov decision model that is described next.

10.1. Finite-Horizon Model

A Markov decision model is said to have a *finite horizon* if there is a maximum number N of state-to-state transitions that might occur, no matter what state is observed initially and what policy is employed. The medfly model has a finite horizon. Figure 10 indicates that it has $N = 2$; decision making is guaranteed to stop after two state-to-state transitions.

If a finite-horizon model, decision making stops after a predictable number of transitions. Finite-horizon models are the subject of

Proposition 6: Consider a finite Markov decision model that has a finite horizon. It has an optimal policy, and that policy's return function is the unique solution to functional equation (24). Furthermore, Program 3, below, has a unique optimal solution v, and $v(i) = f(i)$ for each i.

Program 3: Minimize $\{\sum_{i \in S} v(i)\}$ subject to the constraints

$$v(i) \geq R_i^k + \sum_{j \in S} P_{ij}^k v(j), \quad \text{each } i \in S \text{ and } k \in A_i \tag{25}$$

This section omits proofs. For a proof of Proposition 6, see Denardo (1982, pp. 114–115).

For a finite-horizon model, functional equation (24) can be solved by starting at the end of the planning interval and "rolling back" toward the beginning, exactly as has been done in Figure 10.

Proposition 6 introduces a new way to find f and an optimal policy, which is by linear programming. Program 3 is a linear program. Its decision variables are the $v(i)$'s. Its data are the R_i^k's and the P_{ij}^k's. It has one constraint for each state-action pair.

Is this linear program important in practice? In some cases, yes. On one hand, executing a rollback of (24) takes less computer time than applying a general-purpose linear programming code to Program 3. On the other hand, writing and debugging a computer program that executes this rollback can take more effort than using a commercial linear programming code to solve Program 3. Also, most linear programming codes have extensive capability for sensitivity analysis, which is always required in practice.

Recorded below are the minor changes that adapt Proposition 6 to a minimization problem.

Corollary 1: To adapt Proposition 6 to a cost minimization problem, replace maximum by minimum in (23) and in (24), replace minimize by maximize in the objective of Program 3, and replace $\geq$ by $\leq$ in constraint (25).

Proposition 6 and its corollary conclude that an optimal policy δ does exist and that this policy's return function v^δ equals f, where f is the unique solution to a functional equation. Proposition 6 also shows that f is the unique optimal solution to a linear program. Later in this section, each of these conclusions will be seen to hold more generally, for models that do not have finite horizons.

10.2. Terminal Rewards and Expected Utility

Proposition 6 shows how to maximize expected income in a finite-horizon Markov decision model. Its corollary shows how to minimize expected cost. They encompass the medfly example and all other finite Markov decision problems.

Do they encompass the amniocentesis example and other utility maximization problems? For the case of a finite horizon, they do. Let us see how. A Markov decision model is said to have *terminal rewards* if each

Dynamic Programming

state-action pair (i, k) falls into one of two categories: either $s_i^k = 1$ or $s_i^k = R_i^k = 0$. In such a model, the only occasions at which nonzero rewards can be earned are those for which termination is guaranteed, namely the state-action pairs (i, k) having $s_k^j = 1$. For a model having terminal rewards, interpret R_i^k as the utility of stopping by observing state i and selecting action k.

In a finite-horizon model with terminal rewards, an optimal policy maximizes the expectation of the terminal reward; equivalently, it maximizes expected utility. Hence, Proposition 6 shows how to maximize expected utility in a decision tree. Its corollary shows how to minimize expected disutility. Figure 14 exhibits the decision tree for a Markov decision problem that is finite and has terminal rewards; a rollback has minimized its expected disutility.

10.3. Shortest- and Longest-Path Problems

Let us now observe that the Markov decision model encompasses shortest-path problems and that they too can be solved by linear programming. To establish this connection, consider a directed network (S, A) whose set of nodes is S and whose set of directed arcs is A. Suppose that each arc (i, k) in A has length C_{ik}. To interpret this as a Markov decision problem, define the decision set A_i for state i by $A_i = \{k : (i, k) \in A\}$. Similarly, define the costs and transition probabilities by $R_i^k = C_{ik}$ and $P_{ik}^k = 1$ for each (i, k) in A.

Proposition 7: Consider a directed network (S, A) having these properties: First, no arcs emanate from node j. Second, for each node $i \neq j$, a path exists from node i to node j. Third, no cycle has negative length. With $f(j) = 0$, let $f(i)$ denote the length of the shortest path from node i to node j. Then the unique solution to Program 4, below, is $v = f$.

Program 4: Maximize $\{\sum_{i \in S} v(i)\}$ subject to $v(j) = 0$ and

$$v(i) \leq C_{ik} + v(k) \quad \text{for each } (i, k) \in A \tag{26}$$

For a proof, see Denardo (1967). The hypothesis of Proposition 7 is satisfied if the network is acyclic. This hypothesis is also satisfied if the network is cyclic but has no cycle whose length is negative. As mentioned in an earlier section, the shortest-path problem is ill-posed if the network has a negative cycle.

In effect, Program 4 computes the tree of shortest paths to node j. To compute the tree of longest paths to node j, replace Maximize by Minimize in the objective of Program 4, and replace $\leq$ by $\geq$ in (26). A variant of

Program 4 computes the tree of shortest paths from node i. Can you see what it is?

Examples 7 through 16 are dynamic programs. Nearly all of them have now been interpreted as Markov decision problems and solved by linear programming. Can you find an example that is not a Markov decision problem? Can you guess how to solve it by linear programming?

10.4. The Time Value of Money: Discrete Time Case

A Markov decision model can represent a planning activity whose duration is measured in years or decades. In such cases, one must account for the time value of money, namely for the fact that money received earlier is preferable to the same amount of money received later.

For convenient exposition, we first treat the time value of money for a simple case that is known in the literature as the "discrete" time model. In the *discrete time* model, each transition takes the same length of time, and this interval of time is called a *period*.

Let us suppose that there exists a nonnegative number r with this property: The decision maker is indifferent between receiving any number X of units of income at the start of any period and $X(1 + r)$ units of income one period later. Thus, each unit of money earned at the start of a period has the same value as $(1 + r)$ units of money earned one period later. The number r is called the per-period *interest rate*. In effect, each unit of money (e.g., dollar) on hand for one period earns simple interest of r units of money, paid at the end of the period.

The value that r should take varies with the context in which decisions are being made. When a company makes investment decisions, it may use as r its "hurdle rate," namely the rate that the company can earn on investments that incur an acceptable level of risk. For public expenditures, one may take as r the "real" rate of interest, which equals the nominal interest rate less the anticipated rate of inflation. It is assumed throughout that r is nonnegative.

The same "interest rate" r applies to positive *and* negative values of X. Hence, the decision maker is assumed to be indifferent between paying C units of money now and paying $C(1 + r)$ units of money one period from now.

Each unit of money received one period in the future is equivalent to α units of money received now where

$$\alpha(1 + r) = 1$$

The number $\alpha = 1/(1 + r)$ is called the per-period *discount factor*. The assumption that r is nonnegative implies $0 < \alpha \le 1$. Revenue X received

Dynamic Programming

one period in the future is equivalent to revenue αX received now. Revenue Y received two periods in the future is equivalent to $\alpha^2 Y$ units of money earned now. And so forth.

In the discrete time model, an *income stream* is a sequence $(X_1, X_2, \ldots, X_k, \ldots)$, where X_j denotes the revenue received at the start of period j. Each income stream has a *present value* at the start of each period, which is defined as the number of *current* units of money to which that income stream is equivalent. For example, the income stream $(X_1, X_2, \ldots, X_k)$ has $(X_1 + \alpha X_2 + \cdots + \alpha^{k-1} X_k)$ as its present value at the start of period 1. What is the present value of the same income stream at the start of period 2?

An *undiscounted* model has $r = 0$. A *discounted* model has $r > 0$. To allow for discounting in the Markov decision model, the datum R_i^k must be reinterpreted:

R_i^k = the present value at the current moment of the income earned between now and the time at which the next state is observed given that state i has just been observed and action k in A_i has just been selected

Similarly, the reward function $v^\delta(i)$ must be reinterpreted:

$v^\delta(i)$ = the expectation of the present value at the current moment of all current and future income given that state i has just been observed and if, for each state j, action $\delta(j)$ is selected whenever state j is observed

For an undiscounted model, a recursion for policy δ's return function was presented as Eq. (22). That equation is now modified to accommodate discounting. For the discounted discrete-time model, (22) becomes

$$v^\delta(i) = R_i^{\delta(i)} + \sum_{j \in S} \alpha P_{ij}^{\delta(i)} v^\delta(j), \quad \text{each } i \in S \tag{27}$$

In (27), interpret $v^\delta(j)$ as the present value at the start of period 2 of the income earned from then on if state j is observed then and if policy δ is used. Therefore, $\alpha v^\delta(j)$ equals the present value at the start of period 1 of that income stream. Thus, the prior justification for (22) applies to (27).

10.5. Time Value of Money: General Discussion

Let us turn our attention to a more general situation. Suppose that different transitions can take different lengths of time, that the transition times can be uncertain, and that they can take any values. In this context, r remains the interest rate per period but with "continuous" compounding.

Continuous compounding is the limit of compound interest as the period is chopped into finer and finer pieces, as follows: For a positive integer N, divide each period into N short periods that have equal lengths. Assume that the interest rate of r/N applies to each short period. Wealth grows by the factor of $[1 + r/N]$ during each short period. With compound interest, wealth grows by the factor of $[1 + r/N]^N$ per period. Finally, let the number N of short periods approach infinity. With e as the base of the natural logarithms (so $e = 2.718 \ldots$), one has

$$e^r = \lim_{N \to \infty} \left[1 + \frac{r}{N}\right]^N$$

Consequently, with continuous compounding, each unit of money received now is equivalent to e^r units of money received one period from now. Similarly, each unit of money received now is equivalent to $(e^r)^t = e^{rt}$ units of money received t periods from now. And each unit of money received t periods in the future is equivalent to e^{-rt} units of money received now.

10.6. Markov Renewal Model with Discounting

In the discrete-time model, each transition takes the same length of time. The *Markov renewal* model is the elaboration upon the discrete-time model whose transition times can vary with the state and action in the following manner: For each state-action pair (i, k),

$T_{ij}^k =$ the (possibly random) time that elapses between now and the moment at which transition to the next state occurs, given that state i has just been observed, given that action k in A_i has just been selected, and given that transition will occur to state j

The transition time T_{ij}^k can be uncertain. The notation indicates that the transition time can vary with the state i, the action k, and even with the state j to which transition will occur.

In the discounted discrete-time model, each transition takes the same length of time, and the same discount factor α applies to all transitions. In the discounted Markov renewal model, the transition times vary and so do the discount factors. To account for this, we define

$$\alpha_{ij}^k = E[e^{-rT_{ij}^k}]$$

where E denotes expectation. For the discounted Markov renewal model, (22) is replaced not by (27) but by

$$v^\delta(i) = R_i^{\delta(i)} + \sum_{j \in S} \alpha_{ij}^{\delta(i)} P_{ij}^{\delta(i)} v^\delta(j), \quad \text{each } i \in S \qquad (28)$$

Can you see why? Also, functional equation (24) is replaced by

Dynamic Programming

$$f(i) = \underset{k \in A_i}{\text{maximum}} \{R_i^k + \sum_{j \in S} \alpha_{ij}^k P_{ij}^k f(j)\}, \quad \text{each } i \in S \tag{29}$$

The discrete-time model is the special case of the Markov renewal model each of whose transitions takes one unit of time. Similarly, the so-called *continuous-time* model is the special case of the Markov renewal model in which each transition time has the exponential distribution.

Proposition 6 concerns undiscounted Markov decision models whose horizons are finite. Its conclusions are now shown to hold for discounted models, whether or not their horizons are finite.

Proposition 8: Consider a discounted finite Markov renewal model for which each state-action pair (i, k) has $\Sigma_{j \in S} \alpha_{ij}^k P_{ij}^k < 1$. That model has an optimal policy, and that policy's return function is the unique solution to (29). Furthermore, Program 5, below, has a unique optimal solution v, and $v(i) = f(i)$ for each i.

Program 5: Minimize $\{\sum_{i \in S} v(i)\}$ subject to the constraints

$$v(i) \geq R_i^k + \sum_{j \in S} \alpha_{ij}^k P_{ij}^k v(j), \quad \text{each } i \in S \text{ and } k \in A_i \tag{30}$$

For a proof, see Denardo (1967). Proposition 8 shows that the conclusions of Proposition 6 remain true when P_{ij}^k is replaced by $\alpha_{ij}^k P_{ij}^k$ throughout. The hypothesis of Proposition 8 is that the interest rate r is positive and that no transition takes 0 time with probability of 1. That can be weakened; it suffices that no sequence of $|S|$ transitions takes 0 time with probability of 1, where $|S|$ denotes the number of states.

A Markov decision model is said to have an *infinite horizon* if $s_i^k = 0$ for each $i \in S$ and each $k \in A_i$. In an infinite-horizon model, decision making continues forever. Proposition 8 applies to discounted models with both finite and infinite horizons. It also applies to discounted models in which the decision maker can choose to stop or not.

For a finite-horizon model, f can be computed by starting at the end of the planning interval and executing a rollback. For a discounted infinite-horizon model, no rollback is possible because there exists no "end" at which to initiate it. Proposition 8 shows that f can still be found by solving Program 5.

10.7. Turnpike Theorems

Why study an infinite-horizon model? Nothing continues forever. The planning interval may be long. Its length may be uncertain, and many decisions may be taken before it ends, but end it will. Yet, there are *turnpike theorems* (see Shapiro, 1968), which are statements of this sort:

If the length of the planning interval is long enough, it is best to act now as though the planning interval were infinite.

The precise statements of these turnpike theorems lie beyond this chapter's scope. It suffices for our purposes to observe that turnpike theorems justify the use of infinite-horizon models. In particular, an "optimal" policy found by Proposition 8 is a good one for all but the final stages of a long planning horizon.

10.8. Small Interest Rates, Numerical Instability, and Abelian Theorems

For a discounted infinite-horizon model, (29) has f as its unique solution, and Program 5 has f as its unique solution. But when the interest rate r is small, computing f can be difficult. To suggest why, we suppose that actions must be taken daily and that the interest rate equals 10% per year. The per-period interest rate r is approximately $.1/365 = .00027$. The discount factor α is approximately .99973. With α this close to 1, the matrices with which the simplex method of linear programming deals are *ill-conditioned* because their determinants are close to 0. Numerical instability will make it difficult to find f from (29) or from Program 5.

Fortunately, "Abelian" theorems come to the rescue. Loosely put, an Abelian theorem shows that the behavior of the infinite-horizon undiscounted model determines the behavior of the infinite-horizon discounted model when the interest rate is small. That is good news because the undiscounted model is not afflicted by numerical instability.

Before describing the role played by Abelian theorems, we must pause to examine income streams. As in the discrete-time model, we divide the future into periods of equal length and number these periods 1, 2, 3, ..., without limit. For each period i, we interpret the number a_i as the income earned during period i, and we assume that a_i is received at the *end* of period i. Define S_n by

$$S_n = a_1 + a_2 + \cdots + a_n, \quad \text{for } n = 1, 2, \ldots \tag{31}$$

and interpret S_n as the total income earned during the first n periods. The income stream (sequence) $(a_1, a_2, \ldots)$ is said to *sum* to S if $S_n \to S$ in the normal sense.

Let us assume that income is discounted in accord with an interest rate of r per period, therefore with a discount factor $\alpha = 1/(1 + r)$ per period. Designate by $S(r)$ the present value at the *start* of period 1 of the entire income stream with interest rate r, so that

$$S(r) = \alpha a_1 + \alpha^2 a_2 + \cdots + \alpha^n a_n + \cdots, \tag{32}$$

with $\alpha = 1/(1 + r)$.

Dynamic Programming

The income stream (sequence) $(a_1, a_2, \ldots)$ is said to be *Abel summable* to S if $S(r) \to S$ as $r \to 0+$, that is, as r decreases to 0.

To contrast these two notions of summability, consider the income stream (sequence) $(1, -1, 1, -1, \ldots)$. Evidently, S_n alternates between 1 and 0, while $S(r)$ satisfies the recursion $S(r) = \alpha - \alpha^2 + \alpha^2 S(r) = \alpha(1 - \alpha)/(1 - \alpha^2) = \alpha/(1 + \alpha)$, which converges to $1/2$ as $r \to 0$, equivalently, as $\alpha \to 1$. Thus, the income stream $(1, -1, 1, -1, \ldots)$ is Abel summable to $1/2$ but is not summable in the normal sense. If a sequence is summable in the normal sense, can it fail to be summable in the sense of Abel? No. Consider:

Theorem 1 (Abel) (see Rudin, 1976, p. 174): For the sequence $(a_1, a_2, \ldots)$, define S_n by (31) and $S(r)$ by (32). Then:

$$\lim_{n \to \infty} S_n = S \quad \text{implies} \quad \lim_{r \to 0+} S(r) = S \tag{33}$$

Let us paraphrase this theorem: If the total income S_n over n periods approaches S as n becomes large, the present value of the income stream approaches the same limit, S, as the interest rate r vanishes. Alternatively, if the sequence $(a_1, a_2, \ldots)$ sums to S in the normal sense, then the same sequence sums to S in the sense of Abel.

This Abelian theorem is not quite sufficient for our purposes because it does not distinguish between income streams whose totals approach infinity at different *rates*, e.g., between $(1, 1, \ldots)$ and $(2, 2, \ldots)$. Interpret S_n/n as the *rate* at which income is earned over the first n periods. Consider:

Theorem 2 (*Abelian*): For the sequence $(a_1, a_2, \ldots)$, define S_n by (31) and $S(r)$ by (32). Then

$$\lim_{n \to \infty} \frac{S_n}{n} = g \quad \text{implies} \quad \lim_{r \to 0+} \frac{S(r)}{r} = g \tag{34}$$

$$\lim_{n \to \infty} [S_n - ng] = b \quad \text{implies} \quad \lim_{r \to 0+} \left[S(r) - \frac{g}{r} \right] = b \tag{35}$$

To illustrate this Abelian theorem, consider the sequence $(0, 1, 2, 2, 2, \ldots)$. Notice that $S_n = 2n - 3$ for $n \geq 3$. A routine calculation verifies that $[S(r) - 2/r] \to -3$ as $r \to 0$, exactly as predicted by expression (35).

Theorem 2 shows how the undiscounted income stream determines the discounted stream for small interest rates. Designate as $V_T^\delta(i)$ the expectation of the undiscounted income received by time T if state i is observed initially and if policy δ is used. One can demonstrate that, under a mild regularity condition, there exist numbers $g^\delta(i)$ and $w^\delta(i)$ for which

$$w^\delta(i) = \lim_{T \to \infty} [V_T^\delta(i) - g^\delta(i) T], \quad \text{each } i \tag{36}$$

Equation (36) lets us interpret $g^\delta(i)$ as a *gain rate*, namely as the average rate per unit time at which income is earned if one starts at state i and uses policy δ. Equation (36) also lets us interpret $w^\delta(i)$ as the *bias* for starting in state i and using policy δ. Policy δ is called *gain-optimal* if $g^\delta(i) \geq g^\pi(i)$ for every policy π and for every state i. Policy δ is called *bias-optimal* if $w^\delta(i) \geq w^\pi(i)$ for every gain-optimal policy π and every state i. Under suitable regularity conditions (see Denardo and Rothblum, 1979), there exists a bias-optimal policy.

Equation (36) and Theorem 2 determines the behavior of the discounted infinite-horizon income as r vanishes. Let v_r^δ denote the return function for policy δ, with the interest rate r shown explicitly. Theorem 2 shows that

$$w^\delta(i) = \lim_{r \to 0+} \left\{ v_r^\delta(i) + \frac{g^\delta(i)}{r} \right\}, \quad \text{each } i \qquad (37)$$

Expression (37) shows that a bias-optimal policy maximizes the expected discounted income as the interest rate r vanishes. Indeed, one can show (see Blackwell, 1962) that a bias-optimal policy is optimal when the interest rate is close enough to zero. And bias-optimal policies are found by studying the undiscounted problem, which is not afflicted by the numerical instabilities of the discounted problem.

10.9. Infinite Horizon with No Discounting

Let us now study a Markov decision model with no discounting and an infinite horizon. For it, we seek a gain-optimal policy. (Finding a bias-optimal policy lies somewhat beyond the scope of this discussion.) Curiously, the easy way to find a gain-optimal policy is to look for it among the policies that are stationary and "randomized."

Until now, we have focused on policies that are stationary and nonrandomized. *Stationary* means that the action $\delta(i)$ taken when state i is observed depends only on i, e.g., not on the number of times that state i has been observed. *Nonrandomized* means that this action $\delta(i)$ is a fixed element of A_i, that $\delta(i)$ is not influenced by chance. A stationary *randomized* policy π is defined as a set of number π_i^k such that

$$\pi_i^k \geq 0, \quad \text{each } i \in S, \text{ each } k \in A_i$$

$$\sum_{k \in A_i} \pi_i^k = 1, \quad \text{each } i \in S$$

Interpret π_i^k as the conditional probability of selecting action k if state i is observed. The stationary randomized policies include the stationary nonrandomized policies because the latter have

Dynamic Programming

$$\pi_i^k = \begin{cases} 1 & \text{for } k = \delta(i) \\ 0 & \text{for } k \neq \delta(i) \end{cases}$$

Program 6, below, concerns the discrete-time case. Interpret the decision variable x_i^k in Program 6 as the *joint* probability of observing state i and selecting action k.

Program 6: Maximize h subject to

$$\sum_i \sum_k x_i^k R_i^k = h \tag{38}$$

$$\sum_j \sum_k x_j^k P_{ji}^k = \sum_k x_i^k \quad \text{each } i \in S \tag{39}$$

$$\sum_i \sum_k x_i^k = 1 \tag{40}$$

$$x_i^k \geq 0, \quad \text{each } i \in S \text{ and } k \in A_i \tag{41}$$

The constraints in Program 6 relate to steady-state probabilities. Constraint (41) keeps each probability nonnegative. Constraint (40) equates the sum of the probabilities to 1, as must be. Constraint (39) equates the steady-state probability of a transition to state i to the probability of observing state i. Constraint (38) equates h to the average rate at which income is earned. Recall that $g(i)$ is the largest attainable gain rate while starting in state i. Program 6 seems to maximize $g(i)$ over i.

Proposition 9: Consider a finite discrete-time Markov decision model with an infinite horizon and no discounting. Program 6 is feasible and bounded. Let (h, x) be any basic optimal solution to Program 6, and set $\hat{S} = \{i \in S : \sum_k x_i^k > 0\}$. Then:

1. For each state i in S, $h = g(i) = \text{maximum}\{g^\delta(j) : j \in S, \delta \in \Delta\}$.
2. Each state i in $\hat{S}$ has $x_i^k > 0$ for exactly one value of k.
3. To earn income at average rate of h, start in any state i in $\hat{S}$ and on each visit to each state j select the action k having $x_j^k > 0$.

For a proof, see Denardo (1970). To interpret Proposition 9, consider any optimal basic solution (h, x) to Program 6. Use this basic solution to specify part or all of a policy δ, as follows: If state i has $\Sigma_k x_i^k > 0$, set $\delta(i) = k$ for the unique k having $x_i^k > 0$. If state i has $\Sigma_k x_i^k = 0$, leave $\delta(i)$ undefined. Now, start in any state i for which $\delta(i)$ is defined, and each time a state j is visited, use decision $\delta(j)$. This will never cause transition to a state j for which $\delta(j)$ is undefined, and it will cause income to be earned at the largest possible rate, which equals h.

Program 6 has been developed for the discrete-time model. To adapt it to the Markov renewal model, we must account for the fact that its transition times can vary with the state and the action. For each state i in S and for each action k in A_i, designate v_i^k by

$$v_i^k = \sum_j P_{ij}^k (ET_{ij}^k)$$

Interpret v_i^k as the expectation of the time to transition, given that state i has just been observed and given that action k has been selected.

To generalize Program 6 to the Markov renewal model, replace constraint (38) by

$$h = \frac{\sum_i \sum_k x_i^k R_i^k}{\sum_i \sum_k x_i^k v_i^k} \qquad (42)$$

This redefines h as the ratio of the average income per transition to the average time per transition. The discrete-time case reduces (42) and (38) because it has each transition time T_{ij}^k equal to 1.

For the Markov renewal model, the analogue of Program 6 is to maximize h as defined by (42) subject to constraints (39)–(41). This is a *linear fractional program*. A simple change of variables converts a linear fractional program to an equivalent linear program. This change of variables happens to preserve (38) and to alter (40), as follows:

Corollary 2: For the Markov renewal model with no discounting, the conclusions of Proposition 9 remain true when constraint (40) is replaced by

$$\sum_i \sum_k x_i^k v_i^k = 1$$

provided that each state-action pair has $v_i^k > 0$.

The proviso in Corollary 2 can be weakened. Individual transitions can have expected transition times v_i^k that equal zero, but no policy δ can have a closed (escape-proof) set of states whose expected transition times equal zero.

Let us summarize the discussion of infinite-horizon models. They cannot be solved by rollbacks. They can be solved by linear programing, e.g., by Programs 5 and 6. The value of an infinite-horizon model rests on a turnpike theorem, which states that an infinite horizon is a reasonable surrogate for a long finite horizon. The value of an undiscounted infinite-horizon model is strengthened by an Abelian theorem, which states that its optimal solution remains optimal for the discounted model when the interest rate is close to zero.

Dynamic Programming

11. INVENTORY CONTROL: OPTIMAL (s, S) POLICIES

This section concerns the control of inventory of an item whose demand in each period is uncertain and whose ordering cost includes a fixed charge. This control problem is a Markov decision model. Its optimal policy is described by two parameters, a "reorder point" s and an "order-up-to quantity" S. We shall see that optimal values of s and S can be found quickly. This inventory control problem is specified as

Example 18: A product exists in integer quantities. At the start of each of infinitely many periods, the manager observes the quantity on hand and decides how many units to order, if any. The order is delivered immediately. The demand D that will occur during a period is uncertain. The demands in different periods are independent of each other and have the same probability distribution, which is known to the manager. The quantity sold in a period is the smaller of the demand D during that period and the number y of units on hand at the start of the period, after receipt of any order. The excess of D over y, if any, is assumed to be backlogged, which means that it must be filled from future orders. A value $y < 0$ indicates a backlog of $-y$ units.

The cost of placing an order is $K > 0$, independent of the size of the order. The expectation of the cost incurred during the period is given by the function $G(y)$ of the quantity y on hand at the start of the period, after receipt of any order. The function $G(y)$ is assumed to be quasi-convex and to have a global minimum. Let us designate by y^* the smallest integer that minimizes $G(y)$:

$$y^* = \text{minimum}\{y : G(y) \leq G(z) \text{ for every integer } z\}$$

[Quasi-convexity means that the function $G(y)$ is nonincreasing for $y \leq y^*$ and is nondecreasing for $y \geq y^*$.] It is also assumed that $G(y)$ rises above $G(y^*)$ by more than K as $y \to \infty$ and as $y \to -\infty$. The manager wishes to minimize the average cost per unit time over an infinite planning horizon.

Remarks: Per-unit ordering costs and per-unit sales receipts have been absorbed into the holding cost function, $G(y)$. The results developed here hold not just for Example 18 but for models that include (1) a fixed delay between the time an order is placed and the moment when that order is received, (2) discounting, (3) lost sales, i.e., the case in which any excess of demand over supply is lost rather than backlogged, and (4) continuous review of inventory.

For this inventory control problem, what type of policy makes economic sense? Intuitively, orders below a certain size should be avoided because their benefit cannot offset the fixed charge K for placing an order.

Also, if any order is placed, it should restore the inventory position to its most desired level.

An inventory control rule is called an (s, S) *policy* if the parameters s and S have this interpretation: if the quantity on hand is below s, order the quantity needed to restore the inventory to S units. An (s, S) policy has $s \leq S$. Each order quantity exceeds $(S - s)$, and each order restores the inventory level to S. In an (s, S) policy, the number S is called the *order-up-to quantity*, and s is called the *reorder point*.

Scarf (1960) first showed that an (s, S) policy is optimum; see also Denardo (1982, pp. 150–154). Zheng and Federgruen (1991) introduced a quick method for computing the parameters s and S of the least-cost (s, S) policy. We present their method, adapted to include an insight from Denardo et al. (1995). Their method may seem simple, but it had eluded researchers for three decades.

Designate by $c(s, S)$ the rate per unit time at which cost is incurred by the (s, S) policy. Each moment at which an (s, S) policy causes an order to be placed is a *regeneration point* because each order restores the inventory to S, so that what happens after regeneration is independent of what happened before. A well-known theorem from regenerative processes (see Section 7.4 of Ross, 1994) is that

$$c(s, S) = \frac{\text{expected cost between successive regenerations}}{\text{expected time between successive regenerations}} \qquad (43)$$

We seek formulae for the numerator and denominator of (43). The demand D that occurs during a period can take the values 0, 1, 2, ... Designate by $p(j)$ the probability that D equals j:

$$p(j) = \Pr\{D = j\}, \quad \text{for } j = 0, 1, \ldots$$

The manager is assumed to know the demand distribution, the $p(j)$'s. To avoid a triviality, assume that the demand is not identically zero, that $p(0) = \Pr\{D = 0\} < 1$.

The denominator in (43) is the expectation of the number of periods that elapse before the *total* demand over those periods exceeds $(S - s)$. Aiming for a recursion, we designate, for $j = 0, 1, 2, \ldots$,

$M(j) = $ the expectation of the number of periods that elapse until the total demand exceeds j units

Conditioning this expectation on the demand D in the first period gives

$$M(j) = \sum_{k=0}^{j} \Pr\{D = k\}[1 + M(j - k)] + \sum_{k=j+1}^{\infty} \Pr\{D = k\}[1 + 0]$$

Dynamic Programming

To justify this formula, notice that, if the demand D in the first period equals k and if $k \leq j$, then $[1 + M(j - k)]$ is the expected number of periods that will elapse until the total demand exceeds j. This formula for $M(j)$ simplifies:

$$M(j) = 1 + \sum_{k=0}^{j} p(k)M(j - k) \quad \text{for } j = 0, 1, 2, \ldots \quad (44)$$

Equation (44) is a recursion; it can be solved in ascending j. As suggested earlier, $M(S - s)$ is the expectation of the time between successive regenerations. In other words, $M(S - s)$ equals the denominator of (43).

Evidently, $M(j) \geq M(j - 1)$. Let us designate the difference as $m(j)$. Specifically, with $M(-1) = 0$,

$$m(j) = M(j) - M(j - 1), \quad \text{for } j = 0, 1, \ldots \quad (45)$$

To interpret $m(j)$, suppose that regeneration has just occurred, so the inventory level has just been restored to S. For each $j \leq S - s$, the number $m(j)$ equals the expectation of the number of periods, prior to the next regeneration, during which the inventory level equals $S - j$.

Designate by $H(s, S)$ the expected holding cost between successive regenerations. From the prior interpretation of $m(j)$,

$$H(s, S) = \sum_{j=0}^{S-s} m(j)G(S - j) \quad (46)$$

The expected cost between regenerations is $K + H(s, S)$, and (43) gives

$$c(s, S) = \frac{K + H(s, S)}{M(S - s)} \quad (47)$$

Equation (47) lets one compute $c(s, S)$ for a particular pair, s and S. Equation (47) also reveals information about the values of s and S that minimize $c(s, S)$. Consider:

Proposition 10: $c(s, S)$ is minimized by parameters s and S that satisfy

$$s \leq y^* \leq S \quad \text{and} \quad G(S) \leq c(s, S) \quad (48)$$

Proof: For $x > y^*$, the inequality $H(s - 1, S - 1) \leq H(s, S)$ is evident from (46) and from the fact that $G(y)$ is nondecreasing for $y \geq y^*$, so (47) gives $c(s - 1, S) \leq c(s, S)$ in this case. Similarly, for $S < y^*$, the inequality $H(s + 1, S + 1) \leq H(s, S)$ is evident from (46) and from the fact that $G(y)$ is nonincreasing for $y \leq y^*$, so (47) gives $c(s, S) \geq c(s + 1, S + 1)$ in this case. In short, at least one cost-minimizing (s, S) policy has $s \leq y^* \leq S$.

Conditioning the expected holding cost $H(s,S)$ on the demand in the first period gives

$$H(s, S) = G(S) + \sum_{j=0}^{S-s} p(j)H(s, S - j)$$

To complete the proof, consider an (s, S) policy having $G(S) > c(s, S)$. Rearrange (47) as $c(s, S)M(S - s) = K + H(s, S)$ and use the above to get

$$c(s, S)M(S - s) > K + c(s, S) + \sum_{j=0}^{S-s} p(j)H(s, S - j)$$

The above combines with the recursion in (44) for $M(S - s)$ to give

$$c(s, S) > \frac{K + \sum_{j=0}^{S-s} p(j)H(s, S - j)}{\sum_{j=0}^{S-s} p(j)M(S - j)} \qquad (49)$$

Now, set $q = \Pr\{D \leq S - s\}$, and consider this randomized policy: Whenever the quantity on hand drops below s, restore it to $S - j$ with probability $p(j)/q$ for $j = 0, 1, \ldots, S - s$. The regeneration theorem applies to this policy; its cost rate c equals the right-hand side of (49). So $c(s, S) > c$. Proposition 9 shows that the cost c of this randomized policy cannot be lower than the least of the costs of the nonrandomized policies $c(s, j)$ for $s \leq j \leq S$. In short, no (s, S) policy having $G(s) > c(s, S)$ can minimize cost. □

The next step in our analysis of (s, S) policies is to examine the formula in (46) for the holding cost $H(s, S)$. This formula is a sum whose addends are $m(j)G(S - j)$. The discrete analog of integration by parts relates this formula to a sum whose addends are $M(j)g(S - j)$, where the (difference) function $g(y)$ is defined by

$$g(y) = G(y - 1) - G(y), \qquad y = 0, \pm 1, \pm 2, \ldots \qquad (50)$$

That the function

$$B(s, S) = \sum_{j=0}^{S-s-1} M(j)g(S - j) \qquad (51)$$

plays a central role is evident in

Proposition 11:

$$H(s, S) = M(S - s)G(s) - B(s, S) \qquad (52)$$

$$c(s, S) = G(s) + \frac{K - B(s, S)}{M(S - s)} = G(s - 1) + \frac{K - B(s - 1, S)}{M(S - s)} \qquad (53)$$

Dynamic Programming

Proof: To verify (52), note from (45) and (50) that $m(j)G(S - j) = [M(j) - M(j - 1)][G(S) + g(S) + g(S - 1) + \cdots + g(S - j - 1)]$, substitute the latter for the former in (56), and collect coefficients of $g(j)$ and of $G(S)$. to verify the leftward equation in (53), substitute (52) into (47). Finally, to verify the rightward equation, substitute $B(s, S) = B(s - 1, S) - M(S - s)g(s)$ and $G(s) = G(s - 1) - g(s)$. □

Note from (53) that $c(s, S)$ lies between $G(s - 1)$ and $G(s)$ if and only if K lies between $B(s, S)$ and $B(s - 1, S)$. This motivates the first of the following two definitions:

$$r(S) = \min\{s | B(s, S) \leq K\} \qquad (54)$$

$$c^*(S) = \underset{s<S}{\text{minimum}}\ c(s, S) \qquad (55)$$

Interpret $c^*(S)$ as the least cost of the policies that restore inventory to S when ordering. If $c(s, S) = c^*(S)$, we interpret s as an *optimum* reorder point for order-up-to quantity S. That $r(s)$ is an optimum reorder point is demonstrated in:

Proposition 12: For each S, the functions $B(\cdot, S)$ and $c(\cdot, S)$ of s are quasi-convex. Also, for $S \geq y^*$,

$$c[r(S), S] = c^*(S) \qquad (56)$$

Finally, $s = r(S)$ if and only if

$$G(s - 1) > c(s, S) \geq G(s) \qquad (57)$$

Proof: Since $G(y)$ is quasi-convex with a minimum at y^*, the function $g(y)$ is nonnegative for $y \leq y^*$ and is nonpositive for $y > y^*$. Since $M(j)$ is nonnegative, it is clear from (51) that $B(\cdot, S)$ is quasi-convex, that $B(s, S)$ increases as s decreases below y^*, and that $B(S, S) \leq 0$ for $S \geq y^*$.

To see that $c(\cdot, S)$ is quasi-convex, set $d = S - s$ and use both equations (53) and $m(d + 1) = M(d + 1) - M(d)$ to get

$$[c(s, S) - c(s - 1, S)]M(d)M(d + 1)$$
$$= [K - B(s - 1, S)]m(d + 1)$$

Since $m(j) \geq 0$, $[c(s, S) - c(s - 1, S)]$ and $[K - B(s - 1, S)]$ have the same sign. Thus, $c(\cdot, S)$ is quasi-convex; moreover, $c(s, S)$ is minimized by the smallest s having $B(s, S) \leq K$, which verifies (56). And (53) shows that $s = r(S)$ if and only if (57) holds, completing the proof. □

For a given $S \geq y^*$, Proposition 12 shows that $r(S)$ is an optimal reorder point, moreover, that $r(S)$ and $c^*(S)$ can be found from the following recursion.

Subroutine: compute $r(S)$ and $c^*(S)$:

$x \leftarrow 0$ and $B \leftarrow M(0)g(S)]$.
While $B \leq K$ do
  begin
    $x \leftarrow (x + 1)$,
    $B \leftarrow [B + M(x)g(S - x)]$.
  end
$r(S) \leftarrow S - x$.
$c^*(X) \leftarrow G[r(S) - 1] + \dfrac{K - B}{M[S - r(S)]}$.

To find the parameters of the least-cost (s, S) policy, we could begin with $S = y^*$ and apply the preceding subroutine to compute $r(S)$ and $c^*(S)$ in ascending S, ending when S has increased to the point where $G(S) \geq c^*(S)$.

To see how knowledge of $r(S)$ and $c^*(S)$ can accelerate computation, consider any $T > S$. If $c^*(T) \leq c^*(S)$, then (57) gives $r(T) \geq r(S)$. On the other hand, if $c^*(T) > c^*(S)$, then we need not compute $c^*(T)$. Consequently, for $T > S$, we can adapt the subroutine to compute $\hat{t} = \max\{r(S), r(T)\}$ by terminating its "while" statement after $T - x$ decreases to $r(S)$. The search that results is as follows.

Search for S, $r(S)$, and $c^*(S)$ of the least-cost (s, S) policy:
Set $S \leftarrow y^*$. Compute $r(S)$ and $c^*(S)$.
Set $T \leftarrow (S + 1)$.
While $G(T) < c^*(S)$ do
  begin
    Compute $\hat{t} = \max\{r(S), r(T)\}$ and $c(\hat{t}, T)$.
    If $c(\hat{t}, T) < c^*(S)$ then
      $S \leftarrow T$, $r(S) \leftarrow r(T)$, and $c^*(S) \leftarrow c(\hat{t}, T)$.
    $T \leftarrow (T + 1)$.
  end

This search is one-directional; no update decreases S or $r(S)$. One can show that the work entailed by this search exceeds the work required to compute $M(\cdot)$ by a factor that lies between 2 and 2.4.

12. CONCLUSION

This chapter has acquainted the reader with dynamic programming and with its uses in industrial and systems engineering. In deterministic settings, dynamic programming handles diseconomies of scale for which the direct application of linear programming fails. In a stochastic setting, dynamic programming provides the simplest and the most flexible model of decision making.

Dynamic Programming

One goal of this chapter has been to help the reader learn how to formulate an optimization problem as a dynamic program. The key to formulation is to identify the states. A state is a regeneration point. A state summarizes what has happened in enough detail to evaluate actions and to allow transitions to occur from state to state. A policy assigns an action to each state. A policy is called optimal if, for each starting state, the policy earns the greatest income possible (or incurs the least cost possible) while starting in that state. The principle of optimality asserts that there exists an optimal policy. The functional equation of dynamic programming springs from an optimal policy when only its initial action is changed.

This chapter has surveyed the types of dynamic program that are more likely to occur. These include eight deterministic dynamic programs, each of which can be interpreted as a shortest- or longest-route problem in a network. In most cases, the path's length equals the sum of the lengths of its arcs. In one example, the path's length equals the sum of the lengths of its nodes. In one, the path's length equals the longest of the lengths of its arcs. These definitions of path length satisfy the monotonicity condition, as they must if they are to lead to functional equations.

This chapter has also presented four stochastic dynamic programs, of which the third generalizes the others. The third is actually a family of models of decision making under uncertainty—with finite or infinite planning intervals, with discounting or without. Certain of these stochastic dynamic programs can be solved by rolling back a functional equation. All can be solved by linear programming.

The subject matter of this chapter has concerned industrial and systems engineering. The subjects include project management, the grouping of jobs into labor grades, aggregate planning, production scheduling with economies of scale, inventory control, and expansion of capacity. They include a problem of pest control and a medical decision.

In many applications, the nemesis of dynamic programming is this—too many states. This difficulty was dubbed the *curse of dimensionality* by Richard Bellman (1952, 1957). In this context, "dimensionality" refers to the number of variables that must be kept track of. Suppose that state s keeps track of p variables: $s = (x_1, x_2, \ldots, x_p)$. If the state variables x_1 through x_p can take 10 values each, the number of states equals 10^p, which grows rapidly in p. When $p = 25$, there's no hope for efficient solution on any computer, current or future.

Actually, there are two curses of dimensionality. A model that has 10^{25} states cannot be solved. If a model has 10^{10} states, one might be able to compute its solution economically, but one cannot, in this writer's opinion, convince a manager to use that solution. He or she will want to know why that solution works, not merely what it is. A model must be simple enough to be understood. In addition, its robustness must be demonstrable.

In this writer's view, the most exciting current research in dynamic programming circumvents both curses by restricting attention to a class of policies that are comprehensible, relatively few in number, and tractable. The goal to devise rules that make sense to managers, rules whose operating characteristics can be computed and optimized, rules that give up little by way of optimality. Successful work of this type can be found in the control of systems of inventories, in the scheduling of production flows through plants, and in the routing of traffic through telecommunications networks. The focus of this work shifts from exact solutions to effective ones.

REFERENCES

Aggarwal, A., Park, J. K., "Improved Algorithms for Economic Lot Size Problems," *Opns. Res.*, Vol. 41, pp. 549–571, 1993.

Ahuja, R. K., Magnanti, T. L., Orlin, J. B., *Network Flows: Theory, Algorithms and Applications*. Prentice Hall, Englewood Cliffs, NJ, 1993.

Allais, M., "Le Comportement de l'Homme Rationnel devant le Risque, Critique des Postulates de l'École Américaine," *Econometrica*, Vol. 21, pp. 503–546, 1953.

Bell, D. E., "Regret in Decision Making Under Uncertainty," *Opns. Res.*, Vol. 30, pp. 961–981, 1982.

Bellman, R. E., "On the Theory of Dynamic Programming," *Proc. Natl. Acad. Sci. USA*, Vol. 38, pp. 716–719, 1952.

Bellman, R. E., *Dynamic Programming*. Princeton University Press, Princeton, NJ, 1957.

Blackwell, D., "Discrete Dynamic Programming," *Ann. Math. Statist.*, Vol. 33, pp. 719–726, 1962.

Blackwell, D., "Discounted Dynamic Programming," *Ann. Math. Statist.*, Vol. 36, pp. 226–235, 1965.

Chen, C.-D., Lee, C.-Y., "Error Bound for the Dynamic Lot Size Model Allowing Speculative Motive," *IIE Trans.*, Vol. 27, pp. 683–688, 1995.

Çinlar, E., "Superposition of Point Processes," in *Stochastic Point Processes: Statistical Analysis, Theory, and Applications*, ed. P. A. W. Lewis. Wiley-Interscience, New York, pp. 549–602, 1972.

Denardo, E. V., "Sequential Decision Processes," Ph.D. dissertation, Northwestern University, Evanston, IL, 1965.

Denardo, E. V., "Contraction Mappings in the Theory Underlying Dynamic Programming," *SIAM Rev.*, Vol. 9, pp. 165–177, 1967.

Denardo, E. V., "On Linear Programming in a Markov Decision Problem," *Mgmt. Sci*, Vol. 16, pp. 281–288, 1970.

Denardo, E. V., *Dynamic Programming: Models and Applications*. Prentice Hall, Englewood Cliffs, NJ, 1982.
Denardo, E. V., Hoffman, A. J., MacKensie, T., Pulleyblank, W. R., "A Nonlinear Allocation Problem," *IBM J. Res. Dev.*, Vol. 36, pp. 301–306, 1994.
Denardo, E. V., Huberman, G. R., Rothblum, U. G., "Optimal Locations on a Line Are Interleaved," *Opns. Res.*, Vol. 30, pp. 745–759, 1982.
Denardo, E. V., Rothblum, U. G., "Overtaking Optimality for Markov Decision Chains," *Math. Opns. Res*, Vol. 4, pp. 144–152, 1979.
Denardo, E. V., Feinberg, E., and Kella, O., "Threshold Policies in Clearing and Queueing Processes," Technical Report, Yale, New Haven, CT, 1995.
Ellsberg, D., "Risk, Ambiguity and the Savage Axioms," *Quart. Jour. Econ.*, Vol. 75, pp. 643–669, 1961.
Federgruen, A., Tzur, M., "A Simple Forward Algorithm to Solve General Dynamic Lot Sizing Models with n Periods in $O(n \log n)$ or $O(n)$ Time," *Mgmt. Sci.*, Vol. 37, pp. 909–925, 1991.
Graves, S. C., "Using Lagrangian Techniques to Solve Hierarchical Production Planning Problems," *Mgmt. Sci.*, Vol. 28, pp. 260–275, 1982.
Kahneman, D., Tversky, A., "Prospect Theory: An Analysis of Decision Under Risk," *Econometrica*, Vol. 47, pp. 263–291, 1979.
Lee, C.-Y., Denardo, E. V., "Rolling Planning Horizons: Error Bounds for the Dynamic Lot Size Model," *Math. Opns. Res.*, Vol. 11, pp. 423–432, 1986.
Luss, H., "Operations Research and Capacity Expansion Problems: A Survey," *Opns. Res.*, Vol. 30, pp. 907–947, 1982.
Luss, H., "A Heuristic for Capacity Expansion Planning with Multiple Facility Types," *Naval Res. Logist. Quart.*, Vol. 33, pp. 685–701, 1986.
Manne, A. S., Veinott, A. F., Jr., "Optimal Plant Size with Arbitrary Increasing Time Paths of Demand," in *Investments for Capacity Expansion: Size, Location, and Time Phasing*, ed. A. S. Manne, pp. 178–190. MIT Press, Cambridge, MA, 1967.
Mitten, L. G., "Composition Principles for Synthesis of Optimal Multi-Stage Processes," *Opns. Res.*, Vol. 12, pp. 610–619, 1964.
Mitten, L. G., "Preference Order Dynamic Programming," *Mgmt. Sci.*, Vol. 21, pp. 43–46, 1974.
Pauker, S. P., Pauker, S. G., "The Amniocentesis Decision: Ten Years of Decision Analytic Experience," *Birth Defects: Original Article Series*, Vol. 23, pp. 151–169, 1987.
Ramsey, F. P., "Truth and Probability," in *The Foundations of Mathematics and Other Logical Essays*, ed. F. P. Ramsey. Harcourt Brace, New York, 1931.
Ross, S. M., *An Introduction to Probability Models*, 5th ed. Academic Press, San Diego, 1994.
Roundy, R. O., "Efficient Effective Lot Sizing for Multistage Production Systems," *Opns. Res.*, Vol. 41, pp. 371–385, 1993.
Rudin, W., *Principles of Mathematical Analysis*, 3rd ed. McGraw-Hill, New York, 1976.
Savage, L. J., *The Foundations of Statistics*. Wiley, New York, 1954.

Scarf, H., "The Optimality of (S, s) Policies in the Dynamic Inventory Problem," in *Mathematical Methods in the Social Sciences 1959*, eds. K. Arrow, S. Karlin, and P. Suppes, chapter 13. Stanford University Press, Stanford, CA, 1960.

Shapiro, J. F., "Turnpike Planning Horizons for a Markovian Decision Model," *Mgmt. Sci.*, Vol. 14, pp. 292–300, 1968.

Shapley, L. S., "Stochastic Games," *Proc. Natl. Acad. Sci. USA*, Vol. 39, pp. 1095–1100, 1953.

Von Neumann, J., Morgenstern, O., *The Theory of Games and Economic Behavior*, 2nd ed. Princeton University Press, Princeton, NJ, 1947.

Wagelmans, A., van Hoesel, S., Kolen, A., "Economic Lot Sizing: An $O(n \log n)$ Algorithm That Runs in Linear Time in the Wagner-Whitin Case," *Opns. Res.*, Vol. 40, pp. S145–S156, 1992.

Wagner, H. M., Whitin, T. M., "Dynamic Version of the Economic Lot Size Formula," *Mgmt. Sci.*, Vol. 5, pp. 89–96, 1958.

Zangwill, W. I., "A Deterministic Multi-Period Production Model with Backlogging," *Mgmt. Sci.*, Vol. 14, pp. 429–450, 1966.

Zheng, Y.-S., Federgruen, A., "Finding Optimal (s, S) Policies Is About as Simple as Evaluating a Single Policy," *Opns. Res.*, Vol. 39, pp. 654–665, 1991.

6

Nonlinear Programming

Leon Lasdon
The University of Texas at Austin, Austin, Texas

John Plummer
Southwest Texas State University, San Marcos, Texas

Allan D. Waren
Cleveland State University, Cleveland, Ohio

1. INTRODUCTION

1.1. Problem Statement

A general form for a nonlinear program (NLP) is

$$\text{Minimize or maximize } f(x)$$
$$\text{subject to } \quad a_i \leq g_i(x) \leq b_i, \quad i = 1, \ldots, m \quad (1.1)$$
$$\text{and} \quad l_j \leq x_j \leq u_j, \quad j = 1, \ldots, n$$

In this problem statement, x is a vector of n decision variables $(x_1, \ldots, x_n)$, f is the objective function, and the g_i are constraint functions. The a_i and b_i are specified lower and upper bounds on the constraint functions satisfying $a_i \leq b_i$, and l_j, u_j are lower and upper bounds on the variables with $l_j \le u_j$. If $a_i = b_i$ the ith constraint is an equality constraint, and only upper and only lower limits on g_i correspond to the cases $a_i = -\infty$ and $b_i = +\infty$. Similar comments apply to the variable bounds, with $l_j = u_j$ corresponding to a variable x_j whose value is fixed and $l_j = -\infty$ and $u_j = +\infty$ specifying a free variable.

In applications, equality constraints often arise as equations describing the static or dynamic operation of the system being modeled, and upper bounds enforce budget limits, unit capacities, or safety requirements like limits on pressures or temperatures. Lower limits often impose system requirements; e.g., demands must be satisfied, and system units cannot operate efficiently below specified levels. The objective function is often a cost to be minimized or a profit or system efficiency measure to be maximized. The decision variables can be operating conditions in a physical system like temperature, pressure, or flow rates, design variables like pipe diameters or beam dimensions, or allocations of limited resources to various activities, or they may specify a production schedule, e.g., inventory and production levels of several products in multiple time periods.

The problem (1.1) is nonlinear if one or more of the problem functions $f, g_1, \ldots, g_m$ are nonlinear. It is *unconstrained* if there are no constraint functions g_i and no bounds on the x_i, and *bound constrained* if only the x_i are bounded. In *linearly constrained* problems all constraint functions g_i are linear and the objective f is nonlinear. There are special nonlinear programming (NLP) algorithms and software for unconstrained and bound constrained problems, and we describe these first. Methods and software for solving constrained NLPs use many ideas from the unconstrained case. Most modern software can handle nonlinear constraints and is especially efficient on linearly constrained problems. A linearly constrained problem with a quadratic objective is called a quadratic program (QP). Special methods exist for solving QPs, and these are often faster than general-purpose procedures.

A vector x is *feasible* if it satisfies all the constraints. The set of all feasible points is called the feasible region, F. If F is empty the problem is *infeasible*, and if feasible points exist at which the objective f is arbitrarily large in a max problem or arbitrarily small in a min problem, the problem is *unbounded*. A vector x^* is a *global solution* to the problem if

$$f(x^*) \le f(x) \quad \text{for all } x \in F \tag{1.2}$$

and is a *local solution* if (1.2) holds for all $x \in F$ that are sufficiently close to x^*.

1.2. NLP Geometry

A typical feasible region for a problem with two variables and the constraints

Nonlinear Programming

$$x_j \geq 0, \quad g_i(x) \leq 0, \quad i = 1, 2, \quad j = 1, 2$$

is shown as the unshaded region in Figure 1. Its boundaries are the straight or curved lines $x_j = 0$, $g_i(x) = 0$ for $i = 1, 2, j = 1, 2$. If a nonlinear equality constraint $g_3(x) = 2.6$ were added, the new feasible region would be the intersection of the curve $g_3(x) = 2.6$ with the original set, as shown in Figure 1.

As an example, consider the problem

Minimize $z = (x_1 - 3)^2 + (x_2 - 4)^2$

subject to the linear constraints

$$x_1 \geq 0$$
$$x_2 \geq 0$$
$$5 - x_1 - x_2 \geq 0$$
$$-2.5 + x_1 - x_2 \leq 0$$

This problem is shown graphically in Figure 2. The feasible region is of the linear programming type, having a finite number of corner points. The objective function, being nonlinear (quadratic), has contours of constant value that are not parallel lines, as in the linear case, but concentric circles.

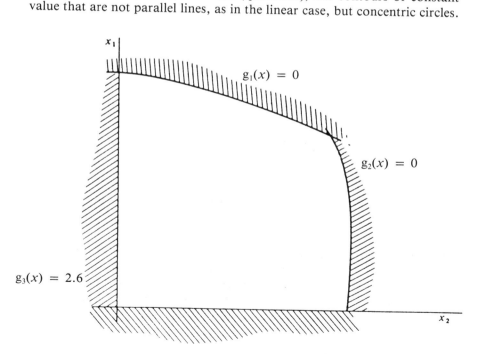

Figure 1 Feasible region.

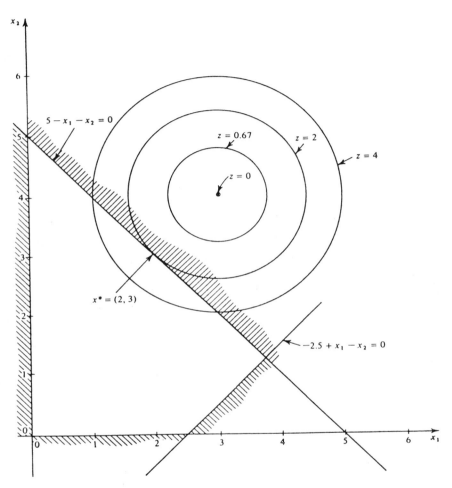

Figure 2 Minimum on boundary of constraint set.

The minimum value of z corresponds to the contour of lowest value having at least one point in common with the feasible region, i.e., at $x_1^* = 2$, $x_2^* = 3$. This is not an extreme point of the set, although it is a boundary point. (Recall that for linear programs the minimum is always at an extreme point.) Furthermore, if the objective function of the previous problem is changed to

$$z = (x_1 - 2)^2 + (x_2 - 2)^2$$

Nonlinear Programming

the situation is as depicted in Figure 3. The minimum is now at $x_1 = 2$, $x_2 = 2$, which is not even a boundary point of the feasible region. Here the unconstrained minimum of the nonlinear function satisfies the constraints.

Neither of the previous problems had local optima that are not global. It is easy, however, to construct nonlinear programs in which such local optima occur. For example, if the objective function of the previous problem had two minima and at least one was interior to the feasible region, then the constrained problem would have two local minima. Contours of

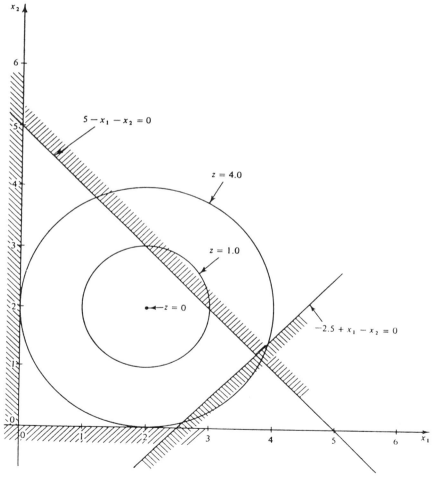

Figure 3 Unconstrained minimum interior to constraint set.

such a function are shown in Figure 4. Note that the minimum at the boundary point $x_1 = 3$, $x_2 = 2$ is the global minimum, because the value of the objective function is $z = 3$, whereas at the interior local minimum $z = 4$.

Although the examples thus far have had linear constraints, the chief nonlinearity of an optimization problem often appears in the constraints. The feasible region will then have curved boundaries. A problem with non-

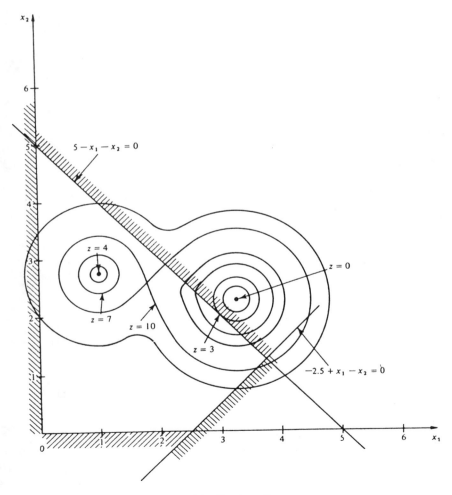

Figure 4 Local optima due to objective function.

Nonlinear Programming

linear constraints may have local optima, even if the objective function has only one unconstrained minimum. Consider a problem with a quadratic objective function and a feasible region as shown in Figure 5. The problem has local optima at the two points a and b, since no point of the feasible region in the immediate vicinity of either point will yield a smaller value of z.

In summary, the optimum of a nonlinear programming problem will in general not be at an extreme point of the feasible region and may not even be on the boundary. Also, the problem may have local optima distinct from the global optimum. These properties are direct consequences of the nonlinearity. However, a class of nonlinear problems can be defined that are guaranteed to be free of distinct local optima. These are called convex programming problems and are now considered.

1.3. Convex and Nonconvex Problems

A set of points is said to be *convex* if, given any two points in the set, the line segment joining them is also in the set. In the examples shown in Figure 6, the two sets A and B are convex, while C is not. It is easily seen that a

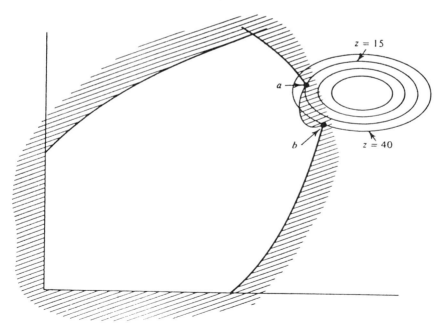

Figure 5 Local optima due to feasible region.

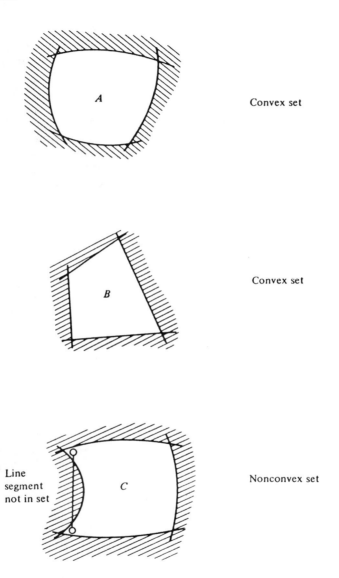

Figure 6 Convex and nonconvex sets.

Nonlinear Programming

convex set is one whose boundary does not bulge inward. Note that the feasible region of a linear programming problem is convex.

In many dimensions, these geometrical ideas must be formulated in algebraic terms. In particular, one must define what is meant by "the line segment between two points." Let the two points be x_1 and x_2 and consider the expression

$$x_3 = \lambda x_1 + (1 - \lambda)x_2$$

where λ is a scalar, $0 \leq \lambda \leq 1$. Obviously, if $\lambda = 0$, $x_3 = x_2$, and if $\lambda = 1$, $x_3 = x_1$. If $0 < \lambda < 1$, then x_3 takes on all values on the line segment between x_1 and x_2. (This is easily verified in two or three dimensions.) Thus the line segment between two points is defined as the set

$$S = \{x | x = \lambda x_1 + (1 - \lambda)x_2, 0 \leq \lambda \leq 1\}$$

A function, $f(x)$ is said to be *convex* if the line segment drawn between any two points on the graph of the function never lies *below* the graph and *concave* if the line segment never lies *above* the graph. Examples of convex and concave functions are shown in Figure 7. The first function is strictly convex, because the line segment is always above the function; the second function is strictly concave. Note that a linear function is both convex and concave but neither strictly convex nor strictly concave.

Algebraically, a function $f(x)$ is convex if

$$f(\lambda x_1 + (1 - \lambda)x_2) \leq \lambda f(x_1) + (1 - \lambda)f(x_2)$$

for all x_1, x_2 in the (convex) domain of definition of f and for all $0 \leq \lambda \leq 1$. The function is strictly convex if the strict inequality holds.

A convex programming problem is one of minimizing a convex function (or of maximizing a concave function) over a convex feasible region. The main theorem of convex programming is the following:

Theorem 1: Any local minimum of a convex programming problem is a global minimum.

Note that the problem may have a number of points at which the global minimum is taken on, but the set of all such points is convex. No distinct (i.e., separated) local minima may exist with different function values. This is a nice property for a problem to have. It serves as motivation for the following results, which enable one to both characterize and recognize convex programming problems more easily.

Theorem 2: If $f(x)$ is convex, then the set

$$R = \{x | f(x) \leq k\}$$

is convex for all scalars k.

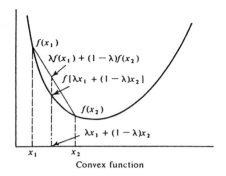
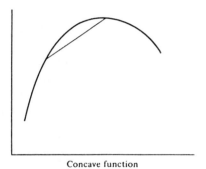
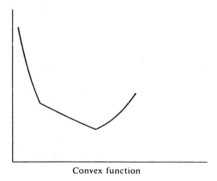

Figure 7 Convex and concave functions.

Nonlinear Programming

The theorem is illustrated in Figure 8, where a convex quadratic function is cut by the plane $f(x) = k$. The set R above is an ellipse plus its interior, which is convex.

Theorem 3: The intersection of any number of convex sets is convex.

This theorem is easily verified geometrically in two dimensions. As a consequence of Theorems 2 and 3, the following holds: Given the problem

Minimize $f(x)$

subject to $g_i(x) \leq b_i$, $i = 1, \ldots, m$

then, if the functions f and g_i are convex, the problem is a convex programming problem. This is true because (1) each of the sets

$$R_i = \{x \mid g_i(x) \leq b_i\}$$

is convex by Theorem 2 and (2) the feasible region R is the intersection of the sets R_i and is convex by Theorem 3. Note that, because a linear function is convex, a linear program is a special case of a convex programming problem. This establishes more firmly the geometrically evident fact that a linear program cannot have local optima distinct from the global optimum.

The previous result implies that convex programs can be identified by

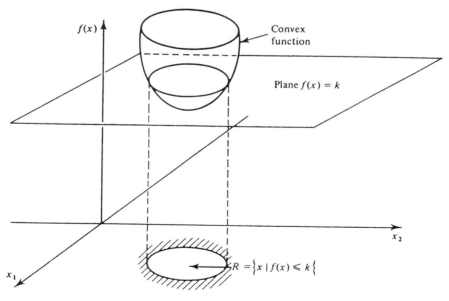

Figure 8 Illustration of Theorem 2.

finding out if the objective and constraint functions of the problem are convex. It is thus of interest to better characterize convex functions. The main results along this line are given below, where ∇f is the gradient of f, with components $\partial f / \partial x_j$.

Theorem 4: If $f(x)$ has continuous first and second partial derivatives, the following statements are equivalent:

(a) $f(x)$ is convex.
(b) $f(x_1) \geq f(x_2) + \nabla f^T(x_2)(x_1 - x_2)$ for any two points x_1, x_2.
(c) The matrix of second partial derivatives of $f(x)$, denoted by $\nabla^2 f(x)$, is positive semidefinite for all points x. (This means that the quadratic form $z^T \nabla^2 f(x) z$ is nonnegative for all vectors z.)

Part (b) of the theorem states that the function evaluated at any point, $f(x_1)$, never lies below its tangent plane passing through any other point x_2.

Theorem 5: A positive semidefinite quadratic form is convex.
This theorem is a direct consequence of part (c) of Theorem 4.

Theorem 6: A positive linear combination of convex functions is convex.

Theorem 7: A function $f(x)$ is convex if and only if the one-dimensional function $g(\alpha) = f(x + \alpha d)$ is convex for all fixed x and d.

Since $f(x + \alpha d)$ is the function evaluated at points along the line d passing through the point x, Theorem 7 says that a convex function is convex along any line. This affords a mean of telling if a given function of n variables is not convex, for if any line in n-dimensional space can be found along which $g(\alpha)$ is not convex, $f(x)$ is not convex either.

1.3.1. "Mixed" Problems

Thus far, problems with both equality and inequality constraints have not been considered. The principal theoretical difficulty in dealing with such problems stems from the fact that, if $g(x)$ is nonlinear, the set

$$R = \{x | g(x) = b\}$$

is generally not convex. This is evident geometrically, since most nonlinear functions have graphs that are curved surfaces. Hence the set R will usually be a curved surface also, and the line segment joining any two points on this surface will generally not lie on the surface.

As a consequence of the above, the problem

Nonlinear Programming

$$\text{Minimize} \quad f(x)$$
$$\text{subject to} \quad g_i(x) \leq 0, \quad i = 1, \ldots, m$$
$$h_k(x) = 0, \quad k = 1, \ldots, r < n \quad (1.3)$$

may not be a convex programming problem in the variables $x_1, \ldots, x_n$ if any of the functions $h_k(x)$ are nonlinear. This, of course, does not preclude efficient solution of such problems, but it does make it more difficult to guarantee the absence of local optima and to generate sharp theoretical results.

In many cases the equality constraints may be used to eliminate some of the variables, leaving a problem with only inequality constraints and fewer variables. Even if the equalities are difficult to solve analytically it may still be worthwhile solving them numerically. This is the approach taken by the generalized reduced gradient method, which is described in Section 4.

1.3.2. Role of Convexity

Although convexity is desirable, many real-world problems turn out to be nonconvex. In addition, there is no simple way to test a nonlinear problem for convexity, because there is no simple way to test a nonlinear function for this property. Why, then, is convex programming studied? The main reasons are:

1. When convexity is assumed, many significant mathematical results have been derived in the field of mathematical programming.
2. Often results obtained under convexity assumptions can give insight into the properties of more general problems. Sometimes, such results may even be carried over to nonconvex problems but in a weaker form.

For example, it is usually impossible to prove that a given algorithm will find the global minimum of a nonlinear programming problem unless the problem is convex. For nonconvex problems, however, many such algorithms will find at least a local minimum. Convexity thus plays a role much like that of linearity in the study of dynamic systems. Many results derived from linear theory are used in the design of, for example, nonlinear control systems.

1.4. Smooth and Nonsmooth Problems

The NLP in (1.1) is *smooth* if f and all g_i have continuous first partial derivatives everywhere; otherwise it is *nonsmooth*. A common type of nonsmooth function is one with a piecewise definition, for example,

$$f(x) = \begin{cases} -x & \text{for } x < 0 \\ x^2 & \text{for } x \geq 0 \end{cases}$$

This is a convex function with continuous derivative everywhere but at the origin, where its minimum lies and where the first derivative is discontinuous. Piecewise linear functions occur often. For example, taxes in most countries are piecewise linear, convex functions of taxable income, and some production processes have costs that exhibit economies of scale as shown in Figure 9. This is a piecewise linear concave function. When included as a term in an overall cost to be minimized, such functions can render a problem nonconvex, and concave minimization problems often have local optima.

Another frequently used discontinuous function is the step function:

$$f(x) = \begin{cases} 0, & x \leq 0 \\ 1, & x > 0 \end{cases}$$

which is often used to model setup costs, arising from work that must be done to prepare a machine to produce a specific item prior to the production process itself.

Almost all available NLP algorithms are designed to solve smooth problems. They all use function and gradient values of the problem functions and can solve a high percentage of the smooth problems posed to them to high accuracy in reasonable time. They may solve a nonsmooth problem, especially if the discontinuities do not occur near or at an optimal point. However, these algorithms can get stuck far from the optimum or

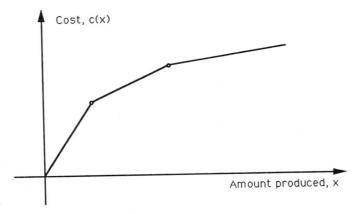

Figure 9 Piecewise linear concave function.

Nonlinear Programming

can oscillate about it. As a partial cure, nonsmooth functions of a single variable can be "smoothed" by approximating them with a smooth function, often a piecewise polynomial. In addition, a nonsmooth problem can often be transformed to a smooth problem by introducing binary variables, as described in Chapter 3 of this book.

1.5. NLP Software and Modeling Systems

Many excellent software systems for posing and solving nonlinear programs are now available and will be described later in this chapter. Some are stand-alone NLP solvers, written in FORTRAN or C. These require the user to write a subroutine that computes the values of his objective and constraint functions for given values of x. Other problem data, including all upper and lower bounds and initial values for x, are either read from a file or passed through an argument list.

Algebraic modeling systems are higher-level programs specially designed for solving optimization problems. They allow a user to define index sets, indexed variables, and indexed collections of constraint functions. The problem is automatically communicated to a solver, and the solution can be output or processed further. Several excellent algebraic modeling systems exist, interfacing to multiple solvers for linear, nonlinear, and mixed-integer programs.

Spreadsheet software now includes facilities called solvers for linear, nonlinear, and mixed-integer programs. For example, the Excel and Quattro Pro solvers can handle problems of up to 200 decision variables and 100 constraints, plus bounds on the variables. Millions of these programs are on desktops worldwide.

2. UNCONSTRAINED PROBLEMS

2.1. Introduction

Here we consider optimization problems of the form

$$\text{Minimize} \quad f(x) \tag{2.1}$$

subject to no constraints. Least squares fitting problems are of this form, where

$$f(x) = \sum_{i=1}^{p} (h(v_i, x) - y_i)^2 \tag{2.2}$$

In (2.2), y_i is a set of observed values of a dependent variable, y, and v_i is a corresponding set of values for a vector of independent variables v that are thought to "cause" y. A model of the form

$$y = h(v,x) \tag{2.3}$$

is constructed, where x is a vector of model parameters or coefficients, to be chosen to make the model fit the observed data as closely as possible. If h is linear in x, i.e.,

$$h(v,x) = \sum_{j=1}^{p} x_j h_j(v) \tag{2.4}$$

then f in (2.2) is a quadratic function of x, and its minimum can be found by setting $\nabla f(x)$ equal to zero. This yields a set of linear equations and is the approach used by statistical software for linear regression problems. When the parameters x do not appear linearly, f is not quadratic and may be minimized using either general-purpose solvers or methods designed specially for least squares problems. Both are described in this chapter.

Aside from curve fitting, unconstrained problems are far less common than constrained ones. The main reason for studying methods for unconstrained problems is that almost all the ideas and procedures involved also apply to constrained problems.

2.2. Geometry and Descent Algorithms

Figure 10 shows contours of constant value of a differentiable nonlinear function, f, of two variables, x_1 and x_2. At a point $\bar{x}$ on the contour $f(x) = 20$, the tangent to the contour and the negative gradient, $-\nabla f(\bar{x})$, are shown. The vector

$$\nabla f(\bar{x}) = \left(\frac{\partial f}{\partial x_1}, \frac{\partial f}{\partial x_2} \right)_{x=\bar{x}}$$

is orthogonal to the tangent line, and $-\nabla f$ points in the direction of maximum rate of decrease of f per unit of distance traveled. $-\nabla f(\bar{x})$ is called the direction of *steepest descent* for f at $\bar{x}$. The function f will decrease along any direction d, emanating from $\bar{x}$ and making an angle of less than 90° with $-\nabla f$. Such a direction is called a descent direction and is defined as any vector, d, such that the directional derivative of f at $\bar{x}$ along d is negative:

$$Df(\bar{x},d) = \nabla f^T(\bar{x})d < 0 \tag{2.5}$$

In Figure 10, all vectors starting at $\bar{x}$ and pointing into the half-space labeled "descent" are descent directions, and those pointing into the opposite half-space are directions of ascent.

Consider the one-dimensional function

$$g(\alpha) = f(\bar{x} + \alpha d) \tag{2.6}$$

Nonlinear Programming

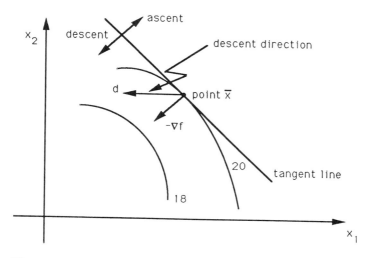

Figure 10 Geometry of unconstrained minimization.

where α is a scalar called the step size. This function gives the value of f as one moves along the line $\bar{x} + \alpha d$, which has $\bar{x}$ as its origin and direction d. The derivative of g at α is

$$g'(\alpha) \equiv \frac{dg(\alpha)}{d\alpha} = d^T \nabla f(\bar{x} + \alpha d) \tag{2.7}$$

and at $\alpha = 0$

$$g'(0) = d^T \nabla f(\bar{x}) \tag{2.8}$$

so the slope of $g(\alpha)$ at the origin is the directional derivative of f at $\bar{x}$ along d. This is illustrated in Figure 11. Clearly, if $g'(0) < 0$, it is possible to choose a positive step size $\bar{\alpha}$ such that

$$g(\bar{\alpha}) < g(0) \tag{2.9}$$

i.e., such that the objective is reduced. The process of choosing $\bar{\alpha}$ satisfying (2.9) is called a *linesearch*. Most algorithms for unconstrained or constrained optimization consist of a sequence of linesearches:

General Descent Algorithm

1. Start with an initial x, say $x = x_0$. Set the iteration counter i to 0.
2. Choose a descent direction d_i.
3. Perform a linesearch to choose a step size α_i such that

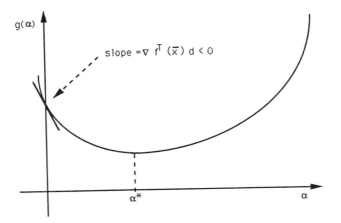

Figure 11 Objective function behavior along a descent direction.

$$g_i(\alpha_i) = f(x_i + \alpha_i d_i) < g_i(0)$$

4. Set $x_{i+1} = x_i + \alpha_i d_i$.
5. Test for convergence. If converged, stop.
6. Replace i by $i + 1$ and go to step 2.

An exact linesearch is one that chooses α_i as the first local minimum of $g_i(\alpha)$, i.e., the one with the smallest α value. This is the point α^* in Figure 11. Finding this minimum to high accuracy is overly time consuming, so modern NLP algorithms use a variety of inexact linesearch procedures, often involving polynomial fits, as described later in this section.

If $f(x)$ has continuous second partial derivatives everywhere, necessary conditions for x^* to be a local minimum of f are

$$\nabla f(x^*) = 0, \quad \nabla^2 f(x^*) \text{ is positive semidefinite} \quad (2.10)$$

A point x^* where $\nabla f(x^*) = 0$ is called a stationary point of f. A stationary point may be a local minimum, local maximum, or a saddle point. Sufficient conditions for x^* to be a local minimum are

$$\nabla f(x^*) = 0, \quad \nabla^2 f(x^*) \text{ is positive definite} \quad (2.11)$$

Stationarity is often used as one of several termination criteria in step 5 of the descent algorithm; i.e., stop when the gradient is small. However, if f is scaled, i.e., multiplied by a positive scale factor c, then

$$\nabla(cf(x)) = c\nabla f(x)$$

Nonlinear Programming

so it is always possible to make a small gradient large by choosing c large, i.e., by changing the units in which f is measured, for example, from dollars to pennies. Hence other criteria are added, for example:

Step 5
If
$$\|\nabla f(x_i)\| < \epsilon$$
or
$$\frac{|f(x_{i+1}) - f(x_i)|}{1 + |f(x_i)|} < \epsilon \quad \text{for } k \text{ consecutive values of } i$$
stop

This stops the iterations whenever some norm of the gradient is less than a user-defined stopping tolerance ϵ, or if the fractional objective change is less than ϵ for k consecutive cycles. The NLP code GRG2 described later uses $\epsilon = 10^{-4}$ and $k = 3$ as default values.

2.3. Choosing a Search Direction

There are several procedures for choosing search directions d_i. The oldest is to choose d_i as the negative gradient, i.e.,

$$d_i = -\nabla f(x_i)$$

which is called the steepest descent method. This leads to an inefficient zigzagging approach to the optimum for functions whose contours differ significantly from concentric circles. This is illustrated in Figure 12, where an exact linesearch is used. If α_i minimizes $g_i(\alpha_i)$ then

$$g'_i(\alpha_i) = d_i^T \nabla f(x_i + \alpha_i d_i) = 0 \tag{2.12}$$

If $d_i = -\nabla f(x_i)$, then (2.12) shows that successive search directions are orthogonal.

A much more efficient approach is to use Newton's method, where d_i is the solution of the system of equations

$$\nabla^2 f(x_i) d_i = -\nabla f(x_i) \tag{2.13}$$

where $\nabla^2 f(x_i)$ is the matrix of second partial derivatives, also called the Hessian matrix of f at x_i. However, second derivatives are usually very tedious to code by hand for functions of any complexity and are not provided by any available modeling system at this time. In addition, d_i in (2.13) is guaranteed to be a descent direction only if $\nabla^2 f(x_i)$ is positive definite. This is likely to be true for a convex function ($\nabla^2 f$ must be positive semidefinite everywhere—see Section 1.3)—but $\nabla^2 f$ may be indefinite otherwise.

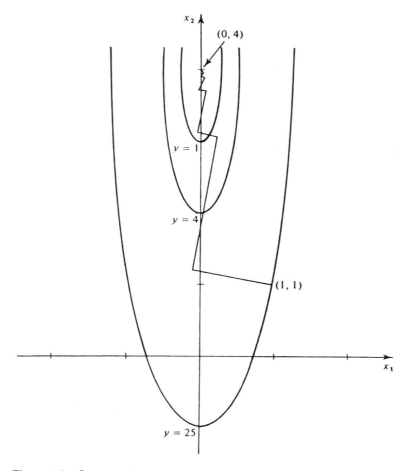

Figure 12 Steepest descent with an exact linesearch.

Modifications of Newton's method exist that completely overcome this problem (see Dennis and Schnabel, 1983, Chapter 6), and such "safeguarded" Newton codes are included in most numerical subroutine libraries. If second derivatives can be computed quickly and accurately, a modern implementation of Newton's method is generally fast, reliable, and locates a local optimum very accurately.

Procedures that compute a search direction using only first derivatives provide an attractive alternative to Newton's method. The most popular of these are the quasi-Newton methods that replace $\nabla^2 f(x_i)$ in (2.13) by a positive definite approximation H_i:

$$H_i d_i = -\nabla f(x_i) \qquad (2.14)$$

H_i is initialized as any positive definite symmetric matrix (often the identity or a diagonal matrix) and is updated after each linesearch using the changes in x and in $\nabla f(x)$ over the last two points, as measured by the vectors

$$s_i = x_{i+1} - x_i \qquad (2.15)$$

and

$$y_i = \nabla f(x_{i+1}) - \nabla f(x_i) \qquad (2.16)$$

The most efficient and widely used updating formula is the BFGS update, where the approximate Hessian is given by

$$H_{i+1} = H_i + \frac{y_i y_i^T}{s_i^T y_i} - (H_i s_i) \frac{(H_i s_i)^T}{s_i^T H_i s_i} \qquad (2.17)$$

If H_i is positive definite and $s_i^T y_i > 0$, it can be shown that H_{i+1} is positive definite (Dennis and Schnabel, 1983, Chapter 9). The condition $s_i^T y_i > 0$ can be interpreted geometrically, since

$$\begin{aligned} s_i^T y_i &= \alpha_i d_i^T (\nabla f(x_{i+1}) - \nabla f(x_i)) \\ &= \alpha_i (d_i^T \nabla f(x_{i+1}) - d_i^T \nabla f(x_i)) \\ &= \alpha_i (\text{slope2} - \text{slope1}) \end{aligned}$$

The quantity slope2 is the slope of the linesearch objective function $g_i(\alpha)$ at $\alpha = \alpha_i$, and slope1 is its slope at $\alpha = 0$, so $s_i^T y_i > 0$ if and only if slope2 > slope1. This condition is always satisfied if f is strictly convex. A good linesearch routine will attempt to meet this condition, and if it is not met then H_i is not updated.

If the BFGS algorithm is applied to a positive definite quadratic function of n variables and the linesearch is exact, it will minimize the function in at most n iterations (Dennis and Schnabel, 1983, Chapter 9). This is also true for some other updating formulas. For nonquadratic functions, a good BFGS code will usually require more iterations than a comparable Newton implementation and may not achieve as high an accuracy. However, each BFGS iteration is generally faster, because second derivatives are not required and the linear system (2.14) need not be solved (H_i^{-1} or the Cholesky factor* of H_i can be updated).

---

*The Cholesky factor of H_i is the positive definite lower triangular matrix L_i such that $H_i = L_i L_i^T$.

For problems with hundreds or thousands of variables, storing and manipulating the matrices H_i or $\nabla^2 f(x_i)$ requires much time and computer memory, and conjugate gradient methods become attractive. These compute d_i using formulas involving no matrices. The Fletcher-Reeves method (Luenberger, 1984, Chapter 8) uses

$$d_0 = -\nabla f(x_0)$$
$$d_i = -\nabla f(x_i) + \beta_i d_{i-1}, \qquad i = 1, 2, \ldots$$

where

$$\beta_i = \frac{\nabla f^T(x_i)\nabla f(x_i)}{\nabla f^T(x_{i-1})\nabla f(x_{i-1})}$$

The one-step BFGS formula is usually more efficient;

$$d_0 = -\nabla f(x_0)$$
$$d_{i+1} = -\nabla f(x_{i+1}) + \frac{1}{s_i^T y_i}\left[-\tau_i s_i^T \nabla f(x_{i+1}) + (y_i^T \nabla f(x_{i+1}))s_i + (s_i^T \nabla f(x_{i+1}))y_i\right]$$

where

$$\tau_i = 1 + \frac{y_i^T H_i y_i}{s_i^T y_i}$$

This formula follows from the BFGS formula for H_i^{-1} by (1) assuming $H_{i-1}^{-1} = I$, (2) computing H_i^{-1} from the update formula, and (3) computing d_i as $-H_i^{-1}\nabla f(x_i)$. Both methods minimize a positive definite quadratic function of n variables in at most n iterations using exact linesearches but generally require signficantly more iterations than the BFGS procedure for general nonlinear functions. A class of algorithms called variable memory quasi-Newton methods (Nocedal, 1980) partially overcomes this difficulty and provide an effective compromise between standard quasi-Newton and conjugate gradient algorithms.

2.4. Linesearch Procedures

All the methods discussed thus far have searched for a minimum in n-dimensional space by performing one-dimensional minimizations down a set of directions d_i. Thus the efficiency of any such procedure depends critically on the efficiency of the method used to solve the single-dimensional search. Two techniques are presented that use polynomial interpolation, one requiring derivatives, the second only function values.

For both interpolative procedures, the variables $x_1, \ldots, x_n$ are scaled so that a unit change in any variable is a significant but not too large

Nonlinear Programming

percentage change in that variable. For example, if a variable is expected to have a value around 100 units, then a 1-unit change would be considered significant, whereas a 10-unit change would be too large.

2.4.1. Cubic Interpolation

This technique, described in Fletcher and Reeves (1964), solves the problem of finding the smallest nonnegative α, α^*, for which the function

$$g(\alpha) = f(x + \alpha d)$$

attains a local minimum in three stages. It uses the derivative $g'(\alpha)$. The first stage normalizes the d vector so that a step size $\alpha = 1$ is acceptable. The second stage establishes bounds on α^*, and the third stage interpolates its value.

Stage 1. Given a search direction d, calculate

$$\Delta = \max_j |(d)_j|$$

where $(d)_j$ is component j of d, and divide each component of d by Δ. This ensures that d is a reasonable change in x.

Stage 2. Evaluate $g(\alpha)$ and $g'(\alpha)$ at the points $\alpha = 0, 1, 2, 4, \ldots, a, b$, where b is the first of these values at which either g' is nonnegative or g has not decreased. It then follows that α^* is bounded in the interval $a < \alpha^* \leq b$. If $g(1)$ is much greater than $g(0)$, divide the components of d by a factor, e.g., 2 or 3, and repeat this stage starting from $\alpha = a$.

Stage 3. A cubic polynomial is now fitted to the four values $g(a)$, $g'(a)$, $g(b)$, $g'(b)$, and its minimum, α_e, is taken to be the value for α^*. It is shown in Fletcher and Reeves (1964) that the cubic has a unique minimum in the interval (a,b) that is given by

$$\alpha_e = b - \frac{g'(b) + w - z}{g'(b) - g'(a) + 2w} (b - a) \tag{2.18}$$

where

$$z = 3\left[\frac{g(a) - g(b)}{b - a}\right] + g'(a) - g'(b) \tag{2.19}$$

$$w = [z^2 - g'(a)g'(b)]^{1/2} \tag{2.20}$$

If neither $g(a)$ nor $g(b)$ is less than $g(\alpha_e)$, then α_e is accepted as the estimate of α^*. Otherwise, according to whether $g'(\alpha_e)$ is positive or negative, the interpolation is repeated over the subinterval (a, α_e) or (α_e, b), respectively.

It is interesting to note that for small values of $g'(a)$ and $g'(b)$, the cubic has the shape that a flat metal spring would assume if fitted to the points a, b with slopes $g'(a)$, $g'(b)$.

2.4.2. Quadratic Interpolation

If derivatives are not available or are difficult to compute, then quadratic interpolation should be used in the one-dimensional minimization. The procedure can again be described in three steps.

Stage 1. This is the same as stage 1 above.
Stage 2. Evaluate $g(\alpha)$ at the points $\alpha = 0, 1, 2, 4, \ldots, a, b, c$, where c is the first of these values at which g has increased. Then

$$a < \alpha^* < c$$

Again, if $g(1) \geqslant g(0)$, divide the components of d by a factor, e.g., 2 or 3, and repeat.

Stage 3. A quadratic polynomial is now fitted to the three values $g(a)$, $g(b)$, $g(c)$, and its minimum, α_e, is

$$\alpha_e = \frac{1}{2} \frac{g(a)(c^2 - b^2) + g(b)(a^2 - c^2) + g(c)(b^2 - a^2)}{g(a)(c - b) + g(b)(a - c) + g(c)(b - a)} \quad (2.21)$$

If

$$g(\alpha_e) < g(b)$$

then α_e is accepted as the estimate of α^*. Otherwise, b is taken as α^*.

2.5. Nonlinear Least Squares Methods

Special-purpose methods exist for minimizing a sum of squares of nonlinear functions [see Eq. (2.2)] and these are often more reliable and efficient for such problems than the general-purpose methods described earlier in this chapter. Good codes implementing these algorithms are contained in most numerical subroutine libraries (see Section 4.5). Dennis and Schnabel (1983) recommend the Levenberg-Marquardt and Gauss-Newton methods for problems with small residuals and secant variations of Newton's method when the residuals are large.*

---

*The residuals are the terms $h(v_i, x) - y_i$ in Eq. (2.2).

Nonlinear Programming

3. CONSTRAINED PROBLEMS— OPTIMALITY CONDITIONS

In this section, it is simplest to pose the problem to be discussed in the form

$$\text{Minimize } f(x) \tag{3.1}$$

subject to both equality and inequality constraints:

$$h_i(x) = b_i, \quad i = 1, \ldots, m \tag{3.2}$$

and

$$g_j(x) \leq c_j, \quad j = 1, \ldots, r \tag{3.3}$$

where all functions have continuous first partial derivatives everywhere. We state necessary conditions, called the Kuhn-Tucker conditions (Luenberger, 1984, Chapter 10), for a point to be a local minimum of this problem. These conditions are used by most NLP software to terminate the algorithm. The Kuhn-Tucker conditions involve new variables called Lagrange multipliers. These provide valuable sensitivity information, because they are equal to the first derivative of the optimal objective value with respect to the constraint right-hand sides b_i and c_j.

3.1. Geometric Interpretation of the Kuhn-Tucker Conditions

3.1.1. Cones

The idea of a cone aids the understanding of the Kuhn-Tucker conditions. A cone is a set of points R such that, if x is in R, λx is also in R for $\lambda \geq 0$. A convex cone is a cone that is a convex set. An example of a convex cone in two dimensions is shown in Figure 13. In two and three dimensions, the definition of a convex cone coincides with the usual meaning of the word.

It is easily shown from the above definitions that the set of all nonnegative linear combinations of a finite set of vectors is a convex cone, i.e., that the set

$$R = \{x \mid x = \lambda_1 x_1 + \lambda_2 x_2 + \cdots + \lambda_m x_m,$$
$$\lambda_i \geq 0, i = 1, \ldots, m\}$$

is a convex cone. The vectors $x_1, x_2, \ldots, x_m$ are called the generators of the cone. For example, the cone of Figure 13 is generated by the vectors (2,1) and (2,4). Thus any vector that can be expressed as a nonnegative linear combination of these vectors lies in the cone. For example, in Figure 13 the vector (4,5) in the cone is given by

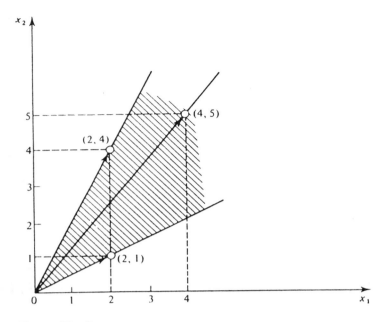

Figure 13 Convex cone.

$$(4,5) = 1 \times (2,1) + 1 \times (2,4)$$

3.1.2. Kuhn-Tucker Conditions — Geometrical Interpretation

The Kuhn-Tucker conditions are predicated on this fact: At any constrained optimum, no (small) allowable change in the problem variables can improve the objective function. To illustrate this consider the nonlinear programming problem

$$\text{Minimize} \quad f(x,y) = (x-2)^2 + (y-1)^2$$
$$\text{subject to} \quad g_1(x,y) = -y + x^2 \leq 0$$
$$g_2(x,y) = x + y \leq 2$$

The problem is shown geometrically in Figure 14. It is evident that the optimum is at the intersection of the two constraints at $(1,1)$. Define a feasible direction as a vector such that a small move along that vector violates no constraints. At $(1,1)$, the set of all feasible directions lies between the line $x + y - 2 = 0$ and the tangent line to $y = x^2$ at $(1,1)$, i.e., the line $y = 2x - 1$. In other words, the set of feasible directions is the cone generated by these lines. The vector $-\nabla f$ points in the direction of

Nonlinear Programming

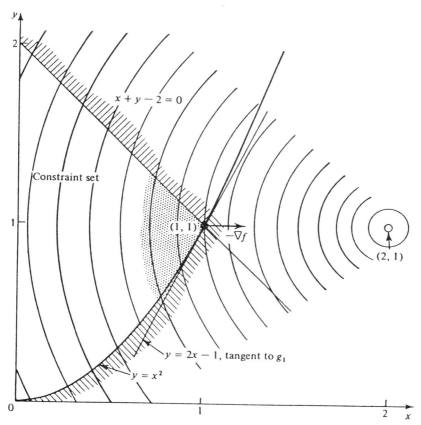

Figure 14 Geometry of constrained optimization problem.

maximum rate of decrease of f and a small move along any direction making an angle of less than 90° with $-\nabla f$ will decrease f. Thus, at the optimum, no feasible direction can have an angle of less than 90° between it and $-\nabla f$.

Consider Figure 15, in which the gradient vectors ∇g_1 and ∇g_2 are drawn. Note that $-\nabla f$ is contained in the cone generated by ∇g_1 and ∇g_2. What if this were not so? If $-\nabla f$ were slightly above ∇g_2, it would make an angle of less than 90° with a feasible direction just below the line $x + y - 2 = 0$. If $-\nabla f$ were slightly below ∇g_1, it would make an angle of less than 90° with a feasible direction just above the line $y = 2x - 1$. Neither case can occur at an optimal point, and both cases are excluded if and only if $-\nabla f$ lies within the cone generated by ∇g_1 and ∇g_2. Of course, this is the

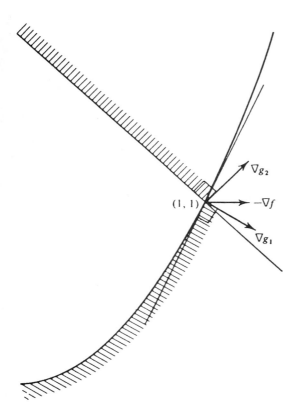

Figure 15 Gradient of objective contained in convex cone.

same as requiring that ∇f lie within the cone generated by $-\nabla g_1$ and $-\nabla g_2$. This is the usual statement of the Kuhn-Tucker conditions; i.e., if f and all g_i are differentiable, a necessary condition* for a point x^0 to be a constrained minimum of the problem

> Minimize $f(x)$
> subject to $g_i(x) \leq c_i$, $i = 1, \ldots, m$

is that, at x^0, ∇f lies within the cone generated by the negative gradients of the binding constraints.†

---

*For this condition to be necessary, a "constraint qualification" must be satisfied; see Luenberger (1984, Chapter 10).
†Binding constraints are those that hold as equalities at x^0.

3.2. Algebraic Statement of the Kuhn-Tucker Conditions

The above results may be stated in algebraic terms. For ∇f to lie within the cone described above, it must be a nonnegative linear combination of the negative gradients of the binding constraints; i.e., there must exist numbers u_i^0 such that

$$\nabla f(x^0) = \sum_{i \in I} u_i^0 (-\nabla g_i(x^0)) \qquad (3.4)$$

where

$$u_i^0 \geq 0, \quad i \in I \qquad (3.5)$$

and I is the set of indices of the binding constraints.

These results may be restated to include all constraints by defining the coefficient u_i^0 to be zero if $g_i(x^0) < c_i$. Note that if this is done, then $u_i^0 \geq 0$ if $g_i(x^0) = c_i$, and $u_i^0 = 0$ if $g_i(x^0) < c_i$; i.e., the product $u_i^0(g_i(x^0) - c_i)$ is zero for all i. Conditions (3.4) and (3.5) then become

$$\nabla f(x^0) = \sum_{i=1}^{m} u_i^0 (-\nabla g_i(x^0)) \qquad (3.6)$$

$$u_i^0 \geq 0, \, u_i^0(g_i(x^0) - c_i) = 0; \qquad (3.7)$$
$$g_i(x^0) \leq c_i, \quad i = 1, \ldots, m$$

Relations (3.6) and (3.7) are the form in which the Kuhn-Tucker conditions are usually stated.

3.2.1. Lagrange Multipliers

The Kuhn-Tucker conditions are closely related to the classical Lagrange multiplier results for equality constrained problems. Form the Lagrangian

$$L(x,u) = f(x) + \sum_{i=1}^{m} u_i g_i(x) \qquad (3.8)$$

where the u_i are viewed as Lagrange multipliers for the inequality constraints $g_i(x) \leq 0$. Then (3.6) and (3.7) state that $L(x,u)$ must be stationary in x at (x^0, u^0) with the multipliers, u^0, satisfying (3.7). The stationarity of L is the same condition as in the equality-constrained case—the additional conditions in (3.7) arise because the constraints here are inequalities.

3.2.2. Equalities and Inequalities

When both equality and inequality constraints are present, as in the problem (3.1)-(3.3), the Kuhn-Tucker conditions are stated as follows. Define Lagrange multipliers λ_i for the equalities and u_j for the inequalities, and form the Lagrangian function

$$L(x,\lambda,u) = f(x) + \sum_{i=1}^{m} \lambda_i h_i(x) + \sum_{j=1}^{r} u_j g_j(x)$$

Then, if x^0 is a local minimum of (3.1)–(3.3), there exist vectors of Lagrange multipliers λ^0 and u^0, such that x^0 is a stationary point of the function $L(x,\lambda^0,u^0)$, i.e.,

$$\nabla_x L(x^0,\lambda^0,u^0) = \nabla f(x^0) + \sum_{i=1}^{m} \lambda_i^0 \nabla h_i(x^0) + \sum_{j=1}^{r} u_j^0 \nabla g_j(x^0) = 0$$

and

$$u_j^0 \geq 0, \quad u_j^0(g_j(x^0) - c_j) = 0; \quad j = 1, \ldots, r$$

3.2.3. Lagrange Multipliers and Sensitivity Analysis

At each iteration, NLP algorithms form new estimates not only of the decision variables x but also of the Lagrange multipliers λ and u. If, at these estimates, all constraints are satisfied and the Kuhn-Tucker conditions are satisfied to within specified tolerances, the algorithm stops. At a local optimum, the optimal multiplier values provide useful sensitivity information. In the NLP (3.1)–(3.3), let $v^0(b,c)$ be the optimal value of the objective f at a local minimum, viewed as a function of the constraint right-hand sides b and c. Then, under mild additional conditions (see Luenberger, 1984, Chapter 10)

$$\lambda_i^0 = \frac{-\partial v^0}{\partial b_i}, \quad i = 1, \ldots, m$$

$$u_j^0 = \frac{-\partial v^0}{\partial c_j}, \quad j = 1, \ldots, r$$

That is, the Lagrange multipliers provide the rate of change of the optimal objective value with respect to changes in the constraint right-hand sides. This information is often of significant value. For example, if the right-hand side of an inequality constraint, c_j, represents the capacity of a process, and this capacity constraint is active at the optimum, then the optimal multiplier value u_j^0 equals the rate of decrease of the minimal cost if the capacity is increased. This is the marginal value of the capacity. In a situation with several active capacity limits, the ones with the largest absolute multipliers should be considered first for possible increases.

Several examples of the use of Lagrange multipliers are given in Section 5 of this chapter.

4. CONSTRAINED OPTIMIZATION: ALGORITHMS

4.1. The Generalized Reduced Gradient Method

The generalized reduced gradient (GRG) algorithm was first developed in the later 1960s by Jean Abadie (Abadie and Carpentier, 1969) and has since been refined by several other researchers. In this section we discuss the fundamental concepts of GRG and describe the version of GRG that is implemented in GRG2, the most widely available nonlinear optimizer (Lasdon et al., 1978; Lasdon and Waren, 1978).

4.1.1. Unconstrained Example

Since many GRG concepts derive from unconstrained optimization, we begin with a very simple unconstrained minimization problem:

Minimize $f(x,y) = x^2 + 4y^2$

Contours of constant value of this function [the sets of (x,y) such that $x^2 + 4y^2 = r^2$ for various values of r] are ellipses centered at the origin, as shown in Figure 16. The minimum of this function also occurs at the origin. Recall that this point can be determined analytically by setting the gradient of f, ∇f, to zero, yielding:

$$\nabla f(x,y) = (2x, 8y) = (0,0)$$

which implies that $x = y = 0$.

Although this simple problem can be solved both analytically and geometrically, more complex ones cannot. Usually the equations that arise when $\nabla \mathbf{f}$ is set to zero cannot be solved analytically, nor, for more than a few variables, can the objective function contours be effectively represented geometrically. Thus a numerical approach must be used. For solving unconstrained problems, either problems originally of this form or unconstrained subproblems that arise during the solution of constrained problems, GRG2 uses the gradient-based iterative method first presented in Section 2.2 and repeated, in a slightly different form, below. Here, in terms of our simple example, $\mathbf{x} = (x,y)$, $\mathbf{x}_c = (x_c, y_c)$, and $\mathbf{x}_n = (x_n, y_n)$ where the subscripts c and n refer to the current and next points.

General descent algorithm
1. Compute the gradient of $f(\mathbf{x})$ at the current point $\mathbf{x}_c$, giving $\nabla \mathbf{f}(\mathbf{x}_c)$.
2. If the current point $\mathbf{x}_c$ is close enough to being optimal, stop.
3. Compute a search direction $\mathbf{d}(\mathbf{x}_c)$ using the gradient $\nabla \mathbf{f}(\mathbf{x}_c)$ and perhaps other information.

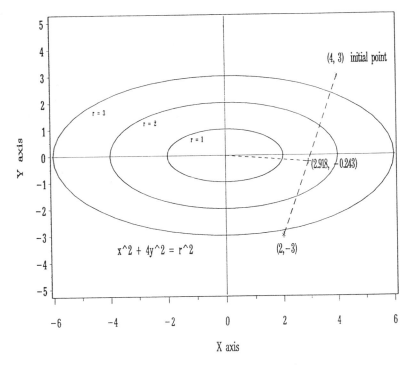

Figure 16 Elliptical objective contours, no constraints.

4. Determine how far to move along the current search direction $d(x_c)$, starting from the current point x_c. This distance, α_c, is most often an approximation of the value of α that minimizes the objective function $f(x_c + \alpha d(x_c))$ and is used to determine the next point $x_n = (x_c + \alpha_c d(x_c))$.
5. Replace the current point x_c by the next point x_n and return to step 1.

For our example we arbitrarily select as our initial current point $x_c = (4,3)$. The gradient of f at this point is

$$\nabla f(x,y) = (2x, 8y) = (8, 24)$$

and is judged not to be sufficiently close to the optimum value of $(0,0)$ in order to stop, so we proceed with step 3. All unconstrained minimization algorithms in GRG2 use the negative gradient direction as the search direction from the initial point, since no other information is yet available and

Nonlinear Programming 417

the negative gradient direction is the direction in which the objective function decreases most rapidly. Thus $\mathbf{d}(\mathbf{x}_c) = -\nabla \mathbf{f}(\mathbf{x}_c) = (-8, -24)$ and this direction is shown in Figure 16 as the dashed line emanating from the initial point. We are now ready for step 4.

Selecting a value for the step size variable, α, determines how far to move in the current search direction. Moving from the current point along the search direction means that new points are given by

$$(x,y) = (x_c, y_c) + \alpha(d_1, d_2)$$
$$= (x_c + \alpha d_1, y_c + \alpha d_2)$$
$$= (4 - 8\alpha, 3 - 24\alpha)$$

At $\alpha = 0$ we start at our initial point, (4,3), and as α increases we move along the search direction reaching, for example, the point $(2, -3)$ when α reaches 0.25. A reasonable approach for selecting the value of α_c is to choose the value of α that reduces the objective function as much as possible. Thus α_c is selected as the value of α that minimizes the following function of one variable:

$$g(\alpha) = f(x_c + \alpha d(x_c))$$
$$= f(4 - 8\alpha, 3 - 24\alpha)$$
$$= (4 - 8\alpha)^2 + 4(3 - 24\alpha)^2$$

In general, the process of finding the minimizing value of α_c cannot be done analytically and this process must also be carried out numerically. This procedure is called the linesearch and was described earlier in Section 2.4. However, for this simple quadratic problem the minimizing value for α can be found by setting the derivative of $g(\alpha)$ to zero, i.e.,

$$\frac{dg(\alpha)}{d\alpha} = 2(4 - 8\alpha)(-8) + 8(3 - 24\alpha)(-24) = 0$$

Solving this Equation gives $\alpha_c = 5/37$ and hence $\mathbf{x}_n = (x,y) = (2.918, -0.243)$. At this point we have completed one iteration of the algorithm.

In subsequent steps GRG2 uses search directions generated by either conjugate gradient or quasi-Newton methods, as discussed in Section 2.3 of this chapter. For this unconstrained quadratic problem, these methods all generate the next search direction as one that passes through the origin. Hence GRG2 terminates at the optimum after two one-dimensional searches.

4.1.2. Equality Constraints

To explain how GRG algorithms handle equality constraints let us consider the problem:

Minimize $x^2 + y^2$
subject to $x + y = 4$

The geometry of this problem is shown in Figure 17. The linear equality constraint plots as a straight line and the contours of constant objective function value are circles centered at the origin. From a geometrical point of view, the problem is to find the point on the line that is closest to the origin at $x = 0$, $y = 0$. Clearly, the solution to the problem is at $x = 2$, $y = 2$, where the objective function value is 8.

GRG takes a direct and natural approach to solve this problem. It uses the equality constraint to solve for one of the variables in terms of the other. For example, if we solve for x the constraint becomes

$$x = 4 - y \tag{4.1}$$

Whenever a value is specified for y the appropriate value for x, which keeps the equality constraint satisfied, can easily be calculated. We call y the

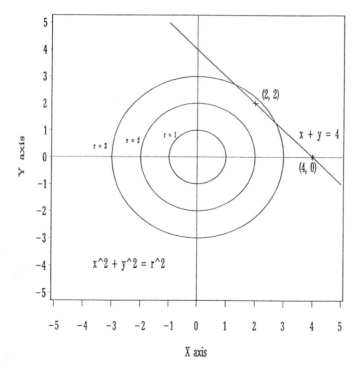

Figure 17 Circular objective contours, linear equality constraint.

Nonlinear Programming

independent or nonbasic variable and x the dependent or basic variable. Since x is now determined by y, we can reduce this problem to one involving only y by substituting $(4 - y)$ for x in the objective function to give

$$F(y) = (4 - y)^2 + y^2$$

The function $F(y)$ is called the reduced objective function and the reduced problem is to minimize $F(y)$ subject to no constraints. Once the optimal value of y is found, the optimal value of x is computed from (4.1).

Since the reduced problem is unconstrained and quite simple, it can be solved both analytically and by the iterative descent algorithm described earlier. First, let us solve it analytically. We set the gradient of $F(y)$, called the *reduced gradient*, to zero, giving

$$\nabla F(y) = \frac{dF(y)}{dy} = -2(4 - y) + 2y$$
$$= -8 + 4y = 0$$

Solving this equation, we get $y = 2$. Substituting this value in (4.1) gives $x = 2$ and $(x,y) = (2, 2)$ is, of course, the same solution as our geometric one.

Let us now apply the steps of the descent algorithm to the reduced problem, starting from an initial $y_c = 0$ for which the corresponding $x_c = 4$. Computing the reduced gradient gives $\nabla F(y_c) = \nabla F(0) = -8$, which is not close enough to zero to be judged optimal, so we proceed with step 3. Our initial search direction will be the negative reduced gradient direction, so $d = 8$ and we proceed to the linesearch of step 4. New points are given by

$$y = y_c + \alpha d \qquad (4.2)$$
$$= 0 + 8\alpha$$

where α is the step size. We start at (4,0) with $\alpha = 0$ and as α increases y also increases. This increase is determined by Eq. (4.2) and keeps (x,y) on the equality constraint shown in Figure 17.

As usual, we select α to minimize $g(\alpha)$, the reduced objective function evaluated along the current search direction, which is given by

$$g(\alpha) = F(y_c + \alpha d)$$
$$= F(0 + 8\alpha)$$
$$= (4 - 8\alpha)^2 + (8\alpha)^2$$

Again, in this simple case, we can proceed analytically to determine α by setting the derivative of $g(\alpha)$ to zero to get

$$\frac{dg(\alpha)}{d\alpha} = -16(4 - 8\alpha) + 128\alpha$$
$$= -64 + 256\alpha = 0$$

Solving for α gives $\alpha = 1/4$. Substituting this value in (4.2) gives $y_n = 2$ and then (4.1) gives $x_n = 2$, which is the optimal solution.

4.1.3. Inequality Constraints

Let us now examine how GRG2 proceeds when some of the constraints are inequalities and there are bounds on some, or all, of the variables. Consider the problem:

$$\begin{aligned}
&\text{Minimize} && (x - 0.5)^2 + (y - 2.5)^2 \\
&\text{subject to} && x - y \geq 0 \\
&\text{and bounds} && 0 \leq x \\
&&& 0 \leq y \leq 2
\end{aligned}$$

The feasible region and some contours of the objective function are shown in Figure 18. Geometrically, the problem is to find the feasible point that is closest to the point (0.5, 2.5) and obviously this is the point (1.5, 1.5).

When there are inequalities, GRG converts them to equalities by introducing slack variables. If we let s be the slack in this case, the inequality above becomes $x - y - s = 0$. We must also add the bound for the slack, $s \geq 0$, giving the new problem

$$\begin{aligned}
&\text{Minimize} && (x - 0.5)^2 + (y - 2.5)^2 \\
&\text{subject to} && x - y - s = 0 \\
&\text{and bounds} && 0 \leq x \\
&&& 0 \leq y \leq 2 \\
&&& 0 \leq s
\end{aligned}$$

Let the starting point be (1,0), at which the objective value is 6.5 and the inequality is satisfied strictly, that is, its slack is positive ($s = 1$). Such constraints are called inactive. At this point the bounds are also all satisfied, although x is at its lower bound. Because all of the constraints (except for bounds) are inactive at our starting point, we have no equalities that must be solved for values for dependent variables. Hence we proceed to minimize the objective subject only to the bounds on the nonbasic variables x and y. There are no basic variables. The reduced problem is simply the original problem ignoring the inequality constraint. In solving this reduced problem, we do, however, keep track of the inequality. If it becomes active or violated, the reduced problem will change.

Nonlinear Programming

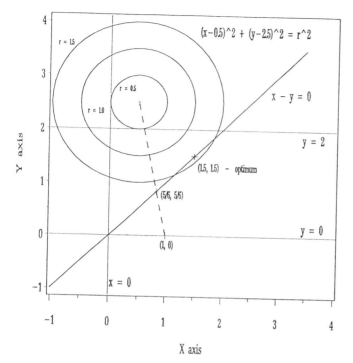

Figure 18 Circular objective contours, linear inequality constraint.

To solve this first reduced problem, we follow the steps of the descent algorithm with some straightforward modifications that account for the bounds on x and y. Whenever a nonbasic variable is at a bound, we must decide whether it should be allowed to leave the bound or whether it should be forced to remain at that bound for the next iteration. The nonbasic variables that are not going to be kept at bounds are called superbasic variables [the term was coined by Murtagh and Saunders (1982)]. Calculating the reduced gradient gives

$$\nabla F(x,y) = \left(\frac{\partial f(x,y)}{\partial x}, \frac{\partial f(x,y)}{\partial y} \right)$$
$$= (2(x - 0.5), 2(y - 2.5))$$
$$\nabla F(1,0) = (1, -5)$$

In this example x is a superbasic variable. To decide whether y should also be a superbasic variable and be allowed to leave its bound, we examine the value of its reduced gradient component. Since this value, -5, is negative, then moving y from its bound into the feasible region, i.e., increasing the value of y, will decrease the objective function value. Therefore we consider letting y leave its bound. In GRG2, a nonbasic variable at a bound is allowed to leave that bound only if (1) doing so improves the value of the objective and (2) the predicted improvement is large compared to the improvement obtained by varying only the current superbasic variables.

We will not discuss the precise logic used by GRG2 in step (2) above because it is fairly complex. References are Lasdon et al. (1978) and Lasdon and Waren (1978). In this example, since the magnitude of the y component of the reduced gradient is five times the magnitude of the x component, it is evident that there is a large incentive to release y from its bound. Thus y is added to our list of superbasic variables.

Continuing with the third step of our descent algorithm (as the gradient is clearly not small enough to stop) the first search direction is chosen as the negative gradient direction:

$$d(x_c, y_c) = -(1, -5) = (-1, 5)$$

In Figure 18, this direction points to the center of the circular objective contours at $(0.5, 2.5)$. In step 4, the line search process moves along $\mathbf{d}$ until either the objective stops decreasing or some constraint or variable bound is reached. In this example the condition that is first encountered is that the constraint $x - y \geq 0$ reaches its bound and GRG2 then selects the intersection of the search direction and the constraint $x - y = 0$ as the next point. This is the point $(5/6, 5/6)$ where $f = 26/9 \approx 2.889$.

Since we now have an active constraint, we use it to solve for one variable in terms of the other, as in the earlier equality-constrained example. Let x be the basic or dependent variable and y and s the nonbasic (independent) ones. Solving the constraint for x in terms of y and the slack s yields

$$x = y + s$$

The reduced objective is obtained by substituting this relation for x in the objective:

$$F(y,s) = (y + s - 0.5)^2 + (y - 2.5)^2$$

The reduced gradient is

$$\nabla F(y,s) = 2((y + s - 0.5) + (y - 2.5), (y + s - 0.5)) \quad (4.3)$$
$$= (4y + 2s - 6, 2y + 2s - 1)$$

which at $(5/6, 0)$ is

Nonlinear Programming

$$\nabla F(\tfrac{5}{6}, 0) = (-\tfrac{8}{3}, \tfrac{2}{3})$$

The variable y is currently superbasic. Since s is at its lower bound of zero, we consider whether s should be allowed to leave its bound, i.e., be a superbasic variable. Since its reduced gradient term is $\tfrac{2}{3}$, increasing s (which is the only feasible change for s) increases the objective value. Since we are minimizing the objective we fix s at zero, and this corresponds to staying on the line $x = y$. The search direction $d = \tfrac{8}{3}$ and new values for y are generated from

$$y = \tfrac{5}{6} + \tfrac{8}{3}\alpha$$

where α is the step size from our current point. The function to be minimized by the line search is

$$g(\alpha) = F(\tfrac{5}{6} + \tfrac{8}{3}\alpha, 0)$$

$$= \left[\tfrac{1}{3} + \tfrac{8}{3}\alpha\right]^2 + \left[\tfrac{-5}{3} + \tfrac{8}{3}\alpha\right]^2$$

The optimal step size of $\alpha = \tfrac{1}{4}$ is determined by setting $dg(\alpha)/d\alpha = 0$, which gives the next point as $y_n = 1.5$. Since s has been fixed at zero, we are on the line $x = y$ and at step 5 we have $(x_c, y_c) = (1.5, 1.5)$, which is the optimal value for this problem. To confirm this, we return to step 1 of our descent algorithm and calculate the reduced gradient at this point using (4.3) to get

$$\nabla F(y, s) = \nabla F(1.5, 0) = (0, 1)$$

First, the reduced gradient with respect to the superbasic variable, y, is zero. Second, because the reduced gradient with respect to s is 1, increasing s (the only feasible change to s) causes an increase in the objective value. These are the two necessary conditions for optimality for this reduced problem and the algorithm terminates at $(1.5, 1.5)$ with an objective value of 2.0.

4.1.4. Nonlinear Constraints

To illustrate how GRG2 handles nonlinear constraints, we replace the linear constraint of the previous example by

$$(x - 2)^2 + y^2 \leq 4$$

Our new problem is

Minimize $\quad (x - 0.5)^2 + (y - 2.5)^2$

subject to $\quad (x - 2)^2 + y^2 \leq 4$

and bounds $0 \leq x$
$0 \leq y \leq 2$

The feasible region is shown in Figure 19. It is bounded by a circle of radius 2 centered at (2, 0) and by the x axis. The point in this region closest to (0.5, 2.5) is optimal. This is the point (0.971, 1.715).

We again start from the point (1, 0). At this point the nonlinear constraint is inactive, y is released from its lower bound to become superbasic (along with x), and progress is along the negative gradient direction until the constraint is encountered. The intersection of the constraint and the negative gradient direction from our starting point is at (0.697, 1.715). Now the nonlinear constraint is active. Adding a slack variable s gives

$$(x - 2)^2 + y^2 + s = 4 \tag{4.4}$$

To form the reduced problem, this equation must be solved for one variable in terms of the other two. The logic in GRG2 for selecting which variables

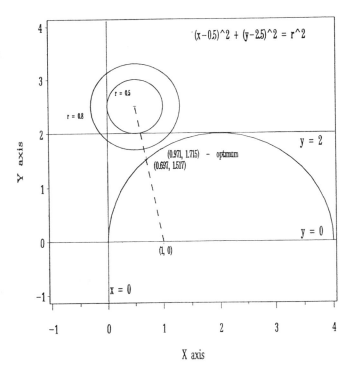

Figure 19 Circular objective contours, nonlinear inequality constraint.

Nonlinear Programming

are to be basic is complex and will not be discussed here [see Lasdon et al. (1978) and Lasdon and Waren (1978) for more information]. In this example at this point GRG2 selects x as basic. Solving (4.4) for x yields

$$x = 2 + \sqrt{4 - y^2 - s}$$

The reduced objective is obtained by substituting this expression into the objective function. The slack, s, will be fixed at its current zero value for the next iteration because moving into the interior of the circle from (0.697, 1.517) increases the objective. Thus, as in the linearly constrained example, y is again the only superbasic variable at this stage.

Since analytic solution of the active constraints for the basic variables is rarely possible, especially when some of the constraints are nonlinear, a numerical procedure must be used. GRG2 uses a variation of Newton's method that, in this example, works as follows. With $s = 0$, the equation to be solved for x is

$$(x - 2)^2 + y^2 - 4 = 0 \tag{4.5}$$

GRG2 determines a new value for y as before, by choosing a search direction d and then a step size α. Since this is the first iteration for the current reduced problem, the direction d is the negative reduced gradient. The linesearch subroutine in GRG2 chooses an initial value for α. At (0.697, 1.517), $d = 1.508$ and the initial value for α is 0.050. Thus the first new value for y, say y_1, is

$$y_1 = y_c + \alpha d = 1.517 + 0.050(1.508) = 1.592$$

Substituting this value into (4.5) gives

$$g(x) = (x - 2)^2 - 1.456 = 0 \tag{4.6}$$

Given an initial guess x_0, for x, Newton's method is used to solve this equation for x by replacing the left-hand side of (4.6) by its first-order Taylor series approximation at x_0:

$$g(x_0) + \left(\frac{\partial g(x_0)}{\partial x} \right)(x - x_0) = 0$$

Solving this equation for x and calling this result x_1 yields

$$x_1 = x_0 - \left(\frac{\partial g(x_0)}{\partial x} \right)^{-1} g(x_0) \tag{4.7}$$

If $g(x_1)$ is close enough to zero, x_1 is accepted as the solution and this procedure stops. "Close enough" is determined by a GRG2 parameter named EPNEWT (which can be set by the user and has a default value of 0.0001) using the criterion:

$$\text{abs}[g(x_1)] \leq \text{EPNEWT} \tag{4.8}$$

If this criterion is not satisfied, x_1 replaces x_0 and a new iteration of Newton's method begins. For this example, the sequence of x and y values generated by GRG2 is

Iteration	x	$g(x)$
Initial point	0.7849	$-0.134\text{E-}01$
1	0.7900	$-0.940\text{E-}03$
2	0.7904	$-0.675\text{E-}04$

In the "pure" Newton's method, $\partial g(x)/\partial x$ is reevaluated at each new value of x. In GRG2, $\partial g(x)/\partial x$ is evaluated only once for each linesearch, at the point from which the linesearch begins. In this example, $\partial g(x)/\partial x$ evaluated at $x = 0.697$ is 2.606, so the GRG2 formula corresponding to (4.7) is

$$x_1 = x_0 - 0.383 g(x_0)$$

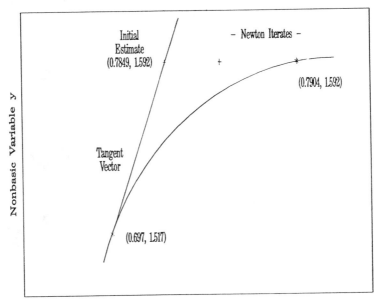

Figure 20 Initial estimate for Newton's method, nonlinear constraint.

Nonlinear Programming

This variation on Newton's method usually requires more iterations than the pure version, but it takes much less work per iteration, especially when there are two or more basic variables. In this case $\partial g(x)/\partial x$ is a matrix (called the basis matrix, as in linear programming), its inverse is $(\partial g(x)/\partial x)^{-1}$, and $g(x_0)$ is the vector of active constraint values at x_0.

Note that the initial guess for x in row 1 of the above table is 0.7849, not its base value of 0.697. GRG2 derives this initial estimate by using the vector that is tangent to the nonlinear constraint at (0.697, 1.517) as shown in Figure 20. Given $y_1 = 1.592$, the x value on this tangent vector is 0.7849. The tangent vector value is used because it usually provides a good initial guess and results in fewer Newton iterations.

Of course, Newton's method does not always converge. GRG2 assumes it has failed if more than ITLIM iterations occur before the Newton termination criterion (4.8) is met or if the norm of the error in the active constraints ever increases from its previous value (an occurrence indicating that Newton's method is diverging). ITLIM has a default value of 10. If Newton's method fails but an improved point has been found, the linesearch is terminated and a new GRG2 iteration begins. Otherwise the step size in the linesearch is reduced and GRG2 tries again. With the GRG2 print control parameter IPR = 2, the output from GRG2 that shows the progress of the linesearch at iteration 4 is

```
              STEP = 5.028E-02    OBJ = 9.073E-01   NEWTON ITERS = 2
              STEP = 1.005E-01    OBJ = 8.491E-01   NEWTON ITERS = 4
              STEP = 2.011E-01    OBJ = 9.128E-01   NEWTON ITERS = 8
QUADRATIC INTERPOLATION
              STEP = 1.242E-01    OBJ = 8.386E-01   NEWTON ITERS = 4
```

Note that as the linesearch process continues and the total step from the initial point gets larger, the number of Newton iterations generally increases. This occurs because the linear approximation to the active constraints, at the initial point (0.697, 1.517), becomes less and less accurate as we move farther from that point.

4.1.5. Infeasible Starting Point

If the initial values of the variables do not satisfy all of the constraints, GRG2 starts with a phase I objective function (which is also done in linear programming) and attempts to find a feasible solution. To illustrate this approach let us consider a problem that has no objective function and has the following three constraints:

$$x^2 + y^2 \leq 4$$
$$x + y \geq 1$$
$$x - y = 0$$

We use a starting point of (0.75, 0). The feasible region is shown in Figure 21 as the dashed line segment. At the initial point constraint 1 is strictly satisfied but constraints 2 and 3 are violated. GRG2 constructs the phase I objective function as the sum of the absolute values of all constraint violations. For this case the sum of the infeasibilities is

$$\begin{aligned}\text{sinf}(x,y) &= (x - y) + [1 - (x + y)]\\ &= 0.75 + 0.25\\ &= 1.0\end{aligned}$$

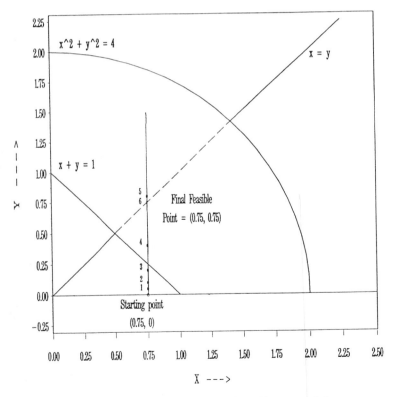

Figure 21 Finding a feasible point, GRG2 with constraints.

Nonlinear Programming

The first term is the violation of constraint 2 and the second term is the violation of constraint 3. Note that both terms are arranged so that the violations are positive.

The optimization problem solved by GRG2 is

Minimize   sinf(x,y)

subject to $x^2 + y^2 \leq 4$

At the initial point, the above nonlinear constraint is inactive, the reduced objective is just sinf(x,y), and the reduced gradient is

$$\nabla \text{sinf}(x,y) = (0,-2)$$

The initial search direction is, as usual, the negative reduced gradient direction, so $d = (0,2)$ and we move from $(0.75,0)$ straight up toward the line $x + y = 1$. The output from GRG2 (with IPR = 2) is shown below.

```
ITERATION   OBJECTIVE       NO.         NO.SUPER    NUMBER      NORM RED.   HESSIAN
NUMBER      FUNCTION        BINDING     BASICS      INFEAS.     GRADIENT    CONDITION

   0        1.0000E+00         1           2           2        2.000E+00   1.000E+00
            STEP = 2.500000E-02       OBJ = 9.000000E-01
            STEP = 5.000000E-02       OBJ = 8.000000E-01
            STEP = 1.000000E-01       OBJ = 6.000000E-01
            STEP = 2.000000E-01       OBJ = 3.500000E-01
            STEP = 4.000000E-01       OBJ = 0.000000E+00
            CONSTRAINT # 3 VIOLATED BOUND
    ALL VIOLATED CONSTRAINTS SATISFIED.  NOW BEGIN TO OPTIMIZE TRUE
    OBJECTIVE

ITERATION   OBJECTIVE       NO.         NO.SUPER    NUMBER      NORM RED.   HESSIAN
NUMBER      FUNCTION        BINDING     BASICS      INFEAS.     GRADIENT    CONDITION

   1        0.0000E+00         1           2           0        2.000E+00   1.000E+00
KUHN TUCKER CONDITIONS SATISFIED
```

As can be seen in this output, at the starting point (iteration 0) there are two infeasible constraints, two superbasics, and sin $f = 1$. Using our usual formula, (x,y) for the first linesearch is calculated as follows:

$$\begin{aligned}
(x,y) &= (x_c, y_c) + \alpha(d_1, d_2) \\
&= (x_c + \alpha d_1, y_c + \alpha d_2) \\
&= (0.75 + 0\alpha, 0 + 2\alpha) \\
&= (0.75, 2\alpha)
\end{aligned}$$

It is clear that the x values remain fixed at 0.75 and the y values are twice the step size at each step. In Figure 21 these steps are labeled 1 through 6. At step 5, GRG2 detects the change in sign of constraint number 3 and backs up until the constraint is binding. Since this point is feasible, GRG2 prints the message ALL VIOLATED CONSTRAINTS SATISFIED.

If the problem had an objective function, GRG2 would begin minimizing the "true" objective, starting from this feasible point. Since we did not specify an objective for this problem, GRG2 stops. Minimizing sinf to find a feasible point, if needed, is phase 1 of the GRG2 algorithm; optimization of the true objective is phase 2. If GRG2 cannot find a feasible solution then phase 1 will terminate with a positive value of sinf($\cdot$) and report that no feasible solution was found.

4.2. Successive Linear Programming

Successive linear programming (SLP) methods solve a sequence of linear programming approximations to a nonlinear programming problem. One of the earliest approaches, the method of approximation programming (MAP), was first suggested by Griffith and Stewart (1961) and we begin with a brief description of their approach.

4.2.1. Method of Approximation Programming

Recall that if $g_i(\mathbf{x})$ is a nonlinear function and $\mathbf{x}^0$ is the initial value for $\mathbf{x}$, then the first two terms in the Taylor's series expansion to $g_i(\mathbf{x})$ around $\mathbf{x}^0$ are

$$g_i(\mathbf{x}) = g_i(\mathbf{x}^0 + \Delta \mathbf{x}) \cong g_i(\mathbf{x}^0) + \Delta g_i(\mathbf{x}^0)^T (\Delta \mathbf{x})$$

and this is the linear approximation to $g_i(\mathbf{x})$ at $\mathbf{x}^0$. Given initial values for the variables, the nonlinear functions are replaced by their linear approximations at this initial point. The variables in the resulting LP are the "change" variables, the Δx_i's, representing changes from the current values. In addition, upper and lower step bounds (also called tolerance bounds) are imposed on the change variables, since the linear approximation is reasonably accurate only in some neighborhood of the initial point.

This resulting LP is solved and if the new point is an improved one it becomes the current point and the process is repeated. If the new point is not an improved one, the step bounds may need to be reduced or we may be close enough to the optimum to stop. Successive points generated by this procedure need not be feasible even if the initial point is. However, the amount of infeasibility generally is reduced as the iterations proceed.

Again we will illustrate the basic concepts with a simple example. Consider the following problem:

Maximize $2x + y$
subject to $x^2 + y^2 = 25$
$x^2 - y^2 \leq 7$
and $x \geq 0$
$y \geq 0$

with an initial starting point of $\mathbf{x}_c = (2, 2)$. Figure 22 shows the two nonlinear constraints and one objective function contour with an objective value

Nonlinear Programming

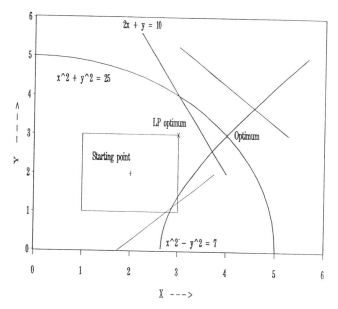

Figure 22 SLP example with linear objective, nonlinear constraints.

of 10. Since the objective function increases with increasing x and y, it is clear that the optimal solution is at the intersection of the two constraint equations where $\mathbf{x}_c = (3, 4)$. Linearizing the problem around the current point and adding tolerance bounds of plus and minus 1 to the change variables leads to the following LP:

Maximize $\quad 2x_c + y_c + 2\Delta x + \Delta y = 2\Delta x + \Delta y + 6$

subject to $\quad x_c^2 + y_c^2 + 2x_c\Delta x + 2y_c\Delta y = 4\Delta x + 4\Delta y + 8 = 25$

$\quad x_c^2 - y_c^2 + 2x_c\Delta x - 2y_c\Delta y = 4\Delta x - 4\Delta y \le 7$

and $\quad \max(0 - x_c, -1) \le \Delta x \le 1$

$\quad \max(0 - y_c, -1) \le \Delta y \le 1$

Lower bounds on the change variables are selected so that adding the changes to the current values of the variables will not violate the original bounds. If there were upper bounds on the variables, they would be handled in a similar fashion. Rearranging terms in the above equations and realizing that a constant in the objective can be ignored gives us the following equivalent problem:

Maximize $2\Delta x + \Delta y$
subject to $\Delta x + \Delta y = 4.25$
$\Delta x - \Delta y \leq 1.75$
and $-1 \leq \Delta x \leq 1$
$-1 \leq \Delta y \leq 1$

Figure 22 also shows these LP constraints. Its optimal solution (recall that the solution to an LP is always at an extreme point) is at $(\Delta x, \Delta y) = (1, 1)$, which gives $(x_n, y_n) = (3, 3)$. This is an improved point, as can be seen by evaluating the original functions, so we set $\mathbf{x}_c = \mathbf{x}_n$ and repeat these steps to get the next LP shown below:

Maximize $2\Delta x + \Delta y$
subject to $\Delta x + \Delta y = 7/6$
$\Delta x - \Delta y \leq 7/6$
and $-1 \leq \Delta x \leq 1$
$-1 \leq \Delta y \leq 1$

The feasible region is shown in Figure 23 and the optimal solution is at $(\Delta x,$

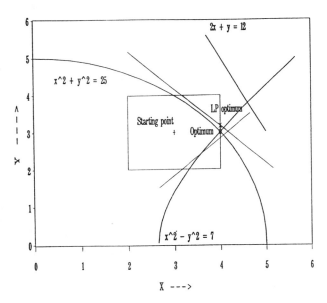

Figure 23 SLP example with linear objective, nonlinear constraints.

$\Delta y) = (1, 1/6)$ or $(x_n, y_n) = 4, 3.167)$. The next step of the algorithm gives $(x_n, y_n) = (4, 3.005)$.

In general, if the optimum is at an extreme point, then convergence to the optimum is quadratic once the current point is sufficiently close to it. However, for optima that are not at an extreme point, convergence can be slow. The rate of convergence will depend on how the tolerance bounds are reduced. Strategies are determining if the tolerance bounds need to be reduced, reducing them, and otherwise improving the performance of SLP methods have been proposed by Beale (1978), Buzby (1974), Boddington (1979), Palacios-Gomez et al. (1982), and Zhang et al. (1985). This last approach, called penalty SLP (PSLP) by the authors, is briefly described below.

4.2.2. Penalty Sequential Linear Programming

All linear programming–based optimization algorithms handle linear constraints particularly well. In order to emphasize this aspect, the general nonlinear programming problem can be restated to show such constraints explicitly. In various formulations it may also be helpful to show linear variables explicitly, although we will not do that here.

To explain the PSLP algorithm we will begin with our general nonlinear programming problem cast into the following form:

Minimize $g_0(x)$
subject to $lb_i \le g_i(x) \le ub_i,$ $i = 1, \ldots, m$
$Ax \le b$
and $l_i \le x_i \le u_i,$ $i = 1, \ldots, n$

We will frequently use $g(x)$ to represent the vector of (nonlinear) constraint functions, excluding the objective. For our next step we define an exact penalty problem (Han and Mangasarian, 1979)

Minimize $p(x,w) = g_0(x) + w \sum_{i=1}^{m} \text{viol}(g_i(x))$
subject to $Ax \le b$
and $l_i \le x_i \le u_i$

where the w is a positive weight and viol($\cdot$) is the function that returns the absolute value of the amount by which the constraint violates its bound or zero if the constraint is within bounds. If the penalty weight is large enough, this penalty problem is, under certain mild conditions (Conn, 1982), equivalent to the original NLP in that any local solution of one is a local solution of the other. "Large enough" means that w must be larger than the absolute value of the largest optimal Lagrange multiplier for the nonlinear constraints. Often such a value is known in advance. If not, logic has been developed for adjusting an initial guess until it is large enough. In the following we no longer show the explicit dependence of the objective function on w.

The next step in our discussion of the PSLP algorithm is to linearize the constraint and objective functions at the current point and to add tolerance bounds to the change variables. In order to simplify the presentation we assume that all of the nonlinear constraints are of the form $g_i(x) = 0$ and hence that $\text{viol}(g_i(x)) = |g_i(x)|$. This leads to the following piecewise linear approximation to the exact penalty problem:

Minimize $\quad \nabla g_0(x_c) \Delta x + w \sum_{i=1}^{m} |g_i(x_c) + \nabla g_i(x_c) \Delta x|$

subject to $\quad A\Delta x \leq b - Ax_c$

and $\quad \max(l_i - x_{c_i}, -s_i) \leq \Delta x_i \leq \min(u_i - x_{c_i}, s_i)$

Here the s_i are the tolerance bounds on the change variables, which, for the linear variables, can be quite large (essentially ignored in the current implementation of PSLP).

This problem can be transformed into a linear program using an idea from goal programming (see Chapter 7 is this book). We replace the absolute value terms, which are piecewise linear convex functions, by a sum of nonnegative deviation variables $n_i + p_i$, where n_i and p_i represent the negative and positive values of the ith constraint, respectively. This gives us the following equivalent linear program:

Minimize $\quad pl(\Delta x) = \nabla g_0(x_c) \Delta x + w \sum_{i=1}^{m} (p_i + n_i)$

subject to $\quad \nabla g_i(x_c) \Delta x - p_i + n_i = -g_i(x_c)$

$\quad A\Delta x \leq b - Ax_c$ \hfill (4.9)

and $\quad \max(l_i - x_{c_i}, -s_i) \leq \Delta x_i \leq \min(u_i - x_{c_i}, s_i)$

$\quad p_i \geq 0, n_i \geq 0$

The rectangle defined by the step bounds in (4.9) is called a "trust region" (Sorenson, 1982). Its size may be varied at each step of the PSLP iteration. The idea is to make this region as large as possible and still have the linearized problem remain a good approximation to the original problem. Changes to the step bounds are determined by evaluating the ratio of the actual reduction in the exact penalty function at each step to the predicted reduction for that step as determined from the linearized problem.

Nonlinear Programming

The major steps in the PSLP algorithm are:

1. Select x^1 satisfying the linear contraints and the variable bounds.
2. Solve the linear program LP obtaining an optimal solution Δx^k.
3. Compute the actual change in the exact penalty function

$$\Delta p^k = p(x^k) - p(x^k + \Delta x^k)$$

   and the change "predicted" by its piecewise linear approximation

$$\Delta p l^k = p(x^k) - pl(x^k + \Delta x^k)$$

4. Test the stopping criteria and stop if any one is satisfied. Otherwise compute the ratio r^k of the actual to the predicted change.
5. If $r^k < 0.025$, divide each step bound by 2 and return to step 2. Otherwise set $x^{k+1} = x^k + \Delta x^k$ and update the step bounds as follows:

    If $r^k < 0.25$, divide each step bound by 2.
    If $r^k > 0.75$, multiply each step bound by 2.
    If any nonlinear variable has been equal to the same step bound for three consecutive iterations, multiply that step bound by 2.
6. Update $k = k + 1$ and return to step 2.

PSLP is now widely used in refineries to solve refinery production planning problems (Baker and Lasdon, 1985).

4.3. Successive Quadratic Programming

Successive quadratic programming (SQP) methods solve a sequence of quadratic programming approximations to a nonlinear programming problem. Quadratic programs (QPs) have a quadratic objective function, linear constraints, and bounds on the variables, and there exist efficient procedures for solving them. As in SLP, the linear constraints are linearizations of the actual constraints about the current point. However, the quadratic objective that is used is not just the quadratic approximation to the original objective function but a modified function that uses the Lagrangian function as well.

Let us begin by considering the following problem:

Minimize $g_0(x)$ (4.10)
subject to $g_i(x) = 0$, $i = 1, \ldots, m$

The Lagrangian for this problem is $L(x_c, \lambda_c) = g_0(x_c) - \lambda_c g(x_c)$. Recall that the necessary conditions for the point x_c to be a local minimum of (4.10) are that there exist Lagrange multipliers λ_c such that

$$\nabla_x L(x_c, \lambda_c) = \nabla g_0(x_c) - \lambda_c \nabla g(x_c) = 0 \tag{4.11}$$

and

$$g(x_c) = 0$$

Newton's method, applied to solve the system of equations (4.11), linearizes them at the current point to yield the linear system

$$\begin{bmatrix} \nabla_x^2 L(x_c, \lambda_c) & -\nabla^2 g(x_c) \\ (\nabla^2 g(x_c))^T & 0 \end{bmatrix} \begin{pmatrix} \Delta x \\ \Delta \lambda \end{pmatrix} = \begin{pmatrix} \nabla L(x_c, \lambda_c) \\ g(x_c) \end{pmatrix} \tag{4.12}$$

It is easily shown that if $(\Delta x, \Delta \lambda)$ satisfy Eqs. (4.12), then $(\Delta x, \lambda_c + \Delta \lambda)$ will satisfy the necessary conditions for the optimality of the following quadratic program:

$$\text{Minimize} \quad \nabla g_0(x_c) \Delta x + \frac{1}{2} \Delta x^T \nabla_x^2 L(x_c, \lambda_c) \Delta x \tag{4.13}$$

$$\text{subject to} \quad g(x_c) + \nabla g(x_c) \Delta x = 0$$

On the other hand, if $\Delta x^* = 0$ is the solution to this quadratic program, then we can show that x_c satisfies the necessary conditions (4.11) for a local minimum of the original problem. First, since $\Delta x^* = 0$, $g(x_c) = 0$ and x_c is a feasible point for the original problem. Since Δx^* solves (4.13), there exist λ^* such that the gradient of the Lagrangian function for (4.13), evaluated at $\Delta x^* = 0$, is also equal to zero, i.e.,

$$\nabla g_0(x_c) - \lambda^* \nabla g(x_c) = 0$$

The Lagrange multipliers λ^* can clearly serve as the Lagrange multipliers for the original problem and hence the necessary conditions (4.11) are satisfied by (x_c, λ^*).

The extension to inequality constraints is straightforward; linearize them also and include them in the Lagrangian function when computing the Hessian matrix (represented below by H) of the Lagrangian. Linear constraints and variable bounds in the original problem are just included directly in the quadratic program (4.13).

Of course, the matrix H need not be positive definite, even at the optimal solution of the NLP problem, so the QP may not have a minimum. Fortunately, positive definite approximations, Q, of H can be used and then the QP will have an optimal solution if it is feasible. Such approximations can be obtained by a slight modification of the popular BFGS updating formula used in unconstrained minimization. This formula requires only the gradient of the Lagrangian function so second derivatives of the problem functions do not need to be computed.

Since the QP can be derived from Newton's method applied to the necessary conditions for the optimum of the NLP, if one simply accepts the

Nonlinear Programming

solution of the QP as defining the next point, the algorithm behaves like Newton's method; it converges rapidly near an optimal solution but may not converge at all from a poor initial point. If Δx is viewed as a search direction, the convergence properties can be improved. However, since both objective function improvement and reduction of the constraint infeasibilities need to be accounted for, the function to be minimized in the linesearch process must incorporate both. Two possibilities that have been suggested are the exact penalty function and the Lagrangian function. The Lagrangian is suitable because

1. On the tangent plane to the active constraints, it has a minimum at the optimal solution to the NLP.
2. It initially decreases along the direction Δx.

If the penalty weight in the exact penalty function is large enough, the exact penalty function also has property (2) and it is minimized at the optimal solution to the NLP.

4.4. Relative Advantages and Disadvantages

Table 1 summarizes the relative merits of SLP, SQP, and GRG algorithms, focusing on their application to problems with many nonlinear equality constraints. One feature appears as both an advantage and a disadvantage—whether or not the algorithm can violate the nonlinear constraints of the problem by relatively large amounts during the solution process.

SLP and SQP usually generate points yielding large violations. This can cause difficulties, especially in models with log or fractional power expressions, since negative arguments for these functions may be generated. Such problems have been documented in reference to complex chemical process examples (Sarma and Reklaitis, 1979) in which SLP and some exterior penalty-type algorithms failed, while the GRG2 code succeeded and was quite efficient. On the other hand, algorithms that do not attempt to satisfy the equalities at each step can be faster than those that do (Berna et al., 1980). The fact that SLP and SQP will satisfy any linear constraints at each iteration should ease the difficulties referred to above but will not eliminate them.

There are situations in which the optimization process must be interrupted before the algorithm has reached optimality and the current point must be used or discarded. These cases usually arise in on-line process control where time constraints force timely decisions. In such cases, maintaining feasibility during the optimization process may be a requirement for the optimizer inasmuch as constraint violations make a solution unusable.

Clearly, all three algorithms have advantages that will dictate their use in certain situations. For large problems, SLP software is used most widely,

Table 1 Relative Merits of SLP, SQP, and GRG Algorithms

Algorithm	Relative advantages	Relative disadvantages
SLP	1. Easy to implement 2. Widely used in practice 3. Rapid convergence when optimum is at a vertex 4. Can handle very large problems 5. Does not attempt to satisfy equalities at each iteration 6. Can benefit from improvements to LP solvers	1. May converge slowly on problems with nonvertex optima 2. Will usually violate nonlinear constraints until convergence to optimum, often by large amounts
SQP	1. Usually requires fewest function and gradient evaluations of all three algorithms (by far) 2. Does not attempt to satisfy equalities at each iteration	1. Will usually violate nonlinear constraints until convergence, often by large amounts 2. Harder than SLP to implement—requires a good QP solver
GRG	1. Probably most robust of all three methods 2. Versatile—especially good for unconstrained or linearly constrained problems but also works well for nonlinear constraints 3. Can utilize existing process simulators employing Newton's method 4. Once it reaches a feasible solution it remains feasible and then can be stopped at any stage with an improved solution	1. Hardest to implement 2. Needs to satisfy equalities at each step of the algorithm

because it is relatively easy to implement given a good LP system. However, large-scale versions of GRG and SQP are increasingly popular.

4.5. Available NLP Software

In this section we survey implementations of the algorithms described in Sections 4.1–4.4. Although a large (and increasing) proportion of NLP users employ systems with higher-level user interfaces to optimizers, such

Nonlinear Programming

as spreadsheets and algebraic modeling systems, all such systems have at their core adaptations of one or more stand-alone optimization packages. (By "stand-alone" we mean software that is designed specifically to accept the specification of a nonlinear program, attempt to solve it, and return the results of that attempt to the user or to an invoking application.) The NLP capabilities and characteristics of those higher-level systems therefore naturally derive from those of their incorporated optimizers. As a result, we begin our discussion with an overview of significant stand-alone NLP optimizers. At the end of this chapter we provide a series of appendices that illustrate, for a simple NLP problem, the inputs and outputs for some of the optimizers described below. A comprehensive list of vendors and sources for the products discussed in this section (as well as for a large number of linear, unconstrained, and discrete optimization products) may be found in More and Wright (1993). Advertisements for many of the systems described below can be found in the monthly magazine *OR/MS Today*, published by INFORMS (Institute for Operations Research and the Management Sciences). This magazine is an excellent source of information on analytical software of all kinds. The April 1995 issue contains an excellent NLP software survey (Nash, 1995).

4.5.1. Optimizers for Stand-Alone Operation or Embedded Applications

Most existing NLP optimizers are FORTRAN based, although C versions are becoming more prevalent. Most are capable of operation as true stand-alone systems (the user must usually code or modify main programs and routines that return function values) or as subsystems that are embedded in larger systems and solve problems generated by or posed through those systems. Some vendors supply source code; others supply only object code for the customers' target platform. Details are available from the vendors as noted below or in More and Wright (1993). All NLP optimizers require that the user supply the following:

- A specification of the NLP problem to be solved—at a minimum, the number of functions, the number of variables, which function is the optimization objective, bounds on the functions and variables (if different from some default scheme), and initial values of some or all variables (the system may supply default values, but using these is recommended only as a last resort).
- One or more subprograms that supply to the optimizer, on demand, the values of the functions for a specified set of variable values. Some systems also allow the user the option of supplying derivative values.

GRG-Based Optimizers

GRG2. Presently, the most widely distributed GRG code is GRG2, whose operation is explained in Section 4.1. In addition to its use as a stand-alone system, it is the optimizer employed by the Solver optimization option within the spreadsheet packages Excel, Quattro Pro, and the GINO interactive solver.

In stand-alone operation, GRG2 requires the user to code a calling program in FORTRAN that allocates working storage and passes through its argument list the problem specifications and any nondefault values for user-modifiable options (an option using text files for problem specifications also exists). In addition, the user must code a subroutine that accepts as input a vector of variable values and returns a vector of function values calculated from the inputs. All constraints are assumed to be of the form

$$l_i \leq g_i \leq u_i$$

where l_i and u_i are (constant) lower and upper bounds on $g(i)$.

GRG2 represents the problem Jacobian (i.e., the matrix of first partial derivatives) as a dense matrix. As a result, the effective limit on the size of problems that can be solved by GRG2 is a few hundred active constraints (excluding variable bounds). Beyond this size, the overhead associated with inversion and other linear algebra operations begins to severely degrade performance. References for descriptions of the GRG2 implementation are given in Liebman et al. (1986) and Lasdon et al. (1978). Appendix A.1 at the end of this chapter shows a GRG2 calling program, function evaluation routine, and sample output.

LSGRG2 is an extension of GRG2 that employs sparse matrix representations and manipulations and extends the practical size limit to at least 1000 variables and constraints. The interfaces to LSGRG2 are very close to those described above for GRG2. LSGRG2 has been interfaced to the GAMS algebraic modeling system. Performance tests and comparisons on several large models from the GAMS library are described in Smith and Lasdon (1992).

CONOPT is the other widely used implementation of the GRG algorithm. It, like LSGRG2, is designed to solve large, sparse problems. CONOPT is available as a stand-alone system, callable subsystem, or as one of the optimizers callable by the GAMS system. Description of the implementation and performance of CONOPT is given in Drud (1985).

SQP-Based Optimizers.

Implementations of the SQP algorithm described in Section 4.3 are the following:

SQP is a sister code to GRG2 and is available from the same source. The interfaces to SQP are very similar to those of GRG2. SQP is useful for small problems as well as large sparse ones, employing sparse matrix structures throughout. The implementation and performance of SQP are documented in Fan et al. (1988).

Nonlinear Programming

NPSOL is a dense matrix SQP code developed at Stanford University. It is available from the same source as MINOS (see below). Additional details are available in More and Wright (1993).

NLPQL is another SQP implementation, callable as subroutine and notable mainly by its use of reverse communication (the called subsystem returns codes to the calling program indicating what information is required upon reentry). A reference for NLPQL is More and Wright (1993).

MINOS. MINOS employs a modified augmented Lagrangian algorithm described in Murtagh and Saunders (1982). MINOS uses sparse matrix representations throughout and is capable of solving nonlinear problems exceeding 1000 variables and rows. MINOS is also capable of exploiting, to the greatest extent possible, the presence of purely linear variables and functions (although, because the user must communicate this structure to the optimizer, the greatest utility of this feature results from coupling MINOS to higher-level modeling systems that are able to determine problems structure). As a stand-alone system, problem specifications and user options are supplied to MINOS via an external text file; problem coefficient structure is supplied through an MPS file (a structured text file). As with the optimizers described above, the user must supply FORTRAN routines that compute function values and, optionally, derivatives. MINOS is the default optimizer option under the GAMS system for both linear and nonlinear problems. Details of stand-alone use of MINOS and additional references are given in Murtagh and Saunders (1982).

Mathematical Software Libraries. Many of the major callable libraries of mathematical software include at least one general NLP (i.e., capable of solving problems with nonlinear constraints) component. IMSL provides individual callable routines for most variations of linear and nonlinear constraints and objectives. The NAG Fortran Library contains an SQP method for constrained problems and a variety of routines for unconstrained or specialized optimization problems. In addition, most such libraries, even those without specific constrained NLP solvers, contain routines that perform such tasks as equation solving, unconstrained optimization, and various linear algebra operations. These routines can be used as subalgorithm components to build customized NLP solvers. References for the IMSL and NAG libraries and their vendors may be found in More and Wright (1993).

4.5.2. Spreadsheet Optimizers

In the 1980s, a major move away from the necessity of coding calling programs and function evaluation routines in FORTRAN began as optimizers; first LP solvers and then NLP solvers were interfaced to spreadsheet systems for desktop computers. The spreadsheet has become, de facto, the universal user interface for entering and manipulating numeric data. Spreadsheet vendors are increasingly incorporating analytic tools accessible from the spreadsheet interface and able, through that interface, to access external databases; examples include statistical packages, optimizers, and equation solvers.

The Excel Solver. Microsoft Excel, beginning with version 3.0, incorporates an NLP solver that operates on the values and formulas of the spreadsheet model. Versions 4.0 and later include an LP solver and mixed-integer programming (MIP) capability for both linear and nonlinear problems. The user specifies a set of cell addresses to be independently adjusted (the decision variables), a set of formula cells whose values are to be constrained (the constraints), and a formula cell designated as the optimization objective. The solver uses the spreadsheet interpreter to evaluate the constraint and objective functions and differences those computations to generate derivatives. The NLP solution engine for the Excel solver is GRG2 (see Section 4.1 and the section on GRG-based optimizers). Available as add-on products for Excel are large-scale LP, MIP, and NLP solvers as well as an enhanced quadratic integer programming capability (Frontline*).

The Quattro Pro Solver. The same team that packaged and developed the Excel solver also interfaced the same NLP engine (GRG2) to the Quattro Pro spreadsheet. Solver operation and problem specification mechanisms are similar to those of Excel.

Lotus 1-2-3. The Lotus 1-2-3 Windows-based products incorporate linear and nonlinear solvers that operate in a fashion similar to those described above. The solver engines are Lotus proprietary developments and are not well documented stand-alone optimizers as in Excel and Quattro Pro. Hence the performance characteristics, strengths, and weaknesses of the Lotus solvers are not at this time documented in the literature.

4.5.3. Algebraic Modeling Systems

An algebraic modeling system normally accepts the specification of a model in text as a system of algebraic equations. The system parses the equations and generates a representation of the expressions that can be numerically evaluated by its interpreter. In addition, some analysis is done to determine the structure of the model and, in some cases, to generate symbolic expressions for evaluating the Jacobian matrix. The processed model is then available for presentation to an equation solver or optimizer. Three modeling systems with NLP capabilities are described below.

---

*Frontline Systems, Inc., P.O. Box 4288, Incline Village, NV 89450.

Nonlinear Programming

GAMS — General Algebraic Modeling System. The GAMS system allows specification and solution of large-scale optimization problems. The modeling language is algebraic with a FORTRANish flavor. The default NLP solver for GAMS is MINOS with ZOOM-XMP available for mixed-integer programming. Optional solver interfaces are availabe, with some platform restrictions, for MPSX, OSL, CONOPT, and DICOPT (for nonlinear MIP). An interface to LSGRG2 has been developed that is not presently supported by the vendor. GAMS is available on a wide variety of platforms ranging from the lowest-level 80x8x DOS-based PC to workstations and mainframes. Examples of GAMS models and GAMS solution output are given in Section 5 of this chapter. General references, system details, and user procedures are given in Brooke et al. (1992).

AMPL — A Mathematical Programming Language. AMPL is a modeling language developed by Robert Fourer and others at Northwestern University. Its main features include an interactive environment for setting up and solving mathematical programs, the ability to select among several solvers (including MINOS, OSL, and CPLEX), and a powerful set construct that allows indexed, named, and nested sets. This set construct allows large-scale optimization problems to be stated tersely and in a form close to their natural algebraic expression. AMPL is described in Fourer et al. (1993).

GINO. GINO is an interactive interface to GRG2. It allows the user to enter an algebraic representation of his model, define an optimization or equation-solving problem on that model, and submit that problem to the solver. Results are communicated to the terminal screen. GINO uses the same user interface and syntax as does LINDO, a sister product for LPs. It does not contain the indexing features of GAMS or AMPL, so is not suitable for large problems. See Liebman et al. (1986) for further information.

4.6. Using NLP Software

In this section we address some of the issues that affect the user of NLP solution software. The primary determinant of solution reliability with LP solvers is numerical stability and accuracy. If the linear algebra subsystem of an LP solver is strong in these areas, the solver will always terminate with one of three conditions — optimal, infeasible, unbounded — or will run up against a time or iteration limit set by the user prior to detecting one of those conditions. In contrast, there are many additional factors that affect NLP solvers and their ability to obtain and recognize a solution.

4.6.1. Evaluation of Derivatives — Issues and Problems

All major NLP algorithms require first derivatives of the problem functions in order to move toward a solution and to evaluate the optimality conditions. If the derivatives are computed inaccurately, the algorithm may progress very slowly, choose poor directions for movement, and terminate due to lack of progress or iteration limits at points far from the actual optima, or, in extreme cases, actually declare optimality at nonoptimal points.

Finite Difference Derivatives. When the user, whether working stand-alone or through a spreadsheet or modeling system, supplies only the values of the problem functions at a proposed point, the solver computes the derivatives by *finite differences*. Each function is evaluated at a base point and then at a perturbed point. The difference between the function values is then divided by the perturbation distance to obtain an approximation of the first derivative at the base point. If the perturbation is in the positive direction from the base point we call the resulting approximation a *forward difference approximation*. For highly nonlinear functions, derivative accuracy may be improved by using *central differences*; here, the base point is perturbed both forward and backward and the derivative approximation is formed from the difference of the function values at those points. The price for this increased accuracy is that central differences require twice the computational effort (and consequently time) of forward differences. If the functions are inexpensive to evaluate, the additional effort may be modest, but for large problems with complex functions, the use of central differences may dramatically degrade solution times. Most NLP solvers possess options that enable the use to specify the use of central differences. Some solvers attempt to assess derivative accuracy as the solution progresses and switch to central differences automatically if the switch seems warranted.

A critical factor in the accuracy of finite difference derivatives is the value of the perturbation step. The default values employed by all NLP solvers (generally 1.E-6 to 1.E-7 times the value of the variable) yield good accuracy when the problem functions can be evaluated to reasonable accuracy. When problem functions cannot be evaluated to "normal" accuracy (perhaps because the functions themselves are the result of iterative computations), the default step is often too small and the resulting derivative approximations may be erroneously determined to be zero or some other invalid value resulting from the difference being only that of the "noise" represented by the insignificant digits of the function value. If the functions are highly nonlinear in the neighborhood of the base point, the default perturbation step may be too large to approximate accurately the tangent to the function at that point. Special care must be taken in derivative computation if the problem functions are not closed-form functions in compiled code or a modeling language (or, equivalently, a sequence of simple computations in a spreadsheet). If each function evaluation involves convergence of some sort of simulation, solution of simultaneous equations, or convergence of some empirical model, the interaction between the derivative per-

Nonlinear Programming

turbation step and the convergence criteria of the functions will strongly affect the derivative accuracy and, as a result, solution progress and reliability. In such cases, increasing the perturbation step by two or three orders of magnitude may aid the solution process.

Most NLP solvers allow the stand-alone user to supply derivative values himself if he chooses to. A user will take advantage of this if he has access to the analytic derivatives of the functions or believes himself able to compute derivative values more quickly and/or more accurately than the solver's finite difference procedures. Modeling languages such as GAMS determine the analytic derivatives of the problem functions and supply evaluations of these to the solver. Analytic derivatives (symbolic derivatives or values computed through automatic differentiation techniques) are faster and more accurate than finite difference approximations and represent an attractive option when available. Analytic derivatives are not presently computed by spreadsheet solvers, but this can be done and may appear in future versions.

What to Do When an NLP Algorithm Is Not "Working" Probably the most common mode of failure of NLP algorithms is termination due to "fractional change" (i.e., when the difference in successive objective function values is a small fraction of the value itself over a set of consecutive iterations) at a point where the Kuhn-Tucker optimality conditions are far from satisfied. Sometimes this criterion is not present and a simple iteration limit is present. Termination at a significantly nonoptimal point is an indication that the algorithm is unable to make any further progress. Such lack of progress is often associated with poor derivative accuracy, which can lead to search directions that do little or nothing to improve the objective function. In such cases, the user should analyze his problem functions and perhaps experiment with different derivative steps.

Algorithm Parameters and "Tuning." Most NLP solvers utilize a set of tolerances and parameters that control the algorithm's determination of which values are "nonzero," when constraints are satisfied, when optimality conditions are met, and other items.

Feasibility Tolerance. Most NLP solvers evaluate the first-order optimality conditions and declare optimality when a feasible solution meets these conditions to within a specified tolerance. Problems that reach what amount to optimal solutions in a practical sense but require many additional iterations to declare optimality may be sped up by increasing the optimality tolerance. Conversely, problems that terminate at points close but not close enough to optimality may often be moved to improved solutions by decreasing the optimality tolerance if derivative accuracy is high enough.

Other Tuning Issues. A critical parameter for GRG2 and systems that employ GRG2 as a solver is the convergence tolerance for the Newton iterations (see Section 4.1 for details of the GRG algorithm). Increasing this from its default value may speed convergence of slow problems, and decreasing it may yield a more accurate solution (at some sacrifice of speed) or "unstick" a sequence of iterations that are going nowhere. MINOS requires specification of a penalty parameter that penalizes constraint violations. Penalty parameter values affect the balance between the seeking of feasibility and improvement of the objective function. In addition, because MINOS employs an infeasible path algorithm, the user can enforce a quasi-trust region through specification of the maximum number of "minor" iterations per "major" iteration.

Scaling. The performance of most NLP algorithms (particularly on large problems) is greatly influenced by the relative scale of the variables, function values, and Jacobian elements. In general, NLP problems in which the values of these quantities lie within a few orders of magnitude (say in the range 0–10) will tend to be solved (if solutions exist) faster and with fewer numerical problems. Most solvers either scale problems by default or allow the user to specify that the problem be scaled. Users can leverage these scaling procedures by building models that are reasonably scaled to begin with.

Model Formulation. Users can enhance the reliability of the solutions of their NLP problems via consideration of some simple model formulation issues:

> Avoid constructs that may result in discontinuities or infinities. Use exponential functions rather than logs. Avoid denominator terms that may tend toward zero [e.g., $1/x$ or $1/(x-1)$], multiplying out these denominators where possible.
>
> Be sensitive to possible "domain violations," that is, the potential for the optimizer to move variables to values where functions are not defined (negative log arguments, negative square roots, negative bases for fractional exponents) or where the functions that make up the model are not valid expressions of the systems being modeled.

Starting Points. The performance of NLP solvers is strongly influenced by the point from which the solution process is started. Points such as the origin (0,0, . . .) should be avoided as there may be a number of zero derivatives at that point (as well as problems with infinities). In general, any point where a substantial number of zero derivatives are possible is undesirable, as is any point where tiny denominator values are possible. Finally, for models of physical systems such as chemical processes, the

Nonlinear Programming

user should avoid starting points that do not represent realistic operating conditions. Such points may cause the solver to move toward points that are stationary points of the NLP functions but are not acceptable configurations of the physical system.

Local and Global Optima. All NLP algorithms and solvers here are capable only of finding local optima. For convex programs, any local optimum is also global. Unfortunately, most general NLPs—especially large ones—are not convex or cannot be determined to be convex and hence we must consider any solution returned by an NLP solver to be local. The user should examine his solution for reasonableness, perhaps resolving his problem from several starting points to investigate what local optima exist and how these solutions differ from one another.

5. APPLICATIONS OF NONLINEAR PROGRAMMING

5.1. Fitting Models to Data

Here we reconsider the problem, introduced in Section 2.1, of fitting a model of the form

$$y = h(v,x) \tag{5.1}$$

to a set of observed data. The vector v contains the independent variables thought to determine the dependent variable y, and x is a vector of model parameters to be chosen to cause the model to match a given set of observed data as closely as possible. There are p sets of observations, (y_i, v_i), $i = 1, \ldots, p$. We will define and compare several criteria for determining a best fit.

5.1.1. Least Squares

It is convenient to define the error between the observed value y_i and the value predicted by the model as

$$e_i(x) = y_i - h(v_i, x)$$

In Section 2 we noted that the most common procedure is to minimize the sum of squares of the errors e_i. In some problems it is desirable to weight some errors more heavily than others, using nonnegative weights w_i, leading to the problem

$$\text{Minimize} \quad n_2(x) = \sum_{i=1}^{p} w_i e_i^2(x) \tag{5.2}$$

For example, in fitting a model to time series data, the index i represents the number of time periods from the present time. Then it makes sense to have the weights decrease as i increases, so that older data receive a smaller weight.

There are several special-purpose algorithms for minimizing $n_2(x)$ in the case in which the model parameters x appear nonlinearly in the function h. These were mentioned in Section 2.5. General-purpose software may also be used. Sometimes there are constraints on the parameters x; for example, some components of x may be known to be nonnegative. In such cases, constrained NLP solvers are required.

5.1.2. Least Absolute Value

If some model errors are much larger than others, perhaps due to errors in obtaining the measurements y_i, the squares of these large errors will dominate the sum of squares. In this case, the parameters that minimize $n_2(x)$ will lead to a fit that tries to make the large errors small but tends to ignore the errors at the other points. In other words, the "outliers" in the data are overemphasized. An error norm that eases this difficulty is the weighted sum of absolute errors:

$$n_1(x) = \sum_{i=1}^{p} w_i |e_i(x)|$$

Minimizing n_1 directly is difficult because it is not differentiable at points x, where one or more errors $e_i(x)$ are zero, and an optimal solution is likely to have several small errors. However, this nonsmooth unconstrained problem is equivalent to the smooth constrained problem shown below:

$$\begin{aligned}
\text{Minimize} \quad & \sum_{i=1}^{p} w_i(p_i + n_i) \\
\text{subject to} \quad & e_i(x) = p_i - n_i, \quad i = 1, \ldots, p \\
\text{and} \quad & p_i \geq 0, n_i \geq 0, \quad i = 1, \ldots, p
\end{aligned} \quad (5.3)$$

Here, p_i and n_i are additional nonnegative variables representing, respectively, positive and negative errors. Assume that all weights w_i are positive [if not, the ith error term $e_i(x)$ may be dropped]. Then, in any optimal solution of (5.3), p_i and n_i will never both be positive. This is because, if they were, say, $p_2 = 4$ and $n_2 = 2$, we could choose $p_2 = 2$ and $n_2 = 0$ and still satisfy the condition $e_2(x) = p_2 - n_2 = 2$. These new values yield $p_2 + n_2 = 2$, which is smaller than the previous value of 6, so the values $p_2 = 4$, $n_2 = 2$ could not have been optimal. Since at least one member of each pair (p_i, n_i) is zero at optimality, and these satisfy $e_i(x) = p_i - n_i$, then $p_i + n_i = |e_i(x)|$. Any of the methods described in Section 4 may be used to solve the problem (5.3).

Nonlinear Programming

5.1.3. Minimax Fit

In some approximation problems, it is important to focus on the largest error and ignore all others. For example, automobile and aircraft manufacturers often use computer-aided design (CAD) software systems to help them design portions of car or aircraft bodies. Data points representing the coordinates in 3-space of the desired surface shape are entered, and a smooth, mathematically defined surface is then fit to these points. These surfaces are often chosen as polynomials of low degree, defined with different coefficients over adjacent regions, and structured so that they join smoothly at region boundaries. Such approximating functions are called *splines*. In such applications, it is typical to choose the spline coefficients so that the largest absolute error is minimized. This is stated mathematically as

$$\text{Minimize}_x \max_i |e_i(x)| \tag{5.4}$$

As with the sum of absolute errors, the above objective function is nonsmooth — it is generally not differentiable at points x where two or more errors $e_i(x)$ are equal. Since optimal solutions tend to have several equal errors, the objective will not be differentiable at the optimum, and the unconstrained methods described earlier will often stop short of the minimum or oscillate about it. However, as with the norm $n_1(x)$, the problem (6.4) is equivalent to the following smooth constrained problem:

$$\text{Minimize} \quad z \tag{5.5}$$
$$\text{subject to} \quad -z \le e_i(x) \le z, \quad i = 1, \ldots, p \tag{5.6}$$

Here, z is an additional nonnegative variable. As z decreases, the constraints (5.6) cause each error $e_i(x)$ to be squeezed down both from above and from below. At an optimal solution (z^0, x^0), $e_i(x^0) = z^0$ or $e_i(x^0) = -z^0$ for at least one index i, because, if not, z^0 could be reduced and the point (z^0, x^0) could not be optimal. Hence z^0 must be the maximum absolute error, since (5.6) forces all errors to lie between z^0 and $-z^0$. Although there are special-purpose algorithms for solving minimax problems, any general-purpose solver can be applied to the formulation (5.5)–(5.6).

As an example, consider the model

$$y = b_1 + b_2 \exp(b_3 * x) \tag{5.7}$$

where y and x are scalars and b_1, b_2, and b_3 are adjustable parameters. A set of six observed (x,y) values are contained in the Excel spreadsheet of Figure 24, in columns B and C. The spreadsheet fits this model to the data using the Excel solver, which was discussed in Section 4.5. Note that here the

	A	B	C	D
1		Nonlinear	Regression	with Three Objectives
2				
3		Approximating	function is	y = b1+b2*exp(x)
4				
5		observed	observed	predicted
6		Independent	Dependent	dependent
7	index,I	variable,x	variable,y	variable,y
8	1	-5	127	=$H$3+$H$4*(EXP($H$5*B8))
9	2	-3	151	=$H$3+$H$4*(EXP($H$5*B9))
10	3	-1	379	=$H$3+$H$4*(EXP($H$5*B10))
11	4	1	426	=$H$3+$H$4*(EXP($H$5*B11))
12	5	3	460	=$I$3+$I$4*(EXP($I$5*B12))
13	6	5	421	=$H$3+$H$4*(EXP($H$5*B13))
14				
15				
16	Problem 1			Problem 2
17	Min N1(x)			Min N2(x)
18				
19	Objective	=G14+H14		Objective
20		Constraints		
21	Index	e(I)-p(I)+n(I)		
22	1	=E8-G8+H8		
23	2	=E9-G9+H9		
24	3	=E10-G10+H10		
25	4	=E11-G11+H11		
26	5	=E12-G12+H12		
27	6	=E13-G13+H13		
28	Optimal Model			Optimal Model
29	Parameters			Parameters
30	b1	480.091		b1
31	b2	-73.946		b2
32	b3	-0.312		b3
33				
34	Optimal y values			Optimal y values
35	y1	127		y1
36	y2	291.16		y2
37	y3	379		y3
38	y4	426		y4
39	y5	451.15		y5
40	y6	464.6		y6
41				
42	Optimal errors,e(I)			Optimal errors,e(I)
43	e(1)	0		e(1)
44	e(2)	-140.16		e(2)
45	e(3)	0		e(3)
46	e(4)	0		e(4)
47	e(5)	8.85		e(5)
48	e(6)	-43.6		e(6)
49				
50				
51	Saved Solver specs			Saved Solver specs
52	=MIN($B$19)			=MIN($E$19)
53	=COUNT($I$3:$I$5,$G$8:			=COUNT($I$3:$I$5)
54	=$G$8:$G$13>=0			={100,100,0.000001,0.05,FALSE,FALS
55	=$H$8:$H$13>=0			
56	=$B$22:$B$27=0			
57	={100,100,0.000001,0.05,FA			

Figure 24 Spreadsheet formulation of nonlinear fitting problem.

Nonlinear Programming

E	F	G	H	I
		Model Coefficients		
exp(x)		b1	575.588	
		b2	-217.787	
		b3	-0.17	
error	squared			
e(i)	error	p(i)	n(i)	
=C8-D8	=E8*E8	0	0	
=C9-D9	=E9*E9	0	140.161	
=C10-D10	=E10*E10	0	0	
=C11-D11	=E11*E11	0	0	
=C12-D12	=E12*E12	8.851	0	
=C13-D13	=E13*E13	0	43.604	
Sum	=SUM(F8:F13)	=SUM(G8:G13)	=SUM(H8:H13)	
		Problem 3		
		Minimax		
=F14		Objective,z	61.623	
			Constraints	
		Index	z+e(i)	z-e(i)
		1	=$H$19+E8	=$H$19-E8
		2	=$H$19+E9	=$H$19-E9
		3	=$H$19+E10	=$H$19-E10
		4	=$H$19+E11	=$H$19-E11
		5	=$H$19+E12	=$H$19-E12
		6	=$H$19+E13	=$H$19-E13
		Optimal Model		
		Parameters		
516.651		b1	575.588	
-149.351		b2	-217.787	
-0.206		b3	-0.17	
		Optimal y values		
96.96		y1	65.38	
239.04		y2	212.62	
333.02		y3	317.38	
395.18		y4	391.9	
436.3		y5	444.91	
463.5		y6	482.62	
		Optimal errors, e(i)		
30.04		e(1)	61.62	
-88.04		e(2)	-61.62	
45.98		e(3)	61.62	
30.82		e(4)	34.1	
23.7		e(5)	15.09	
-42.5		e(6)	-61.62	
		Saved Solver specs		
		=MIN($H$19)		
		=COUNT($I$3:$I$5,$H$19)		
		=$H$22:$H$27>=0		
		=$I$22:$I$27>=0		
		=(100,100,0.000001,0.05,FA		

variables to be optimized are b_1, b_2, and b_3. Three criteria are used: least absolute error, squared error, and maximum error.

In Figure 24, column D contains formulas for the function on the right-hand side of (5.7), with references to the cells headed Model Coefficients, in column G, when the parameters b_1, b_2, and b_3 are needed (the values shown yield the minimax fit). The errors in column E are simply the observed y minus the predicted y, and squared error is $e(i) * e(i)$. The columns headed $p(i)$ and $n(i)$ contain the deviation variables p_i and n_i in problem (5.3).

The simplest of the three problems is the least squares one, shown in column D under Problem 2. Cell E19 contains the sum of squared error, by referring to the sum in F14. This is the solver target cell to be minimized. The changing cells are the parameters b_1, b_2, and b_3 in H3 : H5 and there are no constraints. These solver specifications are shown in cells D52 : D54. The argument of the MIN function is the objective, and the argument of the COUNT function is the range(s) containing the changing cells.

The problem of the sum of absolute errors has as its changing cells the model coefficients and the cells under the $p(i)$ and $n(i)$ headings in columns G and H. These latter cells are summed in cells G14 and H14, and the overall sum is formed in B19. The quantities $e_i(x) - p_i + n_i$ are formed directly below this objective. To satisfy (6.3), they are required to be equal to zero as one of the sets of solver constraints. The other constraints are nonnegativities on the $p(i)$ and $n(i)$ cells. These solver specifications are saved in A52 : A57.

To understand the setup for Problem 3, note that the constraints (5.6) can be written

$$z + e_i(x) \geq 0, \quad z - e_i(x) \geq 0, \quad i = 1, \ldots, p \qquad (5.8)$$

The variable z is contained in H19, and the quantities $z + e_i(x)$ and $z - e_i(x)$ are directly below. These are required to be nonnegative in the solver constraint box.

Optimal solutions to these three problems are shown numerically in rows 30 through 48 and are graphed in Figure 25. The least absolute error fit has three points with zero error but has a large error of -140.16 at $x = -3$. The least squares fit reduces this large error to -88.04, simultaneously allowing the errors at most other points to increase. The minimax fit reduces this largest error to 61.62 by allowing errors at $x = -5, -1$, and 5 to increase to this same value. It is typical of minimax fits to have several points at which the maximum error occurs.

5.2. Inventory Models

Most organizations, from households to manufacturers to retailers, maintain some type of inventory. The major cost of holding inventory is an

Nonlinear Programming

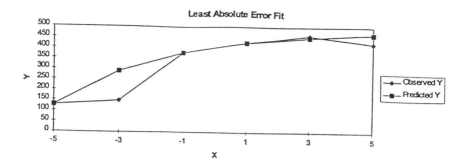

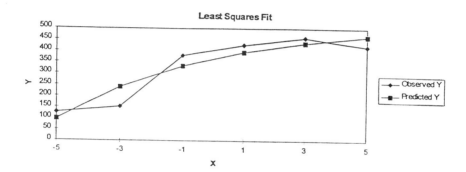

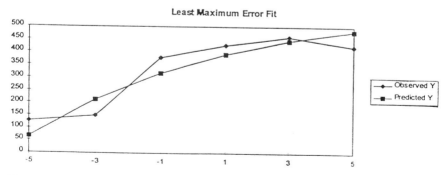

Figure 25 Graphs of three optimal fits.

opportunity cost: money tied up in inventory could be invested elsewhere at some rate of return. However, if there is too little inventory of raw materials or of intermediate products, then additional costs may be incurred due to stopped production or "emergency" purchases at higher prices (e.g., running down to the corner store to pick up milk at 5¢ a gallon more than at the supermarket). Insufficient finished goods inventory can lead to lost sales.

Inventory models can become quite complex, especially when uncertainty in demand or in supply lead time is included. Here, we consider some basic situations with deterministic demand and lead time. For a more complete treatment see Liebman et al. (1986) or Silver and Peterson (1985).

5.2.1. Balancing Storage and Ordering Costs (Economic Order Quantity)

Suppose we are managing the supply department for a manufacturer and that we have two types of products to keep in inventory. Costs include:

1. Purchase cost, assessed per unit. There are no discounts for large orders.
2. Storage or holding cost, assessed per unit stored per unit time. The largest component of this cost is the income lost by having the value of a unit tied up as unsold inventory.
3. Ordering or replenishment cost: a fixed charge per order, independent of the amount ordered or produced.

Future demand for product i is assumed to be known and constant, at a rate of d_i units per unit time, and the replenishment rate is instantaneous; i.e., an order arrives all at once with known lead time, just as inventory hits zero. Under these assumptions, inventory of each item varies with time as shown in Figure 26. If q_i units of product i are ordered, inventory jumps

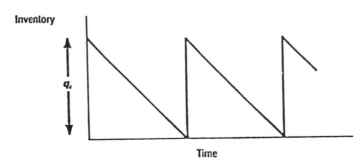

Figure 26 Inventory level versus time for instantaneous replenishment.

Nonlinear Programming

instantaneously to a level q_i just as it hits zero, then declines at a constant rate d_i.

For product i, let the ordering cost be k_i \$/order and the holding cost be h_i \$/order/week. Then the average inventory is $q_i/2$, and the holding cost per week is $h_i(q_i/2)$. The number of orders per week is d_i/q_i, and the ordering cost per week is $k_i(d_i/q_i)$. The total cost for both products is

$$TC = \sum_{i=1}^{2} \left(h_i \frac{q_i}{2} + k_i \frac{d_i}{q_i} \right) \tag{5.9}$$

The optimal order quantity q_i^*, is determined by minimizing the above function independently, subject to no constraints. Since TC is separable, each term of the sum can be minimized independently. As described in Section 2.2, the minimum occurs where the derivative is zero:

$$\frac{d}{dq_i}\left(h_i \frac{q_i}{2} + k_i \frac{d_i}{q_i} \right) = \frac{h_i}{2} - \frac{k_i d_i}{q_i^2} = 0 \tag{5.10}$$

Solving for q_i:

$$\frac{k_i d_i}{q_i^2} = \frac{h_i}{2} \tag{5.11}$$

or

$$\frac{k_i d_i}{q_i} = \frac{h_i q_i}{2} \tag{5.12}$$

The above shows that, at the optimum, holding cost per week equals ordering cost per week. Solving for q_i^* (often called the economic order quantity or EOQ):

$$q_i^* = \left(\frac{2 k_i d_i}{h_i} \right)^{1/2}$$

Note that q_i^* increases as ordering cost or demand increases and decreases as holding cost increases, as it should. Doubling any of these items multiplies or divides q_i^* by a factor of $\sqrt{2} = 1.414$.

A spreadsheet model of this problem is shown in Figure 27. The optimal order quantities q_i are in E8 : E9, holding and ordering costs are computed just to the right of these cells, and total cost is in H8 : H9. This is summed in H10, which contains the objective, TC. The saved solver specifications for minimizing TC subject to no constraints are in A22 : A23. The optimal values are $q_1^* = 38.72$, $q_2^* = 51.64$, with total cost of \$15.49/week (see E24 : E26).

Figure 28 shows a graph of the total cost versus order quantity, q_1, for

	A	B	C	D	E	F	G	H
1								
2	EOQ model with	2 products with	and without	constraints				
3								
4		Per Unit	Per Order					
5		Holding	Ordering	Demand	Order	Holding	Ordering	Total
6		Cost	Cost	Rate	Quantity	Cost	Cost	Cost
7	Product	($/unit/wk)	($/order)	(units/wk)	(units)	($/wk)	($/wk)	($/wk)
8	1	0.2	5	30	11.74	=0.5*B8*E8	=C8*D8/E8	=F8+G8
9	2	0.15	5	40	20.63	=0.5*B9*E9	=C9*D9/E9	=F9+G9
10							Total	=SUM(H8:H9)
11				Unit				
12				Purchase	Storage		Storage	
13				Cost	Space		Space	
14				($/unit)	(sq ft/unit)	($)	(sq ft)	
15				5	7	=D15*E8	=E15*E8	
16				2	10	=D16*E9	=E16*E9	
17	Saved Solver				Total	=SUM(F15:F16)	=SUM(G15:G16)	
18	Specs				Limits	100	350	
19								
20	Unconstrained	Constrained			Optimal	Solutions		
21	Problem	Problem			Unconstrained	Constrained		
22	=MIN($H$10)	=MIN($H$10)			Problem	Problem		
23	=COUNT(E8:E9)	=COUNT(E8:E9)		q1	38.72	11.74		
24		=F17:G17<=F18:G18		q2	51.64	20.63		
25					15.491	25.192		
26				Cost				
27				investment	296.08	100		
28				space	787.44	288.48		

Figure 27 EOQ model.

Nonlinear Programming

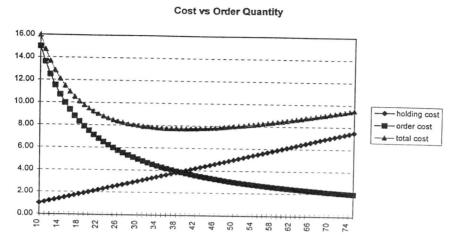

Figure 28 Cost versus order quantity.

product 1. Holding cost increases with q_1 while ordering cost decreases. The minimum value of 7.7460 $/week is achieved at $q_1^* = 38.72$, where the two costs are equal, and the portion of the cost curve near this minimum is very flat. This allows us to round the q_1^* value to 39 with little extra cost: the new cost is 7.7462.

5.2.2. Adding Investment and Space Constraints

Now suppose that there is a limited amount of storage space available for our two products, say 350 square feet, and that we wish to limit the total value of an order to $100. The space and value requirements of the EOQ solution are about 787 ft^2 and 297 $/order, so smaller order quantities are needed to meet these limits. Cells D15 : G18 contain per-unit purchase cost and space requirements and formulas to compute total cost and space used. Cells F17 : F18 contain these totals, which are used as constraint cells by the solver. The saved solver specifications are in B22 : B24 and the constrained solution is in F24 : F28. It has sharply lower order quantities and constraint values, at an additional cost of about $10/week. Note that, even though the unconstrained solution violates both constraints, only one (investment) is active at the constrained optimum.

Since the constraints cause the cost to rise by about 66%, it is sensible to explore the effects of relaxing the investment limit of $100. The Excel solver sensitivity report, shown in Figure 29, provides useful information. The Lagrange multiplier for the purchase cost constraint is -0.197. This means that an increase of the investment limit from $100 to $101 will reduce total cost by about 19.7 cents/week. This is an excellent rate of return of this extra dollar invested (19.7% per week), so we are naturally led to explore further the effects of increasing the investment limit. This is easily done in Excel: simply change the limit and resolve.

Figure 30 shows a scenario summary of the results of solving this problem for investment constraint limits ranging from 100 to 150 in steps of 10. This summary was created by the following sequence of steps. Set the value of the investment limit in cell F18 of the spreadsheet in Figure 27 to 100; for example, invoke the solver and click Solve. When the problem is solved, a Solver Results dialogue box appears telling you that a solution has been found. Click the Save Scenario button in this dialogue box, and another dialogue box will appear asking you to enter a Scenario name (INV100 would be a good choice); after doing so, the Solver Results box reappears; click OK, it vanishes, and you are out of the solver, back in Excel. Change the investment constraint limit to 110, and repeat the process until all values for this limit have been done. Then, from Excel, select Scenarios from the Tools menu, and select Summary from the Scenario Manager dialogue box that appears. A Scenario Summary dialogue box will appear, with the Scenario Summary report type selected. Click OK, and the Scenario Summary table of Figure 30 will be written to a sheet of the current workbook named Scenario Summary.

Now let us examine the scenario summary of Figure 30, because it exhibits some interesting behavior. As the investment limit increases from 100, cost decreases, and inventory investment and storage space used increase. When the investment limit is 130, space has reached its limit of 350 (see if you can determine at about what investment limit space reaches 350). Between 130 and 140, both constraints are active, but at 150 the investment

Changing Cells

Cell	Name	Final Value	Reduced Gradient
$E$8	(units)	11.75	0.000000
$E$9	(units)	20.63	0.000000

Constraints

Cell	Name	Final Value	Lagrange Multiplier
$F$17	Total ($)	100	-0.197404977
$G$17	Total (sq ft)	288.5546345	0

Figure 29 Excel solver sensitivity report.

Nonlinear Programming

	Unconstrained	limit = 100	limit = 110	limit = 120	limit = 130	limit = 140	limit = 150
Changing Cells:							
q1	38.73	11.75	12.96	14.19	16.67	19.44	20.17
q2	51.64	20.63	22.59	24.51	23.33	21.39	20.88
Result Cells:							
Cost	15.49	25.18	23.41	21.98	20.99	20.61	20.60
Investment	296.93	100.00	110.00	120.00	130.00	140.00	142.62
Space	787.50	288.55	316.64	344.50	350.00	350.00	350.00

Figure 30 Scenario summary.

constraint is inactive, and space becomes the limiting factor. The Cost Decrease row shows that the rate of improvement of cost decreases sharply as the Investment constraint is loosened. This illustrates the fact that, in any convex minimization problem, the optimal objective value is a convex nonincreasing function of the right-hand side of any $\leq$ constraint.

Figure 31 illustrates the situation geometrically, displaying contours of constant value of the objective (labeled z) and the initial locations of the investment constraint (#1) and the storage constraint (#2). As the investment constraint is relaxed, it will "slide" over toward the unconstrained solution. Assuming that we are seeking a solution that costs no more than $17 per week, we should investigate what capital investment and storage limits will be required. If we examine a number of solutions on the $17 per week contour and evaluate their investment and storage requirements, we

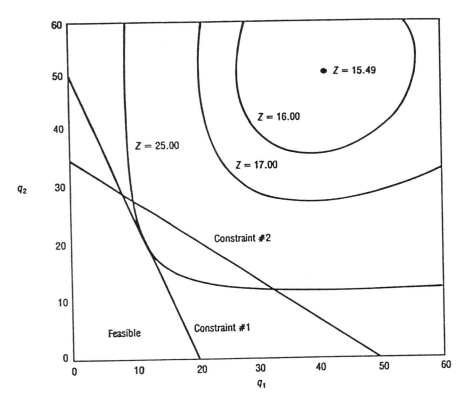

Figure 31 Cost contours for inventory problem.

Nonlinear Programming

find that the solutions on the $17 per week contour that have the smallest order quantities will need the least storage and investment.

5.3. A Simple Pooling Problem

The blending of liquid chemicals (with no chemical reactions) having differing concentrations of impurities leads to nonlinear models if pooling of some chemicals is involved. In the petroleum industry this leads to what is called a pooling problem. Consider the simple pooling model taken from Haverly (1978) and shown in Figure 32. A, B, and C are input chemicals containing sulfur as an impurity. These chemicals are to be blended to provide two output products, X and Y, which must meet sulfur content restrictions as shown on the figure. Customers will buy all of X and Y produced, up to a maximum of 100 units of X and 200 units of Y, at the given prices. The problem is to operate the process in order to maximize profit.

A complicating factor in this blending process is the fact that products A and B are not available separately. Such a situation can arise in a number of ways. The two chemicals may be produced at the same location but, for economic reasons, are transported together, for instance, in one tank car or via one pipeline. Another possibility is that the two chemicals are delivered separately but only a single holding facility is available at the blending site. In any case, until the amounts of A and B are determined, the pool sulfur content is unknown. However, it is the pool sulfur content together with the amounts of pool material and of chemical C used in blending X and Y that determines the X and Y sulfur contents. The sulfur constraints on X and Y affect the amounts of A and B that are needed, and it is this "circularity" that causes a nonlinearity.

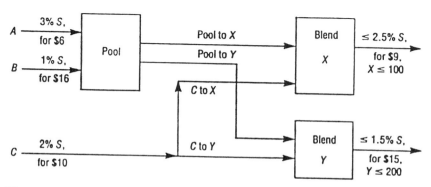

Figure 32 A pooling problem.

The constraint equations defining this system involve material balances together with sulfur constraints on the output products. Consider the material balance equations first.

The mass balance for the pool (assuming that all of the pool material is to be used up) is simply

$$\text{Amount\_A} + \text{amount\_B} = \text{pool\_to\_X} + \text{pool\_to\_Y}$$

For the output products, the balance equations are

$$\text{Pool\_to\_X} + \text{C\_to\_X} = \text{amount\_X}$$
$$\text{Pool\_to\_Y} + \text{C\_to\_Y} = \text{amount\_Y}$$

and for the total amount of C, the equation is

$$\text{C\_to\_X} + \text{C\_to\_Y} = \text{amount\_C}$$

Introducing the pool sulfur percent, POOLS, as a new variable makes it easy to write the X and Y sulfur constraints. If we let POOLS have a value between 0 and 100 and express all other percentages on the same scale, these constraints are

$$\text{POOLS} * \text{pool\_to\_X} + 2 * \text{C\_to\_X} \leq 2.5 * \text{amount\_X}$$
$$\text{POOLS} * \text{pool\_to\_Y} + 2 * \text{C\_to\_Y} \leq 1.5 * \text{amount\_Y}$$

The left-hand side of each inequality represents the actual sulfur content of the appropriate product and the right-hand side is the maximum amount of sulfur permitted in that product. The pool sulfur balance equation is

$$3 * \text{Amount\_A} + 1 * \text{amount\_B} = \text{POOLS} * (\text{amount\_A} + \text{amount\_B})$$

This defines POOLS as the amount of sulfur in the pool divided by the total amount of material in the pool.

As mentioned earlier, product demand sets upper bounds on production as

$$\text{Amount\_X} \leq 100$$
$$\text{Amount\_Y} \leq 200$$

and physical considerations restrict all variables to be nonnegative quantities. Clearly, the pool sulfur can never be less than 1% or more than 3%; thus:

$$1 \leq \text{POOLS} \leq 3$$

Nonlinear Programming

Finally, the profit function must be formulated. If cost_A, cost_B, cost_C, cost_X, and cost_Y are the appropriate cost coefficients, the profit can be written as

Cost_X * amount_X + cost_Y * amount_Y
− cost_A * amount_A − cost_B * amount_B
− cost_C * amount_C

A GAMS model of this problem is shown in Figure 33. It defines two sets: ST for the inputs or stocks and BL for the outputs or blends. The cost and sulfur content of the stocks are parameters indexed over the set ST, and the blend data are the table BDATA(BL,*), whose rows are indexed over the blends and whose columns are not indexed over any set previously defined. Aside from the indexed variables USE(ST) and PRD(BL), the GAMS variables have the same names as in the above formulation, and the constraints are the same as well. The PROFIT equation uses the GAMS SUM verb. Note that, even if the sets BL and ST each had 1000 elements, this equation would be unchanged (but all elements of BL and ST would have to be listed).

This problem is tricky in that it has (as we shall see) two local optima and a third point that satisfies the Kuhn-Tucker conditions but is not a local optimum. These three solutions are found by the three SOLVE statements in Figure 33. These statements all solve the same model, but from different starting points. The first SOLVE uses the default starting point provided by GAMS and converges to the global optimum, where OBJ = 400. The second SOLVE has initial values for the variables equal to the final values from the previous SOLVE, except for the new values established by the block of statements beginning with POOLS.L = 3 (which sets the current value or Level of POOLS to 3) and ending with USE.L ("C") = 50. The next statement, OPTION NLP = CONOPT, establishes CONOPT as the NLP solver. This SOLVE converges to a local optimum where OBJ = 100. The third SOLVE has all variables but POOLS initialized to a value of 0 and terminates at the same point. This is a Kuhn-Tucker point where OBJ = 0 and is not even a local optimum.

These three solutions are shown in the edited GAMS output of Figure 34, where some lines of no interest have been deleted. In the first solution, using MINOS, only product Y is produced, using equal amounts of B and C. The cost per unit of output is $(16 + 10)/2 = $13, and the sale price is $15, giving a profit of $2 per unit. Since all 200 units are produced and sold, the profit is $400.

The second solution, using CONOPT, produces 100 units of X and no Y, using equal amounts of A and C. The average per-unit production

```
$TITLE   A BLENDING PROBLEM WITH POOLING
$OFFUPPER
* This problem finds an optimal allocation of stocks to blends
* when two stocks must be mixed prior to blending
*
   SETS
       ST   stocks              / A, B, C /
       BL   blends              / X, Y / ;

   PARAMETERS

       COST(ST)   per unit costs of stocks
       /   A      6
           B     16
           C     10    /

       SULF(ST)   sulfur percentages of stocks
       /   A      3
           B      1
           C      2     / ;

   TABLE  BDATA(BL,*)   blend data

                    PRICE        MAXSULF      MAXPROD
           X          9            2.5          100
           Y         15            1.5          200    ;


   POSITIVE VARIABLES
       USE(ST)    amount of each stock used
       PRD(BL)    amount of each blend produced
       CTOX
       CTOY
       POOLTOX
       POOLTOY
       POOLS        ;
   VARIABLES
       OBJ          ;

   EQUATIONS
       PROFIT        define objective function
       BALX          material balance for X
       BALY          material balance for Y
       BALPOOL       material balance for pool
       BALC          material balance for C
       SULFPOOL      sulfur balance for pool
       SULFLIMX      max sulfur percentage for X
       SULFLIMY      max sulfur percentage for Y     ;

   PROFIT ..     OBJ =E= SUM(BL, BDATA(BL,"PRICE")*PRD(BL))
                         -SUM(ST,COST(ST)*USE(ST))  ;

   BALX ..     PRD("X") =E= POOLTOX + CTOX ;
   BALY ..     PRD("Y") =E= POOLTOY + CTOY ;

   BALPOOL ..     USE("A") + USE("B") =E= POOLTOX + POOLTOY ;

   BALC ..     USE("C") =E= CTOX + CTOY ;
```

Figure 33   GAMS model of pooling problem.

```
   SULFPOOL ..     SULF("A")*USE("A") + SULF("B")*USE("B") =E=
                   POOLS*(USE("A") + USE("B") )    ;
   SULFLIMX ..     POOLS*POOLTOX + SULF("C")*CTOX =L=
                   BDATA("X","MAXSULF")*PRD("X")    ;
   SULFLIMY ..     POOLS*POOLTOY + SULF("C")*CTOY =L=
                   BDATA("Y","MAXSULF")*PRD("Y")    ;
*
*  Set Bounds, Including upper limits on Blends
*
   PRD.UP("X") = BDATA("X","MAXPROD") ;
   PRD.UP("Y") = BDATA("Y","MAXPROD") ;
   POOLS.LO = 1 ;   POOLS.UP = 3 ;

   MODEL POOL /ALL/ ;
*
*  First Solve-Goes to Global Optimum
*
   SOLVE POOL USING NLP MAXIMIZING OBJ ;
*
*  Second Solve- Obtains Local Optimum
*
   POOLS.L = 3 ;
   USE.L("A") = 50.0 ;  PRD.L("X") = 100.0; PRD.L("Y") = 0;
   USE.L("B") = 0; USE.L("C")=50;

   OPTION NLP = CONOPT ;

   SOLVE POOL USING NLP MAXIMIZING OBJ ;
*
*  Third Solve-Obtains Saddle Point
*
   POOLS.L = 2 ;
   USE.L(ST) = 0.0 ;  PRD.L(BL) = 0.0 ;
   CTOX.L=0; CTOY.L=0; POOLTOX.L=0; POOLTOY.L=0; OBJ.L=0;
   OPTION NLP = CONOPT ;

   SOLVE POOL USING NLP MAXIMIZING OBJ ;
```

cost is $(6 + 10)/2 = \$8$, and the sales price is \$9, yielding a profit of \$1 per unit of X sold. Since only 100 units are called for, the final profit is \$100. This solution is locally optimal; that is, small changes from this operating point reduce the profit. There are no feasible operating conditions close to this one that yield a better solution.

Our earlier solution, yielding a profit of \$400, is also a local optimum. However, there is no other feasible point with a larger profit, so the \$400 solution is a global optimum. The reader is invited to see if there are other local optima, for example, by increasing the use of A and decreasing B and C.

```
GAMS 2.25   386/486 DOS
A BLENDING PROBLEM WITH POOLING

MODEL STATISTICS

BLOCKS OF EQUATIONS        8        SINGLE EQUATIONS      8
BLOCKS OF VARIABLES        8        SINGLE VARIABLES     11
NON ZERO ELEMENTS         30        NON LINEAR N-Z        7
DERIVATIVE POOL            6        CONSTANT POOL         2
CODE LENGTH               51

GENERATION TIME      =      0.113 SECONDS
EXECUTION TIME       =      0.160 SECONDS         VERID MW2-00-037


                 S O L V E      S U M M A R Y

          MODEL    POOL               OBJECTIVE   OBJ
          TYPE     NLP                DIRECTION   MAXIMIZE
          SOLVER   MINOS5             FROM LINE   79

****  SOLVER STATUS      1 NORMAL COMPLETION
****  MODEL STATUS       2 LOCALLY OPTIMAL
****  OBJECTIVE VALUE          400.0000

      RESOURCE USAGE, LIMIT       0.220      1000.000
      ITERATION COUNT, LIMIT      4          1000
      EVALUATION ERRORS           0          0

          M I N O S    5.3    (Nov 1990)         Ver: 225-386-02

      EXIT -- OPTIMAL SOLUTION FOUND
      MAJOR ITNS, LIMIT              8        200
      FUNOBJ, FUNCON CALLS           0         22
      SUPERBASICS                    0
      INTERPRETER USAGE           0.00
      NORM RG / NORM PI      0.000E+00

                         LOWER       LEVEL       UPPER      MARGINAL

      ----  EQU PROFIT       .           .           .        -1.000
      ----  EQU BALX         .           .           .         9.000
      ----  EQU BALY         .           .           .        22.000
      ----  EQU BALPOOL      .           .           .       -16.000
      ----  EQU BALC         .           .           .       -10.000
      ----  EQU SULFPOOL     .           .           .         5.000
      ----  EQU SULFLIMX   -INF          .           .           .
      ----  EQU SULFLIMY   -INF          .           .         6.000

      ----  VAR USE        amount of each stock used

              LOWER       LEVEL       UPPER     MARGINAL

      A         .            .        +INF         .
      B         .         100.000     +INF         .
      C         .         100.000     +INF         .

      ----  VAR PRD        amount of each blend produced
```

Figure 34 Solution of GAMS pooling model.

Nonlinear Programming

	LOWER	LEVEL	UPPER	MARGINAL
X	.	.	100.000	.
Y	.	200.000	200.000	2.000

	LOWER	LEVEL	UPPER	MARGINAL
---- VAR CTOX	.	.	+INF	-1.000
---- VAR CTOY	.	100.000	+INF	.
---- VAR POOLTOX	.	.	+INF	-7.000
---- VAR POOLTOY	.	100.000	+INF	.
---- VAR POOLS	1.000	1.000	3.000	-100.000
---- VAR OBJ	-INF	400.000	+INF	.

Solution Report   SOLVE POOL USING NLP FROM LINE 90

```
                S O L V E      S U M M A R Y

    MODEL    POOL              OBJECTIVE  OBJ
    TYPE     NLP               DIRECTION  MAXIMIZE
    SOLVER   CONOPT            FROM LINE  90

**** SOLVER STATUS     1 NORMAL COMPLETION
**** MODEL STATUS      2 LOCALLY OPTIMAL
**** OBJECTIVE VALUE         100.0000

 RESOURCE USAGE, LIMIT      0.110       1000.000
 ITERATION COUNT, LIMIT     6           1000
 EVALUATION ERRORS          0           0


     C O N O P T    version 2.027
     Copyright (C)  ARKI Consulting and Development A/S
                    Bagsvaerdvej 246 A
                    DK-2880 Bagsvaerd, Denmark


 ** Optimal solution. There are no superbasic variables


 Function calls:      6     Gradient calls:      3
 CONOPT Time             0.110 (Interpreter    0.000)
```

	LOWER	LEVEL	UPPER	MARGINAL
---- EQU PROFIT	.	.	.	1.000
---- EQU BALX	.	.	.	18.000
---- EQU BALY	.	.	.	30.000
---- EQU BALPOOL	.	.	.	-6.000
---- EQU BALC	.	.	.	-10.000
---- EQU SULFPOOL	.	.	.	5.000
---- EQU SULFLIMX	-INF	.	.	4.000
---- EQU SULFLIMY	-INF	.	.	10.000

---- VAR USE   amount of each stock used

	LOWER	LEVEL	UPPER	MARGINAL

A	.	50.000	+INF	.	
B	.	.	+INF	.	
C	.	50.000	+INF	.	

---- VAR PRD   amount of each blend produced

	LOWER	LEVEL	UPPER	MARGINAL
X	.	100.000	100.000	1.000
Y	.	.	200.000	.

		LOWER	LEVEL	UPPER	MARGINAL
---- VAR CTOX		.	50.000	+INF	.
---- VAR CTOY		.	.	+INF	.
---- VAR POOLTOX		.	50.000	+INF	.
---- VAR POOLTOY		.	.	+INF	-6.000
---- VAR POOLS		1.000	3.000	3.000	50.000
---- VAR OBJ		-INF	100.000	+INF	.

Solution Report   SOLVE POOL USING NLP FROM LINE 99

S O L V E   S U M M A R Y

```
    MODEL    POOL              OBJECTIVE  OBJ
    TYPE     NLP               DIRECTION  MAXIMIZE
    SOLVER   CONOPT            FROM LINE  99

**** SOLVER STATUS     1 NORMAL COMPLETION
**** MODEL STATUS      2 LOCALLY OPTIMAL
**** OBJECTIVE VALUE              0.0000

    RESOURCE USAGE, LIMIT       0.110      1000.000
    ITERATION COUNT, LIMIT      3          1000
    EVALUATION ERRORS           0          0
```

C O N O P T   version 2.027

** Optimal solution. Reduced gradient less than tolerance.

```
Function calls:     2    Gradient calls:     1
CONOPT Time         0.060 (Interpreter   0.000)
```

	LOWER	LEVEL	UPPER	MARGINAL
---- EQU PROFIT	.	.	.	1.000
---- EQU BALX	.	.	.	10.000
---- EQU BALY	.	.	.	30.000
---- EQU BALPOOL	.	.	.	-11.000
---- EQU BALC	.	.	.	-10.000
---- EQU SULFPOOL	.	.	.	5.000
---- EQU SULFLIMX	-INF	.	.	.
---- EQU SULFLIMY	-INF	.	.	10.000

---- VAR USE   amount of each stock used

Figure 34 Continued.

Nonlinear Programming

```
          LOWER         LEVEL        UPPER      MARGINAL
A           .             .          +INF          .
B           .             .          +INF          .
C           .             .          +INF          .

---- VAR PRD            amount of each blend produced

          LOWER         LEVEL        UPPER      MARGINAL
X           .             .         100.000      -1.000
Y           .             .         200.000         .

                        LOWER       LEVEL       UPPER    MARGINAL
---- VAR CTOX             .           .         +INF        .
---- VAR CTOY             .           .         +INF        .
---- VAR POOLTOX          .           .         +INF      -1.000
---- VAR POOLTOY          .           .         +INF      -1.000
---- VAR POOLS          1.000       2.000       3.000      EPS
---- VAR OBJ            -INF          .         +INF        .
```

Generally speaking, an initial guess should not set variable values to zero. Because zero multiplied by any quantity is still zero, such values can lead to unusual behavior of the optimization algorithm. In this example, if our initial values are all zero but with POOLS = 2, the third SOLVE terminates at this point. Actually, this point is not even a local optimum, but it does satisfy the Kuhn-Tucker necessary conditions for an optimum. However, if the starting point is perturbed by some small amount, the system should find an actual local optimum and perhaps the global one. In fact, setting all variables previously at zero value to 0.1 does lead to the global maximum solution with profit of $400.

For this problem, all solutions obtained have the property that many constraints are active, in other words, hold as equalities. Of course, the five equality constraints are always active. In addition, in the globally optimal solution, all other constraints are at limits (eight in all), and six variables are either at lower or upper limits A, X, Y, CTOX, POOLTOX, and POOLS. Hence there are 14 active constraints but only 11 variables. When there are at least as many active constraints as there are variables, this is called a vertex solution. In linear programming, any LP having an optimal solution has a vertex optimum. This is not true in NLP, but vertex optima are not uncommon and seem to occur frequently in models involving blending and processing.

When there are more active constraints than variables, the vertex is called degenerate. In the global solution to this problem, there are three "extra" active constraints. One could be removed by dropping the upper and lower limits on POOLS; these are redundant, because they are implied by the SULFPOOL constraint and the nonnegativity of the variables. The lower limits on PRD("X") and PRD("Y") could also be dropped, because they are implied by rows 3 and 4 and the lower limits on CTOX, CTOY, POOLTOX, and POOLTOY. Doing this would lead to the same vertex solution, but with exactly as many active constraints as variables. (Some other constraints are redundant too; the reader is invited to find them.)

Degeneracy can sometimes cause difficulties for NLP algorithms. With GRG algorithms, the problems are very similar to those encountered in the simplex method for linear programming. At a degenerate vertex, one or more basic variables will be at a limit, and it may not be possible to take a step that improves the objective. Some variables at a bound then leave the basis, and (hopefully) the next step will not be "blocked" and progress will again be made. If a solver takes many degenerate steps or stops at a degenerate point that is not optimal, the problem can often be alleviated by seeking redundant constraints and removing them from the model.

The other solution to this problem is also a vertex solution. Verify this, and see if it too is degenerate.

5.4. Alkylation Process Optimization

Crude oil and natural gas as they come from the ground contain a variety of hydrocarbon molecules. Some of the hydrocarbons such as methane, ethane, propane, and butane contain only one to four carbon atoms per molecule, whereas others contain a dozen or more. For the purposes of making gasoline, one of the most desirable hydrocarbons is octane. It contains eight carbon atoms per molecule. An efficient gasoline refinery will "resculpt" some of the undesirable molecules into more desirable ones. For example, molecules with too few carbon atoms will be combined or formed into larger molecules. Molecules with too many carbon atoms will be split or cracked into smaller ones. This reshaping of molecules is done by applying temperature and pressure to the undesirable molecules in the presence of other facilitating chemicals. One such reshaping process is alkylation.

Figure 35 shows a simplified alkylation process model similar to those common in the petroleum industry. The model consists of a reactor and a fractionator. Olefins and isobutane are fed into the reactor, along with fresh acid. The acid is used to catalyze the chemical reactions. The hydrocarbon reactor product is fed to the fractionator.

The fractionating column separates the final alkylate product, which is withdrawn form the bottom of the fractionator, from unreacted isobutane, which is drawn from the top of the fractionator and recycled to the reactor. Total isobutane input to the reactor consists of the recycled isobu-

Nonlinear Programming

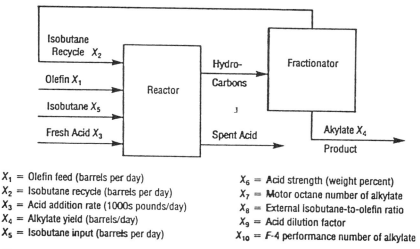

X_1 = Olefin feed (barrels per day)
X_2 = Isobutane recycle (barrels per day)
X_3 = Acid addition rate (1000s pounds/day)
X_4 = Alkylate yield (barrels/day)
X_5 = Isobutane input (barrels per day)
X_6 = Acid strength (weight percent)
X_7 = Motor octane number of alkylate
X_8 = External isobutane-to-olefin ratio
X_9 = Acid dilution factor
X_{10} = F-4 performance number of alkylate

Figure 35 Alkylation process.

tane along with fresh isobutane that is added to maintain a specified isobutane-to-olefin ratio for the reactor input.

The following simplifying assumptions are made:

1. The olefin feed is pure butylene.
2. Isobutane input and isobutane recycle are pure isobutane.
3. Fresh acid strength is 98% by weight.
4. Spent acid is withdrawn from the reactor and discarded rather than recycled.

The variables for the problem are also shown in Figure 35. There are 10 variables of interest, three of which, X_1 (olefin feed), X_2 (isobutane recycle), and X_3 (fresh acid), are considered the controllable (independent) variables. These three variables can be adjusted by the operator by changing set points on automatic control instruments. Such changes, of course, affect the other (dependent) variables. The objective is to determine values for the independent variables that maximize profit. Note that this simplified model does not account for other variables such as the relative humidity of the outside air and the temperature of cooling water in the process.

To complete our model of this system we must now determine the equations relating the dependent variables to the three independent variables. Some of these equations are easy to derive, for example, those representing volume balance or definitional quantities. Others are more complicated and analytic relations are not available. In these cases we resort to experimentation and use regression analysis to fit polynomial expressions to measured data.

The alkylate yield, X_4, is a function of both the olefin feed, X_1, and the external isobutane-to-olefin ratio, X_8. Nonlinear regression analysis, for measurements obtained by holding reactor temperature between 80 and 90°F and reactor acid strength by weight between 85 and 93%, gives

$$X_4 = X_1(1.12 + 0.13167X_8 - 0.00667X_8^2)$$

The motor octane number of the alkylate, X_7, is a function of the external isobutane-to-olefin ratio, X_8, and the acid strength by weight percent, X_6. Its nonlinear regression equation, determined under the same operating conditions as for alkylate yield, is

$$X_7 = 86.35 + 1.098X_8 - 0.038X_8^2 + 0.325(X_6 - 89)$$

Analysis shows that the acid dilution factor, X_9, can be adequately expressed as a linear function of the F-4 performance number, X_{10}. This linear regression equation is

$$X_9 = 35.82 - 0.222X_{10}$$

Similarly, the F-4 performance, X_{10}, can be expressed as a linear function of the motor octane number, X_7, by the equation

$$X_{10} = 3X_7 - 133$$

Turning now to exact relationships, the external isobutane-to-olefin ratio is simply

$$X_8 = \frac{X_2 + X_5}{X_1}$$

The isobutane input, X_5, can be determined by a straightforward volume balance for the system. The input (in barrels per day) is $X_1 + X_5$ and the output is X_4 barrels per day. However, unlike mass, the resulting volume for liquids mixed together in the presence of chemical reactions can be less than the sum of the volumes of the inputs. This volumetric shrinkage is constant for our system, and the output volume loss is 22% of the alkylate output volume. Thus:

$$X_4 = X_1 + X_5 - 0.22X_4$$

or

$$X_5 = 1.22X_4 - X_1$$

The acid strength by weight percent is related to the acid addition rate, the acid dilution factor, and the alkylate yield by the equation

$$1000 X_3 = \frac{X_4 X_6 X_9}{(98 - X_6)}$$

where we have assumed that the fresh acid has an acid strength of 98. This yields

$$X_6 = \frac{98{,}000 X_3}{X_4 X_9 + 1000 X_3}$$

on rearrangement.

The final step in model formulation is to determine the profit function. We assume that the profit depends on the value of the output product minus the cost of the inputs and that all other operating costs are independent of our variables and can be ignored. [It should be noted that this model formulation takes the reactor and fractionator operating conditions (temperature and pressure) and characteristics as given and constant over the range of feed rates used to develop the regression equations.] The specific cost figures used in the profit function are as follows:

Alkylate product value = $0.063 per octane barrel
Olefin feed cost = $5.04 per barrel
Isobutane feed cost = $3.36 per barrel
Isobutane recycle cost = $0.035 per barrel
Acid addition cost = $10.00 per barrel

Hence the profit function is

$$0.063 X_4 X_7 - 5.04 X_1 - 0.035 X_2 - 10 X_3 - 3.36 X_5$$

A GAMS model of this problem is shown in Figure 36. It uses names for decision variables, e.g., OLEFIN and ACID. The first eight equations, from YIELD to DPROFIT, are identical to the algebraic model just presented except for some minor restatement of the equations. Ignore the equations whose names start with RNG for now; these will be explained shortly. Our current model is named PROCESS in the MODEL statement, which also defines a model called RPROC, to be discussed below. The block of statements starting with STRENGTH.LO = 85 and ending with ISOM.UP = 2000 define upper and lower bounds for the variables, and the set of statements starting with OLEFIN.L = 1745 define initial values. In GAMS, any variable, say X, has four attributes: X.LO and X.UP are its lower and upper bounds, X.L sets its level or value (to be adjusted by

```
$TITLE ALKYLATION PROCESS OPTIMIZATION   (PROCESS,SEQ=20)
*
* Turn off equation and cross reference listings
*
$OFFSYMXREF
$OFFSYMLIST
*
* OPTIMIZATION OF A ALKYLATION PROCESS.
*
* REFERENCE: BRACKEN J AND MC CORMICK I P, SELECTED APPLICATIONS OF
*            NONLINEAR PROGRAMMING, JOHN WILEY AND SONS, NEW YORK, 1968.
*            (CHAPTER 4)

 POSITIVE
 VARIABLES  OLEFIN      OLEFIN FEED (BPD)
            ISOR        ISOBUTANE RECYCLE (BPD)
            ACID        ACID ADDITION RATE (1000LB PER DAY)
            ALKYLATE    ALKYLATE YIELD (BPD)
            ISOM        ISOBUTANE MAKEUP (BPD)
            STRENGTH    ACID STRENGTH (WEIGHT PCT)
            OCTANE      MOTOR OCATANE NUMBER
            RATIO       ISOBUTANE MAKEUP TO OLEFIN RATIO
            DILUTE      ACID DILUTION FACTOR
            F4          F-4 PERFORMANCE NUMBER
 VARIABLES  PROFIT, RANGEY, RANGEM, RANGED, RANGEF

 EQUATIONS  YIELD       ALKYLATE YIELD DEFINITION
            RNGYIELD    RANGED ALKYLATE YIELD DEFINITION
            MAKEUP      ISOBUTANE MAKEUP DEFINITION
            SDEF        ACID STRENGTH DEFINITION
            MOTOR       MOTOR OCTANE NUMBER
            RNGMOTOR    RANGED MOTOR OCTANE NUMBER
            DRAT        ISOBUTANE TO OLEFIN RATIO
            DDIL        DILUTION DEFINITION
            RNGDDIL     RANGED DILUTION DEFINITION
            DF4         F-4 DEFINITION
            RNGDF4      RANGED F-4 DEFINITION
            DPROFIT     PROFIT DEFINITION;

 YIELD..   ALKYLATE =E= OLEFIN*(1.12+.13167*RATIO-.00667*SQR(RATIO)) ;

 MAKEUP..  ALKYLATE =E= OLEFIN + ISOM -.22*ALKYLATE ;

 SDEF..    ACID =E= ALKYLATE*DILUTE*STRENGTH/(98-STRENGTH)/1000 ;

 MOTOR..   OCTANE =E= 86.35 + 1.098*RATIO -.038*SQR(RATIO) -.325*(89-STRENGTH);

 DRAT..    RATIO =E= (ISOR+ISOM)/OLEFIN ;

 DDIL..    DILUTE =E= 35.82 - .222*F4 ;

 DF4..     F4 =E= -133 + 3*OCTANE ;

 DPROFIT.. PROFIT =E= .063*ALKYLATE*OCTANE - 5.04*OLEFIN - .035*ISOR - 10*ACID
                     -3.36*ISOM;
*
* Ranged Model Equations
*
 RNGYIELD.. RANGEY*ALKYLATE =E= OLEFIN*(1.12+.13167*RATIO-.00667*SQR(RATIO)) ;
```

Figure 36 GAMS model of alkylation problem.

the optimization algorithm), and X.M is the marginal value or Lagrange multiplier associated with X.

In Figure 36, the statement SOLVE PROCESS MAXIMIZING PROFIT USING NLP solves the model called PROCESS using whatever NLP solver has been designated. Here none is designated, so the GAMS

Nonlinear Programming

```
RNGMOTOR..  RANGEM*OCTANE =E= 86.35 + 1.098*RATIO - .038*SQR(RATIO)
                              -.325*(89-STRENGTH);
RNGDDIL..   RANGED*DILUTE =E= 35.82 - .222*F4 ;

RNGDF4..    RANGEF*F4 =E= -133 + 3*OCTANE ;
*
* Model With Regression Equations as Equalities
*
MODEL PROCESS PROCESS MODEL WITH EQUALITIES / YIELD, MAKEUP, SDEF, MOTOR,
                                             DRAT,DDIL, DF4, DPROFIT / ;
*
* Model with Regression Equations as Ranges
*
MODEL RPROC    RANGED PROCESS MODEL  / RNGYIELD, MAKEUP, SDEF, RNGMOTOR,
                                      DRAT,RNGDDIL,RNGDF4, DPROFIT / ;
*
* Bounds and Initial Values for RANGE Variables
*
RANGEY.LO = .99; RANGEY.UP = 1.01; RANGEY.L = 1;
RANGEM.LO = .99; RANGEM.UP = 1.01; RANGEM.L = 1;
RANGED.LO = .9; RANGED.UP = 1.1; RANGED.L = 1;
RANGEF.LO = .99; RANGEF.UP = 1.01; RANGEF.L = 1;
*
* Bounds and Initial Values for other Variables
*
STRENGTH.LO = 85;   STRENGTH.UP = 93;  OCTANE.LO = 90;   OCTANE.UP = 95;
RATIO.LO = 3;       RATIO.UP = 12;     DILUTE.LO = 1.2;  DILUTE.UP = 4;
F4.LO = 145;  F4.UP = 162;
OLEFIN.LO = 10; OLEFIN.UP = 2000; ISOR.UP = 16000; ACID.UP = 120;
ALKYLATE.UP = 5000; ISOM.UP = 2000;

OLEFIN.L = 1745; ISOR.L = 12000; ACID.L = 110; ALKYLATE.L = 3048;
ISOM.L = 1974; STRENGTH.L = 89.2; OCTANE.L = 92.8;
RATIO.L = 8; DILUTE.L = 3.6; F4.L = 145; PROFIT.L = 872;

SOLVE PROCESS MAXIMIZING PROFIT USING NLP;

SOLVE RPROC MAXIMIZING PROFIT USING NLP;
```

default NLP solver, MINOS, has been used. The statement OPTION NLP = CONOPT would cause the solver named CONOPT (a GRG code) to be used instead. This illustrates a major advantage of modeling languages — the ability to change solvers with minimal effort. This is especially useful in nonlinear optimization, where any solver can fail to solve a given problem. One can simply try another solver, starting from either the same initial point or the final point produced by the solver that failed, by adding OPTION and SOLVE statements.

The solution produced by GAMS using MINOS to solve the model PROCESS is shown in the first part of Figure 37. The maximum profit is 1161.33 $/day. Note, however, the large values for the Marginal values (i.e., Lagrange multipliers) for rows MOTOR, DDIL, and DF4. These rows, along with row YIELD, are the regression relations. The large associated Marginals indicate that relaxing these constraints could pay large dividends in terms of increased profit. Because these equations are "best fit"

```
GAMS 2.25   386/486 DOS
ALKYLATION PROCESS OPTIMIZATION   (PROCESS,SEQ=20)
MODEL STATISTICS

BLOCKS OF EQUATIONS          8     SINGLE EQUATIONS       8
BLOCKS OF VARIABLES         11     SINGLE VARIABLES      11
NON ZERO ELEMENTS           28     NON LINEAR N-Z        11
DERIVATIVE POOL              7     CONSTANT POOL         10
CODE LENGTH                165

GENERATION TIME     =    0.168 SECONDS
EXECUTION TIME      =    0.438 SECONDS           VERID MW2-00-037

STEP SUMMARY:      0.109 STARTUP
                   0.113 COMPILATION
                   0.438 EXECUTION
                   0.051 CLOSEDOWN
                   0.711 TOTAL SECONDS
Solution Report      SOLVE PROCESS USING NLP FROM LINE 91

              S O L V E      S U M M A R Y
      MODEL    PROCESS            OBJECTIVE  PROFIT
      TYPE     NLP                DIRECTION  MAXIMIZE
      SOLVER   MINOS5             FROM LINE  91

**** SOLVER STATUS       1 NORMAL COMPLETION
**** MODEL STATUS        2 LOCALLY OPTIMAL
**** OBJECTIVE VALUE            1161.3366

RESOURCE USAGE, LIMIT         0.440       1000.000
ITERATION COUNT, LIMIT          45           1000
EVALUATION ERRORS                0              0

     M I N O S    5.3    (Nov 1990)       Ver: 225-386-02
     = = = = =

     B. A. Murtagh, University of New South Wales
        and
     P. E. Gill, W. Murray, M. A. Saunders and M. H. Wright
     Systems Optimization Laboratory, Stanford University.

Work space allocated            --       0.04 Mb

EXIT -- OPTIMAL SOLUTION FOUND
MAJOR ITNS, LIMIT              13         200
FUNOBJ, FUNCON CALLS           96          96
SUPERBASICS                     1
INTERPRETER USAGE            0.16
NORM RG / NORM PI       7.041E-08

                        LOWER      LEVEL      UPPER     MARGINAL

---- EQU YIELD             .          .          .        1.228
---- EQU MAKEUP            .          .          .        3.594
---- EQU SDEF              .          .          .      -10.000
---- EQU MOTOR          57.425     57.425     57.425    442.373
---- EQU DRAT              .          .          .      120.383
---- EQU DDIL           35.820     35.820     35.820   -375.096
---- EQU DF4          -133.000   -133.000   -133.000     83.271
```

Figure 37 Solution of GAMS alkylation model.

Nonlinear Programming

```
----  EQU DPROFIT         .          .          .        -1.000

                       LOWER       LEVEL       UPPER    MARGINAL

----  VAR OLEFIN      10.000    1728.921    2000.000        .
----  VAR ISOR            .    16000.000   16000.000      0.035
----  VAR ACID            .       98.161     120.000        .
----  VAR ALKYLATE        .     3056.493    5000.000        .
----  VAR ISOM            .     2000.000    2000.000      0.304

                       LOWER       LEVEL       UPPER    MARGINAL

----  VAR STRENGTH    85.000      90.616      93.000        .
----  VAR OCTANE      90.000      94.188      95.000        .
----  VAR RATIO        3.000      10.411      12.000        .
----  VAR DILUTE       1.200       2.617       4.000        .
----  VAR F4         145.000     149.563     162.000       EPS
----  VAR PROFIT       -INF     1161.337       +INF         .


****  REPORT SUMMARY :      0    NONOPT
                            0    INFEASIBLE
                            0    UNBOUNDED
                            0    ERRORS
MODEL STATISTICS

BLOCKS OF EQUATIONS       8     SINGLE EQUATIONS       8
BLOCKS OF VARIABLES      15     SINGLE VARIABLES      15
NON ZERO ELEMENTS        32     NON LINEAR N-Z        19
DERIVATIVE POOL           7     CONSTANT POOL         10
CODE LENGTH             201

GENERATION TIME     =      0.168 SECONDS
EXECUTION TIME      =      0.219 SECONDS          VERID MW2-00-037

STEP SUMMARY:     0.109 STARTUP
                  0.000 COMPILATION
                  0.219 EXECUTION
                  0.051 CLOSEDOWN
                  0.379 TOTAL SECONDS
Solution Report     SOLVE RPROC USING NLP FROM LINE 93

                S O L V E      S U M M A R Y

        MODEL    RPROC              OBJECTIVE   PROFIT
        TYPE     NLP                DIRECTION   MAXIMIZE
        SOLVER   MINOS5             FROM LINE   93

****  SOLVER STATUS      1  NORMAL COMPLETION
****  MODEL STATUS       2  LOCALLY OPTIMAL
****  OBJECTIVE VALUE            1763.3466

RESOURCE USAGE, LIMIT      0.270     1000.000
ITERATION COUNT, LIMIT       25         1000
EVALUATION ERRORS             0            0

     M I N O S    5.3     (Nov 1990)     Ver: 225-386-02
     = = = = =
```

B. A. Murtagh, University of New South Wales
and
P. E. Gill, W. Murray, M. A. Saunders and M. H. Wright
Systems Optimization Laboratory, Stanford University.

```
EXIT -- OPTIMAL SOLUTION FOUND
MAJOR ITNS, LIMIT            7       200
FUNOBJ, FUNCON CALLS        51        51
SUPERBASICS                  1
INTERPRETER USAGE         0.05
NORM RG / NORM PI    5.798E-09
```

	LOWER	LEVEL	UPPER	MARGINAL
---- EQU RNGYIELD	.	.	.	0.680
---- EQU MAKEUP	.	.	.	4.207
---- EQU SDEF	.	.	.	-10.000
---- EQU RNGMOTOR	57.425	57.425	57.425	231.208
---- EQU DRAT	.	.	.	59.443
---- EQU RNGDDIL	35.820	35.820	35.820	-314.431
---- EQU RNGDF4	-133.000	-133.000	-133.000	70.509
---- EQU DPROFIT	.	.	.	-1.000

	LOWER	LEVEL	UPPER	MARGINAL
---- VAR OLEFIN	10.000	1698.296	2000.000	.
---- VAR ISOR	.	15843.614	16000.000	EPS
---- VAR ACID	.	54.559	120.000	.
---- VAR ALKYLATE	.	3031.390	5000.000	.
---- VAR ISOM	.	2000.000	2000.000	0.882

	LOWER	LEVEL	UPPER	MARGINAL
---- VAR STRENGTH	85.000	90.103	93.000	.
---- VAR OCTANE	90.000	95.000	95.000	173.608
---- VAR RATIO	3.000	10.507	12.000	.
---- VAR DILUTE	1.200	1.577	4.000	.
---- VAR F4	145.000	153.535	162.000	.
---- VAR PROFIT	-INF	1763.347	+INF	.
---- VAR RANGEY	0.990	0.990	1.010	-2060.413
---- VAR RANGEM	0.990	0.990	1.010	-2.196E+4
---- VAR RANGED	0.900	1.100	1.100	495.987
---- VAR RANGEF	0.990	0.990	1.010	-1.083E+4

```
**** REPORT SUMMARY :     0     NONOPT
                          0     INFEASIBLE
                          0     UNBOUNDED
                          0     ERRORS

EXECUTION TIME      =     0.059 SECONDS    VERID MW2-00-037
```

Figure 37 Continued.

Nonlinear Programming

solutions to measurements, it is wise to examine the accuracy of both the measurements and the fit to determine how much these constraints could be relaxed.

Assume that such an analysis indicates that rows YIELD, MOTOR, and DF4 are quite good but that equation DDIL is not a tight fit. Accordingly, the problem is reformulated with these regression equations redefined as range rows, using new variables RANGEY, RANGEM, RANGED, and RANGEF. Referring to Figure 36 again, examine the equations RNGYIELD through RNGDF4, which are the modified regression equations. The four variables whose names begin with RANGE have upper and lower bounds and initial values assigned in the group of statements starting with RANGEY.LO = 0.99. These variables multiply the left-hand sides of the four equations whose names start with RNG. This means that, for example, the right-hand side of the RNGYIELD equation can take on a value anywhere in the range .99 · ALKYLATE to 1.01 * ALKYLATE. The three tight-fitting regression equations have been relaxed within narrow ranges in this way, and the loose-fitting equation DDIL has the wide range .9 * DILUTE to 1.1 * DILUTE.

Although this is a compact way to relax an equality into a range, it does introduce an additional nonlinearity (the products RANGEY * DILUTE, ETC.) that was not present originally. Taking the F4 equation as an example, the direct approach would be to rewrite it as .99 * F4 ≤ −133 − 3 * OCTANE ≤ 1.01 * F4. This formulation is recommended especially when the original model is difficult to solve.

The solution to this ranged model is in the second half of Figure 37, right after the first STEP SUMMARY. The maximum profit has risen from 1161 to 1763 $/day, a major increase. However, this profit may not be achievable, just potentially achievable depending on the true form of the regression equations. This analysis is still very useful, because it indicates that more accurate modeling of the process could yield significant additional profits.

APPENDIX A.1: GRG2 CALLING PROGRAM AND OUTPUT

The code below represents the function evaluation routine for GRG2 (GCOMP) together with a main program that calls GRG2 through its callable interface with problem specifications for the nonlinear regression example of Section 5.1. Solved here is the least squares version of the problem started from the optimal solution of the least absolute value form.

```
cc----------------------------------------------------------------
cc   This program calls GRG2 through its callable interface
cc   The problem is the nonlinear regression example from this
cc   chapter. Here we solve the least squares form of the problem
cc----------------------------------------------------------------
      program nlpdemo
      implicit none
      integer maxvar,maxrow,maxz
      parameter(maxvar=100,maxrow=100,maxz=10000)
cc----------------------------------------------------------------
cc    GRGSUB argument declarations
cc----------------------------------------------------------------
      integer ncore,nnvars,nfun,maxbas,maxhes,nnobj,
     *   nnstop,iitlim,llmser,iipr,iipn4,iipn5,
     *   iipn6,iiper,iidump,iiquad,lderiv,mmodcg,nbind,
     *   nnonb,inform,inbind(maxrow),nonbas(maxvar),i
cc
      double precision blvar(maxvar),buvar(maxvar),blcon(maxrow),
     *   bucon(maxrow),xx(maxvar),fcns(maxrow),rmults(maxrow),
     *   redgr(maxvar),defaul(20),z(maxz),ttitle(19),blank,big,
     *   fpnewt,fpinit,fpstop,pphlep,fpspiv
cc
      logical inprnt,otprnt
cc
      character*8 ramcon(maxrow),ramvar(maxvar)
cc
      character*76 atitle
      character*4  ttitle(19)
      data atitle/'Nonlinear Regression Example'/
      data blank/'        '/
cc----------------------------------------------------------------
cc   big is the value for +infinity and is used to set
cc   infinite bounds on variables and functions
cc----------------------------------------------------------------
      data big/1.D31/
cc----------------------------------------------------------------
cc   blank out name arrays
cc----------------------------------------------------------------
      read(atitle,'(19a4)') ttitle
cc
      do i=1,maxvar
        ramvar(i) = blank
      enddo
      do i=1,maxrow
        ramcon(i) = blank
      enddo
cc----------------------------------------------------------------
cc   set up problem dimensions and objective identification
cc----------------------------------------------------------------
      NCORE  = maxz
      NNVARS = 3
      NFUN   = 1
      NNOBJ  = 1
      MAXBAS = 3
      MAXHES = 3
CC
CC  SET VARIABLE BOUNDS
CC
      do i=1,nnvars
        BLVAR(i) = -big
        BUVAR(i) =  big
      enddo
CC
CC SET ROW BOUNDS
```

Nonlinear Programming

```
cc
      BLCON(1) = -big
      BUCON(1) =  big
cc
cc    SET INTIAL VALUES
cc
      XX(1) =  490.091
      XX(2) =  -73.947
      XX(3) =   -0.3126
cc------------------------------------------------------------
cc    set grgsub user options
cc    . inprnt/otprnt control tabular output on entry/exit
cc    . defaul controls whether default (1.0) or user specified
cc           value (defaul(i) not = 1.0) will be used for ith
cc           user specifiable option (see GRGSUB documentation)
cc------------------------------------------------------------
      INPRNT = .TRUE.
      OTPRNT = .TRUE.
cc
      DO 10 I=1,19
10    DEFAUL(I) = 1.0
cc
cc    set GRG2 print level
cc
      defaul(9) = 0.0
      iipr      = 1
cc
cc    set derivative method (0 = forward differences)
cc
      defaul(16) = 0.0
      lderiv     = 1
cc------------------------------------------------------------
cc    GRG2 writes its output to FORTRAN unit 6
cc    on systems where it is more convienent to open files
cc    explicitly within the program (PC's and VMS, etc.) , the open
cc    statement below is useful.  On many mainframes, the assignment
cc    of logical unit 6 to a physical file or device is more easily
cc    done externally via the system command language.
cc------------------------------------------------------------
      OPEN(UNIT = 6,FILE = 'nlpdemo.out')
cc------------------------------------------------------------
cc    call grgsub to solve problem
cc------------------------------------------------------------
      CALL GRGSUB(INPRNT,OTPRNT,NCORE,NNVARS,NFUN,MAXBAS,
     * MAXHES,NNOBJ,TTITLE,BLVAR,BUVAR,BLCON,BUCON,DEFAUL,FPNEWT,FPINIT,
     * FPSTOP,FPSPIV,PPH1EP,NNSTOP,IITLIM,LLMSER,IIPR,IIPN4,IIPN5,
     * IIPN6,IIPER,IIDUMP,IIQUAD,LDERIV,MMODCG,
     * RAMCON,RAMVAR,XX,FCNS,INBIND,RMULTS,NONBAS,REDGR,
     * NBIND,NNONB,INFORM,Z)
cc
      STOP
      END
      subroutine gcomp(g,b)
      implicit none
cc------------------------------------------------------------
cc    this gcomp computes the least squares objective for the
cc    nonlinear regression example of section 5.1
cc
cc    the approximating function is of the form
cc    y = b1 + b2 * exp( b3 * x)
cc
cc    independent variables are unknown parameters b
```

```
cc------------------------------------------------------------------
      integer nbr_points
      parameter(nbr_points=6)
cc
      double precision g(*),b(*),x_observed(nbr_points),
     *                 y_observed(nbr_points),predicted,error
cc
      integer i
cc
      data x_observed/ -5.0, -3.0, -1.0,  1.0,  3.0 ,5.0/
      data y_observed/127.0,151.0,379.0,426.0,460.0,421.0/
cc------------------------------------------------------------------
      g(1) = 0.0
      do i=1,6
          predicted = b(1) + b(2) * exp(b(3) * x_observed(i))
          error     = y_observed(i) - predicted
          g(1) = g(1)  + error * error
      enddo
cc
      return
      end
```

Below is the contents of the output generated by GRG2 when called by the main program above.

```
NUMBER OF VARIABLES IS      3
NUMBER OF FUNCTIONS IS      1
SPACE RESERVED FOR HESSIAN HAS DIMENSION    3
LIMIT ON BINDING CONSTRAINTS IS    1
ACTUAL LENGTH OF Z ARRAY IS     87
    0.4900910E+03 -0.7394700E+02 -0.3126000E+00
EPNEWT = 0.1000E-03 EPINIT = 0.1000E-03 EPSTOP = 0.1000E-03 EPPIV = 0.1000E-02 PH1EPS = 0.00000E+00
NSTOP =    3  ITLIM =   10  LIMSER =    10000
IPR =    1  PN4 =    0  PN5 =    0  PN6 =    0  PER =    0  DUMP =    0
TANGENT VECTORS WILL BE USED FOR INITIAL ESTIMATES     BASIC VARIABLES
THE FINITE DIFFERENCE PARSH USING CENTRAL DIFFERENC    WILL BE USED
OBJECTIVE FUNCTION WILL BE MINIMIZED.
LIMIT ON HESSIAN IS    3


                        OUTPUT OF INITIAL VALUES


Nonlinear Regression Example

SECTION 1 -- FUNCTIONS
      FUNCTION                   INITIAL        LOWER        UPPER
NO.   NAME     STATUS  TYPE      VALUE          LIMIT        LIMIT
 1                     OBJ     2.5738121E+04



SECTION 2 -- VARIABLES
      VARIABLE          INITIAL         LOWER           UPPER
NO.   NAME     STATUS   VALUE           LIMIT           LIMIT
 1                    4.9009100E+02  -1.0000000E+31  1.0000000E+31
 2                   -7.3946999E+01  -1.0000000E+31  1.0000000E+31
 3                   -3.1259999E-01  -1.0000000E+31  1.0000000E+31


ITERATION  OBJECTIVE    NO.BINDING   NO.SUPER-  NUMBER      NORM RED.   HESSIAN      HESSIAN                DEGENERATE
NUMBER     FUNCTION     CONSTRAINTS  BASICS     INFEASIBLE  GRADIENT    CONDITION    UPDATE    STEPSIZE     STEP
   0       2.573812E+04     0           3         0         1.984E+05   1.000E+00      F       0.000E+00
   1       2.298267E+04     0           3         0         1.984E+05   1.000E+00      F       1.260E-07
   2       1.543715E+04     0           3         0         1.033E+04   3.065E+04      T       5.557E-02
   3       1.493508E+04     0           3         0         7.224E+04   4.148E+04      T       1.377E+00
   4       1.440973E+04     0           3         0         1.284E+04   3.961E+04      T       6.594E-01
   5       1.414525E+04     0           3         0         3.518E+04   4.488E+04      T       2.684E+00
   6       1.410506E+04     0           3         0         2.312E+04   7.754E+04      T       6.689E-01
```

```
     7      1.408914E+04     0       3       0    3.702E+03  7.580E+04   T    8.301E-01
     8      1.408519E+04     0       3       0    2.102E+03  5.160E+04   T    1.990E+00
     9      1.408514E+04     0       3       0    9.969E+02  5.464E+04   T    8.952E-01
    10      1.408514E+04     0       3       0    1.004E+01  5.455E+04   T    1.049E+00
    11      1.408514E+04     0       3       0    4.721E-01  5.080E+04   T    1.036E+00

TOTAL FRACTIONAL CHANGE IN OBJECTIVE LESS THAN    1.00000E-04 FOR    3 CONSECUTIVE ITERATIONS
1                     FINAL RESULTS

Nonlinear Regression Example
SECTION 1 -- FUNCTIONS

                                                     DISTANCE
                    INITIAL      FINAL                 FROM          LAGRANGE
NO.    NAME         VALUE        VALUE      STATUS    NEAREST        MULTIPLIER
                                                      BOUND
  1                 2.57381E+04  1.40851E+04  OBJ

SECTION 2 -- VARIABLES

                                                     DISTANCE
                    INITIAL      FINAL                 FROM          REDUCED
NO.    NAME         VALUE        VALUE      STATUS    NEAREST        GRADIENT
                                                      BOUND
  1                 4.90091E+02  5.16651E+02 SUPBASIC  NO BOUND      -1.80842E-05
  2                -7.39470E+01 -1.49352E+02 SUPBASIC  NO BOUND      -8.33978E-06
  3                -3.12600E-01 -2.06643E-01 SUPBASIC  NO BOUND       3.48921E-03


RUN STATISTICS

Nonlinear Regression Example

NUMBER OF ONE-DIMENSIONAL SEARCHES =     11

NEWTON CALLS =    0   NEWTON ITERATIONS =    0   AVERAGE =   0.00

FUNCTION CALLS =    64   GRADIENT CALLS =    12

  ACTUAL FUNCTION CALLS (INC. FOR GRADIENT) =    136

NUMBER OF TIMES BASIC VARIABLE VIOLATED A BOUND =    0

NUMBER OF TIMES NEWTON FAILED TO CONVERGE =    0

TIMES STEPSIZE CUT BACK DUE TO NEWTON FAILURE =    0
```

REFERENCES

Abadie, J., Carpentier, J., "Generalization of the Wolfe Reduced Gradient Method to the Case of Nonlinear Constraints," in *Optimization*, ed. R. Fletcher, pp. 37–47. Academic Press, New York, 1969.

Baker, T. E., Lasdon, L. S., "Successive Linear Programming at Exxon," *Management Science*, Vol. 31, pp. 264–274, 1985.

Beale, E. M. L., "Nonlinear Programming Using a General Mathematical Programming System," in *Design and Implementation of Optimization Software*, ed. H. J. Greenberg, pp. 259–281. Sijthoff and Noordhoff, Holland, 1978.

Berna, T. J., Locke, M. H., and Westerberg, A. W., "A New Approach to Optimization of Chemical Processes," *AIChE J.*, Vol. 26, pp. 37–43, 1980.

Boddington, C. E. "Nonlinear Programs for Product Blending," Joint National TIMS/ORSA meeting, New Orleans, April/May 1979.

Brooke, A., et al., *GAMS: A User's Guide*. Boyd and Fraser, Danvers, MA, 1992.

Buzby, B. R., "Techniques and Experience Solving Really Big Nonlinear Programs," in *Optimization Methods*, eds. R. W. Cottle and J. Krarup. English Universities Press, London, 1974.

Conn, A. R., "Penalty Function Methods," in *Nonlinear Optimization, 1981*, ed. M. J. D. Powell. Academic Press, New York, 1982.

Dennis, J. E., Schnabel, R. F., *Numerical Methods for Unconstrained Optimization and Nonlinear Equations*. Prentice-Hall, Englewood Cliffs, NJ, 1983.

Drud, A. "CONOPT—A GRG-Code for Large Sparse Dynamic Nonlinear Optimization Problems," *Mathematical Programming*, Vol. 31, pp. 153–191, 1985.

Fan, Y., Sarkar, S., Lasdon, L., "Experiments with Successive Quadratic Programming Algorithms," *Journal of Optimization Theory and Applications*, Vol. 56, No. 3, pp. 359–383, 1988.

Fletcher, R., Reeves, C. M., "Function Minimization by Conjugate Gradients," *British Computer Journal*, Vol. 7, pp. 149–154, 1964.

Fourer, R., Gay, G., Kernighan, B., *AMPL: A Modeling Language for Mathematical Programming*. Boyd and Fraser, Danvers, MA, 1993.

Griffith, R. E., Stewart, R. A. "A Nonlinear Programming Technique for the Optimization of Continuous Processing Systems," *Management Science*, Vol. 7, pp. 379–392, 1961.

Han, S. P., Mangasarian, O. L., "Exact Penalty Functions in Nonlinear Programming," *Mathematical Programming*, Vol. 17, pp. 251–269, 1979.

Haverly, C. A., "Studies of the Behavior Recursion for the Pooling Problem," *SIGMAP Bulletin*, No. 25, pp. 19-28.

Lasdon, L. S., Waren, A. D., "Generalized Reduced Gradient Software for Linearly and Nonlinearly Constrained Problems," in *Design and Implementation of Optimization Software*, ed. H. J. Greenberg, pp. 363–397. Sijthoff and Noordhoff, Holland, 1978.

Lasdon, L., Waren A., et al., "Design and Testing of a Generalized Relaxed Gradient Code for Nonlinear Programming," *ACM Transactions on Mathematical Software*, Vol. 4, No. 1, pp. 34–50, 1978.

Liebman, J. F., et al., *Modeling and Optimization with GINO*. Boyd and Fraser, Danvers, MA, 1986.

Luenberger, D. G., *Linear and Nonlinear Programming*, 2nd ed. Addison-Wesley, Menlo Park, CA, 1984.

More, J. J., and Wright, S. J. *Frontiers in Applied Mathematics: Optimization Software Guide*. Society for Industrial and Applied Mathematics (SIAM), 1993.

Murtagh, B. A., Saunders, M. A., "A Projected Lagrangian Algorithm and Its Implementation for Sparse Nonlinear Constraints," *Mathematical Programming Study*, Vol. 16, pp. 84–117, 1982.

Nash, S. G., "NLP Software Survey," *OR/MS Today*, pp. 60–63, April 1995.

Nocedal, J., "Updating Quasi-Newton Matrices with Limited Storage," *Mathematics of Computation*, Vol. 35, pp. 773–782, 1980.

Palacios-Gomez, F. E., Lasdon, L. S., Engquist, M., "Nonlinear Optimization by Successive Linear Programming," *Management Science*, Vol. 28, pp. 1106–1120, 1982.

Sarma, P. V. L. N., Reklaitis, G. V., "Optimization of a Complex Chemical Process Using an Equation Oriented Model," *Mathematical Programming Study*, Vol. 20, pp. 113–160, 1982.

Silver, A. E., Peterson, R. *Decision Systems for Inventory Management and Production Planning*, J. Wiley & Sons, New York, 1985.

Smith, S., Lasdon, L., "Solving Large Sparse Nonlinear Programs Using GRG," *ORSA Journal on Computing*, Vol. 4, No. 1, pp. 3–15, 1992.

Sorenson, D. C., "Trust Region Methods for Unconstrained Optimization," in *Nonlinear Optimization, 1981*, ed. M. J. D. Powell. Academic Press, New York, 1982.

Zhang, J., Kim, N., Lasdon, L., "An Improved Successive Linear Programming Algorithm," *Management Science*, Vol. 31, pp. 1312–1331, 1985.

7

Multiobjective Programming for Industrial Engineers

Mario T. Tabucanon
Asian Institute of Technology, Bangkok, Thailand

1. INTRODUCTION

The industrial engineer (IE) is more and more becoming the centerpiece of an organization. With his knowledge of the basic elements of a productive system—man, material, machine, money, and information—he is able to conceptualize an approach that will systematically encompass the broad range of interlocking activities of these elements. With this knowledge comes the need to make alternative choices from among the complex qualitative and quantitative factors that make up the intricate fabric of a production system.

The IE identifies alternative methods, courses of action, or designs and makes specific time-oriented estimates of the consequences of choosing each of the possible alternatives. The time value of money is usually used as a basis for determining which alternative is most attractive. This, however, does not mean that the economic viewpoint always prevails. The advantages and disadvantages of each alternative that cannot be reduced to monetary terms have to be bidded against the quantitative factors.

Multiobjective programming is an area of research in management science and a decision-making tool for engineers, managers, and planners. It is a domain of a broader scientific discipline in management better known as multiple-criteria decision making (MCDM).

2. THEORETICAL CONCEPTS AND APPROACHES

2.1. What Is MCDM?

In simple terms, multiple criteria decision making is the process of selecting an act or course of action from among a set of alternatives such that it will produce "satisficing" results under some criteria of optimization. Satisficing is a jargon word expressing the notion of *satis*faction through fair sacri-*fices*. One would notice that from the traditional definition of decision making, the terms "optimal results" and "a criterion" are respectively replaced by "satisficing results" and "some criteria." Optimality, as it is defined in the traditional sense, is valid only for the monocriterion case. One has to think of a satisficing solution in dealing with the case of multiple criteria.

2.2. The Process

In the theoretical sense, there are two actors involved in the process—the decision maker (DM) and the analyst. The DM has the power to make decisions. He can be an individual, like the CEO; or individuals acting as a group, like the board of directors of a company; or an organization, like a government ministry.

There is also the analyst who serves as information and fact-finding agent or technical supporter of the DM. The analyst can be a machine (computer), a person, a group of persons, or an agency, depending on the level at which the decision-making process is focused. It is the DM who provides information on preferences or importance of criteria. The analyst inputs this preference information into the model. Results are then presented by the analyst to the DM for selection. The DM-analyst interaction is something worth scrutinizing. In fact, one way of classifying MCDM techniques is via the nature of this interaction.

In practice, one must look at a longer horizon beyond that of the decision-making process alone. The real "value-adding" operation is in the implementation phase. Doing nothing but decision making does not add value to anything. Only when these decisions are sucessfully implemented can one claim an added value. Therefore, one must not ignore the importance of the implementor as the third actor in the decision-making process. Many decisions have failed or have encountered extreme difficulty in implementation because the implementors were not involved in the decision-making process. A worthy example is that of the Japanese "bottom-up" decision-making style, where implementors are involved throughout the process. Despite the slow pace of the decision-making process, successful implementation is virtually certain. Like the DM and the analyst, the implementor can be an individual, a group of individuals, or an organization.

Multiobjective Programming for Industrial Engineers

It is understandable that theoretical considerations and advances in MCDM focus attention on the bipartite relationship of the DM and analyst. This is with the assumption that the implementors are a group of robots that are programmed to follow instructions from the DM. In the distant past, this was true; but now this notion is history. A tripartite relationship, DM-analyst-implementor, is a must.

2.3. Set of Alternatives

The feasible set of alternatives ranges in number from two to infinity. The lower bound of two already includes the "do nothing" option. A large set of alternatives are usually expressed in an analytical form in which mathematical relationships can be derived among decision variables and parameters. For example, the feasible set of alternatives for a product mix problem can be defined by a set of mathematical expressions for resource limitations and resource requirements of products. If the decision variables are continuous, there can be an infinite number of alternatives. Even for integer-valued decision variables, the feasible region necessitates the use of mathematical programming techniques and a computer—the problem is that of design. MCDM problems of this sort are specifically known as MODM, meaning multiple-objective decision making.

There are many real-life problems in which the number of feasible alternatives is relatively small—e.g., choosing the best from a list of candidate sites for a new factory, choosing the most suitable type of power plant, buying a car. This is a problem of choice in which the alternatives are enumerative. MCDM problems of this kind are known as MADM or multiple-attribute decision making.

MODM problems can become MADM problems if one discretizes the continuous feasible domain. By trying to select, a priori, a manageable number of alternatives from the original infinite set, or in other words short-listing, one reduces the MODM problem to MADM. MADM problems need special outranking methods to arrive at a solution. MODM problems, which are the thrust of this chapter, need mathematical programming to derive a satisficing or compromise solution, thus the use of the term multiobjective programming.

2.4. Criteria

In traditional optimization, one maximizes or minimizes a criterion function, say maximizes profit or minimizes cost or their surrogates. Linear programming, dynamic programming, and nonlinear programming are well developed for the purpose of deriving an optimal solution from a large (often infinitely large) set of alternatives. For a small, finite number of alternatives, one merely evaluates the value of each alternative with respect to the figure of merit; the highest or the lowest, as the case may be, is chosen. For example, economists use internal rate of return or net present worth in selecting investment alternatives.

For a monocriterion case, the problem is straightforward and the point of optimality is objectively determined and in many cases unique. If the criteria are conflicting, meaning that the full satisfaction of one precludes or impairs the full satisfaction of others, then the individual optimal solutions of the criteria are different points. In the case of MODM, unlike the traditional monocriterion optimization, efficiency is no longer unique. To recall, an efficient (noninferior, nondominated, or pareto optimal) solution is one in which no increase can be obtained in any of the objectives without causing a simultaneous decrease in at least one of the other objectives (see Figure 1 for an illustration of a biobjective case). The terms "criteria" and "objectives" are used interchangeably when referring to MODM problem.

If the objectives are nonconflicting, their individual optima (points a and b in Figure 1) are on a common point; thus the problem is in fact non-MODM, because picking up one of the objectives would just give the same common point. A real MODM problem, with conflicting objectives, would resemble the situation described in Figure 2. The elements along the diagonal of the payoff matrix are maxima (or minima, in case of a minimizing problem) of the individual objectives, also known as the ideal solutions,

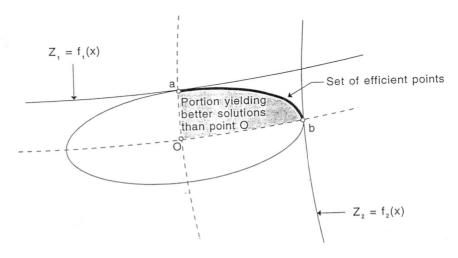

Figure 1 A graphical representation of a biobjective optimization problem.

	x^{1*}	x^{2*}	. . .	x^{h*}	. . .	x^{k*}
z_1	$f_1(x^{1*})$	$f_1(x^{2*})$	. . .	$f_1(x^{h*})$	. . .	$f_1(x^{k*})$
z_2	$f_2(x^{1*})$	$f_2(x^{2*})$	. . .	$f_2(x^{h*})$	. . .	$f_2(x^{k*})$
:	:	:	. : .	:	. : .	:
z_l	$f_l(x^{1*})$	$f_l(x^{2*})$	. . .	$f_l(x^{h*})$	. . .	$f_l(x^{k*})$
:	:	:	:	:	. : .	:
z_k	$f_k(x^{1*})$	$f_k(x^{2*})$	. . .	$f_k(x^{h*})$	. . .	$f_k(x^{k*})$

Figure 2 Payoff matrix (or payoff table).

which are unachievable. The Z_i's are the objective functions and the x^{h*}'s are the individual optima. When $l = h$, the element in the payoff matrix is an ideal value.

3. TECHNIQUES

Solution procedures for MODM-formulated problems rely on mathematical programming. The most well-known MODM technique is goal programming. MADM-type problems are typically solved by outranking/paired comparison methods such as the multiattribute utility theory or MAUT, the analytic hierarchy process or AHP, the ELECTRE method, and the PROMETHEE, among others. Classification of techniques may be based on whether the technique is of outranking, distance-based, value or utility, direction-based, or mixed type. The *outranking type* compares discrete alternatives as to their relative preference. The *distance-based type*, in which distance is viewed as a proxy of measure of human preference, determines a solution in reference to some point (often the ideal) in decision space. The *value* (for the deterministic case) or *utility* (for the probabilistic case) *type* represents the DM's preference through a value or utility function in mathematical terms. The *direction-based type* is of an interactive nature where at each step the DM is asked to state a preferred direction in the search for the compromise solution. Apart from these four types of techniques, there are those that cannot be categorically classified under any one of the above, thus named *mixed type*.

For more details on MODM techniques, the following books are suggested for further readings: Tabucanon [1], Lieberman [2], Steuer [3], Zeleny [4], Chankong and Haimes [5], Osyczka [6], Bogetoft and Pruzan [7], Lee and Shim [8], Ignizio [9], Lee [10], and Hwang and Masud [11]. For a typology of MCDM techniques see Duckstein [12].

The most popular techniques from the standpoint of applications are goal programming (GP), MAUT, AHP (all developed in the United States), and perhaps ELECTRE (French developed) if one looks at the European region. The reason for their practical attractiveness is mainly their ability to accommodate situations that indeed happen in practice. For example, GP features targets or goals, which is also the case in practice, as organizations set goals in the foreseeable future. It also features priorities and differential weights, which are common in practice. An added attraction of GP is that the algorithm is simplex based. MAUT is based on ranking and rating, and AHP operates on hierarchies and paired comparisons of criteria. ELECTRE is also based on paired comparison. These common features made these four techniques useful in practical applications.

Another classification of techniques based on information flow as given by the DM would be as follows: (1) no articulation of preference information, (2) a priori articulation of cardinal information, (3) a priori articulation of ordinal and cardinal information, (4) progressive articulation of explicit trade-off information, (5) progressive articulation of implicit trade-off information, and (6) a posteriori articulation of preference information. As the classification implies, the DM can state his preferences, if at all required by the technique, either before (ordinally/cardinally), during (implicitly/explicitly), or after the generation of a solution(s). A survey of techniques based on this classification may be found in Hwang and Masud [11], and a tutorial on some of these techniques is presented in Hwang et al. [13].

Many multiobjective mathematical programming methods have been reported. Due to space limitations, only a few are discussed here. The selection does not reflect any particular bias other than mere representations of the various MCDM approaches discussed earlier. The techniques described in the subsequent sections are the global criterion method and the maximum effectiveness method representing the "no a priori articulation" approach, goal programming as an example of the "a priori articulation" approach, the step method showing the "progressive articulation of explicit preference information" approach, the two-person-zero-sum game method showing the "progressive articulation of implicit preference information"

Multiobjective Programming for Industrial Engineers

approach, and the parametric method illustrating the "a posteriori articulation" approach.

Some of the techniques discussed will be used to solve selected industrial engineering problems later in the chapter. Multiobjective programming applications discussed are those related to problems of production planning, product mix, job sequencing and scheduling, maintenance, and project selection.

3.1. Global Criterion Method

The global criterion method (also known as the ideal distance minimization method) evolved from the works of Yu [14], Yu and Leitmann [15], Hwang [16], and Salukvadze [17, 18]. A global function is developed that is the sum of the distances of the individual objective functions from the ideal solution taken as a proportion to the ideal value. The k-objective vector maximization problem can then be reformulated as

$$\text{Minimize} \quad F = \sum_{l=1}^{k} \left[\frac{f_l(x^{l*}) - f_l(x)}{f_l(x^{l*})} \right]^p$$

subject to $x \in X$

F is a dimensionless function that reflects the total "distance" of the compromise solution from the ideal solutions of the individual objectives. The compromise solution would depend on the p-value. The value $p = 1$ implies equal weights for all distances or deviations, and $p = 2$ implies that these deviations are weighted proportionately with the longest deviation having the largest weight. As p approaches a very large number, minimizing F is tantamount to minimizing only the maximum deviation, which in effect is equivalent to the minimax objective function.

This method is appealing for cases in which input from the decision maker is not readily available. Except for the prespecified value of p, it does not require a priori articulation of preference information.

3.2. Maximum Effectiveness Method

Another distance-based technique requiring no articulation of preference information is the maximum effectiveness method, which was developed by V. V. Khomenynk (see Lieberman [2]). The method maximizes the minimum attainment by any objective of its respective ideal value by making use of the natural normalization. The lth objective's relative attainment of its ideal value is denoted by λ_l, and assuming all objectives are being maximized, λ_l is defined as

$$\lambda_l(x) = \frac{f_l(x) - Z_l^{\text{worst}}}{Z_l^* - Z_l^{\text{worst}}}, \quad l = 1, 2, \ldots, k$$

where Z_l^* and Z_l^{worst} are, respectively, the ideal and worst values of the objective taken from the payoff matrix. The final model is then

> Maximize λ
> subject to $\quad x \in X \quad$ (original constraint set)
> $\quad\quad\quad\quad \lambda_l(x) \geq \lambda, \quad l = 1, 2, \ldots, k$

The method is also known as the λ-criteria optimization.

3.3. Goal Programming

Goal programming (GP) is by far the most popular among the multiobjective mathematical programming techniques. It has its roots in the seminal work of Charnes and Cooper [19] and the early works of Lee [10] and Ignizio [9]. Its popularity is driven by its resemblance to traditional linear programming (LP). GP extends the LP formulation to accommodate mathematical programming with multiple objectives.

GP's objective function is always of a minimizing character. It minimizes the deviations of the compromise solution from target goals, weighted and prioritized by the decision maker. There are two categories of constraints, namely structural/system constraints (strict and rigid as in traditional LP) and the so-called goal constraints, which are expressions of the original objective functions with target goals set a priori and negative and positive deviational variables built into them to accommodate possible over- or underachievement. A general GP model can be written as follows:

Minimize $\quad\quad Z = \sum_{k=1}^{K} P_k \sum_i (w_{i,k}^+ d_i^+ + w_{i,k}^- d_i^-)$

subject to

Goal constraints: $\quad \sum_{j=1}^{n} a_{ij} x_j + d_i^- - d_i^+ = b_i, \quad i = 1, 2, \ldots, m$

Structural/
system constraints: $\quad \sum_{j=1}^{n} a_{ij} x_j \begin{bmatrix} \leq \\ = \\ \geq \end{bmatrix} b_i, \quad i = m+1, \ldots, m+p$

$$x_j, d_i^+, d_i^- \geq 0, \quad \begin{matrix} j = 1, \ldots, m \\ i = 1, \ldots, m \end{matrix}$$

where there are m goals, p structural constraints, K priority levels, and n decision variables. P_k is the priority coefficient for the kth priority, $w_{i,k}^+$ is the relative weight of the d_i^+ variable in the kth priority level, and $w_{i,k}^-$ is the relative weight of the d_i^- variable in the kth priority level. The deviational

Multiobjective Programming for Industrial Engineers 495

variables, d_i^- and d_i^+ represent, respectively, the underachievement and overachievement of the objectives at a compromise solution with reference to the target goals (b_i).

The objective function is the lexicographic sum of all the deviational variables that are to be minimized; some deviational variables may not enter into the objective function if their minimization is not desired. The weighted sum of the deviational variables at a certain priority level must first be minimized as much as possible before going to a lower priority level. If P_k is the value of the priority at level k and P_{k+1} is one corresponding to the lower priority level $k + 1$, then $P_k \gg MP_{k+1}$ even if M is a very big number. The expression implies that P_k is much bigger than P_{k+1}; hence it is assured that minimization is carried out lexicographically. In some cases, it is desirable to put differential weights among the deviational variables within a priority level.

The goal constraints are derived from the original objective functions whose target values are set a priori. In real life, it is often the case that managements of organizations set aspiration levels. For instance, a profit objective is not stated as "to maximize profit as much as possible"; rather its statement is usually given in the form "to increase profit by x percent." The quantitative statements of objectives thus reflect reality. Since objective goals are imposed, they may not be achieved on the mark—they may be under- or overachieved. To accommodate situations like this, it is imperative always to include positive and negative deviational variables in all goal constraints.

System or structural constraints are usually nonnegotiable resource and technological considerations. As in traditional LP, these constraints must be adhered to without exception. The feasible region formed by this set of constraints is fixed.

To illustrate a GP formulation numerically, take a simple example. Consider a case in which a firm produces two types of products. Each product type must pass through two departments. Product type I requires 20 hours in department 1 and 10 hours in department 2; product type II requires 1.5 hours each in both departments. Department 1 is limited to 60 hours of operation, and department 2 is limited to 40 hours of operation per week. Contributions to profit of the two products are $40 and $80, respectively. Material availability restricts production of product types I and II to 90 and 100, respectively, for the week under consideration. The first priority goal is the maximization of profit. Since product type II contributes twice as much to profit as product type I, it seems logical to put twice as much weight on product type II as on product type I, in the first priority level. The profit goal is $3000 for the week. The two product types are aimed at two different markets, and because of market prioritization, a preference is given to these markets by establishing minimum production goals of 20 units of product type I and 30 units of product type II. The meeting of these minimum production goals constitutes the second priority level. The third priority goal is to limit total overtime in the two production operations to 20 hours. The GP formulation of the problem is

Minimize $Z = P_1 d_1^- + P_2(d_2^- + 2d_3^-) + P_3 d_6^+$

subject to

Profit goal: $\quad 40x_1 + 80x_2 + d_1^- - d_1^+ = 3000$

Production goals: $\quad x_1 + d_2^- - d_2^+ = 20$

$ x_2 + d_3^- - d_3^+ = 30$

Overtime goals: $\quad 2x_1 + x_2 + d_4^- - d_4^+ = 60$

$ 1.5x_1 + 1.5x_2 + d_5^- - d_5^+ = 40$

$ d_4^+ + d_5^+ + d_6^- - d_6^+ = 20$

Material constraints: $\quad x_1 \leq 90$

$ x_2 \leq 100$

$ x_1, \; x_2, \; d^-\text{'s}, \; d^+\text{'s} \geq 0$

For the two-variable case, the graphical solution, as in LP, can be used to solve a GP problem. For a bigger problem, one could still use available LP software, using it sequentially. The sequential LP procedure (SLP) to solve a GP formulation runs this way: The first LP problem of the sequence minimizes the first-priority deviational variables, subject to the constraints corresponding to P_1. The second LP problem minimizes the second-priority deviational variables, subject to the constraints corresponding to P_2, as well as the values of the deviational variables in priority P_1 that were found in the preceding phase (first LP). This sequential procedure continues until the last linear program is solved or until in one of the problems of the sequence there are no alternative or multiple solutions. It is highly possible that a compromise solution will be obtained before reaching the lowest priority level. The SLP procedure demands a large number of computations as one has to run, one at a time, several LP programs. Ignizio and Perlis [20] tried to cut down the computational effort by linking all LP problems in a single computer run. Software to solve GP is also available, based on modified simplex (see Lee and Shim [8]). It should be noted that in GP a priori articulation of the DM's cardinal (for the differential weights) and ordinal (the priority structure) preference information is required. However, GP has also attracted some criticism from various researchers. For a review of such criticism and an excellent guide on where and how to use GP see Hannan [30].

Multiobjective Programming for Industrial Engineers

3.4. The Step Method

The step method (STEM), introduced by Benayoun et al. [21], allows the decision maker to learn to recognize good satisficing solutions and relative importance of the objectives. Phases of computation alternate, interactively, with phases of decision. For a k-objective problem with the payoff table discussed in Section 2.4, the interactive procedure can be described in the following fashion:

1. Calculation phase. At the mth iteration, the feasible solution is sought that is nearest, in the minimax sense, to the ideal solution Z_l^*. The model becomes

   Minimize $z = y$

   subject to $y \geq [f_l(x^{l^*}) - f_l(x)]\pi_l$, $\quad l = 1, 2, \ldots, k$

   $\quad x \in X^m$

   $\quad y \geq 0$

   where $\pi_l = \dfrac{\alpha_l}{\sum_{i=1}^{k} \alpha_i}$, $\quad l = 1, 2, \ldots, k$

   $\alpha_l = \dfrac{M_l - m_l}{M_l} \left[\dfrac{1}{\sqrt{\sum_{j=1}^{n} c_{lj}^2}} \right]$, $\quad l = 1, 2, \ldots, k$ (for $M_l > 0$)

   $\alpha_l = \dfrac{m_l - M_l}{m_l} \left[\dfrac{1}{\sqrt{\sum_{j=1}^{n} c_{lj}^2}} \right]$, $\quad l = 1, 2, \ldots, k$ (for $M_l \leq 0$)

   π_l gives the relative importance of the distances of the optima and c_{lj} are the coefficients of the lth objective.

2. Decision phase. The compromise solution obtained in the calculation phase, x^m, is presented to the decision maker, who compares its objective vector f^m (or Z^m) with Z^*, the ideal one. If some of the objectives are satisfactory and others are not, the decision maker must relax a satisfactory objective f_l^m enough to allow an improvement of the unsatisfactory objectives in the next iteration. The decision maker specifies Δ_{f_l} as the amount of acceptable relaxation. Then, in the next cycle, the following constraints are added to modify the feasible region.

   $f_l(x) \geq f_l(x^m) - \Delta_{f_l}$

   $f_i(x) \geq f_i(x^m)$, $\quad i \neq l, i = 1, 2, \ldots, k$

   The weight π_l is set to zero and the calculation phase of cycle $m + 1$ starts.

3. The process stops when all the values obtained for the objective function are acceptable to the decision maker. At times, the decision maker's acceptable vector may be infeasible. Infeasibility

occurs when no $f_i(x^m)$ can be relaxed anymore while there are still unsatisfactory $f_i(x^m)$ or when the $m + 1$ cycle is a repetition of the mth cycle.

STEM requires progressive articulation of preference information. The DM has to express, in explicit terms, the allowable trade-offs among the objectives at each cycle of the process.

3.5. Game Theory: Two-Person-Zero-Sum Method

This is an interactive method that requires the decision maker's implicit trade-off information in the form of a yes or no answer to the dialogue at decision-making and calculation phases. It was first introduced by Belenson and Kapur [22].

The method starts from the payoff table (or payoff matrix), where the decision maker is asked whether any of the individual optimal solutions is a satisfactory compromise solution. If the answer is yes, the problem is solved. A yes answer is sometimes possible, especially when there is a good number of objectives and one of them has an individual optimum falling in between other individual optima. If the decision maker is not satisfied with any one in the payoff table, then the algorithm progresses by normalizing the payoff table.

The normalization of the payoff table is done to overcome differences in dimensions, and hence incommensurable magnitudes, of objective functions. Without it, one may obtain unfair dominance of some objectives over others. For example, a profit or revenue-maximizing objective (which is expressed in monetary values) would undoubtedly dominate a reliability-maximizing objective (which is usually expressed as a percentage). The normalization is done by dividing each element of the row by the maximum element of that row (M_l), and this is done for all the rows. The normalized matrix is considered as the payoff table of a two-person-zero-sum game. The strategies in the form of probabilities taken as relative weights of criteria are calculated in the usual way. But since these probabilities are those of the normalized matrix, the equivalent probabilities corresponding to the original matrix are to be determined. If w_l's are weights of the normalized matrix, the weights for the unnormalized matrix are

$$w_l^* = \frac{n_l}{\sum_{l=1}^{k} n_l}, \quad \text{for } l = 1, 2, \ldots, k$$

where $n_l = w_l/M_l$, for $l = 1, 2, \ldots, k$.

Game theory stipulates that to be able to use LP to solve a two-person-zero-sum game problem, the value of the game must be positive. To ensure this condition, the unnormalized matrix must first be made positive by adding a certain constant to all elements of the matrix to eliminate negative elements, if any.

An equivalent linear program is formed using the weights obtained from the game, as follows:

Maximize $\quad Z = \sum_{l=1}^{k} w_l^* f_l(x)$

subject to $\quad x \in X$

The solution to the above problem is efficient and considered as a potential compromise solution x^1. If x^1 was previously considered the multiobjective problem, as solved by this method, does not have a satisfactory solution. Otherwise, the decision maker will be asked if x^1 is acceptable. If so, the problem is solved and x^1 is the compromise solution.

If the decision maker is not satisfied, a new game is formed in which x^1 is substituted for x^{h*}, which is the individual optimum for the least preferred objective. The new game is solved for the new set of weights and then a second linear program is formulated and solved. The solution x^2 is a new potential compromise solution. If x^2 was previously considered, then another efficient solution must be generated. This is done by developing another game via substitution of x^1 for x^{i*} (instead of x^{h*}), where x^{i*} is the individual optimum of the next least preferred objective. There will be another x^2 solution using the process discussed earlier. If this new x^2 is satisfactory from the decision maker's point of view, the problem is solved. Otherwise, the substitution and reiteration process continues until $(k - 1)$ lesser preferred objectives have been considered. Thus, either a satisfactory compromise solution will be obtained or there does not exist a solution that satisfies the decision maker.

In summary, the method is less burdensome for the decision maker as his preference information, unlike that in the step method, is only implicit, or of a yes-no type. Game theory is used to derive the weights of objectives, and a new compromise solution, which of course is efficient, is generated by solving a new linear program. The generation of new games is done by replacing columns, from less to more preferred objectives, of the payoff table by the column corresponding to the new efficient solution generated.

3.6. Parametric Approaches

Methods in which articulation of preference information is required a posteriori involve determination of a set (or subset) of efficient solutions. The decision maker chooses the most satisfactory solution from this set, making implicit or explicit trade-offs between objectives based on certain criteria.

The trade-off information is received from the decision maker only after the set of efficient solutions has been generated.

Parametric methods do not require any assumption or information regarding the decision maker's preference. However, they can generate a large number of efficient solutions such that it becomes extremely difficult, if not impossible, to choose clearly one that is the most satisfactory. To overcome this drawback, parametric methods are generally incorporated in some of the interactive methods. Parametric methods also provide links between MODM and MADM problems. The efficient solutions that are generated from an MODM problem can then be considered as discrete alternatives for a sequel MADM problem.

The simplest version of the parametric methods is the weighted-sum model as follows:

$$\text{Maximize} \quad Z = \sum_{l=1}^{k} w_l f_l(x)$$

subject to $\quad x \in X$

where $w_l \geq 0$ and $\sum_{l=1}^{k} w_l = 1$. The w_l's do not reflect the relative importance of the objectives in the proportional sense; rather, they are mere parameters that are varied to locate efficient points. For a linear case, if the feasible region is a convex set, then a systematic variation of w_l's will result in the generation of a complete efficient solution set. LP software with good postoptimality computational capability is a good tool for this purpose.

A vector maximum problem attempts to "maximize" a vector (as defined by the multiple objectives in this case). A vector maximization algorithm is used to compute all efficient extreme points. A vector maximization problem

$$\text{eff}\{z = Cx | x \in X\}$$

where eff signifies that the set of all solutions is the efficient set, requires no information at all about the decision maker's preference, but it generates an inundation of efficiency information (for more details see Steuer [3]). With the weighted-sum LP problem

$$\text{Max}\{z = \lambda^T Cx | x \in X\}$$

$\lambda \in \Lambda$, λ fixed

one generates enough (but not all) efficiency information. ADBASE is a FORTRAN program used for enumerating efficient extreme points and unbound efficient edges (see Steuer [3]). It incorporates a number of features, including the ability to reduce the criterion cone via the specification of internal criterion weights.

4. APPLICATIONS

4.1. Production Planning*

Aggregate production planning is one of the major decision problems facing production managers. It concerns the decision maker's response to a changing demand pattern over time. For real-life problems in business organizations, the factory may wish to find an optimal production level not only to minimize the total cost but also to maintain a balanced level of workforce, among other things. The conflict arises because an improvement in meeting one objective can be made only to the detriment of meeting the others.

In material requirements planning (MRP) or manufacturing resources planning (MRP II) in particular, one would need an aggregate production planning model to convert the sales forecasts (from the marketing department) to a master production schedule (MPS) for use by the production department. The MPS in turn, together with information on inventory and bill of materials (BOM), is used to generate scheduled releases of production orders to outside suppliers or to internal production stations (see Figure 3).

In this section, the particular problem being considered is that of a monthly production planning in a small diesel engine assembly plant. The market situation is highly competitive and oligopolistic. The company tries to develop an efficient (in resource utilization) and effective (in meeting targets) production planning model to achieve the goals of management.

Three categories of decision variables are identified:

1. Inventory level I_{lt}: amount of holding inventory of product l in period t
2. Production level P_{lt}: amount of product l produced in period t
3. Workforce level M_{lt}: workforce level in period t to produce product l

For the inventory level, lower and upper limits are introduced as the factory does not allow back orders and the delivery date of a spare part from a local supplier is never reliable. Shortage of this spare part and fluctuating demand force the factory to increase the inventory level.

4.1.1. Cost Goal

The cost goal (G_1) is considered with the financial view of minimizing the amount by which the total cost, consisting of costs due to inventory holding, materials, and workers' salaries, exceeds the target level beyond the

---

*This is an edited version of materials taken, with the publisher's permission, from Tabucanon and Mukyangkoon [23].

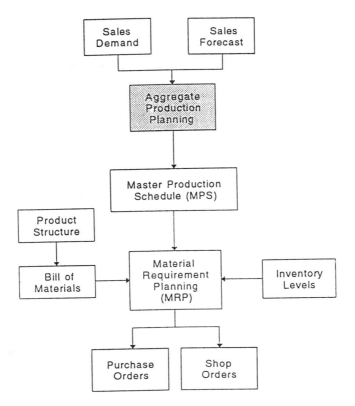

Figure 3 MRP information flow.

planning horizon. The holding cost excludes the fixed cost for the management and maintenance of the warehouse.

$$i \sum_l \sum_t C1_{lt} I_{lt} + \sum_l \sum_t C2_{lt} P_{lt} + \sum_t C3_t M_t + d_{11}^- - d_{11}^+ = G_1$$

where $C1_{lt}$ is the selling price of engine model 1, $C2_{lt}$ the material cost, and $C3_{lt}$ the average salary; d_{11}^+ and d_{11}^- are, respectively, the overachievement and underachievement of the cost goal, G_1; i is the discounting rate used to estimate the holding cost, which is mainly attributed to the cost of capital.

4.1.2. Manpower Goal

Efficient utilization of manpower is an important consideration. The management of the factory has a policy of avoiding both over- and underemployment of their workers. The other consideration is the hiring and firing of workers. The management does not want, as much as possible, either to hire new workers or to fire existing ones. Hiring new workers requires a large training budget and results in a fluctuation in quality, due to learning effects. Firing existing workers would lower morale. The factory has therefore developed a policy of maintaining a balanced workforce. Mathematically, it tries to minimize the absolute deviation in the fluctuation of the manpower requirement in every period, as shown below:

$$M_{lt} + d_{2t}^- - d_{2t}^+ = G_2, \quad t = 1, \ldots, T$$

where d_{2t}^- is the underachievement of manpower goal in period t; d_{2t}^+ is the overachievement of manpower goal, G_2, in period t.

4.1.3. Production Goal

By industry agreement, specific levels for the production of various models of small diesel engines that should be achieved in a year are set. The factory has therefore established constraints for the production goal as shown below:

$$\sum_{t=1}^{T} P_{lt} + d_3^- - d_3^+ = r_l G_3, \quad l = 1, \ldots, L$$

where d_3^- and d_3^+ are the underachievement and overachievement, respectively, of production goals; r_l is the proportion of engine model l in this goal.

4.1.4. Demand Constraints

For each period, there are four balance equations for the production level and demand forecast. Inventory levels provide the link between successive periods. Each equation is formulated according to each model of a small diesel engine. The expression below ensures that (at least) the demand forecast for any period is fulfilled by the production in that period and the remaining inventory. Two assumptions are made about the inventory level: first, the initial inventory level is assumed to be zero for every product and second, there is no final inventory for any model of the product. Every unit of the product must therefore be sold out within the planning horizon:

$$P_{l1} - I_{l0} \geq D_{l1}$$
$$I_{lt} + P_{lt} - I_{lt-1} \geq D_{lt}, \quad t = 2, \ldots, T, \quad l = 1, \ldots, L$$

where D_{lt} is the demand forecast for product l at period t.

4.1.5. Capacity Constraints

The total production during any period is limited by the available production capacity:

$$P_{1t} + P_{2t} + P_{3t} + P_{4t} \leq A_{1t}, \quad t = 1, \ldots, T$$

where A_{1t} is the available production capacity in period t.

At least A_{0t} units of production are processed in every period. This lower limit can be expressed by

$$P_{1t} + P_{2t} + P_{3t} + P_{4t} \geq A_{0t}, \quad t = 1, \ldots, T$$

4.1.6. Workforce Constraints

For each period t, the balance of the workforce level is formulated as

$$M_1 - h_1 P_{1t} - h_2 P_{2t} - h_3 P_{3t} - h_4 P_{4t} = 0, \quad t = 1, \ldots, T$$

where h_l is the manpower requirement for producing one unit of engine model l.

4.1.7. Complete Planning Model Structure

The complete planning model structure, based on the goals, constraints, and other relevant information that have been presented, is shown below. This model is presented in a GP formulation using some assumptions about the cardinal preference information of the goal structure. The goals are ranked by the management, the first priority being the cost goal, followed by manpower and then production goals:

$$\text{Min} \quad P_1(d_{11}^+) + P_2\left(\sum_{t=1}^{6} d_{2t}^- + d_{2t}^+\right) + P_3\left(\sum_{l=1}^{4} d_{3l}^- + d_{3l}^+\right)$$

The model is subject to resource and goal constraints.

Resource Constraints

1. *Demand constraints*

$$P_{11} - I_{11} = -1473 \qquad P_{21} - I_{21} = -2137$$
$$I_{11} + P_{12} - I_{12} = 501 \qquad I_{21} + P_{22} - I_{22} = 1534$$
$$I_{12} + P_{13} - I_{13} = 569 \qquad I_{22} + P_{23} - I_{23} = 1359$$
$$I_{13} + P_{14} - I_{14} = 660 \qquad I_{23} + P_{24} - I_{24} = 1210$$
$$I_{14} + P_{15} - I_{15} = 863 \qquad I_{24} + P_{25} - I_{25} = 1810$$
$$I_{15} + P_{16} - I_{16} = 1114 \qquad I_{25} + P_{26} - I_{26} = 2144$$

$$P_{31} - I_{31} = -2634 \qquad P_{41} - I_{41} = -1527$$
$$I_{31} + P_{32} - I_{32} = 957 \qquad I_{41} + P_{42} - I_{42} = 655$$
$$I_{32} + P_{33} - I_{33} = 814 \qquad I_{42} + P_{43} - I_{43} = 853$$
$$I_{33} + P_{34} - I_{34} = 666 \qquad I_{43} + P_{44} - I_{44} = 565$$
$$I_{34} + P_{35} - I_{35} = 999 \qquad I_{44} + P_{45} - I_{45} = 470$$
$$I_{35} + P_{36} - I_{36} = 1012 \qquad I_{45} + P_{46} - I_{46} = 526$$

2. *Capacity constraints*

$$P_{11} + P_{21} + P_{31} + P_{41} \leq 3740$$
$$P_{12} + P_{22} + P_{32} + P_{42} \leq 3740$$
$$P_{13} + P_{23} + P_{33} + P_{43} \leq 3740$$
$$P_{14} + P_{24} + P_{34} + P_{44} \leq 3740$$
$$P_{15} + P_{25} + P_{35} + P_{45} \leq 3740$$
$$P_{16} + P_{26} + P_{36} + P_{46} \leq 3740$$

$$P_{11} + P_{21} + P_{31} + P_{41} \geq 2220$$
$$P_{12} + P_{22} + P_{32} + P_{42} \geq 2220$$
$$P_{13} + P_{23} + P_{33} + P_{43} \geq 2220$$
$$P_{14} + P_{24} + P_{34} + P_{44} \geq 2220$$
$$P_{15} + P_{25} + P_{35} + P_{45} \geq 2220$$
$$P_{16} + P_{26} + P_{36} + P_{46} \geq 2220$$

3. *Workforce constraints*

$$M_1 - 2.97P_{11} - 2.97P_{21} - 3.08P_{31} - 3.08P_{41} = 0$$
$$M_2 - 2.97P_{12} - 2.97P_{22} - 3.08P_{32} - 3.08P_{42} = 0$$
$$M_3 - 2.97P_{13} - 2.97P_{23} - 3.08P_{33} - 3.08P_{43} = 0$$
$$M_4 - 2.97P_{14} - 2.97P_{24} - 3.08P_{34} - 3.08P_{44} = 0$$
$$M_5 - 2.97P_{15} - 2.97P_{25} - 3.08P_{35} - 3.08P_{45} = 0$$
$$M_6 - 2.97P_{16} - 2.97P_{26} - 3.08P_{36} - 3.08P_{46} = 0$$

Goal Constraints

1. *Cost goals*

$$0.6144\,(P_{11} + P_{12} + P_{13} + P_{14} + P_{15} + P_{16})$$
$$+\ 0.7000\,(P_{21} + P_{22} + P_{23} + P_{24} + P_{25} + P_{26})$$
$$+\ 0.7474\,(P_{31} + P_{32} + P_{33} + P_{34} + P_{35} + P_{36})$$
$$+\ 0.8541\,(P_{41} + P_{42} + P_{43} + P_{44} + P_{45} + P_{46})$$

$$+ 0.00094 (M_1 + M_2 + M_3 + M_4 + M_5 + M_6)$$
$$+ 0.008775 (I_{11} + I_{12} + I_{13} + I_{14} + I_{15} + I_{16})$$
$$+ 0.010 (I_{21} + I_{22} + I_{23} + I_{24} + I_{25} + I_{26})$$
$$+ 0.01092 (I_{31} + I_{32} + I_{33} + I_{34} + I_{35} + I_{36})$$
$$+ 0.0122 (I_{41} + I_{42} + I_{43} + I_{44} + I_{45} + I_{46})$$
$$+ d_{11}^- - d_{11}^+ = 15{,}000$$

2. *Manpower goals*

$$M_1 + d_{21}^- - d_{21}^+ = 10{,}560$$
$$M_2 + d_{22}^- - d_{22}^+ = 10{,}560$$
$$M_3 + d_{23}^- - d_{23}^+ = 10{,}560$$
$$M_4 + d_{24}^- - d_{24}^+ = 10{,}560$$
$$M_5 + d_{25}^- - d_{25}^+ = 10{,}560$$
$$M_6 + d_{26}^- - d_{26}^+ = 10{,}560$$

3. *Production goals*

$$P_{11} + P_{12} + P_{13} + P_{14} + P_{15} + P_{16} + d_{31}^- - d_{31}^+ = 3687$$
$$P_{21} + P_{22} + P_{23} + P_{24} + P_{25} + P_{26} + d_{32}^- - d_{32}^+ = 8114$$
$$P_{31} + P_{32} + P_{33} + P_{34} + P_{35} + P_{36} + d_{33}^- - d_{33}^+ = 4581$$
$$P_{41} + P_{42} + P_{43} + P_{44} + P_{45} + P_{46} + d_{34}^- - d_{34}^+ = 4159$$

This model was run using the MPSX/MIP package, which is IBM mainframe software for LP/ILP. The software was run on a sequential basis for the GP formulation. The results are shown in Table 1.

4.2. Product Mix Problem*

Given limitations in resources available to produce a product and the need to satisfy some objectives, the problem is to determine the optimal mix of production levels. In product mix problems, the company is faced with the decision of how much of each type of product to produce. It is not enough just to produce the largest possible quantity of the most profitable product that current resources permit. When there is a large number of products and resources, the interactive nature of these elements makes the consequences of selecting a particular product mix uncertain.

---

*This is an edited version of materials taken, with the publisher's permission, from parts of Section 5.9 of Tabucanon [1].

Multiobjective Programming for Industrial Engineers

Table 1 Results of the GP Model

Period	Model A	Model B	Model C	Model D
		Production level		
Jan.	30.	2465.	0.	374.
Feb.	0.	1009.	765.	1045.
Mar.	1322.	912.	370.	257.
Apr.	1029.	295.	1225.	280.
May	410.	2311.	0.	156.
Jun.	1486.	850.	528.	0.
		Inventory level		
Jan.	1503.	4602.	2634.	1901.
Feb.	1002.	4077.	2442.	2291.
Mar.	1755.	3630.	1998.	1695.
Apr.	2124.	2715.	2557.	1410.
May	1671.	3216.	1558.	1096.
Jun.	2043.	1922.	1074.	570.

	Workforce level	
	No. of workers	
Jan.	8574	
Feb.	8574	
Mar.	8574	
Apr.	8574	
May	8574	
Jun.	8574	

4.2.1. Production Situation

The company, a certain Chocolate Manufacturer, Inc. (Chocoman), produces various types of chocolate bars, candy and wafer. It has both production and marketing capabilities to produce and sell all or a mixture of the following products: (1) milk chocolate bars, 250 g weight; (2) milk chocolate bars, 100 g weight; (3) crunchy chocolate bars, 250 g weight; (4) crunchy chocolate bars, 100 g weight; (5) chocolate with nuts, 250 g weight; (6) chocolate with nuts, 100 g weight; (7) chocolate candy, packed in 300 g weight each; (8) chocolate wafer, packed in 12 pieces at 10 g per piece.

For production of these products, the following materials are used: (1) cocoa, (2) milk, (3) nuts, (4) confectionery sugar, (5) flour (for wafers), (6) aluminum foil for packaging, (7) paper for packaging, (8) plastic sheets for packaging.

Usage of these raw materials varies for each product. Cocoa, milk, nuts, sugar, and flour are measured by weight in kilograms. For packaging the end products, the use of aluminum foil, paper, and plastic sheets is measured by area in square feet. Details of raw material usage are presented in Table 2.

The following major facilities are used for production: (1) cooking and melting vats, (2) mixing machines, (3) forming machines, (4) grinding machines, (5) wafer-making machines, (6) cutting machines, (7) packaging type 1 machine for paper and foil wrappings, (8) packaging type 2 machine for sealing plastic.

Facilities (1) to (5) are expressed in ton-hours. These machines are limited by both tonnage weight capacity and hours of operation. Cutting and packaging facilities are expressed in hours. Operations of these facilities are independent of the product weight. Details of the facility usage are presented in Table 3.

Labor is also an input in production. Indirect labor, such as administrative, supervisory, and maintenance, is excluded from consideration because the associated costs are fixed in nature. Direct labor required for manufacturing the products, such as labor required in packaging, is considered and details of the requirements are included in Table 3.

Chocoman is in the process of preparing a monthly plan for the next manufacturing period. Available raw materials and facilities are presented in Table 4. Prices of end products are given in Table 5. Prices of the raw materials and labor are given in Table 6.

The company undertakes periodic planning, taking into consideration various alternatives available and with objectives geared toward maximizing the benefit of the company as a whole.

Table 2 Material Requirements for the Production of Chocoman Products

Materials required (per 1000 units)	Product types							
	Milk 250	Choco 100	Crunchy 250	Crunchy 100	Choco 250	W/nuts 100	Candy	Wafer
Cocoa (kg)	87.5	35	75	30	50	20	60	12
Milk (kg)	62.5	25	50	20	50	20	30	12
Nuts (kg)	0	0	37.5	15	75	30	0	0
Conf. sugar (kg)	100	40	87.5	35	75	30	210	24
Flour (kg)	0	0	0	0	0	0	0	72
Alum. foil (ft^2)	500	0	500	0	500	0	0	250
Paper (ft^2)	450	0	450	0	450	0	0	0
Plastic (ft^2)	60	120	60	120	60	120	1600	250

Table 3 Facility Usage in the Production of Chocoman Products

Facility usage required (per 1000 units)	Product types							
	Milk 250	Choco 100	Crunchy 250	Crunchy 100	Choco 250	W/nuts 100	Candy	Wafer
Cooking (ton-hours)	0.5	0.2	0.425	0.17	0.35	0.14	0.6	0.096
Mixing (ton-hours)	0	0	0.15	0.06	0.25	0.1	0	0
Forming (ton-hours)	0.75	0.3	0.75	0.3	0.75	0.3	0.9	0.36
Grinding (ton-hours)	0	0	0.25	0.1	0	0	0	0
Wafer making (ton-hours)	0	0	0	0	0	0	0	0.3
Cutting (hours)	0.1	0.1	0.1	0.1	0.1	0.1	0.2	0
Packaging 1 (hours)	0.25	0	0.25	0	0.25	0	0	0.1
Packaging 2 (hours)	0.05	0.3	0.05	0.3	0.05	0.3	2.5	0.15
Labor (hours)	0.3	0.3	0.3	0.3	0.3	0.3	2.5	2.5

Table 4 Material and Facility Availability

Material/facility (units)	Availability
Cocoa (kg)	100,000
Milk (kg)	120,000
Nuts (kg)	60,000
Confectionery sugar (kg)	200,000
Flour (kg)	20,000
Aluminum foil (ft^2)	500,000
Paper (ft^2)	500,000
Plastic (ft^2)	500,000
Cooking (ton-hours)	1,000
Mixing (ton-hours)	200
Forming (ton-hours)	1,500
Grinding (ton-hours)	200
Wafer making (ton-hours)	100
Cutting (hours)	400
Packaging 1 (hours)	400
Packaging 2 (hours)	1,200
Labor (hours)	1,000

Table 5 Prices of End Products

Product	Price/100 pcs.
Milk chocolate, 250 g	$375
Milk chocolate, 100 g	150
Crunchy chocolate, 250 g	400
Crunchy chocolate, 100 g	160
Chocolate with nuts, 250 g	420
Chocolate with nuts, 100 g	175
Chocolate candy	400
Chocolate wafer	150

4.2.2. Product Mix Model Formulation:

The linear programming model can be formulated with the following variables:

MB = milk chocolate of 250 g to be produced (in '000)
MS = milk chocolate of 100 g to be produced (in '000)
CB = crunchy chocolate of 250 g to be produced (in '000)
CS = crunchy chocolate of 100 g to be produced (in '000)
NB = chocolate with nuts of 250 g to be produced (in '000)
NS = chocolate with nuts of 100 g to be produced (in '000)
CD = chocolate candy to be produced (in '000 packs)
WF = chocolate wafer to be produced (in '000 packs)

The objective of the company is to maximize profit, which is, alternatively, equivalent to maximizing gross contribution. Since the prices of the end products, usage and cost of raw materials are given, the variable cost

Table 6 Prices of Raw Materials

Material (units)	Price ($/unit)
Cocoa (kg)	1.20
Milk (kg)	0.50
Nuts (kg)	2.00
Confectionery sugar (kg)	0.15
Flour (kg)	0.25
Aluminum foil (ft^2)	0.06
Paper (ft^2)	0.02
Plastic (ft^2)	0.03
Labor (hours)	10.00

Multiobjective Programming for Industrial Engineers

can be computed. For example, for milk chocolate of 250 g size, the total variable cost and the gross contribution are computed as follows:
For every 100 units of milk chocolate, 250 g,

Price		$375.00
Material usage and costing		
Cocoa (kg)	(87.5) ($1.20) =	$105.00
Milk (kg)	(62.5) ($0.50) =	31.25
Sugar (kg)	(100.0) ($0.06) =	15.00
Foil (ft^2)	(500.0) ($0.06) =	30.00
Paper (ft^2)	(450.0) ($0.02) =	9.00
Plastic (ft^2)	(60.0) ($0.03) =	1.80
Labor (hour)	(0.3)($10.00) =	3.00
Total variable cost	=	$195.05
Gross contribution ($375.00 − $195.05)		$179.95

After computing the gross contribution for all other products, the entire objective function can be written as

Maximize: $Z = 179.95\text{MB} + 82.9\text{MS} + 153.08\text{CB} + 72.15\text{CS}$
$+ 129.95\text{NB} + 69.9\text{NS} + 208.5\text{CD} + 83\text{WF}$

There are two sets of constraints, raw material availability and facility capacity constraints. Based on the material consumption of each product as shown in Table 2, facility usage shown in Table 3, and the resource availability shown in Table 4, the constraints can be developed for each material and facility. These are presented, together with the objective function in Model 1.

The following conditions and requirements were established by the marketing department of Chocoman:

1. Product mix requirements. Large-sized products (250 g) of each type should not exceed 60% of the small-sized product (100 g), such that: MB ≤ 60% MS, CB ≤ 60% CS, and NB ≤ 60% NS.
2. Main product line requirement. Chocoman wants to preserve its image as the leading chocolate bar manufacturer. Marketing and top management have specified that total sales from candy and wafer products should not exceed 15% of the total revenues from the chocolate bar products, such that:

    400CD + 150WF ≤ 0.15(375MB + 150MS + 400CB
    + 160CS + 420NB + 175NS)

3. Demand lower limits for each product have been established as follows:

Product	Limit ('000 units)
Milk chocolate, 250 g	500
Milk chocolate, 100 g	800
Crunchy chocolate, 250 g	400
Crunchy chocolate, 100 g	600
Chocolate with nuts, 250 g	300
Chocolate with nuts, 100 g	500
Chocolate candy	200
Wafer	400

The problem is to find the optimal product mix under the technical, raw material, and market considerations as discussed. The complete set of constraints and the objective function are shown in Model 1. Results are presented in Table 7.

Model 1: Product Mix Model — Single-Objective Case

Maximize:
$$Z = 179.95MB + 85MS + 153.08CB + 72.15CS + 129.95NB + 69.9NS + 208.5CD + 83WF$$

subject to:

Material constraints

Cocoa
$$87.5MB + 35MS + 75CB + 30CS + 50NB + 20NS + 60CD + 12WF \leq 100{,}000$$

Table 7 Results of the Single Objective Model

Variable	Solution value
Profit contribution (objective)	266,157
Sales revenue	592,533
Variable cost	326,376
Budget	400,000
Product mix:	
Milk chocolate, 250 g	239
Milk chocolate, 100 g	800
Crunchy, 250 g	0
Crunchy, 100 g	600
Choco with nuts, 250 g	300
Choco with nuts, 100 g	500
Chocolate candy	80
Wafer	278

Multiobjective Programming for Industrial Engineers

Milk
62.5MB + 25MS + 50CB + 20CS + 50NB + 20NS + 30CD + 12WF ≤ 120,000

Nuts
37.5CB + 15CS + 75NB + 30NS ≤ 60,000

Sugar
100MB + 40MS + 87.5CB + 35CS + 75NB + 30NS + 210CD + 24WF ≤ 200,000

Flour
72WF ≤ 20,000

Foil
500MB + 500CB + 250WF ≤ 500,000

Paper
450MB + 450CB + 450NB ≤ 500,000

Plastic
60MB + 120MS + 60CB + 120CS + 60NB + 120NS + 1600CD + 250WF ≤ 500,000

Facility constraints

Cooking
0.5MB + 0.2MS + 0.425CB + 0.17CS + 0.35NB + 0.14NS + 0.6CD + 0.096WF ≤ 1000

Mixing
0.15CB + 0.06CS + 0.25NB + 0.1NS ≤ 200

Forming
0.75MB + 0.3MS + 0.75CB + 0.3CS + 0.75NB + 0.3NS + 0.9CD + 0.36WF ≤ 1500

Grinding
0.25CB + 0.1CS + ≤ 200

Wafer making
0.3WF ≤ 100

Cutting
0.5MB + 0.1MS + 0.1CB + 0.1CS + 0.1NB + 0.1NS + 0.2CD ≤ 400

Packaging 1
0.25MB + 0.25CB + 0.25NB + ≤ 400

Packaging 2
0.05MB + 0.3MS + 0.05CB + 0.3CS + 0.05NB + 0.3NS + 2.5CD + 0.15WF ≤ 1000

Labor
0.3MB + 0.3MS + 0.3CB + 0.3CS + 0.3NB + 0.3NS + 2.5CD + 0.25WF ≤ 1000

Volume mix constraints
MB ≤ 0.60 MS, CB ≤ 0.60 CS, NB ≤ 0.60 NS

Product mix constraint (for company image)
400CD + 150WF ≤ 0.15 (375MB + 150 MS + 400CB + 160CS + 420NB + 175NS)

Demand constraints

MB ≤ 500	NB ≤ 300
MS ≤ 800	NS ≤ 500
CB ≤ 400	CD ≤ 200
CS ≤ 600	WF ≤ 400

All variables ≥ 0

Consider that the management of the company has five objectives: simultaneous maximization of revenue, profit, market share of chocolate bar products, units of products produced, and plant machinery utilization. Each of these objectives is formulated as follows:

1. Maximize revenue. Revenue is equal to price multiplied by units produced. Therefore,

   Maximize $Z_1 = 375\text{MB} + 150\text{MS} + 400\text{CB} + 160\text{CS}$
   $+ 420\text{NB} + 175\text{NS} + 400\text{CD} + 150\text{WF}$

2. Maximize contribution to profit. Profit is equivalent to gross contribution. The variable cost can be calculated, and profit would be selling price less variable cost. Therefore, as formulated in the single-objective model, the profit contribution is as follows:

   Maximize $Z_2 = 179.95\text{MB} + 82.90\text{MS} + 153.08\text{CB}$
   $+ 72.15\text{CS} + 129.95 + 69.90\text{NS}$
   $+ 208.5\text{CD} + 83\text{WF}$

3. Maximize market share of chocolate bar products. Maximizing market share is equivalent to maximizing the tonnage of chocolate bars produced. For every 1000 units of 250 g chocolate bar, the weight would be 0.25 tons. Therefore,

   Maximize $Z_3 = 0.25\text{MB} + 0.10\text{MS} + 0.25\text{CB} + 0.10\text{CS}$
   $+ 0.25\text{NB} + 0.10\text{NS}$

4. Maximize the units of products produced. The advertising department would like to maximize the exposure of the company's brand name. For this to be achieved, the consumers have to be constantly reminded of the product through the appearance of the packaging. In order to maximize on the repeated exposure of the brand, the units of products sold must be maximized. This, therefore, would be

   Maximize $Z_4 = \text{MB} + \text{MS} + \text{CB} + \text{NB} + \text{NS} + \text{CD} + \text{WF}$

5. Maximize plant machinery utilization. To maximize plant utilization, the machines should be loaded to the maximum tonnage for the maximum number of hours. Coefficients of usage are given in Table 3, and the total machine capacity utilization can be calculated by adding usage of each machine per product. For the cutting and packaging machines, as there are no restrictions on weight, the loading can be assumed to be unity and simply added to the others. The objective function would be

Multiobjective Programming for Industrial Engineers 515

$$\text{Maximize} \quad Z_5 = 1.65\text{MB} + 0.9\text{MS} + 1.975\text{CB} + 1.03\text{CS}$$
$$+ 1.75\text{NB} + 0.94\text{NS} + 4.2\text{CD} + 1.006\text{WF}$$

Objectives 1 and 2 can be considered as overall company goals. Objectives 3 and 4 are marketing subsystem considerations, and objective 5 is of interest to the production subsystem.

For the product mix model, the following are the materials and labor availability: cocoa, 100,000 kg; milk, 120,000 kg; nuts, 60,000 kg; sugar, 200,000 kg; flour, 20,000 kg; aluminum foil, 500,000 ft^2; paper, 500,000 ft^2; plastic, 500,000 ft^2; labor, 1000 hours.

The resulting multiobjective product mix model is presented as Model 2.

The payoff matrix for the multiobjective product mix model is given in Table 8. This table is presented as a transpose form of the payoff matrix format exhibited in Fig. 2, Section 2.4.

Model 2: Multiobjective Product Mix Model

Maximize:
Revenue
$Z_1 = 375\text{MB} + 150\text{MS} + 400\text{CB} + 160\text{CS} + 420\text{NB} + 175\text{NS} + 400\text{CD} + 150\text{WF}$

Profit
$Z_2 = 179.95\text{MB} + 82.9\text{MS} + 153.08\text{CB} + 72.15\text{CS} + 129.95\text{NB} + 69.9\text{NS} + 208.5\text{CD} + 83\text{WF}$

Market share for chocolate bars
$Z_3 = 0.25\text{MB} + 0.1\text{MS} + 0.25\text{CB} + 0.1\text{CS} + 0.25\text{NB} + 0.1\text{NS}$

Units produced
$Z_4 = \text{MB} + \text{MS} + \text{CB} + \text{CS} + \text{NB} + \text{NS} + \text{CD} + \text{WF}$

Plant utilization
$Z_5 = 1.65\text{MB} + 0.9\text{MS} + 1.975\text{CB} + 1.03\text{CS} + 1.75\text{NB} + 0.94\text{NS} + 4.2\text{CD} + 1.006\text{WF}$

Table 8 Payoff Matrix for Multiobjective Product Mix Model

	Z_1	Z_2	Z_3	Z_4	Z_5
X^{1*}	641,110	273,590	343.92	2610.44	3410.72
X^{2*}	636,852	275,202	349.03	2829.00	3410.44
X^{3*}	598,745	247,040	371.43	2259.60	2840.99
X^{4*}	614,212	273,421	349.83	3254.20	3448.94
X^{5*}	603,472	261,968	323.17	2700.45	3508.33
Optimal	641,110	275,202	371.43	3254.20	3508.33
Minimum	598,745	247,040	323.17	2259.60	2840.99
Range	42,365	28,162	48.26	994.60	667.34
% deviation	6.6%	10.2%	13.0%	30.6%	19.0%
Total deviation		79.4%			
Average deviation		15.8%			

Subject to:

Facility constraints

Cooking
$0.5MB + 0.2MS + 0.425CB - 0.17CS + 0.35NB + 0.14NS + 0.6CD + 0.096WF \le 1000$

Mixing
$0.15CB - 0.06CS + 0.25NB + 0.1NS \le 200$

Forming
$0.75MB + 0.3MS + 0.75CB + 0.3CS + 0.75NB + 0.3NS + 0.9CD + 0.36WF \le 1500$

Grinding
$0.25CB + 0.1CS \le 200$

Wafer making
$0.3WF \le 100$

Cutting
$0.1MB + 0.1MS + 0.1CB + 0.1CS + 0.1NB + 0.1NS + 0.2CD \le 400$

Packaging 1
$0.25MB + 0.25CB + 0.25NB + 0.1WF \le 400$

Packaging 2
$0.05MB + 0.3MS + 0.05CB + 0.3CS + 0.05NB + 0.3NS + 2.5CD + 0.15WF \le 1000$

Labor
$0.3MB + 0.3MS + 0.3CB + 0.3CS + 0.3NB + 0.3NS + 2.5CD + 0.25WF \le 1000$

Material constraints

Cocoa
$87.5MB + 35MS + 75CB + 30CS + 50NB + 20NS + 60CD + 12WF \le 100{,}000$

Milk
$62.5MB + 25MS + 50CB + 20CS + 50NB + 20NS + 30CD + 12WF \le 120{,}000$

Nuts
$37.5CB + 15CS + 75NB + 30NS \le 60{,}000$

Sugar
$100MB + 40MS + 87.5CB + 35CS + 75NB + 30NS + 210CD + 24WF \le 200{,}000$

Flour
$72WF \le 20{,}000$

Foil
$500MB + 500CB + 500NB + 250WF \le 500{,}000$

Paper
$450MB + 450CB + 450NB \le 500{,}000$

Plastic
$60MB + 120MS + 60CB + 120CS + 60NB + 120NS + 1600CD + 250WF \le 500{,}000$

Volume size constraints
$MB \le 0.60MS, \quad CB \le 0.60CS, \quad NB \le 0.60NS$

Product mix constraint (for company image)
$400CD + 150WF \le 0.15(375MB + 150MS + 400CB + 160CS + 420NB + 175NS)$

Demand constraints
$MB \le 800, \quad CB \le 750, \quad NB \le 500, \quad CD \le 400$
$MS \le 1000, \quad CS \le 800, \quad NS \le 800, \quad WF \le 1000$
All variables ≥ 0

Multiobjective Programming for Industrial Engineers

4.2.3. Solution Procedure: The Step Method

Suppose the objective of the decision maker is to be within 5% of each objective's optimal values; we now explore the possibility of attaining this objective by the step method.

First Iteration Using the formula outlined in the theoretical discussion (Section 3.4) and the values in Table 8, the following factors are computed:

	Max	Min	$(\Sigma\, c_{kj}^2)^{1/2}$	α_k	π_k
Z_1	641,110	598,745	859.2	7.69×10^{-5}	0
Z_2	275,202	247,040	374.2	2.73×10^{-4}	0
Z_3	371.43	323.17	0.466	0.278	0.66
Z_4	3354.2	2259.6	2.828	0.108	0.26
Z_5	3508.3	2841.0	5.576	0.034	0.08

The STEM model is formulated as follows:

Minimize y

subject to

$y \geq [371.43 - (0.25\text{MB} + 0.1\text{MS} + \cdots + 0.1\text{NS})]\, 0.66$

$y \geq [3254.2 - (\text{MB} + \text{MS} + \cdots + \text{WF})]\, 0.26$

$y \geq [3508.33 - (1.65\text{MB} + 0.9\text{MS} + \ldots + 1.006\text{WF})]\, 0.08$

and the original constraints in Model 2

The solution set obtained is

$Z_k = (607901;\ 269677;\ 354.11;\ 3214.96;\ 3365.87)$

which, when compared to the ideal solution

$Z^* = (641110;\ 275202;\ 371.43;\ 3254.20;\ 3508.33)$

has the following deviation:

% deviation = 5.18%; 2.01%; 4.66%; 1.21%; 4.06%)

Second Iteration From the deviations, only objective 1 is not yet satisfied; objectives 2 to 5 are satisfied and could be relaxed. The second iteration would, therefore, be

Minimize y

subject to

$y = [641110 - (375\text{MB} + 150\text{MS} + \cdots + 150\text{WF})]\ 1$

$375\text{MB} + 150\text{MS} + \cdots + 150\text{WF} \geq 607901$

$179.95\text{MB} + 82.90 + \ldots + 83\text{WF} \geq 0.95\ Z_2^*$

$0.25\text{MB} + 0.1\text{MS} + \ldots + 0.1\text{NS} \geq 0.95\ Z_3^*$

$\text{MB} + \text{MS} + \ldots + \text{WF} \geq 0.95\ Z_4^*$

$1.65\text{MB} + 0.9\text{MS} + \ldots + 1.006\text{WF} \geq 0.95\ Z_5^*$

and the original Model 2 constraints

The new solution set is obtained as

$Z_k = (619781;\ 272427;\ 352.86;\ 3091.49;\ 3375.26)$

with % deviation = (3.33%; 1.01%; 5%; 5%; 3.79%) from the individual optimum and acceptable. The compromise solutions in decision space is the following product mix: MB = 131, MS = 1000, CB = 0, CS = 506, NB = 359, NS = 800, CD = 20, and WF = 278.

4.3. Job Sequencing and Scheduling: Jobs on One Machine

Job sequencing problems deal with the determination of the best ordering of jobs given one or more criteria. Most bicriterion machine sequencing problems minimize any of the following: weighted flow time and maximum tardiness, total flow time and maximum penalty cost, total flow time and number of tardy jobs, maximum tardiness and crashing costs, or some combination thereof. The sequencing problem discussed in this section is a bicriterion case that minimizes both earliness and tardiness of the jobs.

A job finished before its due date that cannot be shipped out incurs an inventory holding cost. Since a penalty is imposed on jobs that are completed early, machine idle time may be inserted into a schedule so that jobs are completed when they are needed. The concept of inserted idle time is contrary to the conventional wisdom of keeping workers busy all the time and not allowing downstream workstations to become idle. Just in time (JIT), zero inventory, kanban, and stockless production systems, however, suggest that the time (thus money) lost in having workstations sit idle could be offset by savings in reduced inventory carrying cost (see Adulbhan and Tabucanon [31]).

JIT is more generally known as the pull system. Transition to such a system can be facilitated by recognizing earliness as an explicit penalty in the optimizing process. The incorporation of earliness as a penalty greatly

Multiobjective Programming for Industrial Engineers

complicates the sequencing problem. Earliness is not a regular measure of performance, making the optimal solution a nonpermutation schedule (see Fry and Leong [24]). Although $N!$ permutation schedules may be quite large, they are finite. The number of nonpermutation schedules becomes infinite due to the continuous variable representing the inserted machine idle time.

The problem illustrated is a single-machine job shop where all jobs are simultaneously available for processing. Due dates are known in advance, and the machine can process only one job at a time. The processing of a job, once started, cannot be stopped until its completion.

The problem is modeled using GP with a structure similar to that of Fry and Leong [24]. The following notation is used: The number of jobs available for processing is represented by N; C_k is the completion time of the job in the kth position; W_k is the amount of machine idle time directly preceding the job in the kth position; T_k is the processing time of the job in the kth position; X_{jk} is equal to 1 if job j is assigned to position k in the sequence and is 0 otherwise; S_k is the identification of the job occurring in position k in the sequence; and d_k is the due date of the job in the kth position. The Gantt chart in Figure 4 shows the model variables.

Since a job can be assigned to only one position in the sequence, the following equations hold:

$$\sum_{j=1}^{N} X_{jk} = 1, \quad k = 1, 2, \ldots, N$$

$$\sum_{k=1}^{N} X_{jk} = 1, \quad j = 1, 2, \ldots, N$$

The due date goal constraints reflecting earliness and tardiness variables are presented as

$$C_k + l_k^- - l_k^+ = d_k, \quad k = 1, 2, \ldots, N$$

where l_k^- and l_k^+ are the earliness and tardiness deviational variables, respectively, of the job in position k. Since

$$C_k = \sum_{l=1}^{k} W_l + \sum_{l=1}^{k} T_l$$

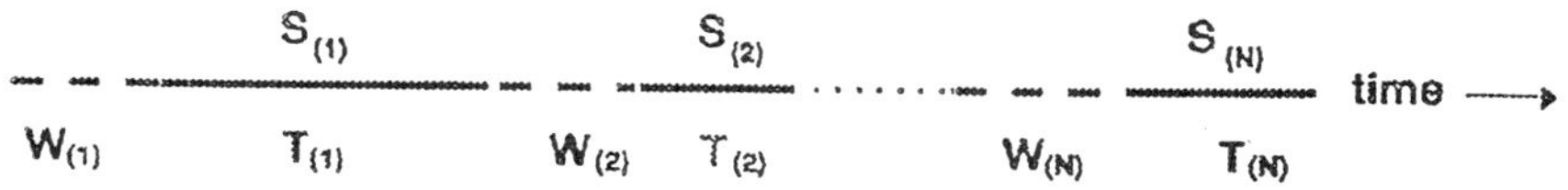

Figure 4 Gantt chart of the sequencing problem.

$$T_k = \sum_{j=1}^{N} X_{jk} P_j$$

$$d_k = \sum_{j=1}^{N} X_{jk} d_j$$

the goal constraints can be written as

$$\sum_{l=1}^{k} W_l + \sum_{j=1}^{N} \sum_{l=1}^{k} X_{jl} T_j + l_k^- - l_k^+ = \sum_{j=1}^{N} X_{jk} d_j, \quad k = 1, 2, \ldots, N$$

The objective function is a result of a weighted sum of earliness and tardiness penalties. Thus,

$$\text{Min} \quad z = v_h \sum_{j=1}^{N} l_j^- + v_t \sum_{j=1}^{N} l_j^+$$

where v_h is the penalty per unit of time associated with holding an item in inventory and v_t is the penalty per unit of time associated with finishing a job tardy.

The model is illustrated using a sample problem given in Table 9.

Three runs are made to illustrate cases in which earliness minimization is more important that lateness minimization (run 1), both are equally important (run 2), and earliness is less important than lateness goal (run 3). This is illustrated in Figure 5. Run 1 shows that two jobs are tardy and there are no early jobs. When the cost of holding inventory ($v_h = 4$) is higher than the cost of finishing a job late ($v_t = 1$), more jobs are expected to be finished after their due dates. The makespan is 70 and total machine idle time is 32 time units.

In the second run, when $v_h = v_t = 1$, makespan decreases to 66 because the total inserted idle time decreases by four time units. There are equal numbers of tardy and early jobs. The sequence remains the same (4-3-5-2-1). In the third run, where $v_h = 1$ and $v_t = 3$, the sequence changes from (4-3-5-2-1) to (4-3-2-5-1), and the resulting schedule has one tardy job and two early jobs. The total inserted idle time becomes 22.

The total penalty over the three runs is also different, with values of $Z = 16$, $Z = 12$, and $Z = 18$, respectively, for the three runs.

Table 9 Sample Problem

Job	1	2	3	4	5
T_j	12	10	4	8	4
d_j	58	54	36	20	48

Multiobjective Programming for Industrial Engineers

CPU time grows in an almost exponential manner as the problem size is increased. Combinatorial techniques such as branch and bound or dynamic programming may be used to curtail enumeration for larger problems. Curtailment can be accomplished using strong bounding or dominance theorems that could determine the ordering of jobs before the enumeration begins. Implementing these techniques, however, would require some procedure to insert idle time as the enumeration is performed. Because of the computational difficulty in solving a mathematical programming-based sequencing problem, many resort to heuristics, especially for multimachine cases (see Chapter 9). One example is presented in the next section.

4.4. Job Sequencing and Scheduling: Jobs on Parallel Machines

Production scheduling is the allocation of resources over time to perform a collection of tasks. The resources are machines. Pure sequencing is a specialized scheduling problem in which the ordering of jobs completely determines a schedule. The effectiveness of a specific sequence may be measured

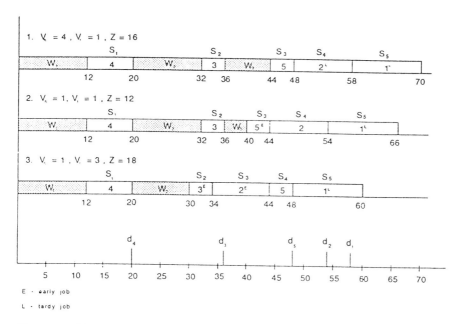

Figure 5 Model results of job sequencing and scheduling.

in terms of average completion time, due date performance, inventory of jobs in process, labor, capacity, etc. Two commonly used measures of performance are mean flow time and maximum tardiness.

A manufacturing firm engaged in batch or small lot production is usually faced with scheduling problems. Objectives in scheduling vary from firm to firm and often from time to time. Traditionally, the aim would be either to achieve certain contractual target dates or simply to finish all the work as early as possible. Although research on the subject has continued to deal predominantly with the single-objective case, real-world managers often perform the scheduling function under multiple objectives.

Traditionally, measures of job shop performance such as makespan, flow time, tardiness, and number of jobs engaged are considered one at a time in classical single-objective job shop sequencing. Heuristics to solve this type of problems involve the use of dispatching rules such as shortest processing time (SPT) for mean flow time minimization, earliest due date (EDD) for maximum tardiness minimization, minimum slack time (MST) for minimizing total tardiness, and largest processing time (LPT) for minimizing mean flow time for parallel operations.

Smith [25] dealt with the problem of minimizing total flow time while meeting due dates of jobs. Wassenhove and Gelders [26] extended Smith's idea to solving a bicriterion scheduling problem. They developed an algorithm considering two criteria, namely minimizing total flow time and minimizing maximum tardiness. The concept of bicriterion scheduling, as a development of Smith's algorithms, was further extended by Chand and Schneeberger [27] for the problem of minimizing the weighted completion time and the maximum allowable tardiness.

In this section, an algorithm is developed for two or more parallel machines in single-stage production. The algorithm deals with two criteria: total flow time and maximum tardiness.

4.4.1. Notations, Assumptions, and Concepts

The following notation will be used throughout the extent of this section: n is the number of jobs available in the shop; m is the number of machines working in each product family (in line with the group technology concept); p_i is the processing time of job i; d_i is the due date of job i; C_i is the completion time of job i; F_i is the flow time of job i (total processing time of jobs completed earlier plus the processing time of its own); F_{ave} is the average flow time; c is the cost function of the schedule; S is a schedule; π is a permutation schedule; Π is the set of permutation schedules; $T(\pi)$ is the tardiness of schedule π; $H(\pi)$ is the total flow time of schedule π; τ is also the total processing time of all available jobs; U is the set of all jobs; J is a job; Δ is the allowable maximum tardiness; k is the position of a job in a set of jobs; and T_{max} is the maximum tardiness.

The following assumptions are valid unless otherwise specified: (1) no preemption; (2) no cancellation; (3) the processing times are independent of schedules; (4) in-process inventory is allowed; (5) no machine can handle more than one job at a time; (6) all machines are available throughout the scheduling horizon; (7) the technological constraints are known in advance and are immutable during the process; (8) the number of jobs, the number of machines, the processing times, and the times at which the jobs are available for assignment are known and fixed.

French [28] considers a single-machine problem in which T_{max} is the suitable indicator of the penalty cost arising from the late completion of jobs and that F_{ave} is an equally important indicator of in-process inventory cost. It is assumed that any other cost component can be neglected. Thus it is confirmed that the total cost is a function of T_{max} and F_{ave} alone, say $c(T_{max}, F_{ave})$. Now it is reasonable to suppose that if either T_{max} or F_{ave} increases, the total cost does, too. Therefore, $c(T_{max}, F_{ave})$ is an increasing function of both its arguments.

Now, suppose that S and S' are two schedules for which $T_{max} < T'_{max}$ and $F_{ave} < F'_{ave}$, where F'_{ave} and T'_{max} refer to S'.

In this case it is clear that S is more acceptable than S' because we have $c(T_{max}, F_{ave}) < c(T'_{max}, F'_{ave})$.

Indeed, a little thought shows that we can always say that S is better than S' if $T_{max} \leq T'_{max}$ and $F_{ave} \leq F'_{ave}$ with strict inequality holding in at least one case.

We shall say that a schedule S' is efficient if there does not exist a schedule S that dominates it.

Then, the problem (P) can be written as

$$\text{(P)} \quad \underset{\pi \in \Pi}{\text{Min}} \sum_{i=1}^{n} C_t = \underset{\pi \in \Pi}{\text{Min}} H(\pi)$$

$$\underset{\pi \in \Pi}{\text{Min}} \underset{i=1,\ldots,n}{\text{Max}} \{\text{Max}(0, C_i - d_i)\} = \underset{\pi \in \Pi}{\text{Min}} T(\pi)$$

We know that the first objectie above can be reached by ordering the jobs according to nondecreasing processing times (SPT) and the second objective can be achieved by ordering the jobs according to nondecreasing due dates (EDD). But if we wish to consider both objectives, the rules work rather poorly. For this we want to have a sequence that does well on both objectives. To define such sequences the concept of efficiency was used by Wassenhove and Gelders [26].

A sequence $\pi^* \in \Pi$ is efficient in problem (P) if there exists no $\pi \in \Pi$ such that $H(\pi) \leq H(\pi^*)$ and $T(\pi) \leq T(\pi^*)$ where at least one relation holds with strict inequality. Similarly, a sequence π_1 dominates a sequence π_2 when $H(\pi_1) \leq H(\pi_2)$ and $T(\pi_1) \leq T(\pi_2)$ where at least one relation is a strict inequality.

4.4.2. Algorithm for the Sequencing of Jobs with Two Performance Measures

Smith [25] considered a single-machine sequencing problem with $T_{max} = 0$; that is, all the due dates can be met if the jobs are sequenced with the EDD rule. He developed an algorithm that gives other sequences that minimize flow time subject to the condition that $T_{max} = 0$. Wassenhove and Gelders [26] modified Smith's algorithm to minimize F_{ave} subject to $T_{max} \leq \Delta$, that is, subject to no job being finished more than Δ after its due date. Their algorithm generates all the possible efficient sequences. This algorithm can be described as follows:

Step 1: Put $\Delta = \Sigma_i^n p_i$.
Step 2: Let $D_i = d_i + \Delta$ for all i.
Step 3: Solve the new problem using Smith's algorithm as modified by Wassenhove and Gelders [26]. If a solution exists, it is efficient. Else go to step 5.
Step 4: Compute $T(\pi^*) = \max_{i=1\ldots n}\{\max(0, C_i - d_i)\}$. Put $\Delta = T(\pi^*) - 1$. Go to step 2.
Step 5: Stop.

4.4.3. Algorithm for Parallel Processors with Bicriterion Performance Measures*

The problem now addressed is to sequence N independent jobs on m parallel processors with two criteria, all jobs being available at an arbitrary time zero. The problem can be considered in two stages, namely (1) allocation of n jobs on m processors and (2) sequencing of jobs to achieve both criteria.

An algorithm that can deal with parallel processors and yet generates efficient sequences meeting the two criteria, namely minimizing the maximum tardiness and minimizing the total flow time, is as follows:

Step 1: Start with the given information on the jobs.
Step 2: Generate an efficient sequence of the jobs using Wassenhove and Gelders' algorithm.

---

*This is an edited version of materials taken, with the publisher's permission, from Cenna and Tabucanon [29].

Step 3: Distribute the jobs among the machines following the one-at-a-time (H_1) heuristic developed by Baker and Merten [32]. In this heuristic, the machine with the smallest amount of scheduled jobs is assigned the first job on the waiting list. This step is repeated until all jobs are assigned.
Step 4: Generate all the efficient sequences at each machine level as suggested by Wassenhove and Gelders.
Step 5: Determine the efficient sequences at the product family level.
Step 6: Go to step 2 for next sequence. If no such sequence is possible go to step 7.
Step 7: Compare all the efficient sequences at product family level to get the overall efficient sequences.
Step 8: Stop.

Using Wassenhove and Gelder's algorithm, several sequences may be created in a single queue before the allocation of jobs among the machines. For each of these sequences efficient sequences are created at the machine level. Finally, the overall efficient sequences are taken. This technique is illustrated at Figure 6.

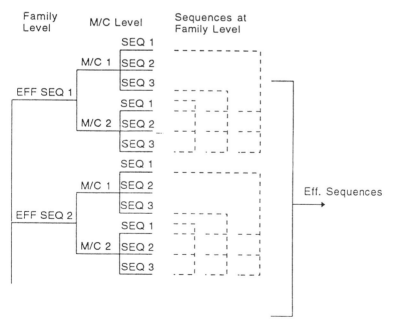

Figure 6 Generation of efficient sequences.

4.4.4. The Case Study*

In order to illustrate the practicality of the model, a real-life case study was conducted in an electronic manufacturing plant. While dealing with smaller components, it has to deal with a large variety of products. The specification of the products changes with different orders, and thus keeping raw material inventory does not seem logical. So the plant has to produce on order with a large lead time. The scheduling time horizon is 1 month.

The production department comprises a number of machines. It has to handle a large number of orders with a large variety of product specifications. The concept of product family has been introduced to take advantage of product classification according to the technological constraints. When the products are divided into families it is found that each family of products is handled by two or more machines and the machines are operating in parallel. Most of the products require only single-stage operation. Machines are mostly semiautomatic. So, for each family of products there is a single queue that is served by two or more parallel machines. Here, only the two-machine case is considered.

The model was run with the actual data set. The model gives the possible efficient sequences with respect to total flow time and maximum tardiness. The program also calculates the total tardiness for each of the efficient sequences. Actually, the selection of a sequence depends on the weights given to the parameters of maximum tardiness and total flow time. A third criterion, total tardiness, helps the decision maker to make a decision. Results are shown in Tables 10 and 11.

4.5. Maintenance of Machines

To maintain an efficient production system and reduce the possibility of work being suspended due to breakdown of machines, the necessity of a maintenance system is evident. Although maintenance does not actually eliminate deterioration, it does prolong the life of the machine by fast corrective action while the problem is still manageable. A regular inspection routine, more commonly called preventive maintenance, is most desirable to uncover minor faults before the damage becomes major. Periodic checks allow production to be planned around these times when machines and equipment are necessarily pulled out of work circulation; unexpected production downtime can be reduced, equipment life expectancy is increased, safety is improved, large expenditures for major repairs are minimized, and machine efficiency is maximized.

---

*This is an edited version of materials taken, with the publisher's permission, from Cenna and Tabucanon [29].

Multiobjective Programming for Industrial Engineers

Table 10 Output Parameters for the Input Data of Product Family 1

Input Information on Jobs											
Job Number	1	2	3	4	5	6	7	8	9	10	11
Processing time	8	7	5	7	7	4	3	2	10	1	2
Due date: June	21	6	4	10	12	21	14	22	17	22	16
Due date: July	21	6	5	11	11	21	15	22	18	22	16
Due date: August	22	7	4	10	12	22	15	23	19	23	17

		Output Information on Processing Orders								
Month	Total Flowtime	Maximum Tardiness	Total Tardiness	Job Numbers						
			Efficient Sequence 1							
June	136	10	48	m/c 1	10	8	6	2	5	1
				m/c 2	11	7	3	4	9	–
July	136	10	46	m/c 1	10	8	7	3	5	9
				m/c 2	11	6	2	4	1	–
August	136	9	44	m/c 1	10	8	7	3	4	9
				m/c 2	11	6	2	5	1	–
			Efficient Sequence 2							
June	153	9	48	m/c 1	10	7	3	4	9	8
				m/c 2	11	6	2	5	1	–
July	142	9	38	m/c 1	10	8	2	4	6	1
				m/c 2	11	7	3	5	9	–
August	142	8	36	m/c 1	10	8	2	5	6	1
				m/c 2	11	7	3	4	9	–
			Efficient Sequence 3							
June	174	8	49	m/c 1	7	3	4	9	10	8
				m/c 2	11	6	8	5	1	–
July	159	8	38	m/c 1	10	7	3	5	9	8
				m/c 2	11	2	4	6	1	–
August	169	7	35	m/c 1	10	7	3	4	9	8
				m/c 2	11	2	5	6	1	–
			Efficient Sequence 4							
June	no production			m/c 1	–	–	–	–	–	–
				m/c 2	–	–	–	–	–	–
July	180	7	39	m/c 1	7	3	5	9	10	8
				m/c 2	11	2	4	6	1	–
August	180	6	35	m/c 1	7	3	4	9	10	8
				m/c 2	11	2	5	6	1	–

Table 11 Output Parameters for the Input Data of Product Family 2

Input Information on Jobs												
Job Number	1	2	3	4	5	6	7	8	9	10	11	–
Processing time	8	10	9	10	7	3	3	4	1	1	–	–
Due date: June	21	21	15	15	7	6	3	4	1	1	–	–
Due date: July	21	21	16	16	8	6	4	5	1	1	–	–
Due date: August	23	23	16	15	9	3	3	6	7	16	6	–

Output Information on Processing Orders										
Month	Total flowtime	Maximum Tardiness	Total Tardiness	Job Numbers						
Efficient Sequence 1										
June	121	11	34	m/c 1	9	7	8	1	4	–	–
				m/c 2	10	6	5	3	2	–	–
July	121	10	29	m/c 1	9	7	8	1	4	–	–
				m/c 2	10	6	5	3	2	–	–
August	124	9	31	m/c 1	11	9	10	6	5	3	1
				m/c 2	7	8	2	4	–	–	–
Efficient Sequence 2										
June	123	9	31	m/c 1	9	7	8	4	1	–	–
				m/c 2	10	6	5	3	2	–	–
July	123	9	26	m/c 1	9	7	8	4	1	–	–
				m/c 2	10	6	5	3	2	–	–
August	119	11	30	m/c 1	11	10	6	2	3	1	–
				m/c 2	9	7	8	5	4	–	–

4.5.1. The Case Study: Copying Machines

Copying machines break down from time to time, not necessarily due to low-quality technology but mainly because they are usually operated extensively in offices. While a copying machine is being repaired, a customer's opportunity cost is incurred. To reduce the frequency of breakdowns, the machine has to be inspected periodically and any minor defects that could eventually cause a major breakdown have to be rectified. These preventive inspections, however, cost money in terms of materials (spare parts, etc.), wages, and loss of production due to scheduled downtime.

A company vending copying machines is faced with a problem of determining an inspection policy for its customers as part of after-sales service. Profit is not the only criterion that customers want to maximize. Those who use the copying machines in offices would mainly desire downtime minimization, not profit. On the other hand, the vendor would necessarily view profit and sales maximization, although these are limited by the availability of resources, such as availability of after-sales service engineers,

Multiobjective Programming for Industrial Engineers

budget (especially for maintenance service), and minimum requirement of after-sales service.

Historical analysis has shown that the machines fail according to a negative exponential distribution with mean time to failure (MTTF) equal to $1/\lambda_k$, where λ_k is the failure arrival rate. The breakdown rate of machine type k is a function of the frequency of inspection, n. Repair times are negative exponentially distributed with mean time to repair (MTTR) equal to $1/\mu_k$. The company's policy is to perform n inspections per 10,000-copy volume regardless of the type of copying machine. The inspection times (for preventive maintenance) are also negative exponentially distributed with mean time to inspect (MTTI) equal to $1/i_k$. The value of output in an uninterrupted 10,000-copy volume has a "profit" contribution of v_k. The average cost of inspection and of repair per uninterrupted 10,000-copy volume are I_k and R_k, respectively. There are four types of copiers:

X_1 machines: low speed (10 pages/min)
X_2 machines: medium speed (30 pages/min)
X_3 machines: medium/high speed (60 pages/min)
X_4 machines: very high speed (90 pages/min)

The average speeds of copy volume (cv) are 4000 cv/month (equivalent to 0.5 page/min), 10,000 cv/month (equivalent to 1.2 pages/min), 30,000 cv/month (equivalent to 3.6 pages/min), and 75,000 cv/month (equivalent to 8.9 pages/min) for X_1, X_2, X_3, and X_4 machines, respectively. The after-sales service engineers work on a regular basis of TM hours/day and the total available budget for maintenance is B per month.

There are two parties involved in this problem: the customer, who would like to maximize its satisfaction, and the vendor company, who, as much as possible, wishes to maximize its maintenance capacity, which leads to maximizing sales capacity. In total there are four objectives: maximizing customer's profit, minimizing customer's downtime for customer's satisfaction, maximizing maintenance capacity based on company's preference, and maximizing maintenance capacity based on customer's preference.

4.5.2. Maximizing Customer's Profit

The profit per unit copy volume from operating a machine of type k is a function of the number of inspections, n, is denoted by $P_k(n)$. This is determined by deducting the opportunity costs due to repair and inspection and the costs of actual repair and inspection from the revenue generated,

$$P_k(n) = v_k - v_k\left[\frac{\lambda_k(n)}{\mu_k}\right] - v_k\left[\frac{n}{i_k}\right] - R_k\left[\frac{\lambda_k(n)}{\mu_k}\right] - I_k\left[\frac{n}{i_k}\right]$$

Thus the first objective of maximizing the customer's total profit can be expressed as

$$\text{Max} \quad Z_1 = \sum_{k=1}^{4} x_k P_k(n)$$

where x_k is the maintenance capacity (in units) for each type of copy machine.

4.5.3. Minimizing Customer's Downtime

The total downtime per 10,000-copy volume is a function of the inspection frequency, n, and is determined by the sum of downtimes for repair and inspection. The downtime function of a machine of type k, $D_k(n)$, can be expressed as

$$D_k(n) = \frac{\lambda_k(n)}{\mu_k} + \frac{n}{i_k}$$

and the second objective can be expressed as

$$\text{Min} \quad Z_2 = \sum_{k=1}^{4} x_k D_k(n)$$

where the variables are as defined above.

4.5.4. Maximizing Maintenance Capacity from Company's Preference

The company's top management prefers to sell the big machines rather than the smaller ones. Therefore the maximum speed of each machine can be used as the weight in the objective function. Thus the third objective can be expressed simply as follows:

$$\text{Max} \quad Z_3 = \sum_{k=1}^{4} w_k x_k$$

where $w_1 = 1$, $w_2 = 3$, $w_3 = 6$, and $w_4 = 9$, taken as the ratios of copying speeds given earlier.

4.5.5. Maximizing Maintenance Capacity from Customer's Preference

This objective is similar to the previous one except that the weights are different. The collective preference of the customers is taken from the distribution of machine types currently owned by the customers. The historical record shows that there is a general customer preference for smaller machines as evidenced by the following patterns of purchases: X_1, X_2, X_3, and X_4 types accounted for 51%, 39.2%, 9.7%, and 0.1% of total sales, respectively. Making use of this empirical distribution, the fourth objective can be expressed as

$$\text{Max} \quad Z_4 = \sum_{k=1}^{4} w'_k X_k$$

where $w'_1 = 51$, $w'_2 = 39.2$, $w'_3 = 9.7$, and $w'_4 = 0.1$.

4.5.6. Constraints

The constraints that form the feasible set of alternatives for the problem are the availability of working hours of after-sales service engineers, budget, minimum requirement of inspection frequency, and bounds due to contractual agreements, if any.

The total time required to do the inspections and repairs cannot exceed the working hours. The available working hours will be adjusted for travel time of the engineers, lunch breaks, and preparation time. The adjustment is in terms of percentage of the total available time. Thus, the working hours constraint can be expressed as

$$\sum_{k=1}^{4} \left(\frac{n}{i_k}\right) x_k + \left[\frac{\lambda_k(n)}{\mu_k}\right] x_k \leq (af)(\text{SE})(\text{TM})$$

where the first term is the total inspection time, the second term is the total repair time, and the right-hand side of the constraint is the available man-time — af is the adjustment factor, SE is the total number of after-sales service engineers, and TM is the monthly available working time of an engineer.

Each inspection for a copying machine k will cost c_{ik} and each repair cost is c_{rk}; thus the budget constraint can be expressed as

$$\sum_{k=1}^{4} [\lambda_k(n) c_{rk} x_k + n c_{ik} x_k] \leq B$$

Following the company policy, the minimum requirement of inspection frequency simply states that for every $cv_k^{\min}$ copy volume the machine should be inspected:

$$n_k \geq cv_k^{\min}$$

Here, n_k is not restricted to be integer because it can take any nonnegative value. Also, due to contractual agreements there are established lower bounds LB_k for x_k:

$$x_k \geq \text{LB}_k$$

The analysis of data taken from the vendor under consideration is summarized in Table 12.

Table 12 Vendor Data

1.	Failure data	$1(n) = 0.48/n_1$ failures per 10,000 cv $2(n) = 0.46/n_2$ failures per 10,000 cv $3(n) = 0.27/n_3$ failures per 10,000 cv $4(n) = 0.27/n_4$ failures per 10,000 cv
2.	Average of real copy volume	for low speed — 0.5 pages/min for medium speed — 1.2 pages/min for medium/high speed — 3.6 pages/min for very high speed — 8.9 pages/min
3.	Time data per 10,000 copy volume	$MTTR_1 = $ 72 minutes $=$ 36 cv $= 0.0036$ $MTTR_2 = $ 80 minutes $=$ 96 cv $= 0.0096$ $MTTR_3 = $ 107 minutes $=$ 386 cv $= 0.0386$ $MTTR_4 = $ 147 minutes $=$ 1308 cv $= 0.1308$ $MTTI_1 = $ 37 minutes $=$ 19 cv $= 0.0019$ $MTTI_2 = $ 49 minutes $=$ 58 cv $= 0.0058$ $MTTI_3 = $ 68 minutes $=$ 245 cv $= 0.0245$ $MTTI_4 = $ 98 minutes $=$ 872 cv $= 0.0872$
4.	Customer profit and cost data per 10,000 cv (estimated)	$v_k = \$4{,}000$
5.	Engineer's data	SE $=$ 220 persons TM $=$ 8 hours/day af $=$ 0.4
6.	Company cost and budget	$C_{ik} = \$16$ $C_{rk} = \$22$ $B = \$600{,}000$

4.5.7. Solution Procedure: The Global Criterion Method

The global criterion method has been chosen to solve this problem. Several values of p are chosen to see the sensitivity of the compromise solution to the different weights that are given to the problem (the values of p are 1, 2, and 5).

The complete problem can be stated as follows:

1. Objectives

$$\text{Max } Z_1 = \left[100 - \frac{0.2938}{n_1} - 0.247 n_1\right] x_1$$
$$+ \left[100 - \frac{0.7507}{n_2} - 0.745 n_2\right] x_2$$

Multiobjective Programming for Industrial Engineers 533

$$\text{Min } Z_2 = \left[\frac{1.7280}{n_1} + 19n_1\right]x_1 + \left[\frac{4.4160}{n_2} + 58n_2\right]x_2$$
$$+ \left[100 - \frac{1.7717}{n_3} - 3.185n_3\right]x_3$$
$$+ \left[100 - \frac{5.3366}{n_4} - 11.336n_4\right]x_4$$
$$+ \left[\frac{10.422}{n_3} + 245n_3\right]x_3$$
$$+ \left[\frac{31.392}{n_4} + 842n_4\right]x_4$$

Max $Z_3 = [x_1 + 3x_2 + 6x_3 + 9x_4]$

Max $Z_4 = [51.0x_1 + 39.3x_2 + 9.7x_3 + 0.1x_4]$

2. Time constraint in 1 month

$$x_1\left[\left(\frac{0.48}{n_1}\right)72 + 37n_1\right]\frac{10,000}{4,000} + x_2\left[\left(\frac{0.46}{n_2}\right)80 + 49n_2\right]\frac{10,000}{10,000}$$
$$+ x_3\left[\left(\frac{0.27}{n_3}\right)107 + 68n_3\right]\frac{10,000}{30,000} + x_4\left[\left(\frac{0.24}{n_4}\right)147 + 98n_4\right]$$
$$\frac{10,000}{75,000} \leq (0.4)(220)(14,400)$$

3. Budget constraint in 1 month

$$x_1\left[\left(\frac{0.48}{n_1}\right)550 + 400n_1\right]\frac{10,000}{4,000} + x_2\left[\left(\frac{0.46}{n_2}\right)550 + 400n_2\right]$$
$$\frac{10,000}{10,000} + x_3\left[\left(\frac{0.27}{n_3}\right)550 + 400n_3\right]\frac{10,000}{30,000} + x_4\left[\left(\frac{0.24}{n_4}\right)550\right.$$
$$\left. + 400n_4\right]\frac{10,000}{75,000} \leq 600,000$$

4. Minimum maintenance frequency requirement for each type of machine

   $n_1 \geq 0.20$
   $n_2 \geq 0.15$
   $n_3 \geq 0.10$
   $n_4 \geq 0.05$

5. Minimum scheduled maintenance (due to contract)

$x_1 \geq 4000$ (scheduled maintenance)
$x_2 \geq 2000$ (scheduled maintenance)
$x_3 \geq 500$ (scheduled maintenance)
$x_4 \geq 10$ (scheduled maintenance)

The solution is obtained by using the GINO software (Lindo Systems, Inc.) for the nonlinear case. Several sensitivity analyses concerning the changing of available resources are also conducted. The sensitivity analysis is done with the following variation:

SE = 275 persons
B = 800,000/month
SE = 300 persons and B = 700,000

After calculating the optimal values for the individual objectives, the payoff matrix for this problem can be depicted as follows:

	x_1^*	x_2^*	x_3^*	x_4^*
Z_1^*	1,498,792			
Z_2^*		168,555		
Z_3^*			81,594	
Z_4^*				409,297

Formulating the problem as a distance function in LP space, the compromise solution can be obtained by substituting a value of p. The values of p that are chosen are 1, 2, and 5. Different values of p are chosen in order to see the effect of different weights to the compromise solution. The results can be seen in Table 13. The fractional values of the n_k's imply that, say for $p = 1$, the inspection frequency for each type of machine is one per about 10,000 cv (10,000 ÷ 0.94) for type 1, about 13,000 cv for type 2, about 29,000 cv for type 3, and about 43,000 cv for type 4 machines. The x_k's indicate the maintenance capacities.

For $p = 1$, the limiting factor is the working hours constraint; for $p = 2$ both the working hours and budget constraints are active; and for $p = 5$, the budget constraint becomes the only one active. This pattern implies that as more weight is given to larger deviations from the ideal, the company will be able to provide maintenance to machines that are still under contract.

Sensitivity runs for budget and sales engineers' availability show interesting results (see Table 14 and Table 15).

Given equal weights to all objectives, if the company is able to increase the budget, then the additional budget will not be so useful because

Multiobjective Programming for Industrial Engineers

Table 13 Results of Using Different Weights on the Compromise Solution

	Parameter		
Variable	$p = 1$	$p = 2$	$p = 5$
n_1	0.94	0.51	0.48
n_2	0.75	0.21	0.29
n_3	0.35	0.20	0.21
n_4	0.23	0.19	0.19
x_1	4000	4000	4000
x_2	3806	2000	2000
x_3	3808	500	500
x_4	3805	10	10

the active constraint is not the budget for this case. But by giving more weight to the longest distance from the ideal point, if the company's budget can be increased, the additional budget can be used to increase the maintenance capacity of copying machine types 2, 3, and 4, as well as increase mainly the inspection frequency of copying machine type 1. It seems that increasing the budget constraint is not so useful here.

Naturally, if the working time is increased, budget will become an active constraint, and therefore more budget will be needed. Increasing both constraints with $SE = 275$ and budget $B = 700,000$ gives the following compromise results: frequency will have a negative correlation with the

Table 14 Compromise Solution for Budget of $800,000

	Parameter		
Variable	$p = 1$	$p = 2$	$p = 5$
n_1	0.94	0.65	0.43
n_2	0.75	0.38	0.30
n_3	0.35	0.22	0.21
n_4	0.23	0.20	0.19
x_1	4000	4000	4000
x_2	3806	2561	2000
x_3	3808	2900	500
x_4	3805	2156	10

Table 15 Compromise Solutions for Different Combinations of Budget and Sales Engineers Availability (for $p = 2$)

Variables	Parameters			
	SE = 20 B = 600,000	SE = 275 B = 600,000	SE = 220 B = 800,000	SE = 300 B = 700,000
n_1	0.51	0.99	0.65	0.54
n_2	0.21	0.20	0.38	0.31
n_3	0.20	0.19	0.20	0.19
n_4	0.19	0.19	0.20	0.19
x_1	4000	4000	4000	4087
x_2	2000	3695	2561	4087
x_3	500	2856	2900	4087
x_4	10	10	2156	4087

number of maintenance capacity. The crucial thing with the present condition is the number of available after-sales service engineers during the available working hours. Increasing the number of engineers while reducing the inspection frequency will result in higher maintenance capacity for lower-speed and medium-speed machines (even reduce the high and very high maintenance capacity).

Given unequal weights to all criteria, it is interesting to see that all constraints are active in a certain range of increase in the resources. Moreover, by increasing the number of after-sales service engineers, the company will be able to give better inspection frequency to low-speed copying machines and at the same time increase the maintenance capacity of medium- and high-speed copying machines. Increasing the budget will enable the company to increase all inspection frequencies and also increase the maintenance capacity for almost all types of copying machines.

4.6. R&D Project Selection

In modern-day manufacturing, shortening of product life cycle is a common phenomenon especially for electronic products. No firm in the industry can survive in the immensely competitive environment without investing in research and development (R&D), continuously trying to develop new products or improve existing ones so as to satisfy the market's appetite for change and variety.

The case in question is a television firm actively engaged in funding its own R&D projects. Two objectives are identified, namely to increase profitability and to increase market share. The problem requires decisions about funding allocations to the particular projects. Six projects are under

Multiobjective Programming for Industrial Engineers

consideration. Projects I, II and III are all related to increasing the size of the TV screen as well as incorporating multiple functions. The distinction among the three "multifunction big-screen TV" projects is based merely on the screen dimensions. Project IV deals with adding improvements to the computer (PC) screens instead of the traditional TV screens, including incorporation of super VGA color monitors. Project V is concerned with research and design of "superthin" TVs for convenient placement on walls like ornaments. Project VI deals with design and development of "super-small" TVs appropriate for watches, calculators, cars, etc.

In assessing the viability of these projects, a 12-year planning horizon is taken in which the firm incurs net cost in the early years and generates net benefit in the later years. Table 16 shows the project cost during the

Table 16 Project's Cash Flow (in Thousand US$)

(a) During Development Phase

Year	I	II	III	IV	V	VI	Budget Available at Time T
0	-325	-300	-315	-450	-350	-300	621
1	-350	-320	-300	-350	-380	-300	672
2	-350	-300	-300	-350	-320	-350	672
3	0	0	-300	-300	-350	-300	672

(b) During Commercialization Phase

Year	Cash Flow of Mid-term Projects						Budget Available at Time T
	I	II	III	IV	V	VI	
3	-7000	-6000	0	0	0	0	10000
4	-5190	-4410	-4000	-4500	-4800	-5000	10000
5	2040	2550	-2000	200	500	-2500	0
6	2040	2550	300	2000	1800	400	0
7	2040	2550	2500	2000	2000	3000	0
8	360	450	2500	2000	2000	3000	0
9	3560	4950	2500	2000	2000	3000	0
10	3560	3950	2500	2000	2000	3000	0
11	3560	2950	0	0	0	0	0
12	4000	3000	0	0	0	0	0

R&D phase (denoted by minus signs) and the incurred benefit during the commercial phase (denoted by plus signs) as well as the budget availability. The projects are anticipated to be put into commercial operation from year 3 onward; thus the negative cash flows in the table from here on signify that some projects during the early stage of commercialization are not profitable.

Let Y_j be the unused money of the project budget in year j, Y'_j the net cash returns in year j coming from investment in selected projects, and X_i the fraction of funding in project i. It is expected that there will be 40, 20, 40, 25, 35, and 40% increases in the domestic market share as a result of fully undertaking projects I to VI, respectively. There are two specific objectives. One is maximization of the total net present value of cash flows over the 12-year period, at a discount rate of 8% for the first 5 years and 10% for the following years. The other is the maximization of a market share function. The constraint set is purely budgetary and is divided into two categories—those at the development phase and those at the commercialization phase. The complete model is as follows:

Objective functions:

$$\text{Max} \quad Z_1 = Y_0 + 0.926Y_1 + 0.867Y_2 + 0.794Y_3 + 0.735Y'_4$$
$$+ 0.680Y'_5 + 0.564Y'_6 + 0.513Y'_7 + 0.467Y'_8$$
$$+ 0.424Y'_9 + 0.386Y'_{10} + 0.350Y'_{11} + 0.320Y'_{12}$$

$$\text{Max} \quad Z_2 = 40X_1 + 20X_2 + 40X_3 + 25X_4 + 35X_5 + 40X_6$$

subject to:

1. Budget constraints at R&D phase

$$325X_1 + 300X_2 + 315X_3 + 450X_4 + 350X_5 + 300X_6 + Y_0 = 621$$
$$350X_1 + 320X_2 + 300X_3 + 350X_4 + 380X_5 + 300X_6 + Y_1 = 672$$
$$350X_1 + 300X_2 + 300X_3 + 350X_4 + 320X_5 + 350X_6 + Y_2 = 672$$
$$300X_3 + 300X_4 + 350X_5 + 300X_6 + Y_3 = 672$$
$$7000X_1 + 6000X_2 + \qquad\qquad\qquad\qquad\qquad Y_4 = 10{,}000$$

2. Budget constraints at commercialization phase

$$5190X_1 + 4410X_2 + 4000X_3 + 4500X_4 + 4800X_5 + 5000X_6 + Y'_4 = 10{,}000$$
$$-2040X_1 - 2550X_2 + 2000X_3 - 200X_4 - 500X_5 + 2500X_6 + Y'_5 = 0$$
$$-2040X_1 - 2550X_2 - 300X_3 - 2000X_4 - 1800X_5 - 400X_6 + Y'_6 = 0$$
$$-2040X_1 - 2550X_2 - 2500X_3 - 2000X_4 - 2000X_5 - 3000X_6 + Y'_7 = 0$$
$$360X_1 - 450X_2 - 2500X_3 - 2000X_4 - 2000X_5 - 3000X_6 + Y'_8 = 0$$
$$-5560X_1 - 4950X_2 - 2500X_3 - 2000X_4 - 2000X_5 - 3000X_6 + Y'_9 = 0$$
$$-3560X_1 - 3950X_2 - 2500X_3 - 2000X_4 - 2000X_5 - 3000X_6 + Y'_{10} = 0$$
$$-3560X_1 - 2950X_2 + \qquad\qquad\qquad\qquad\qquad Y'_{11} = 0$$
$$-4000X_1 - 3000X_2 + \qquad\qquad\qquad\qquad\qquad Y'_{12} = 0$$

Multiobjective Programming for Industrial Engineers

Table 17 Efficient Solutions

Efficient Solution No.	Non-zero Decision Variable	Activity*	Net Present Value	Increase in Market Share
1	X_1 X_3 X_5 X_6	1.000 1.000 0.030 0.030	15427.58	82.32
2	X_1 X_2 X_3 X_5 X_6	1.000 0.007 1.000 0.028 0.029	15431.92	82.32
3	X_1 X_2 X_3 X_5 X_6	1.000 0.012 1.000 0.021 0.032	15434.59	82.31
4	X_1 X_2 X_3 X_6	0.934 0.088 1.000 0.052	15620.77	81.26
5	X_1 X_2 X_6	0.001 1.000 0.438	18195.57	66.67
6	X_2 X_3 X_6	1.000 0.733 0.433	18196.97	66.66
7	X_2 X_3 X_5 X_6	1.000 0.727 0.001 0.438	18197.11	66.66
8	X_2 X_3 X_5 X_6	1.000 0.678 0.039 0.446	18232.51	66.33
9	X_2 X_3 X_6	1.000 0.951 0.176	18450.20	64.18
10	X_2 X_5	1.000 0.926	19273.12	52.42
11	X_2 X_4 X_5	1.000 0.288 0.661	19296.93	50.34
12	X_2 X_4	1.000 0.802	19311.28	40.06

* All figures are fractions of the required budget.

The model was solved using ADBASE to obtain all nondominated or efficient solutions. More details on ADBASE can be found in Steuer [3]. The results are shown in Table 17. With this set of efficient solutions, the decision maker needs to incorporate his preferences so that a final choice can be made.

5. CONCLUDING REMARKS

Multiobjective programming techniques are numerous, but their application to real-life problems is minimal, to say the most. This is rather unfortunate, because reality often faces multicriteria problems.

In the past, when work methods were the principal occupation of industrial engineers, the criterion used in making decisions to improve productivity depended on the focus of analysis, such as minimizing delays for an overall productive system, minimizing motions for a stationary worker at a workstation, and minimizing idle time and interference for workers interacting with each other. The decision-making methodology used was mainly charting. When analytical decision making became heavily involved with operations research, the criterion was either maximizing profit (or revenue) or minimizing cost or their equivalents. The use of mathematical programming models as aids to decision making became more publicized. Today, as industries are compelled to face problems of higher complexity, traditional attempts to view industrial engineering problems as monocriterion problems are likely to be less satisfactory because they do not reflect the heterogeneity and conflicting nature of the many interacting factors. The occurrence of externalities, risks, uncertainties, irreconcilable interests, and soft information reduces the efficacy of the traditional decision and choice techniques. Multiple criteria decision-making models are now more appropriate.

ACKNOWLEDGMENT

Most sincere gratitude is extended to Miss Juliet T. Lim, who kindly assisted the author in the preparation of the manuscript.

REFERENCES

1. Tabucanon, M. T., *Multiple Criteria Decision Making in Industry*. Elsevier, Amsterdam, 1988.
2. Lieberman, E. R., *Multi-Objective Programming in the USSR*. Academic Press, San Diego, 1991.
3. Steuer, R. E., *Multiple Criteria Optimization: Theory, Computation, and Application*. Wiley, New York, 1986.

Multiobjective Programming for Industrial Engineers

4. Zeleny, M., *Multiple Criteria Decision Making*. McGraw-Hill, New York, 1982.
5. Chankong, V., Haimes, Y. Y., *Multiobjective Decision Making*. North-Holland, Amsterdam, 1983.
6. Osyczka, A., *Multicriterion Optimization in Engineering*. Ellis Horwood, Chichester, 1984.
7. Bogetoft, P., Pruzan, P., *Planning with Multiple Criteria*. North-Holland, Amsterdam, 1991.
8. Lee, S. M., Shim, J., *Micro Management Science*. Wm. C. Brown, Dubuque, Iowa, 1986.
9. Ignizio, J. P., *Goal Programming and Extensions*. Lexington Books, Lexington, MA, 1976.
10. Lee, S. M., *Goal Programming for Decision Analysis*. Auerbach Publishers, Philadelphia, Pennsylvania, 1972.
11. Hwang, C. L., Masud, A. S. M., *Multiple Objective Decision Making — Methods and Applications: A State of the Art Survey*. Springer-Verlag, New York, 1979.
12. Duckstein, L., "Selected MCDM Techniques with Non-numerical Criteria," in *Proceedings of the Seminar-Workshop on Conflict Analysis in Reservoir Management*, ed. J. J. Bogardi. Asian Institute of Technology, Bangkok, Thailand, 1989.
13. Hwang, C. L., Masud, A. S. M., Paidy, S. R., Youn, K., "Mathematical Programming with Multiple Objectives: A Tutorial," *Computers & Operations Research*, Vol. 7, No. 1-2, pp. 5-31, 1980.
14. Yu, P. L., "A Class of Solutions for Group Decision Problems," *Management Science*, Vol. 19, No. 8, pp. 936-946, 1973.
15. Yu, P. L., Leitmann, G., "Compromise Solutions, Domination Structures and Salukvadze's Solution," *Journal of Optimization Theory and Applications*, Vol. 13, No. 3, pp. 363-378, 1974.
16. Hwang, S. C., "Note on the Mean-Square Strategy for Vector-Valued Objective Functions," *Journal of Optimization Theory and Applications*, Vol. 9, No. 5, pp. 364-366, 1972.
17. Salukvadze, M. E., "Optimization of Vector Functionals I: The Programming of Optimal Trajectories," *Automation and Remote Control*, Vol. 32, No. 8, pp. 1169-1178, 1971.
18. Salukvadze, M. E., "Optimization of Vector Functionals II: The Analytic Construction of Optimal Controls," *Automation and Remote Control*, Vol. 32, No. 9, pp. 1347-1357, 1971.
19. Charnes, A., Cooper, W. W., *Management Models and Industrial Applications of Linear Programming*. Wiley, New York, 1961.
20. Ignizio, J. P., Perlis, J. H., "Sequential Linear Goal Programming: Implementation Via MPSX," *Computers and Operations Research*, Vol. 6, No. 3, pp. 141-145, 1979.
21. Benayoun, R., De Montgolfier, J., Tergny, J., Larichev, O., "Linear Programming with Multiple Objective Functions: Step Method (STEM)," *Mathematical Programming*, Vol. 1, pp. 366-375, 1971.
22. Belenson, S. M., Kapur, K. C., "An Algorithm for Solving Multicriterion

Linear Programming with Examples," *Operational Research Quaterly*, Vol. 24, pp. 65-78, 1973.
23. Tabucanon, M. T., Mukyangkoon, S., "Multi-objective Microcomputer-based Interactive Production Planning," *International Journal of Production Research*, Vol. 23, No. 5, pp. 1001-1023, 1985.
24. Fry, T. D., Leong, G. K., "Bi-Criterion Single Machine Scheduling with Forbidden Early Shipments," *Engineering Costs and Production Economics*, Vol. 10, pp. 133-137, 1986.
25. Smith, W. E., "Various Optimizers for Single Stage Production," *Naval Research Logistics Quarterly*, Vol. 3, pp. 59-66, 1956.
26. van Wassenhove, L. N., Gelders, L. F., "Solving Bicriterion Scheduling Problem," *Eur. J. Op. Res.*, Vol. 4, pp. 42-48, 1980.
27. Chand, S., Schneeberger, H., "A Note on the Single Machine Scheduling Problem with Minimum Weighted Completion Time and Maximum Allowable Tardiness," *Naval Research Logistics Quarterly*, Vol. 33, No. 3, pp. 551-557, 1986.
28. French, S., *Sequencing and Scheduling, An Introduction to the Mathematics of the Job Shop*. Wiley, New York, 1982.
29. Cenna, A. A., Tabucanon, M. T., "Bicriterion Scheduling Problem in a Job Shop with Parallel Processors," *International Journal of Production Economics*, Vol. 25, Nos. 1-3, pp. 95-102, 1991.
30. Hannan, E. L., "An Assessment of Some Criticisms of Goal Programming," *Computers and Operations Research*, Vol. 12, No. 6, pp. 525-541, 1985.
31. Adulbhan, P., Tabucanon, M. T., eds., Chapter 8 of *Decision Models for Industrial Systems Engineers and Managers*, Asian Institute of Technology (distributed by Pergamon Press, Elmsford, NY), 1980.
32. Baker, K. R., Merten, A. G., "Scheduling with Parallel Processors and Linear Delay Costs," *Naval Research Logistics Quarterly*, Vol. 20, No. 4, pp. 793-804, 1973.

8

Stochastic Programming

John R. Birge
The University of Michigan, Ann Arbor, Michigan

John M. Mulvey
Princeton University, Princeton, New Jersey

1. INTRODUCTION

Industrial engineering decisions often involve the consideration of uncertain or stochastic parameters. Optimization procedures are increasingly helpful as the size and complexity of tractable problems grow. Ignoring fundamentally random characteristics may, however, limit the usefulness of an optimal solution. Stochastic programs that explicitly consider randomness are often much more beneficial for actual operations. This chapter discusses the use of these methods and presents some of the basic techniques in stochastic programming.

The expected objective advantage of using a stochastic programming solution over a deterministic program solution is called the *value of the stochastic solution*. In Section 2, we illustrate this quantity through a model in financial planning in which one seeks to maximize expected utility concerning an engineering project selection problem. Of particular note is that the value of the stochastic solution is different from the expected value of perfect information that represents the expected objective improvement possible with perfect knowledge of the future.

The major difficulty in stochastic programming is evaluating the future effect of a current action. Issues in modeling stochastic parameters are discussed in Section 3. Approximations are presented in Section 4 that can bound this future effect from above or from below. Section 5 goes beyond bounding procedures to discuss efficient methods for generating the scenarios that can be used to replace a stochastic programming model with an equivalent deterministic model. With the bounding approximations or scenarios, stochastic programs become large-scale mathematical programs that have a particular structure. We briefly describe some techniques to take advantage of that structure in Section 6. Section 7 gives several important examples of particular relevance for industrial engineering.

2. FINANCIAL PLANNING AND THE VALUE OF THE STOCHASTIC SOLUTION

Most industrial engineering decisions involve some financial considerations. Financial decision-making problems, in turn, can often be modeled as stochastic programs. This area represents one of the largest application areas of stochastic programming. Many references can be found in, for example, Mulvey and Vladimirou (1992) and Ziemba and Vickson (1975).

We consider a simple example (from Birge, [1992]) that illustrates additional stochastic programming properties. The random variables reflect uncertain investment yields. The role of the stochastic program is to hedge against poor outcomes by maximizing an expected objective function that is concave and represents some aversion to risk.

For the current problem, suppose we wish to provide a project investment plan Y years from now. We currently have \$$w$ to invest in any of K industrial engineering projects. After Y years, we will have a wealth of \$$W$ that we would like to have exceed a goal of \$$G$. We suppose that we can change investments every y years so we have $T = Y/y$ investment periods. For our purposes here, we ignore transaction costs and taxes on income, although these considerations would be important in practice.

In formulating the problem, we must first describe our objective in mathematical terms. We suppose that exceeding \$$G$ after Y years would be equivalent to our having an income of i of the excess, while not meeting the goal would lead to borrowing for a cost q of the amount short. This gives us a concave utility function, although many other forms of nonlinear utility functions are possible [see Kallberg and Ziemba (1983) for a description of their relevance in financial planning].

The major uncertainty in this model is the return on each investment k within each period t. We describe this random variable as $\mathbf{r}(k,t) = r(k,t,\omega)$ where ω is some underlying random element. The decision on investments will be random as well. We describe these decisions as $\mathbf{x}(k,t) = x(k,t,\omega)$. From the randomness of the returns and investment decisions, our final wealth will also be a random variable $\mathbf{W} = W(\omega)$.

A key point about this investment model is that we cannot completely

Stochastic Programming

observe the random element ω when we make all of our decisions $x(k,t,\omega)$. We can observe only the returns that have already taken place. In stochastic programming, we say that we cannot *anticipate* every possible outcome so that our decisions are *nonanticipative* of future outcomes. Before the first period, this restriction corresponds to saying that we must make fixed investments, $x(k,1)$, for all $\omega \in \Omega$, the space of all random elements.

In the next period, we suppose that the random elements ω correspond to N_1 different possibilities for outcomes in the first investment period. We can therefore partition Ω into $\Omega_1^1, \ldots, \Omega_{N_1}^1$ corresponding to these different initial outcomes. Decisions for a given value of ω must be the same for every ω within the set. We can therefore describe our second period decisions just in terms of the Ω_i^1 that occurs in the first period. We write these decision variables as $x(k,2,i)$ where $i = 1, \ldots, N_1$.

We can continue this process by defining $\Omega_{i_1,\ldots,i_t}^t$ as the set of all ω that correspond to outcomes i_j in periods $j = 1, \ldots, t$. We can then describe the nonanticipative decisions as $x(k, t + 1, i_1, \ldots, i_t)$, which depend on the outcomes $r(t, i_1, \ldots, i_t)$. In the following, for simplicity, we assume that each set of outcomes $i_1, \ldots, i_{t-1}$ up to time $t - 1$ leads to N_t outcomes at time t. In all T periods, we would then have $N_1 \cdot N_2 \cdots N_T$ different possible outcomes. To illustrate how quickly these problems can grow, note that, with just 10 outcomes per period, we would obtain 10^{10} outcomes in 10 periods.

We need to attach probabilities to all outcomes. We let the probability of the ith outcome in period t given outcomes i_j in periods $j = 1, \ldots, t - 1$ be $p(t, i_1, \ldots, i_t)$. Note that we must have $\sum_{i=1}^{N_t} p(t, i_1, \ldots, i_{t-1}, i) = 1$ for all t and $i_1, \ldots, i_{t-1}$.

We then can state a formulation for our problem. We wish to find

$$\max \quad z = \sum_{i_1,\ldots,i_T} p(T, i_1, \ldots, i_T)(iv(i_1, \ldots, i_T)$$
$$- qs(i_1, \ldots, i_T)) \qquad (2.1a)$$

subject to (s.t.) $\quad \sum_{k=1}^{K} x(1, k) = w \qquad (2.1b)$

$$\sum_{k=1}^{K} r(k, t, i_1, \ldots, i_{t-1}) x(k, t-1,$$

$$i_1, \ldots, i_{t-2}) - \sum_{k=1}^{K} x(k, t, i_1, \ldots, i_{t-1}) = 0$$

$$\text{for all } (i_1, \ldots, i_{t-1}), t = 2, \ldots, T \qquad (2.1c)$$

$$\sum_{k=1}^{K} r(k, T, i_1, \ldots, i_T) x(k, T, i_1, \ldots, i_{T-1})$$

$$- v(i_1, \ldots, i_T) + s(i_1, \ldots, i_T) = G \qquad (2.1d)$$

$$x(k, t, i_1, \ldots, i_{t-1}) \geq 0 \qquad (2.1\text{e})$$
$$v(i_1, \ldots, i_T) \geq 0$$
$$s(i_1, \ldots, i_T) \geq 0 \quad \text{for all } 1 \leq i_t \leq N_t,$$
$$1 \leq k \leq K, \quad 1 \leq t \leq T$$

where $v(i_1, \ldots, i_T)$ represents any amount above the goal and $s(i_1, \ldots, i_T)$ represents any shortfall from the target value G. The objective function (2.1a) depicts a weighted combination of maximizing expected excess profits (above the target) while minimizing the expected shortfall. Constraint (2.1b) states that all initial wealth, w, is invested in the first period. The constraints in (2.1c) balance the returns on investments from period t with the investments made in period t. Constraints (2.1d) add shortages, s, or subtract excesses, v. The inequalities in (2.1e) keep all variables nonnegative (preventing short sales in this case).

Another approach to multistage problems like this is to consider the possible outcomes over the entire horizon as *scenarios*, σ. We then substitute a scenario set S for the random elements Ω. Probabilities $p(\sigma)$, returns $r(k,t,\sigma)$, and investments $x(k,t,\sigma)$ become functions of the T-period scenarios.

We must still maintain nonanticipativity, but this time we do so explicitly in the formulation via constraints. First, the scenarios that correspond to the same set of past outcomes at each period form groups, $S^t_{i_1,\ldots,i_{t-1}}$, for scenarios at time t. Now, all actions up to time t must be the same within a group. We do this through an explicit constraint. The new formulation of (2.1) becomes:

$$\max \quad z = \sum_\sigma p(\sigma)(iv(\sigma) - qs(\sigma))$$

$$\text{s.t.} \quad \sum_{k=1}^K x(k,1,\sigma) = w, \quad \forall \sigma \in S$$

$$\sum_{k=1}^K r(k,t,\sigma)x(k,t-1,\sigma) - \sum_{k=1}^K x(k,t,\sigma) = 0,$$
$$\forall \sigma \in S, t = 2, \ldots, T$$

$$\sum_{k=1}^K r(k,T,\sigma)x(k,T,\sigma) - v(\sigma) + s(\sigma) = G$$

$$\left(\sum_{\sigma' \in S^t_{I(\sigma,t)}} p(\sigma')x(k,t,\sigma')\right) - \left(\sum_{\sigma' \in S^t_{I(\sigma,t)}} p(\sigma')\right)$$
$$x(k,t,\sigma) = 0, \quad 1 \leq k \leq K, 1 \leq t \leq T, \forall \sigma \in S$$
$$x(k,t,\sigma) \geq 0, \quad v(\sigma) \geq 0, \quad s(\sigma) \geq 0,$$
$$1 \leq k \leq K, 1 \leq t \leq T, \forall \sigma \in S \qquad (2.2)$$

Stochastic Programming

where $I(\sigma,t) = \{i_1, \ldots, i_{t-1}\}$ such that $\sigma \in S^t_{i_1,\ldots,i_{t-1}}$, $v(\sigma)$ is the amount above the goal in scenario σ, and $s(\sigma)$ is the amount below the goal in scenario σ. Note that the last equality constraint indeed forces all decisions within the same group at time t to be the same. Formulation (2.2) has a special advantage for the problem here because these *nonanticipativity* constraints are the only constraints linking the separate scenarios. Without them, the problem would decompose into a separate problem for each σ, maintaining the structure of that problem.

In modeling terms, this simple additional constraint makes it relatively easy to move from a deterministic model to a stochastic model of the same problem. The scenario indicators and nonanticipativity constraints are the only additions to a deterministic model. Given the ease of this modeling effort, standard optimization procedures can be simply applied to this problem. However, as we noted above, the number of scenarios can become extremely large. Standard methods may not be able to solve the problem in any reasonable fashion, necessitating other techniques.

In most practical situations involving financial markets, decision makers display risk-averse behavior. It is therefore important to represent risk aversion within the context of the stochastic programming model. There are several ways to model risk, for example, the von Neumann-Morgenstern expected utility approach (1947) or the Markowitz marginal utility approach (1959).

In this engineering project investment problem, it is particularly worthwhile to try to exploit the underlying structure of the problem without the nonanticipativity constraints. This relaxed problem is, in fact, a generalized network that allows the use of efficient network techniques. Several decomposition algorithms have been developed to take advantage of this fact. One method is the progressive hedging algorithm of a Rockafellar and Wets (1991), which relies on an augmented Lagrangian to force a solution that satisfies the nonanticipativity constraints. This procedure has been applied by Mulvey and Vladimirou (1992) to financial planning problems. Nielsen and Zenios (1993) have used a different decomposition method, the row-action algorithm of Censor and Lent (1981), in their work with financial planning models. This approach allows them to solve stochastic network problems using massively parallel computers.

With either formulation (2.1) or (2.2), in completing the model, some decisions must be made about the possible set of outcomes or scenarios and the coarseness of the period structure, i.e., the number of periods T allowed for investments. We must also find probabilities to attach to outcomes within each of these periods. These probabilities are often approximations that can, as we shall see in Section 3, provide bounds on true values or on uncertain outcomes with incompletely known distributions. A key observation we will make is that the important step is to include stochastic elements at least approximately and that deterministic solutions often give misleading results.

To illustrate the effects of including stochastic outcomes as well as modeling effects from choosing the time horizon Y and the coarseness of the period approximations T, we use a simple example with $K = 2$ possible project investments: the first is relatively uncertain, project A, whereas the second is more stable, project B. We begin by setting Y at 15 years and allow investment changes every 5 years so that $T = 3$.

The other data for this example include $N_1 = N_2 = N_3 = 2$, so we have $|S| = 8$ scenarios. The scenarios correspond to independent and equal likelihoods of having (inflation-adjusted) returns of 1.25 for project A and 1.14 for project B or 1.06 for project A and 1.12 for project B over the 5-year period. This yields probabilities $p(\sigma) = 0.125$ for each scenario. The returns are $r(1,t,\sigma) = 1.25$, $r(2,t,\sigma) = 1.14$ for $t = 1, \sigma = 1, \ldots, 4$, for $t = 2, \sigma = 1,2,5,6$, and for $t = 3, \sigma = 1,3,5,7$. In the other cases, $r(1,t,\sigma) = 1.06, r(2,t,\sigma) = 1.12$.

The remaining data are the initial wealth, $w = 55,000$, the target value, $G = 80,000$, the surplus reward, $i = 5$, and the shortage penalty, $q = 20$. Solving the problem in (2.2) as an ordinary linear program with these parameter values yields an optimal expected utility value of -7.6. We will call this *RP*, for the expected *recourse problem* solution. The optimal solution (in thousands of dollars) appears in Table 1. In this solution, the initial investment is heavily in project A ($41,500) with only $13,500 in project B. Notice the reaction to first-period outcomes, however. In the case of scenarios 1-4, the risky project A is even more prominent, while scenarios 5-8 reflect a more conservative mix. In the last period, notice that the investments are either completely in project A or completely in project B. This is a general trait of one-period decisions. It occurs here because in scenarios 1 and 2, there is no risk of missing the target. In scenarios 3 through 6, the risky investments (A) may cause one to miss the target so they are avoided. In scenarios 7 and 8, the only hope of reaching the target is through project A.

We now compare the results in Table 1 with a deterministic model. If we substitute expected returns for the random variables, then project A, with an expected 5-year yield of 1.155, dominates project B. Everything is invested in the risky project A in each period. If we apply this decision process to the stochastic model with the alternative outcomes and probabilities given above, then the expected utility value is -19.2, which we call EMV for the *expectation of the mean value solution*. The *value of the stochastic solution* [VSS, introduced in Birge (1982)] is the difference between the expected utility of the stochastic programming model (RP =

Stochastic Programming

Table 1 Optimal Solution with Three-Period Stochastic Program

Period, scenario	Project A	Project B
1, 1–8	41.5	13.5
2, 1–4	65.1	2.17
2, 5–8	36.7	22.4
3, 1–2	83.8	0.0
3, 3–4	0.0	71.4
3, 5–6	0.0	71.4
3, 7–8	64.0	0.0
Scenario	Above G	Below G
1	24.8	0.0
2	8.87	0.0
3	1.42	0.0
4	0.0	0.0
5	1.42	0.0
6	0.0	0.0
7	0.0	0.0
8	0.0	12.2

-7.6) and the expected value for using the deterministic model with means substituted for the random variables (EMV $= -19.2$). In this case, that difference is

$$\text{VSS} = \text{RP} - \text{EMV} = -7.6 - (-19.2) = 11.6$$

The expected value of perfect information, on the other hand, compares the recourse problem value (or maximum expected utility, RP) to the expectation of solution values that would be obtained if the future was known perfectly. In this problem, there are eight potential future outcomes (two per period). The perfect information [or, in stochastic programming, following Madansky (1960), the *wait-and-see* solution WS] is to invest in whatever has the highest yield in each period. In this case, we would invest in project A if it increases 25% in a 5-year period (project B increase 14%) and we would invest in project B if it increases 12% (project A increase 6%). Thus, with probability 0.125, we receive a return of $(1.25)^3 = 1.953$; with probability 0.125, we receive a return of $(1.12)^3 = 1.405$; with probability 0.375, we receive a return of $1.12 \times (1.25)^2 = 1.75$; and with probability 0.375, we receive a return of $(1.12)^2 \times 1.25 = 1.568$. The result is that we would have an expected utility of WS $= 52.5$. The expected value of perfect information is

$$\text{EVPI} = \text{WS} - \text{RP} = 52.5 - (-7.6) = 60.1$$

In this case, EVPI > VSS, but in many cases (see Birge, 1982), we may have VSS > EVPI. In fact, since WS > RP > EMV, VSS and EVPI are assured of being the same only when WS = EMV. Either could be zero, while the other is positive.

In closing this section, note that the mathematical form of this problem actually represents a broad class of control problems. In fact, it is basically equivalent to any problem governed by a linear system of state equations. Many industrial engineering examples, especially as we shall see below in relation to production planning, fit this form. Other types of control problems may have nonlinear (e.g., quadratic) costs associated with the control (investment transactions) in each time period. This presents no complication for our purposes. We may include any of these problems as potential applications.

In this problem, we had a limited set of possible outcomes in each period. In practical problems, we could not expect such a small finite set of realizations. One of the main steps in stochastic program modeling involves how to select such sets of scenarios and how to compare their use with what may actually happen. We discuss bounding approximations in Section 4 and scenario generation techniques in Section 5. However, first we need to discuss how we actually model stochastic parameters.

3. MODELING STOCHASTIC PARAMETERS

A critical element in the application of a stochastic program involves dynamic modeling of the stochastic parameters. The quality of the recommendations depends on an accurate portrayal of the random variables over time. There are four basic approaches for carrying out this task: (1) econometrics and related time series statistical methods (see Wonnacott and Wonnacott, 1970), (2) expert opinion (see Eppen et al., 1989), (3) diffusion and stochastic processes (see Çinlar, 1975), and (4) simulation models (see Law and Kelton, 1991). Generally, the first approach relies heavily on historical data and the second relies on expert judgment regarding possible realizations of the random variables, whereas the third and fourth approaches combine expert judgment regarding the structure of the real-world problem with historical data for calibrating the system. Of course, there is much overlap and it is likely that a blend of methods will occur in practice.

For our purposes, we distinguish two forms of applications. The first involves large organizations who plan their affairs using a small number of scenarios for future economic activity. Many industrial companies and governmental organizations employ this form of planning. A typical example with five scenarios might possess a momentum scenario (e.g., current interest rates continuing), an optimistic scenario (slowly dropping interest rates), a highly optimistic scenario (sharply dropping interest rates), a pessimistic scenario (gradually rising interest rates), and a highly pessimistic scenario (quickly rising interest rates). These projections and the accompanying probabilities are determined by elaborate means, such as a series of meetings whose objective is to obtain consensus of the top management. The results may also come from complex simulation models.

Governmental regulatory bodies are beginning to employ this form of planning. For example, New York State requires seven scenarios to be analyzed when insurance companies propose rate increases and other changes. Similar concepts have been employed by businesses for projecting demand for their products.

A special case of stochastic programming, called robust optimization (RO), has been designed for the situation in which there is a small number of alternative scenarios. The goal is to discover recommendations that are relatively insensitive to the choice of the scenario. Robust optimization, developed by Mulvey et al. (1995), attempts to find an optimal solution to a stochastic problem that is both solution robust and model robust. A solution is solution robust if it is "close" to the optimal solution across all possible scenarios; it is model robust if it is "close" to feasible across all scenarios. RO uses a multiobjective approach to balance the solution robustness by allowing the solution to be infeasible in some scenarios but attempting to minimize the number of infeasibilities that exist. The underlying assumption is that the infeasibilities can be handled outside the framework of the optimization model.

The second class of applications involves stochastic models that come from a set of stochastic differential equations. The famous random walk of stock prices falls within this category. Herein, independence across time is assumed. A slightly more elaborate model is the Ornstein-Ulenbeck equation

$$dx = a(b - x)\, dt + c\, dz \qquad (3.1)$$

The term $a(b - x)\, dt$ controls the drift of the process via the mean reversion parameter b, and c represents the weighted volatility. By building processes of this type, a system of differential equations can be developed. The results can become rather complex—with possibly dozens of interrelated equations for economic and other factors [see Mulvey (1995) and Cariño et al. (1994) for examples of these models].

The calibration of a stochastic system requires considerable care. The basic idea is to find a set of parameters that correspond to the modeler's view of the future in terms of broad goals, while at the same time maintaining a reasonable relationship to historical statistical summaries. Much work has been done in the modeling of economic factors (see, for example, Wonnacott and Wonnacott, 1970). Nonlinear optimization provides a systematic method for carrying out this calibration.

Typically, the linkage of the stochastic model and the optimization is performed by some form of sampling. Either the set of scenarios is selected by the sponsoring organization prior to any optimization (Mulvey, 1995) or a sampling technique is used. Whenever possible, variance reduction methods are employed or sampling is done in conjunction with the optimization algorithm itself, as with importance sampling (Dantzig and Glynn, 1990) or with stochastic decomposition (Higle and Sen, 1991).

Another modeling concern is the treatment of choices over time. Because multistage stochastic programs model decision making over time, the proper formulation of the objective function is important. Multiobjective functions are commonly used, with most analysts discounting the utility over time. This modeling issue has not been a subject of much investigation. Loewenstein and Elster (1992) discuss many of the issues of decision making over time, and Wierzbicki (1988) deals with some of the multiobjective modeling aspects.

As we have seen, there are many important modeling issues with stochastic programming, especially regarding scenario generation. Scenario generation may benefit from the research conducted in the stochastic modeling domain. The integration of stochastic modeling and multistage stochastic programming is just beginning, and it will require considerable future research.

4. DISCRETE BOUNDING APPROXIMATIONS

The most common procedures in stochastic programming approximations are to find some relatively low-cardinality discrete set of realizations that somehow represents a good approximation of the true underlying distribution or whatever is known about this distribution. The basic procedures are extensions of Jensen's inequality [(1906), generalization of the midpoint approximation] and an inequality due to Edmundson (1956) and Madansky (1959) (the generalization of the trapezoidal approximation). For convex functions in random parameters, Jensen provides a lower bound and Edmundson and Madansky provide an upper bound. Significant refinements of these bounds appear in Huang et al. (1979), Kall and Stoyan (1982), and Frauendorfer (1988).

We refer to a general integrand $f(x,\xi)$, where $\xi = \xi(\omega) \in \Xi$, ω has an associated probability measure P, and f represents the value of a problem

Stochastic Programming

such as (2.2) for a single realization ω without the nonanticipativity constraint. The true stochastic program can then be written as

$$\min_{x \text{ nonanticipative}} \int_\Omega f(x,\xi) P(d\omega) \tag{4.1}$$

To construct (2.2), we need to determine an appropriate set of scenarios to replace the multiple integration in (4.1). The basic ideas are to partition the support Ξ into a number of different regions (analogous to intervals in one-dimensional integration) and to apply bounds in each of those regions. We let the partition of Ξ be $S^\nu = \{S^l, l = 1, \ldots, \nu\}$. Define $\xi^l = E[\xi | S^l]$ and $p^l = P[\xi \in S^l]$. The basic lower bounding result is the following:

Theorem 1 Suppose that $f(x, \cdot)$ is convex for all $x \in D$; then

$$E_f(x) \geq \sum_{l=1}^{\nu} p^l f(x,\xi^l) \tag{4.2}$$

Proof Write $E_f(x)$ as

$$E_f(x) = \sum_{l=1}^{\nu} \int_{S^l} f(x,\xi) P(d\xi)$$

$$= \sum_{l=1}^{\nu} p^l E[f(x,\xi) | S^l]$$

$$\geq \sum_{l=1}^{\nu} p^l f(x, E[\xi | S^l])$$

where the last inequality follows from Jensen's inequality that the expectation of a convex function of some argument is always greater than or equal to the function evaluated at the expectation of its argument, i.e., $E(f(\xi)) \geq f(E(\xi))$. $\square$

This result applies directly to (2.2) by letting each $\sigma = \xi^l$ and $p(\sigma) = p^l$. The approximating distribution P^ν is the discrete distribution with atoms of probability p^l at each ξ^l for $l = 1, \ldots, \nu$. By choosing $S^{\nu+1}$ so that each $S^l \in S^{\nu+1}$ is completely contained in some $S^{l'} \in S^\nu$, the approximations actually improve, i.e.,

$$E_f(x) \geq E_f^{\nu+1}(x) \geq E_f^\nu(x) \tag{4.3}$$

Various methods can achieve convergence in distribution of the P^ν to P.

In general, the goal of refining the partition from ν to $\nu + 1$ is to achieve as great an improvement as possible. More details appear in Birge and Wets (1986) and Frauendorfer and Kall (1988). These articles also present upper bounds that again use convexity, although in a different way.

Comparisons of upper and lower bounds allow for error bounding and can lead to conditions for stopping a sequential solution procedure.

In many cases, explicit error bounds are not possible either because of difficulties in calculating the bounds or because of lack of information, in particular information related to cost. A convenient approach in these circumstances is to attach penalties to violated constraints. As mentioned, one method that utilizes these penalties is robust optimization.

5. GENERATING SCENARIOS

While bounding procedures can approximate the actual distribution of the stochastic parameters, there are several other modeling approaches based on efficient procedures for generating representative scenarios. Of these, the most important are importance sampling, stochastic decomposition, and cluster analysis.

Importance sampling is a statistical procedure that has been previously applied to simulation modeling (Glynn and Iglehart, 1989). This procedure provides a method for generating observations such that rare events occur more frequently than would happen randomly. In simulation, it allows rare events to occur with enough frequency that they can be analyzed. Dantzig and Glynn (1990) and Infanger (1992) have applied this technique to stochastic programming. Importance sampling is used to generate the scenarios that provide the cuts to the solution algorithm, based on the decomposition principle of linear programming. As in a simulation experiment, the use of importance sampling allows very rare events to occur more often than would happen normally. Thus, solutions to problems that use such scenarios to approximate the underlying distributions are more likely to be better solutions to the original problem than solutions that sample the original distribution directly.

A second approach is to use stochastic decomposition (Higle and Sen, 1991). With this approach, cutting planes (scenarios) are derived that are only asymptotically valid; that is, the cuts are valid cuts only as you generate an infinite number of cuts. The algorithm, unlike most other decomposition algorithms, terminates when no more information is generated by new cuts. With this method, there is no guarantee that an optimal solution will be found.

Another approach to scenario generation is based on cluster analysis. Cluster analysis is a statistical technique for grouping together objects such that all elements in a cluster are similar to each other and are dissimilar to elements in the other clusters (Mulvey and Crowder, 1979). This technique has been used to group together variables without losing too much information about them (Mulvey and Beck, 1981). This technique could be applied

Stochastic Programming

to scenario optimization. Even a problem with a few parameters could result in too many scenarios to evaluate if a scenario were generated for each possible outcome. Cluster analysis may provide a way to group together scenarios in a manner that minimizes the loss of information, just as cluster analysis can group together data points. The resulting problem will be much smaller, while only a relatively small amount of information is lost.

Given sets of scenarios (from bounds or perhaps sampled solutions), the next step in stochastic programming is to solve the resulting problem. The main procedures are based on large-scale mathematical programming procedures constructed to take advantage of the structure of stochastic programs. In the next section, we describe typical methods for stochastic programs such as the financial planning problem in Section 2.

6. SOLUTION PROCEDURES FOR STOCHASTIC PROGRAMS

Many solution procedures have been suggested for stochastic programming problems. For linear programs, these methods are often based on standard large-scale linear programming methods (see Wets, 1988). One of the more common procedures is called *nested decomposition*, first proposed for deterministic models by Ho and Manne (1974), as well as Glassey (1973). These first approaches are essentially inner linearizations that begin treating all previous periods as subproblems of a current period master problem. The previous periods generate columns that can be used by the master problem of the present period.

A difficulty with these inner linearization methods is that the set of inputs may be fundamentally different for different last-period realizations. Because the number of last-period realizations is the total number of scenarios in the problem, these procedures are not well adapted to efficient implementations. Some success has been achieved, however, by Noël and Smeers (1987).

Instead of inner linearization, the general approach in stochastic programming has been to use the outer linearization or generalization of the two-stage L-shaped method given by Van Slyke and Wets (1969). Louveaux (1980) first performed this generalization for multistage quadratic problems. Birge (1985) extended the two-stage method in the linear case as in the following description.

The basic idea of the nested L-shaped or Benders (see Benders, 1962) decomposition method is to decompose a problem in the form of (2.1) into distinct subproblems for each stage and scenario. They are linked by cuts on future objective values as a function of a current decision x^t. Other cuts achieve an x^t that has a feasible completion in all descendant scenarios. The cuts represent successive linear approximations of the future objective including feasibility. If the future objective function is polyhedral, this process converges to an optimal solution in a finite number of steps.

The general version of the multistage stochastic linear program is the following:

$$\min \sum_{t=1}^{T} \sum_{k=1}^{K^t} p_k^t c_k^{tT} x_k^t$$

s.t. $\quad Ax^1 = b$

$\quad W^t x_k^t = h_k^t - T_k^{t-1} x_{a(k)}^{t-1}, \quad t = 2, \ldots, T, k = 1, \ldots, K^t$

$\quad x_k^t \geq 0, \qquad\qquad\qquad\qquad t = 1, \ldots, T, k = 1, \ldots, K^t$

(6.1)

where we have $k = 1, \ldots, K^t (= N^1 \cdots N^t)$ scenarios for every stage $t = 1, \ldots, T$ (with $K^1 = 1$), and $a(k)$ is the ancestor scenario of k at stage $t - 1$.

In decomposition methods, each period and scenario combination corresponds to a separate master problem that generates cuts to stage t and proposals for stage $t + 1$. In our example, this problem has the form:

$$\min \quad c_k^{tT} x_k^t + \theta_k^t \tag{6.2}$$

s.t. $\quad W^t x_k^t = h_k^t - T_k^{t-1} x_{a(k)}^{t-1} \tag{6.3}$

$\quad D_{k,j}^t x_k^t \geq d_{k,j}^t, \quad j = 1, \ldots, r_{k,j}^t \tag{6.4}$

$\quad E_{k,j}^t x_k^t + \theta_k^t \geq e_{k,j}^t, \quad j = 1, \ldots, s_{k,j}^t \tag{6.5}$

$\quad x_k^t \geq 0 \tag{6.6}$

where $x_{a(k)}^{t-1}$ is the current solution from the previous period, and where for $t = 1$ we interpret $b = h^1 - T^0 x^0$ as the initial conditions of the problem. We may refer also to the stage T problem in which θ_k^T and constraints (6.4) and (6.5) are not present. We will describe a basic algorithm for iterating among these stages. We then discuss some enhancements of this basic approach. In the following, $\mathcal{D}^t(j)$, denotes the period t descendants of a scenario j at period $t - 1$. We assume that all variables have finite upper bounds to avoid complications presented by unbounded solutions (although again these can be treated as in Van Slyke and Wets).

Nested Benders Decomposition Method

Step 0. Set $t = 1$, $k = 1$, $r_k^t = s_k^t = 0$, add the constraint $\theta_k^t = 0$ to (6.2)–(6.6) for all t and k, and let DIR = FORE. Go to (1).

Step 1. Solve the current problem (6.3)–(6.6) for t and k. If infeasible and $t = 1$, then stop, problem (6.1) is infeasible. If infeasible and $t > 1$, then let $r_{a(k)}^{t-1} = r_{a(k)}^{t-1} + 1$, let DIR = BACK. Let the infeasibility condition obtained be $\pi_k^t, \rho_k^t \geq 0$ such that $\pi_k^{tT}W^t + \rho_k^{tT}D_k^t \leq 0$ but $\pi_k^{tT}(h_k^t - T_k^{t-1}x_{a(k)}^{t-1} + \rho_k^{tT}d_k^t) > 0$. Let $D_{a(k),r_{a(k)}^{t-1}}^{t-1} = \pi_k^{tT}T_k^{t-1}$, $d_{a(k),r_{a(k)}^{t-1}}^{t-1} = \pi_k^{tT}h_k^t + \rho_k^{tT}d_k^t$. Let $t = t - 1$, $k = a(k)$ and return to (1).

If feasible, update the values of x_k^t, θ_k^t, and store the value of the dual multipliers on constraints (6.3)–(6.5) as $(\pi_k^t, \rho_k^t, \sigma_k^t)$, respectively. If $k < K_t$, let $k = k + 1$, and return to (1). Otherwise, ($k = K_t$), if DIR = FORE and $t < T$, let $t = t + 1$ and return. If $t = T$, let DIR = BACK. Go to (2).

Step 2. For all scenarios $j = 1, \ldots, K^{t-1}$ at $t - 1$, compute

$$E_j^{t-1} = \sum_{k \in \mathcal{D}^t(j)} \frac{p_k^t}{p_j^{t-1}} \pi_k^t T_k^t$$

and

$$e_j^{t-1} = \sum_{k \in \mathcal{D}^t(j)} \frac{p_k^t}{p_j^{t-1}} \left(\pi_k^t h_k^t + \sum_{i=1}^{r_k} \rho_{ki}^t d_{ki}^t + \sum_{i=1}^{s_k} \sigma_{ki}^t e_{ki}^t \right).$$

The current conditional expected value of all scenario problems in $\mathcal{D}^t(j)$ is then $\bar{\theta}_j^{t-1} = e_j^{t-1} - E_j^{t-1}x_j^{t-1}$. If the constraint $\theta_j^{t-1} = 0$ appears in (6.2)–(6.6) for $t - 1$ and j, then remove it, let $s_j^{t-1} = 1$, and add a constraint (6.5) with E_j^{t-1} and e_j^{t-1} to (6.2)–(6.6) for $t - 1$ and j.

If $\bar{\theta}_j^{t-1} > \theta_j^t$, then let $s_j^{t-1} = s_j^{t-1} + 1$, and add a constraint (6.5) with E_j^{t-1} and e_j^{t-1} to (6.2)–(6.6) for $t - 1$ and j. If $t = 2$ and no constraints are added to (6.2)–(6.6) for $t - 1 = 1$ ($j = K^1 = 1$), then stop with x_1^1 optimal. Otherwise, let $t = t - 1$, $j = 1$. If $t = 1$, let DIR = FORE. Go to (1).

Many alternative strategies are possible in this algorithm in terms of determining the next subproblem (6.2)–(6.6) to solve. For feasible solutions, the description above explores all scenarios at t before deciding to move to $t - 1$ or $t + 1$. For feasible iterations, the algorithm proceeds from t in the direction of DIR until it cannot proceed further in that direction. This is the "fast forward–fast back" procedure proposed by Wittrock (1983) for deterministic problems and implemented with success by Gassmann (1990) and Birge et al. (1993) for stochastic problems. One may alternatively enforce a move from t to $t - 1$ ("fast back") or from t to $t + 1$ ("fast forward") whenever it is possible. From experiments conducted by Gassmann and Birge et al., the fast forward–fast back sequencing protocol seems generally to work better than either of these alternatives.

We note that much of this algorithm can also run in parallel with near (or sometimes even) superlinear speedups (see Birge et al., 1993). Adding regularizing terms can also speed convergence (see Ruszczyński, 1986). Proving the convergence of this method is relatively straightforward (see Birge, 1985).

A principal advantage of this decomposition procedure over standard mathematical programming techniques is that several subproblems (6.1)–(6.5) may be quite similar for different realizations of the random variable. After one of these is solved, the others may be solved quite quickly. The basic procedures for doing this efficiently are called *sifting* (see Gartska and Rutenberg, 1973) and *bunching* (see Wets, 1988; Gassmann, 1990). They generally avoid repeated matrix operations while saving only the set of subproblems solved by a given basis. The result is often a dramatic decrease in solution time in comparison with nonspecialized techniques. For example, Birge et al. (1993) report speedups of orders of magnitude for many problems compared to IBM's OSL package (International Business Machines Corp., 1991).

Other techniques have also been suggested for stochastic programming. Some use sampling techniques, whereas others focus on relaxing the nonanticipativity constraints. In sampling procedures, the main methods are stochastic quasi-gradient methods (see Ermoliev, 1983) plus methods based on outer linearizations such as stochastic decomposition (Higle and Sen, 1991) and importance sampling (Dantzig and Glynn, 1990).

The procedures that relax nonanticipativity place these constraints into the objective with a Lagrangian term. One method called progressive hedging (Rockafellar and Wets, 1991) uses the proximal point algorithm (see Rockafellar, 1976) to obtain a Lagrangian saddle point. The other procedure (see Dempster, 1988) uses dual ascent steps in an augmented Lagrangian framework. Both of these procedures are especially advantageous when the original problem has a special structure (such as a network—see Mulvey and Vladimirou, 1992) that would be destroyed by adding the nonanticipativity constraints. Also, these Lagrangian-based methods are able to handle the nonlinear objective functions required by risk-averse decision makers.

The progressive hedging method of Rockafellar and Wets (1991) provides a framework for combining the solutions from the individual scenarios to create an aggregate solution that is robust; that is, the solution will behave reasonably well in the uncertainty region defined by the weighted scenarios. The progressive hedging methodology views two types of solutions: implementable policies, which satisfy the nonanticipativity constraints, and admissible policies, which satisfy the constraints of the subproblems. The goal is to find the best solution that is both implementable and admissible, with the best one determined by the scenario weightings.

The method uses a penalty parameter in the subproblems to generate admissible polices that are closer to being implementable, while implementable policies are created by weighting the scenario solutions. The optimal solution is found when both the primal decision variables and the weights (dual prices) converge to within a user-specified tolerance.

A variation of progressive hedging has been developed by Mulvey and Vladimirou (1991) for multistage stochastic networks. Their technique allows the multistage stochastic model to be decomposed into scenarios that have a generalized network structure. Efficient specialized network codes can then be employed to solve the subproblems. Acceleration techniques can be applied to the algorithm to speed up convergence. In addition, the decomposition permits coarse-grained parallelization of the subproblems, which can greatly speed up the solution times.

Another solution technique is the diagonal quadratic approximation (DQA) method of Mulvey and Ruszczyński (1992). Based on the augmented Lagrangian method, it is designed for problems with a block-diagonal structure and linking constraints, such as multistage stochastic programming problems. The augmented Lagrangian is approximated locally by a linear-quadratic function that is separable. The resulting subproblems have a special form that can be solved using an interior point method, such as the one developed by Carpenter et al. (1993) or Vanderbei's LOQO (1992). Berger et al. (1994) have developed an algorithm based on DQA that can be implemented on a parallel computer.

7. INDUSTRIAL ENGINEERING APPLICATIONS

Stochastic programming models can be applied to many other types of models apart from the investment model described above. In this section, we describe several basic model forms from different application areas: capacity planning, personnel scheduling, and production and inventory control. Each of these applications corresponds to practical models that have been used in industry. We have simplified them somewhat for presentation but add references to some key applications.

In general, most applications to date involve stochastic linear programming. However, future applications will require nonlinear objectives to handle risk aversion and other real-world considerations. Existing technology and algorithms allow you to solve nonlinear stochastic programs almost as easily as linear stochastic programs.

7.1. Capacity Planning Models

Manufacturers face the situation of having certain plants with installed capacity to produce specific products. The question is to determine whether additional capacity should be installed at a plant where no capacity for a product currently exists. This additional capacity would allow the plant to continue production if demand for the new product is higher than existing capacity at other plants and if the demand for products at the new plant is lower than the existing plant capacity.

As an example, consider Figure 1. Here, there are two products, A and B, and three plants, 1, 2, and 3. The solid lines in the diagram indicate that each plant currently produces only a single product. We could assume that each of the plants is built to meet the mean demand exactly. In that case, if demand for a product ever exceeds the mean, potential sales are lost.

By building additional capacity at 2 for product B (the dashed line in Figure 1), one incurs a large, one-time cost; however, if demand for product A is lower than the mean and demand for product B is higher than its mean, then the excess product B demand can be produced at 2 and fewer sales would be lost. That is the basic goal of flexible capacity. Other relevant measures (apart from net expected value gains) are the utilization of the plants (the fraction of installed capacity actually used) and the number of sales lost for each product type.

The decision problem is to trade off the costs of adding additional capacity against the potential revenue from additional sales due to the extra capacity. This basic problem has been considered by Eppen et al. (1989), who applied a mixed-integer, stochastic linear programming model for a large manufacturer.

Given this network characterization, the model determines where to install additional capacity now in order to maximize the value added to the firm by these capacity decisions. We will describe the model in the AMPL formulation that appears in Figure 2. We use the modeling language AMPL (Fourer et al., 1993) as an example for building practical stochastic programming models. We do not include any special stochastic programming features of the language, which are currently under development.

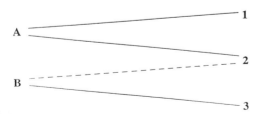

Figure 1 Adding flexible capacity at Plant 2.

Stochastic Programming

```
# This is a capacity planning model for assigning products to plants.
# SETS
set product;
set plant;
# PARAMETERS
param no_scenarios;
param capacity{plant}; # CAPACITY OF PLANT
param ot_capacity{plant}; # OVERTIME CAPACITY
param demand{product,1..no_scenarios}; # DEMAND FOR PRODUCT
# UNDER SCENARIO
param cap_cost{plant,product}; # CAPITAL COST
param cap_budget; # CAPITAL BUDGET
param profit{product}; # PROFIT FROM EACH PRODUCT
param otcost{product}; # OVERTIME OPERATING COSTS
param prob{1..no_scenarios}; # PROBABILITY OF SCENARIO
param amort; # AMORTIZATION FACTOR
#
var y{plant,product}>= 0; # 1 = j IS PRODUCED AT i, 0 OTHERWISE
#integer y
var reg_prod{plant,product,1..no_scenarios}>= 0; #REGULAR
# PRODUCTION
var ot_prod{plant,product,1..no_scenarios} >= 0; #OVERTIME
# PRODUCTION
# OBJECTIVE
maximize opt_val:
- amort*cap_cost[i,j]*y[i,j] + sum{i in plant} sum{j in
product} sum{k in 1..no_scenarios} (prob[k]*profit[j]
*(reg_prod[i,j,k] + ot_prod[i,j,k]) - otcost[j]*prob[k]
*ot_prod[i,j,k]);
# CONSTRAINTS
subject to
#
under_bud{k in 1..no_scenarios}:
sum{i in plant}sum{j in product} y[i,j]*cap_cost[i,j] <=
cap_budget;
#
no_gdemand{j in product, k in 1..no_scenarios}:
sum{i in plant} (reg_prod[i,j,k] + ot_prod[i,j,k]) <= demand[j,k];
#
j_atplant{i in plant,j in product, k in 1..no_scenarios}:
(reg_prod[i,j,k] + ot_prod[i,j,k]) <= y[i,j]*(capacity[i]
+ ot_capacity[i]);
#
r_eq_rcap{i in plant, k in 1..no_scenarios}:
sum{j in product} reg_prod[i,j,k] <= capacity[i];
#
ot_eq_otc{i in plant, k in 1..no_scenarios}:
sum{j in product} ot_prod[i,j,k] <= ot_capacity[i];
```

Figure 2 AMPL model of capacity problem.

The SETS in that model are the names of the products and plants. The parameters include the number of scenarios, the regular and overtime capacity of the plants, the demand for each product under each scenario, the cost of additional fixturing for each product to be produced at each plant, the capital budget, the operating profit on each product, the additional cost for overtime production, and the probabilites of the scenarios. We represent discounting with an amortization factor. This factor is varied according to the lifetime of each product.

The current model assumes that each product will be produced with the same demand pattern through the same time horizon. The model also assumes that capacity decisions occur only now and not in the future. In this way, we essentially build a two-stage model, although it may actually encompass several periods of demand. Sales in each period are discounted by some factor $\gamma(T)$ that depends on the horizon length T and the discount factor r. In this model, we divide the objective by $\gamma(T)$ so that the first-stage decision is multiplied by $amort = 1/\gamma(T)$ and the second-stage decision corresponds to a single year's revenues. This was used to modify as few data items as necessary.

The most fundamental decision is to determine where each product may be produced. We designate this binary variable as $y(plant, product)$. The next variables determine regular and overtime production of each product at each plant and in each scenario.

The objective to maximize is simply the expected revenues with a regular operating profit factor minus the additional costs for overtime production and minus the stage-one capital costs. The constraints state the capital expenditure is under budget. They also limit production of specific products by demand and plant capacity multiplied by the binary variable of whether the product can be produced at a given plant. Total regular and overtime production within a plant are also limited.

For the computational experiments, the basic procedure was to use the integer programming capability of IBM's OSL (1991). This software enabled solutions of relatively large capacity expansion models, although the majority of analyses considered the six-scenario problem in order to have relatively quick turnaround times for varying parameter combinations on the RS6000 processors.

The linear programming relaxations of the basic model in Figure 2 were also solved using the nested decomposition method ND-UM (Birge et al., 1993), which used the OSL solver for each subproblem. Since the OSL solver also follows the integer programming path, we present only the linear programming times for varying numbers of scenarios. For this experiment we also considered multiple stage versions of the problem.

Table 2 indicates a clear advantage for decomposition in these prob-

Stochastic Programming

Table 2 Capacity Planning Solution Times

Stages	Scenarios	Rows	Columns	OSL CPUs	ND-UM CPUs
2	4	478	966	4.7	1.7
2	8	894	1.81E03	11.9	2.5
2	16	1.73E03	3.49E03	35.4	3.6
3	36	4.43E03	8.95E03	230.7	15.2
3	256	2.84E04	5.72E04	12361	140.5
4	4096	4.54E05	9.17E05	Failed	5024

lems. The results for large problems represent orders of magnitude speedups over the straightforward simplex method approach in OSL. Although these results do not occur in all practical problems, they are illustrative of the potential for special-purpose stochastic programming codes.

In terms of modeling value, the stochastic model is also quite valuable. Eppen et al. (1989) report, for example, that the stochastic model obtains a result that no deterministic scenario would be able to find. This is a common characteristic for all optimization problems that necessarily find extreme point optima. The extreme point solution cannot include hedging and, hence, misses the fundamental nature of dealing with uncertainty.

7.2. Personnel Scheduling

Scheduling personnel is another common area in which optimization models are applied. The basic problem in this situation is to consider how to assign permanent staff to work hours to prevent costly overtime or temporary staff or other penalties due to an inability to meet demand. A typical situation would be planning hospital nursing schedules (see Kao and Queyranne, 1985) or scheduling professors to classes (see Mulvey, 1979).

As an example, consider a hospital emergency room (ER). The ER has historical data on the number of cases in each hour over an extended period of time. Permanent staff are used to meet most of the needs in the ER, but when several critical care cases arrive simultaneously, staff may be called from outside the ER to handle the overload. The financial result of this work shift is most likely some overtime cost when the non-ER staff return to their units. It also causes delays for noncritical cases in the ER.

In the general model we consider, the first decision is to fix a planning period (e.g., a week) and the length of intervals within that period (e.g., 1-hour blocks). We assume that possible *shifts*, complete schedules for a given planning period, are also formed by considering minimum work times, typical start times, minimum off times, etc. (Another option is to let the optimization problem fix shifts, but this leads to more complicated models.) The decision is then to determine the number of workers to assign to a shift.

The difficulty in determining this number is related to consideration of the uncertain demand. The most typical method is to consider the probabilistic or chance-constrained form of the constraints. In this approach, for every time interval, one finds a given percentile (e.g., 95%) of the demand and then estimates the number of nurses required to handle that number of average cases. The basic formulation for this type of model appears in Figure 3. This is an example of using a *chance constraint* (see Charnes and Cooper, 1959) or a constraint that is satisfied a given fraction of the time.

In this example, the decisions are the number to assign to each shift in the variable, SCHED{SHIFTS}. The total assigned is then minimized subject to meeting the demand (95%) in each period, j in 1..T. The demand constraint uses a zero-one data matrix ON __DUTY[i,j] with a value of one whenever shift i includes work in period j.

The problem with solving a model such as that in Figure 3 is how to determine the demand values and whether fixed demand values can really capture the decision goal to minimize some form of expected costs. Fixing the quantile targets implies a cost for missing demand. This cost may vary by period and may depend on the entire planning period's demand, not just

```
#This is a model for scheduling hospital workers with fixed
#shifts for each worker.
# SETS:
set SHIFTS; #ALL POSSIBLE SHIFTS IN CYCLE
# PARAMETERS:
param T>0 ; #NO. OF PERIODS
param DEMAND {1..T}; # NO. NEEDED AT TIME T
param ON_DUTY{SHIFTS,1..T}; # PATTERN OF ON-DUTY TIME FOR EACH
#SHIFT
# INDICATOR FOR SHIFT WORKING IN PERIOD
# VARIABLES:
var SCHED{SHIFTS} >= 0 ; # NO. SCHEDULED TO #WORK EACH SHIFT
# SHOULD BE INTEGER
#**************
#OBJECTIVE:
minimize TOTAL_SCHED: #MINIMIZE TOTAL
sum {i in SHIFTS} (SCHED[i]); # SCHEDULED
# CONSTRAINTS:
subject to MEET_REQT {j in 1..T}: sum{i in SHIFTS}
ON_DUTY[i,j]*SCHED[i] >= DEMAND[j];
```

Figure 3 AMPL model of personnel problem.

Stochastic Programming

an individual period. For example, the cost if demand in 1 period out of 20 in a week exceeds the personnel capacity is different from the cost if demand in every period of 1 in 20 weeks is above the capacity. In the former case, noncritical cases are probably shifted to a later period at low cost, but in the latter case no demand can be shifted to a low-demand period. High-cost overtime and probably temporary employees are necessary.

An alternative formulation to that in Figure 3 is to assume a set of demand scenarios (most likely historical data for a given week with some modification for recent trends). These demand scenarios include both critical and noncritical cases. Critical cases must be served in the given interval, while noncritical cases may be queued. Costs are estimated for overtime necessary to accommodate all cases. A typical formulation appears in Figure 4.

In this formulation, additional data requirements are the critical and noncritical demands under each scenario, the probabilities of the scenarios, and the overtime rate. Costs are explicitly charged in the objective at the rate OVERTIME. A simple expected cost objective is used, although some nonlinear form could also be introduced.

```
#This is a model for scheduling hospital workers with fixed
#shifts for each worker.
# SETS:
set SHIFTS; #ALL POSSIBLE SHIFTS IN CYCLE
# PARAMETERS:
param T>0 ; #NO. OF PERIODS
param S>0; # NO OF SCENARIOS
param DEMAND_CRIT {1..T,1..S };
# NO. NEEDED AT TIME T UNDER S - CRITICAL
param DEMAND_NONC {1..T,1..S };
# NO. NEEDED AT TIME T UNDER S - NONCRITICAL
param ON_DUTY{SHIFTS,1..T}; # PATTERN of ON-DUTY TIME FOR EACH
#SHIFT
# INDICATOR FOR SHIFT WORKING IN PERIOD
param OVERTIME; # OVERTIME RATE IN EACH PERIOD
param PROB {1..S }; # PROBABILITY OF EACH SCENARIO
var SCHED{SHIFTS} >= 0 ; # NO. SCHEDULED TO #WORK EACH SHIFT
# SHOULD BE INTEGER
var BACKLOG{1..T,1..S} >= 0; #AMT OF NONCRIT. IN PD. T UNDER S TO CARRY TO T+1
var OVER{1..T,1..S} >= 0 ; #AMT. OVER IN PERIOD T UNDER S
#*************
#OBJECTIVE:
minimize TOTAL_SCHED: #MINIMIZE TOTAL
(sum {i in SHIFTS} (SCHED[i])) + sum {j in 1..T,k in 1..S}
PROB[k]*OVERTIME*OVER[j,k] ; # SCHEDULED
# CONSTRAINTS:
subject to MEET_REQT {j in 1..T,k in 1..S}: sum{i in SHIFTS}
ON_DUTY[i,j]*SCHED[i] + OVER[j,k] + BACKLOG[j,k] - BACKLOG[j,k-1]
>= DEMAND_CRIT[j,k]+DEMAND_NONC[j,k];
PRIORITY{j in 1..T,k in 1..S}: BACKLOG[j,k] <= DEMAND_NONC[j,k];
```

Figure 4 Stochastic AMPL model of personnel problem.

The extra variables are for the number of backlog cases (BACKLOG) and the amount of overtime (OVER) used. The constraints in MEET _REQT apply overtime and backlog to meet the two demand categories. The last constraint set, PRIORITY, forces the backlog to be no greater than the number of noncritical cases in each period. This constraint forces service of critical cases within a period and holds the backlog to at most one period. Longer delays could be allowed with constraints of the form, BACKLOG$[j,k]$ − BACKLOG$[j,k-1]$ ≤ DEMAND _NONC$[j,k]$.

The resulting integer linear program in Figure 4 is substantially larger than the model in Figure 3 but the results generally attain substantial improvements in long-term operating costs. As an example of the potential cost consequences, Kao and Queyranne describe an example in which the commonly used deterministic model produced significant underestimates of total budget. They also show that aggregating demand over periods but maintaining a stochastic model obtains close results for overall workforce level and budget.

7.3. Production and Inventory Control

The basic goal in production and inventory control is to balance holding costs with shortage and overtime penalties. The problem in general practical implementations is that the demands, vendor supplies, actual production, prices, and costs are all unknown quantities. Deterministic models may be used, but care must be taken to ensure that the deterministic model result does indeed reflect the original problem adequately.

The basic form of the stochastic production and inventory control model is that decisions are made to produce at regular or overtime (or with external purchase) in each of several periods. The production is used to meet demand in each of the periods or is carried over as inventory (or shortage) to a subsequent period. Costs are associated with production, holding inventory, shortage penalty, and expedition (overtime, premium shipping, etc.).

In practice, the model is generally solved on a rolling horizon basis so that each period's production decision is made with updated information about current and future demand. This form of review and decision makes the model a true multiperiod decision problem. A basic form of the production planning model appears in Figure 5.

The formulation uses three sets for parameter variable indices, *products*, *resources*, and *scenarios*. The parameters include a zero-one matrix, *scen_links*, which gives *scen_links*$[i,j,k]$ a value of one if scenarios i and j are the same at time period k. This constraint is used for the explicit nonanticipative constraint form of the model. This representation is useful

Stochastic Programming

```
#This model describes a simple production planning problem
set products; #different investment options
set resources; #resources used in production
param scenarios; #no of scenarios (total S)
#The following 0-1 array shows which scenarios are combined
#at period T
param scen_links{1..scenarios,1..scenarios,1..T};
param T; #no. of periods
param capacity{resources}; #resource limits -normal
param cost{products}; #normal production cost of products
param holding_cost{products};
param technology{resources,products};
param penalty{products}; #cost of not meeting demand
param overtime_cost{resources}; #cost for using overtime workers
param overtime_limit{resources}; #max amount of overtime
param demand{products,1..scenarios,1..T}; #amt demanded
param prob {1..scenarios}; #probability of each scenario variables
var amt_made{products,1..scenarios,1..T} >= 0; #actual amounts made
var reg_res{resources,1..scenarios,1..T}>=0;#amt of reg. resource used
var extra_res{resources,1..scenarios,1..T} >= 0;#ext resource used
var inventory{products,1..scenarios,1..T+1}>= 0; #amt above demand
var backorder{products,1..scenarios,1..T+1}>=0;#amt below demand in
each period
minimize exp_value : sum{i in 1..scenarios} #objective
(sum{t in 1..T} ( sum{j in products} prob[i]*(cost[j]*amt_made[j,i,t] +
holding_cost[j]*inventory[j,i,t+1]+penalty[j]*backorder[j,i,t+1]))) +
sum{i in 1..scenarios}(sum{t in 1..T}(sum{j in resources}
prob[i]*overtime_cost[j]*extra_res[j,i,t] ));#cost of prod + extras
subject to capacity_bound{j in resources, i in 1..scenarios,
t in 1..T} : reg_res[j,i,t]<= capacity[j];#keep within bounds
subject to total_resource{j in resources, i in 1..scenarios, t in
1..T} : sum{k in products} technology[j,k]*amt_made[k,i,t]-
reg_res[j,i,t] - extra_res[j,i,t] = 0;
subject to ot_bounds{j in resources, i in 1..scenarios, t in 1..T} :
extra_res[j,1,t]<=overtime_limit[j];#keep within bounds
subject to nonanticip{k in products,j in 1..scenarios,t in 1..T}: (sum{i in 1..scenarios}
scen_links[j,i,t]*prob[i]*amt_made[k,i,t])-(sum{i in 1..scenarios}
scen_links[j,i,t]*prob[i])*amt_made[k,j,t] = 0; # makes nonanticipative
subject to balance{j in products,i in 1..scenarios, t in 1..T} :
amt_made[j,i,t] + inventory[j,1,t] - backorder[j,i,t] - demand[j,i,t] =
inventory[j,i,t+1] - backorder[j,i,t+1];
subject to initial{j in products, i in 1..scenarios} :
inventory[j,i,1] + backorder[j,i,1] = 0;
```

Figure 5 AMPL model of production planning problem.

especially when a deterministic model was developed earlier, as mentioned in Section 2.

The other parameters are somewhat self-explanatory except for the *technology* parameter matrix. The value *technology*[i,j] is used to indicate the amount of resource i used to make a unit of product j.

The variables include *amt_made* for actual production and *reg_res* for resources used under regular time. The variables *extra_res* include overtime or outside purchase. Positive and negative inventories are denoted by *inventory* and *backorder*, respectively.

The objective is again a simple expected cost minimization. The constraints limit regular production with *capacity_bound* and extra resources by *overtime_limit* in *ot_bounds*. Total resource use is equated in *total_resource*. The nonanticipativity constraint is in *nonanticip*. The constraints for conserving inventory appear in *balance* and in *initial* for the start of the planning horizon.

Some examples of solving problems of the form in Figure 5 appear in Escudero et al. (1993). Again, they observe that adding stochastic characteristics to the problem improves the quality of the solution substantially by stabilizing the results in the presence of alternative possible outcomes for the modeling parameters. They also consider other types of models that include service-level constraints, which are a form of probabilistic constraint.

Adding service-level constraints is similar to the deterministic form in Figure 3 for personnel scheduling. The basic idea is to pick a demand percentile to meet in each period. Bitran and Yanasse (1984) include constraints of this form to produce approximate formulations. Their approximations behave well for this type of model with errors generally below 10%.

7.4. Additional Models

The examples given above are only a sample of the industrial engineering problems that stochastic programs can address. Other areas include facility layout and location (see, for example, Louveaux and Thisse, 1985), project scheduling (see Cleef and Gaul, 1982), and production planning for style items (see Bitran et al., 1986). The range of stochastic programming applications is growing rapidly as advanced hardware as well as computational methods continue to expand the set of possible models. Recent implementations (see Birge et al., 1993) now routinely solve problems with up to 5 million variables.

Almost any application that relies on forecasts for model parameters is a candidate for being modeled as a stochastic model. Traditionally, these models were formulated as deterministic models. However, many traditional large-scale integrated models, such as Agrico Chemical Company's production, inventory, and logistic planning system (Glover et al., 1979), Citgo Petroleum Corporation's products planning system (Klingman et al., 1986, 1987a, 1987b), and W. R. Grace's logistic planning system (Klingman et al., 1988) have uncertainty in the demands utilized by these systems. Advances in computer technology and solution techniques since these systems were developed could now handle stochastic variations of these systems. Research into this area can result in the development of robust large-scale integrated planning systems.

8. CONCLUSIONS

This chapter presented a brief discussion of stochastic programming methods and models. We began with a simple example in financial planning that illustrated the value of the stochastic solution over that of a deterministic model solution. We also showed how this quantity is different from the expected value of perfect information.

In the third section, we discussed issues in modeling stochastic parameters. We discussed the four existing methods for generating stochastic parameters—time series modeling, expert opinion, stochastic modeling, and simulation modeling. We also examined the two main applications for stochastic models—organizations that use a small number of scenarios in their planning process and models derived from stochastic difference equations.

We then described the main procedures for creating finite-dimensional problems when the original problem has an infinite range of possibilities. We showed how these approximations could be used to obtain bounds on the optimal objective value.

In the fifth section, we discussed an important issue in solving real-world stochastic programs—scenario generation. Efficient generation techniques permit an analyst to create a deterministic mathematical programming model whose optimal solution is close to the optimal solution for the stochastic program. Importance sampling, scenario decomposition, and cluster analysis are the principal methods for efficiently generating scenarios. This is another major research area in stochastic programming.

In the sixth section, we described the most common solution procedure based on outer linearizations of the value function. This method obtains significant computational advantages over nonspecialized procedures by exploiting the similarities among problems differing only in some random parameters. We also discussed two alternative approaches, progressive hedging and DQA. Progressive hedging is a decomposition method that breaks the problem down into a series of subproblems that can be efficiently solved using specialized algorithms, then updates the weightings of solutions to find an optimal value. DQA uses a local linear-quadratic approximation to the augmented Lagrangian so that an efficient interior point method can be applied to the subproblems. Developing new solution procedures remains an active research area. Overall, these methods allow industrial engineers to build large and practical stochastic programs and, with continued advances in computational capabilities, should lead to widespread solutions of models that explicitly incorporate randomness.

The last section considered several common models for industrial engineering in terms of capacity planning, workforce scheduling, and production and inventory control. In each area, stochastic programming models offer substantial advantages over deterministic approaches. We expect these areas to continue to grow and to aid in continuously improving operational decision making.

ACKNOWLEDGMENT

The research of J. R. Birge was funded in part by NSF grant ECS-92-15921. The research of J. M. Mulvey was funded in part by NSF grant CCR-9102660 and Air Force grant AFOSR-91-0359. We would like to acknowledge the assistance of Michael T. Tapia.

REFERENCES

Benders, J. F., "Partitioning Procedures for Solving Mixed-Variables Programming Problems," *Numerische Mathematik*, Vol. 4, pp. 238–252, 1962.

Berger, A. J., Mulvey, J. M., Ruszczyński, A., "An Extension of the DQA Algorithm to Convex Stochastic Programs," *SIAM Journal of Optimization*, Vol. 4, pp. 735–753, 1994.

Birge, J. R., "The Value of the Stochastic Solution in Stochastic Linear Programs with Fixed Recourse," *Mathematical Programming*, Vol. 24, pp. 314–325, 1982.

Birge, J. R., "Decomposition and Partitioning Methods for Multi-Stage Stochastic Linear Programs," *Operations Research*, Vol. 33, pp. 989–1007, 1985.

Birge, J. R., "Stochastic Programming: Optimizing the Uncertain," in *Optimization Techniques and Applications*, eds. K. Phua, C. Wang, W. Yeong, T. Leong, H. Loh, K. Tan, F. Chou, eds., Vol. 2, pp. 613–632. World Scientific, Singapore, 1992.

Birge, J. R., Donohue, C. J., Holmes, D. F., Svintsitski, O. G., "A Parallel Implementation of the Nested Decomposition Algorithm for Multistage Stochastic Linear Programs," Technical Report, Department of Industrial and Operations Engineering, University of Michigan, Ann Arbor, 1993; also *Mathematical Programming*, to appear.

Birge, J. R., Wets, R. J.-B., "Designing Approximation Schemes for Stochastic Optimization Problems, in Particular, for Stochastic Programs with Recourse," *Math. Progr. Study*, Vol. 27, pp. 54–102, 1986.

Bitran, G., Yanasse, H., "Deterministic Approximations to Stochastic Production Problems," *Operations Research*, Vol. 32, pp. 999–1018, 1984.

Bitran, G. R., Haas, E. A., Matsuo, H., "Production Planning of Style Goods with High Setup Costs and Forecast Revisions," *Operations Research*, Vol. 34, 226–236, 1986.

Cariño, D. R., Kent, T., Myers, D. H., Stacy, C., Sylvanus, M., Turner, A., Watanabe, K., Ziemba, W. T., "The Russell-Yasuda Kasai Financial Planning Model," *Interfaces*, Vol. 24, pp. 29–49, 1994.

Carpenter, T. J., Lustig, I. J., Mulvey, J. M., Shanno, D. F., "Separable Quadratic Programming via Primal-Dual Interior Point Method and Its Use in a Sequential Procedure," *ORSA Journal on Computing*, Vol. 5, pp. 182–191, 1993.

Censor, T., Lent, A., "An Iterative Row-Action Method for Interval Convex Programming," *Journal of Optimization Theory and Applications*, Vol. 34, pp. 321–353, 1981.

Charnes, A., Cooper, W. W., "Chance-Constrained Programming," *Management Science*, Vol. 5, pp. 73–79, 1959.

Çinlar, E., *Introduction to Stochastic Processes*. Prentice-Hall, Englewood Cliffs, NJ, 1975.

Cleef, H. W., Gaul, W., "Project Scheduling via Stochastic Programming," *Math. Operationsforschung Statist. Ser. Optimization*, Vol. 13, pp. 449–468, 1982.

Dantzig, G., Glynn, P., "Parallel Processors for Planning Under Uncertainty," *Annals of Operations Research*, Vol. 22, pp. 1–21, 1990.

Dempster, M. A. H., "On Stochastic Programming II: Dynamic Problems Under Risk," *Stochastics*, Vol. 25, pp. 15–42, 1988.

Edmundson, H. P., "Bounds on the Expectation of a Convex Function of a Random Variable," RAND Corporation Paper 982, Santa Monica, CA, 1956.

Eppen, G. D., Martin, R. K., Schrage, L., "A Scenario Approach to Capacity Planning," *Operations Research*, Vol. 37, pp. 517–527, 1989.

Ermoliev, Y., "Stochastic Quasigradient Methods and Their Applications to Systems Optimization," *Stochastics*, Vol. 9, pp. 1–36, 1983.

Escudero, L. F., Kamesam, P. V., King, A. J., Wets, R. J.-B., "Production Planning via Scenario Modeling," *Annals of Operations Research*, Vol. 43, pp. 311–335, 1993.

Fourer, R., Gay, D. M., Kernighan, B. W., *AMPL: A Modeling Language for Mathematical Programming*. Scientific Press, South San Francisco, CA, 1993.

Frauendorfer, K., "Solving S.L.P. Recourse Problems with Arbitrary Multivariate Distributions—the Dependent Case," *Mathematics of Operations Research*, Vol. 13, pp. 377–394, 1988.

Frauendorfer, K., Kall, P., "A Solution Method for SLP Recourse Problems with Arbitrary Multivariate Distributions—the Independent Case," *Problems Control Inf. Theory*, Vol. 17, pp. 177–205, 1988.

Gartska, S., Rutenberg, D., "Computation in Discrete Stochastic Programs with Recourse," *Operations Research*, Vol. 21, pp. 112–122, 1973.

Gassmann, H. I., "MSLiP: A Computer Code for the Multistage Stochastic Linear Programming Problem," *Mathematical Programming*, Vol. 47, pp. 407–423, 1990.

Glassey, C. R., "Nested Decomposition and Multistage Linear Programs," *Management Science*, Vol. 20, pp. 282–292, 1973.

Glover, F., Jones, D., Karney, D., Klingman, D., Mote, J., "An Integrated Production, Distribution, and Inventory Planning System," *Interfaces*, Vol. 9, No. 5, pp. 21–35, 1979.

Glynn, P. W., Iglehart, D. L., "Importance Sampling for Stochastic Simulations," *Management Science*, Vol. 35, pp. 1367–1392, 1989.

Higle, J., Sen, S., "Stochastic Decomposition: An Algorithm for Two Stage Linear Programs with Recourse," *Mathematics of Operations Research*, Vol. 16, pp. 650–669, 1991.

Ho, J. K., Manne, A. S., "Nested Decomposition for Dynamic Models," *Mathematical Programming*, Vol. 6, pp. 121–140, 1974.

Huang, C. C., Ziemba, W., Ben-Tal, A., "Bounds on the Expectation of a Convex Function of a Random Variable: With Applications to Stochastic Programming," *Operations Research*, Vol. 25, pp. 315–325, 1979.

Infanger, G., "Monte Carlo (Importance) Sampling Within a Benders' Decomposition for Stochastic Linear Programs," *Annals of Operations Research*, Vol. 39, pp. 69–95, 1992.

International Business Machines Corp., "Optimization Subroutine Library Guide and Reference, Release 2," document SC23-0519-02, International Business Machines Corp., 1991.

Jensen, J. L., "Sur les fonctions convexe et les inégalités entre les valeurs moyennes," *Acta. Math.*, Vol. 30, pp. 175–193, 1906.

Kall, P., Stoyan, D., "Solving Stochastic Programming Problems with Recourse Including Error Bounds," *Math. Operationsforsch. Statist. Ser. Optim.*, Vol. 13, pp. 431–447, 1982.

Kallberg, J. G., Ziemba, W. T., "Comparison of Alternative Utility Functions in Portfolio Selection Problems," *Management Science*, Vol. 29, pp. 1257–1276, 1983.

Kao, E., Queyranne, M., "Budgeting Costs of Nursing in a Hosptial," *Management Science*, Vol. 31, pp. 608–621, 1985.

Klingman, D., Phillips, N., Steiger, D., Wirth, R., Young, W., "The Challenges and Success Factors in Implementing an Integrated Products Planning System for Citgo," *Interfaces*, Vol. 16, No. 3, pp. 1–19, 1986.

Klingman, D., Phillips, N., Steiger, D., Young, W., "The Successful Deployment of Management Science Throughout Citgo Petroleum Corporation," *Interfaces*, Vol. 17, No. 1, pp. 4–25, 1987a.

Klingman, D., Phillips, N., Steiger, D., Wirth, R., Padman, R., Krishnan, R., "An Optimization Based Integrated Short-Term Refined Petroleum Product Planning System," *Management Science*, Vol. 33, pp. 813–830, 1987b.

Klingman, D., Mote, J., Phillips, N. V., "A Logistics Planning System at W. R. Grace," *Operations Research*, Vol. 36, pp. 811–822, 1988.

Law, A., Kelton, D., *Simulation Modeling and Analysis*. McGraw-Hill, New York, 1991.

Loewenstein, G., Elster, J., eds., *Choice over Time*. Russell Sage Foundation, New York, 1992.

Louveaux, F. V., "A Solution Method for Multistage Stochastic Programs with Recourse with Application to an Energy Investment Problem," *Operations Research*, Vol. 28, pp. 889–902, 1980.

Louveaux, F. V., Thisse, J.-F., "Production and Location on a Network Under Demand Uncertainty," *Operations Research Letters*, Vol. 4, No. 4, pp. 145–149, 1985.

Madansky, A., "Bounds on the Expectation of a Convex Function of a Multivariate Random Variable," *Ann. Math. Statist.*, Vol. 30, pp. 743–746, 1959.

Madansky, A., "Inequalities for Stochastic Linear Programming Problems," *Management Science*, Vol. 6, pp. 197–204, 1960.

Markowitz, H., *Portfolio Selection: Efficient Diversification of Investments*. Yale University Press, New Haven, CT, 1959.

Mulvey, J. M., "Strategies in Modeling: A Personnel Scheduling Example," *Interfaces*, Vol. 9, No. 3, pp. 66–77, 1979.

Mulvey, J. M., "Generating Scenarios for the Towers Perrin Investment System," *Interfaces*, in press.

Mulvey, J. M., Beck, M. P., "Aggregating Variables via Optimal Cluster Analysis," Technical Report EES 81-2, Department of Civil Engineering and Operations Research, Princeton University, Princeton, NJ, 1981.

Mulvey, J. M., Crowder, H. P., "Cluster Analysis: An Application of Lagrangian Relaxation," *Management Science*, Vol. 25, pp. 329–340, 1979.

Mulvey, J. M., Ruszczyński, A., "A Diagonal Quadratic Approximation Method for Large Scale Linear Programs," *Operations Research Letters*, Vol. 12, pp. 205–215, 1992.

Mulvey, J. M., Vanderbei, R. J., Zenios, S. A., "Robust Optimization of Large Scale Systems," *Operations Research*, Vol. 43, pp. 264–281, 1995.

Mulvey, J. M., Vladimirou, H., "Solving Multistage Stochastic Networks: An Application of Scenario Aggregation," *Networks*, Vol. 21, pp. 619–643, 1991.

Mulvey, J. M., Vladimirou, H., "Stochastic Network Programming for Financial Planning Problems," *Management Science*, Vol. 38, pp. 1642–1664, 1992.

Nielsen, S. S., Zenios, S. A., "A Massively Parallel Algorithm for Nonlinear Stochastic Network Problems," *Operations Research*, Vol. 41, pp. 319–337, 1993.

Noël, M.-C., Smeers, Y., "Nested Decomposition of Multistage Nonlinear Programs with Recourse," *Mathematical Programming*, Vol. 37, pp. 131–152, 1989.

Rockafellar, R. T., "Monotone Operators and the Proximal Point Algorithm," *SIAM J. Control Opt.*, Vol. 14, pp. 877–898, 1976.

Rockafellar, R. T., Wets, R. J.-B., "Scenarios and Policy Aggregation in Optimization Under Uncertainty," *Mathematics of Operations Research*, Vol. 16, pp. 119–147, 1991.

Ruszczyński, A., "A Regularized Decomposition for Minimizing a Sum of Polyhedral Functions," *Mathematical Programming*, Vol. 35, 309–333, 1986.

Vanderbei, R. J., "LOQO User's Manual," Technical Report SOR-92-5, Department of Civil Engineering and Operations Research, Princeton University, Princeton, NJ, 1992.

Van Slyke, R., Wets, R. J.-B., "L-Shaped Linear Programs with Application to Optimal Control and Stochastic Programming," *SIAM Journal on Applied Mathematics*, Vol. 17, pp. 638–663, 1969.

von Neumann, J., Morganstern, O., *Theory of Games and Economic Behavior*. Princeton University Press, Princeton, NJ, 1947.

Wets, R., "Large-Scale Linear Programming Techniques in Stochastic Programming," in *Numerical Techniques for Stochastic Optimization*, eds. Y. Emroliev and R. Wets. Springer, Berlin, 1988.

Wierzbicki, A. P., "Dynamic Aspects of Multiobjective Optimization," in *Lecture Notes in Economics and Mathematical Systems*, No. 351, eds. A. Lewandowski and V. Volkovich. Springer-Verlag, Berlin, 1988.

Wittrock, R. J., "Advances in a Nested Decomposition Algorithm for Solving Staircase Linear Programs," Technical Report SOL 83-2, Systems Optimization Laboratory, Stanford University, Stanford, CA, 1983.

Wonnacott, R. J., Wonnacott, T. H., *Econometrics*. Wiley, New York, 1970.

Ziemba, W. T., Vickson, R. G., eds., *Stochastic Optimization Models in Finance*. Academic Press, New York, 1975.

9

Heuristic Methods

Michael Pinedo
Columbia University, New York, New York

David Simchi-Levi
Northwestern University, Evanston, Illinois

1. INTRODUCTION

Many industrial problems, such as machine scheduling, vehicle dispatching and routing, and facility layout and location problems, can be formulated as linear programs, integer programs, quadratic programs, or disjunctive programs. Unfortunately, many of these problems belong to the class of *NP-hard* problems, which implies that it is unlikely that efficient, polynomial time, algorithms will be developed for their optimal solutions. Hence, it is important to develop effective heuristics for these problems; these heuristics must generate solutions fast with cost close to the best, or optimal, cost.

In this section the various forms of mathematical programs are described. To demonstrate the different formulations, we provide examples of vehicle dispatching and routing, machine scheduling, and facility location problems and show how they are formulated as mathematical programs.

1.1. Linear Programming Formulation

The most basic form of mathematical program is the linear program (LP). An LP refers to an optimization problem in which the objective and the constraints are linear in the variables to be determined. It can be formulated as follows:

Minimize $\quad c_1x_1 + c_2x_2 + \cdots + c_nx_n$

subject to $\quad a_{11}x_1 + a_{12}x_2 + \cdots + a_{1n}x_n \le b_1$

$\qquad\qquad a_{21}x_1 + a_{22}x_2 + \cdots + a_{2n}x_n \le b_2$

$\qquad\qquad\vdots$

$\qquad\qquad a_{m1}x_1 + a_{m2}x_2 + \cdots + a_{mn}x_n \le b_m$

$\qquad\qquad x_j \ge 0 \quad \text{for } j = 1, \ldots, n$

The objective is the minimization of costs. The $c_1, \ldots, c_n$ vector is usually referred to as the cost vector. The variables $x_1, \ldots, x_n$ have to be determined so that the objective function $c_1x_1 + \cdots + c_nx_n$ is minimized. The column vector $a_{1j}, \ldots, a_{mj}$ is referred to as activity vector j. The value of the variable x_j refers to the level at which this activity j is utilized. The $b_1, \ldots, b_m$ is usually referred to as the resources vector.

An important special case of the linear program is the linear assignment problem, commonly referred to as the assignment problem. The assignment problem can be formulated as follows.

Minimize $\quad \sum_{i=1}^{n}\sum_{j=1}^{n} c_{ij}x_{ij}$

subject to $\quad \sum_{i=1}^{n} x_{ij} = 1 \qquad \text{for } j = 1, \ldots, n$

$\qquad\qquad \sum_{j=1}^{n} x_{ij} = 1 \qquad \text{for } i = 1, \ldots, n$

$\qquad\qquad x_{ij} \ge 0 \qquad \text{for } i = 1, \ldots, n, j = 1, \ldots, n$

Note that the assignment problem, as formulated here, has n^2 variables. That the assignment problem is important in the formulation of real-world problems is illustrated by the following scheduling example.

Example 1.1.1 Consider a single machine and n jobs. The n jobs are all available at time 0 and have identical processing times. Let p denote the processing time of a job. Job j has a due date d_j and a weight w_j; the weight of a job usually corresponds to the importance of that particular job in the schedule. If job j is completed at time C_j, its tardiness T_j is equal to $\max(C_j - d_j, 0)$. The objective to be minimized is the sum of the weighted tardiness Σw_jT_j. That this problem is identical to an assignment problem is clear. If job j is assigned to the ith position in the sequence, its weighted tardiness equals $w_j \max(ip - d_j, 0)$, i.e.,

$$c_{ij} = w_j \max(ip - d_j, 0) \qquad \square$$

Heuristic Methods

1.2. Integer Programming Formulations

An integer program (IP) is basically a linear program with the additional requirement that the variables $x_1, \ldots, x_n$ have to be integers. If only a subset of the variables are required to be integer and the remaining ones are allowed to be real, the problem is referred to as a mixed-integer program (MIP). In contrast with the LP, an efficient (polynomial time) algorithm for the IP or MIP does *not* exist.

Many routing and scheduling problems can be formulated as integer programs. In what follows, two examples of integer programming formulations are given. The first example describes an integer programming formulation of the *traveling salesman problem (TSP)*.

Example 1.2.1 A salesman has to visit n cities, starting and finishing in one of them, with as objective to minimize the total distance traveled. Let d_{ij} denote the distance between cities i and j. Assume $d_{ij} = d_{ji}$ for all i and j (in this case the problem is referred to as the symmetric TSP). The TSP can be formulated as follows. For every i and j, $i < j$, let x_{ij} denote a binary variable that takes a value of 1 if the salesman travels between cities i and j and 0 otherwise. Let S denote a subset of the n cities and let $|S|$ denote the number of cities in this subset. Using this notation the TSP is formulated as follows.

$$\text{Minimize} \quad \sum_{i=1}^{n} \sum_{j=i+1}^{n} d_{ij} x_{ij}$$

$$\text{subject to} \quad \sum_{j=1}^{i-1} x_{ji} + \sum_{j=i+1}^{n} x_{ij} = 2 \quad \text{for } i = 1, \ldots, n$$

$$\sum_{i \in S} \sum_{j \in S, j > i} x_{ij} \leq |S| - 1 \quad \text{for every } S, S \neq \emptyset$$

$$x_{ij} \in \{0, 1\} \quad \text{for every } i < j$$

The first set of constraints ensures that each city is visited exactly once. The second set, called the *subtour elimination constraints*, guarantees that no subtour exists. That is, for every S, $S \neq \emptyset$, there is not a cycle in the solution consisting only of cities in S. Observe that if such a subtour exists then

$$\sum_{i \in S} \sum_{i < j, j \in S} x_{ij} = |S|$$

which violates this constraint. Finally, observe that this set of constraints is very large, since every subset S requires a constraint. □

The next example describes an integer programming formulation of a scheduling problem.

Example 1.2.2 Consider a single machine and n jobs. Job j has a processing time p_j and a weight w_j. All jobs are available at time 0. The completion time C_j of job j depends on the sequence and the objective is to minimize the sum of the weighted completion times $\Sigma\, w_j C_j$. It happens that there is an easy solution for this problem, namely sequence the jobs in decreasing order of w_j/p_j. Nevertheless, the formulation of the integer program is of interest as this type of formulation is also applicable to more complicated scheduling problems.

Let x_{jk} denote a 0-1 decision variable that assumes the value 1 if job j precedes job k in the sequence and 0 otherwise. The values x_{jj} have to be 0 for all j. The completion time of job j is then equal to $\sum_{k=1}^{n} p_k x_{kj} + p_j$. The integer programming formulation of the problem thus becomes

$$\text{Minimize} \quad \sum_{j=1}^{n} \sum_{k=1}^{n} w_j p_k x_{kj} + \sum_{j=1}^{n} w_j p_j$$

subject to

$$x_{kj} + x_{jk} = 1 \quad \text{for } j, k = 1, \ldots, n, j \neq k$$
$$x_{kj} + x_{lk} + x_{jl} \geq 1 \quad \text{for } j, k, l = 1, \ldots, n, j \neq k, j \neq l, k \neq l$$
$$x_{jk} \in \{0,1\} \quad \text{for } j, k = 1, \ldots, n$$
$$x_{jj} = 0 \quad \text{for } j = 1, \ldots, n$$

The third set of constraints can be replaced by a combination of (1) a set of linear constraints that require all x_j to be nonnegative, (2) a set of linear constraints requiring all x_j to be less than or equal to 1, and (3) a set of constraints requiring all x_j to be integer. Constraints requiring certain precedences between the jobs can be added easily by specifying the corresponding x_{jk} values. □

There are several ways for dealing with integer programs. These techniques are discussed in Chapter 3. The best known approaches are:

1. Cutting plane (polyhedral) techniques
2. Branch and bound techniques

The first class of techniques focuses on linear program relaxations of the integer program. The techniques aim at generating *additional* linear constraints that have to be satisfied for the variables to be integer. These additional inequalities constrain the feasible set more than the original set of linear inequalities without cutting off integer solutions. Solving the LP relaxation of the IP with the additional inequalities then yields a different solution, which may be integer. If the solution is integer, the procedure stops as the solution obtained is an optimal solution for the original IP. If the variables are not integer, more inequalities are generated.

Heuristic Methods

The second approach, branch and bound, is basically a sophisticated way of doing complete enumeration that can be applied to many combinatorial problems. The branching refers to a partitioning of the solution space. Each part of the solution space is then considered separately. The bounding refers to the development of lower bounds for parts of the solution space. If a lower bound on the objectives in one part of the solution space is larger than a solution already obtained in a different part of the solution space, the corresponding part of the former solution space may be disregarded.

1.3. Quadratic Programming Formulations

A generalization of the linear program described above is the *quadratic program*. The difference between the two is only in the objective function: in the latter the objective is quadratic rather than linear. The constraints, however, remain the same and are linear. The quadratic program with the additional constraints that the variables have to be either 0 or 1 is an important problem in practice. This program, referred to as the *quadratic assignment problem*, is used to formulate many facility layout problems.

Example 1.3.1 Consider the problem of assigning n facilities, such as departments or offices, to a given set of n sites. Let d_{ij} denote the distance between sites i and j and let w_{kl} denote the weight or the amount of flow per unit time between facilities k and l. The problem of assigning facilities to sites so as to minimize total weighted distance, referred to as *material handling cost*, can be formulated as a quadratic assignment problem. For this purpose let x_{ki} be a variable that equals to 1 if facility k is assigned to site i and 0 otherwise.

$$\text{Minimize} \quad \sum_{i=1}^{n} \sum_{j=1}^{n} \sum_{k=1}^{n} \sum_{l=1}^{n} d_{ij} w_{kl} x_{ki} x_{lj}$$

subject to

$$\sum_{k=1}^{n} x_{ki} = 1 \qquad \text{for } i = 1, \ldots, n$$

$$\sum_{i=1}^{n} x_{ki} = 1 \qquad \text{for } k = 1, \ldots, n$$

$$x_{ki} \in \{0,1\} \qquad \text{for } k = 1, \ldots, n, \, i = 1, \ldots, n \qquad \square$$

1.4. Disjunctive Programming Formulations

The last form of mathematical program to be considered in this chapter is the disjunctive program. The class of disjunctive programs is a large class of mathematical programs in which the constraints can be divided into a set of *conjunctive* constraints and one or more sets of *disjunctive* constraints. A set of constraints is called conjunctive if each one of the constraints has to be satisfied. A set of constraints is called disjunctive if at least one of the constraints has to be satisfied but not necessarily all.

In the standard linear program all constraints are conjunctive. The integer program described in Example 1.2.2 in essence contains pairs of disjunctive constraints. The fact that the integer variable x_{jk} has to be either 0 or 1 can be enforced by a pair of disjunctive linear constraints: either $x_{jk} = 0$ or $x_{jk} = 1$. This implies that the scheduling problem in Example 1.2.2 can be formulated as a disjunctive program as well, even when there are precedence constraints.

Example 1.4.1 Consider a generalization of the scheduling problem described in Example 1.2.2. The generalization allows the jobs to be subject to precedence constraints. There is a precedence constraints graph that specifies for any given job which jobs have to be completed before the job can start. Before formulating this scheduling problem as a disjunctive program it is of interest to represent the problem by a disjunctive graph model. For a more general representation of such models see Chapter 4. Let N denote the set of nodes that correspond to the n jobs. Between any pair of nodes (jobs) j and k in this graph exactly one of the following three conditions has to hold:

1. Job j precedes job k.
2. Job k precedes job j.
3. Jobs j and k are independent with respect to one another.

The set of directed arcs A represent the precedence relationships between the jobs. These arcs are the so-called conjunctive arcs. Let set I contain all the pairs of jobs that are independent of one another. Each pair of jobs $(j,k) \in I$ are now connected with one another by two arcs going in opposite directions. These arcs are referred to as disjunctive arcs. The problem is to select from each pair of disjunctive arcs between two independent jobs j and k one arc that indicates which of the two jobs goes first. The selection of disjunctive arcs has to be such that these arcs together with the conjunctive arcs do not contain a cycle. The selected disjunctive arcs together with the conjunctive arcs determine a schedule for the n jobs.

Let the variable x_j in the disjunctive program formulation denote the completion time of job j. The set A denotes the set of precedence constraints $j \to k$ that require job j to be processed before job k. The scheduling problem is formulated as

$$\text{Minimize} \sum_{j=1}^{n} w_j x_j$$

Heuristic Methods

subject to

$$x_k - x_j \geq p_k \quad \text{for all } j \to k \in A$$
$$x_j \geq p_j \quad \text{for } j = 1, \ldots, n$$
$$x_k - x_j \geq p_k \text{ or } x_j - x_k \geq p_j \quad \text{for all } (j,k) \in I$$

The first and second sets of constraints are sets of conjunctive constraints. The third set is a set of disjunctive constraints. □

The same techniques that are applicable to integer programs are also applicable to disjunctive programs. The application of branch and bound to a disjunctive program is straightforward. First the LP relaxation of the disjunctive program has to be solved (i.e., the LP obtained after deleting the set of disjunctive constraints). If the optimal solution of the LP happens to satisfy all disjunctive constraints, then the solution is optimal for the disjunctive program as well. However, if one of the disjunctive constraints is violated, say the constraint

$$(x_k - x_j) \geq p_k \quad \text{or} \quad (x_j - x_k) \geq p_j$$

then two additional LPs are generated. One has the additional constraint $(x_k - x_j) \geq p_k$ and the other has the additional constraint $(x_j - x_k) \geq p_j$. The procedure is in all other respects similar to the branch and bound procedure for integer programming.

As described in the Introduction, most forms of mathematical programs are *NP-hard*, which implies that it is not likely to find a polynomial time algorithm that solves every instances of the problem to optimality. For that reason effective heuristics are often a necessity, especially when solution procedures for problems in practice require an answer in real time. Such is the case in machine scheduling and vehicle routing, where every morning the dispatcher/scheduler receives the assignments for that day and has to perform them effectively.

2. HEURISTICS

This section describes a number of general-purpose techniques that have proved useful in industrial problems. These techniques do not guarantee finding an *optimal* solution but instead aim at finding reasonably good solutions in a relatively short time. Although many such techniques exist, only a few representative techniques are described here in detail.

The first class of techniques are usually referred to as *myopic rules*. The second class are the so-called *local search* procedures. These procedures tend to be fairly generic and can be applied to different problems with only minor customization. The third class of procedures are derivatives of branch and bound. They aim at eliminating branches in an intelligent way so that not all branches have to be examined. The last subsection discusses how several empirical techniques can be combined into a single framework.

There are many other empirical procedures; the selection described here is not meant to be exhaustive but rather aims at providing a flavor of the thinking behind such techniques; the interested reader is referred to Morton and Pentico (1994) or Pinedo (1995).

2.1. Myopic Rules

Myopic rules are used when a solution of a problem is constructed in a progressive manner. At each iteration there is a partial solution and an extension of this solution is constructed by selecting one of a number of options available. Usually, the option with the minimum cost is selected for the extension of the current partial solution.

For example, one well-known rule for constructing an efficient tour for the traveling salesman problem is the *nearest neighbor* (*NN*) rule. Suppose a partial (open) tour has been constructed in previous iterations. This open tour has two ends, i.e., two cities that are currently linked only with a single city. One of these two cities is then linked to the city that is closest by. It has been shown that this heuristic may perform arbitrarily bad. However, this heuristic is often used in practice. This rule is myopic as it considers only the best possible next step. It does not take into consideration any one of the possible steps to be taken after the next step.

Many myopic rules are used in scheduling as well. These myopic rules are often referred to as *dispatching* rules. For example, consider a single machine with n jobs. The processing time of job j is p_j and there are sequence-dependent setup times. If job j is followed by job k, a setup time s_{jk} is incurred. The objective to be minimized is the completion time of the last job, often referred to as the makespan and denoted by C_{max}. It can be shown that this problem is equivalent to the traveling salesman problem. For this purpose, consider this machine scheduling problem and a corresponding traveling salesman problem with $n + 1$ cities $0, 1, 2, \ldots, n$. City i, $i = 1, \ldots, n$, corresponds to job i and city 0 is the starting city where the salesman is initially located. The distance between cities j and k is the setup time s_{jk} for every two cities, $j, k \geq 1$. The distance between cities 0 and k is the time it takes to prepare the machine for processing the first job, say job k. Minimizing the makespan is the same as minimizing the length of the traveling salesman tour on this network. An often used heuristic is the *shortest setup time first* (*SST*), which selects every time the job with the shortest setup time. The above presentation of the makespan problem as a traveling salesman problem implies that the SST rule is equivalent to the NN rule.

Heuristic Methods

Another example in scheduling concerns the sum of the weighted completion times as objective. Consider again a single machine and n jobs. Job j has a processing time p_j and a weight (or priority factor) w_j. The objective is to find a schedule that minimizes the sum of the weighted completion times of the n jobs. A well-known dispatching rule is the so-called *weighted shortest processing time first* (*WSPT*) rule. According to this rule, whenever the machine is idle, the job with the highest ratio of weight over processing time is selected to go next; i.e., the jobs are ranked in decreasing order of w_j/p_j. This dispatching rule is well known and is also used in many stochastic scheduling problems and queueing problems.

Research in myopic rules has been active for several decades, and a number of different rules have been studied for the many different problems in industrial engineering. For comprehensive overviews of dispatching rules see Panwalkar and Iskander (1977) and Bhaskaran and Pinedo (1992).

2.2. Heuristics Based on Local Search

A procedure based on local search, in contrast with a global search procedure, does not guarantee an optimal solution. A procedure based on local search usually attempts to find a solution better than the current one through a search in the *neighborhood* of the current solution. Two solutions are *neighbors* if one can be obtained through a well-defined modification of the other.

Many local search procedures can be viewed as generalizations of a local search procedure developed for the traveling salesman problem in the early 1970s. These procedures are often referred to in the literature as the *2-opt*, *3-opt*, or, in general *k-opt* procedures; see Golden and Stewart (1985).

To illustrate these procedures, consider the *symmetric* traveling salesman problem. The *k-opt* procedure, for $k \geq 2$, starts from a traveling salesman tour

$$j_1, \ldots, j_{k-1}, j_k, j_{k+1}, \ldots, j_{l-1}, j_l, j_{l+1}, \ldots, j_n, j_1$$

where j_i is the ith city visited by the salesman. A *2-opt* exchange takes a tour segment $j_k, \ldots, j_l$ and reverses it. This exchange results in the new tour

$$j_1, \ldots, j_{k-1}, j_l, j_{l-1}, \ldots, j_{k+1}, j_k, j_{l+1}, \ldots, j_n$$

This new tour has a total distance less than the total distance of the original schedule if

$$d_{j_{k-1}, j_k} + d_{j_l, j_{l+1}} > d_{j_{k-1}, j_l} + d_{j_k, j_{l+1}}$$

If the reversal does not result in a reduced cost, an additional reversal is made. A tour segment $j_l, j_{l-1}, \ldots, j_p$ is taken and reversed. Whether the new tour has a smaller total distance than the very first tour can be checked easily. The combination of the two reversals is typically referred to as a *3-opt* exchange. The procedure is usually implemented in such a way that at a given iteration a (variable) number of tour segment reversals are performed in an attempt to obtain a tour with a smaller total distance. All tour segments at a given iteration have as their first city the city immediately following a given city j_{k-1} (which maintains its position in the sequence throughout that iteration). The tour obtained at the end of an iteration is always at least as good as the tour obtained at the end of the previous iteration. If no better tour is found, the search continues in the next iteration with reversals of tour segments following a different city.

The structure of the neighborhood is a very important aspect of a local search procedure. For a single machine scheduling problem a neighborhood can be defined easily. A neighborhood of a particular schedule may be simply defined as all schedules that can be obtained by performing a single adjacent pairwise interchange. This implies that there are $n - 1$ schedules in the neighborhood of the original schedule. A larger neighborhood for a single machine schedule may be defined by taking an arbitrary job in the schedule and inserting it in another position in the schedule. Clearly, each job can be inserted in $n - 1$ other positions. The entire neighborhood contains less than $n(n - 1)$ neighbors as some of these neighbors are identical. The neighborhood of a schedule in a more complicated machine environment is usually more complex.

Neighborhood search procedures that are currently popular are *simulated annealing*, *tabu-search*, and *genetic algorithms*.

Simulated annealing is a search procedure that has its origin in a different field; it was first developed as a simulation model for describing the physical annealing process for condensed matter. The simulated annealing process goes through a number of iterations. At each iteration k there is a current solution, as well as a best solution obtained so far. For example, for a single machine problem these solutions are merely given sequences (permutations) of the jobs, say S_k and S_0. Let $G(S_k)$ and $G(S_0)$ denote the corresponding values of the objective function. The value of the best solution obtained so far is often referred to as the aspiration criterion. The process, in its search for an optimal solution, moves from one solution to another. From S_k, the schedule at iteration k, a search is conducted within its neighborhood for a new solution. First, a so-called *candidate* solution, say S, is selected from the neighborhood. This selection can be done at random or in an organized, possibly sequential, way. If S is a better solution than S_k, a move is made, setting $S_{k+1} = S$. If S is better than the best solution obtained so far, S_0 is set equal to S. However, if S is a worse solution than S_k, a move is made to S with probability

Heuristic Methods

$$P(S_k, S) = \exp\left(\frac{G(S_k) - G(S)}{\beta_k}\right)$$

With probability $1 - P(S_k, S)$ solution S is rejected in favor of the current solution, setting $S_{k+1} = S_k$. The $\beta_1 \geq \beta_2 \geq \beta_3 \geq \cdots > 0$ are control parameters referred to as cooling parameters or temperatures (in analogy with the annealing process mentioned above). Often β_k is chosen to be a^k, where $0 < a < 1$.

From the preceding description of the simulated annealing procedure, it is clear that moves to worse solutions are allowed. This is the major difference with regular neighborhood searches. The reason for allowing these moves is to give the procedure the opportunity to move away from a local minimum and find a better solution later on. Since β_k decreases with k, the acceptance probability for a nonimproving move is lower in later iterations of the search process. From the definition of the acceptance probability it also follows that if a neighbor is significantly worse, the acceptance probability is very low and the move is not likely to be made.

In practice, several stopping criteria are used for this procedure. One way is to let the procedure run for a prescribed number of iterations. Another is to let the procedure run until no improvement has been obtained for a given number of iterations.

The method can be summarized as follows:

Algorithm 2.2.1

> *Step 1*
> Set $k = 1$ and select β_1.
> Select an initial solution S_1 through some heuristic.
> Set $S_0 = S_1$.
> *Step 2*
> Select a candidate solution S from the neighborhood of S_k.
> If $G(S_0) < G(S) < G(S_k)$, set $S_{k+1} = S$ and go to step 3.
> If $G(S) < G(S_0)$, set $S_0 = S_{k+1} = S$ and go to step 3.
> If $G(S) > G(S_k)$ generate a random number U_k from a Uniform $(0,1)$ distribution;
> If $U_k \leq P(S_k, S)$ set $S_{k+1} = S$ otherwise set $S_{k+1} = S_k$ and go to step 3.
> *Step 3*
> Select $\beta_{k+1} \leq \beta_k$.
> Increment k by 1.
> If $k = K$ then STOP, otherwise go to step 2.

The effectiveness of simulated annealing depends on the design of the neighborhood as well as on the manner in which the search is conducted within this neighborhood. If the neighborhood is designed in a way that makes moves to better solutions or moves out of a local minimum easier, the procedure will perform more efficiently.

The search within a neighborhood can be done randomly or in a more organized way. If it is done in a more organized way, the contribution of each job to the objective function may be computed and the job with the highest impact on the objective may be selected as the one to be moved. For an application of simulated annealing to routing see Kirkpatrick et al. (1983). For an application to scheduling, see Matsuo et al. (1988) and to facility layout, see Wilhelm and Ward (1987).

Tabu-search is in many ways similar to simulated annealing. The procedure also moves from one solution to another, with the next solution being possibly worse than the preceding solution. For each solution a neighborhood is defined as in simulated annealing. The search for a neighbor within the neighborhood as a potential candidate to move to is again a design issue. This can, just as in simulated annealing, be done in a random way or in a organized way. The basic difference between tabu-search and simulated annealing lies in the mechanism used for approving candidate moves. In tabu-search the mechanism is not probabilistic but rather of a deterministic nature. At any stage of the process a tabu-list of mutations that the procedure is *not* allowed to make is kept. Mutations on the tabu-list may be, for example, pairs of jobs that may not be interchanged. The list has a fixed number of entries (this number usually lies between 5 and 9). Every time a move is made through a mutation in the current solution the *reverse* mutation is entered at the top of the tabu-list; all other entries are pushed down one position and the bottom entry is deleted. The reverse mutation is put on the list to avoid returning to a local minimum that has been visited before. Actually, at times a reverse mutation that is tabu could have led to a new solution, not visited before, with an objective value lower than any one obtained before. This may happen when the mutation is close to the bottom of the tabu-list and a number of moves have already been made since the mutation was entered in the list. Thus, if the number of entries in the tabu-list is too small, cycling may occur; if the number of entries is too large the search may be constrained unduly.

The method can be summarized as follows:

Algorithm 2.2.2

> *Step 1*
> Set $k = 1$.
> Select an initial solution S_1 through some heuristic.
> Set $S_0 = S_1$.

Heuristic Methods

Step 2
Select a candidate solution S from the neighborhood of S_k.
If the move $S_k \to S$ is prohibited by a mutation on the tabu-list, set $S_{k+1} = S_k$ and go to step 3.
If the move $S_k \to S$ is not prohibited by any mutation on the tabu-list, set $S_{k+1} = S$;
enter reverse mutation at the top of the tabu-list;
push all other entries in the tabu-list one position down;
delete the entry at the bottom of the tabu-list.
If $G(S) < G(S_0)$, set $S_0 = S$;
Go to step 3.
Step 3
Increment k by 1.
If $k = K$ then STOP,
otherwise go to step 2.

The following example illustrates the method.

Example 2.2.3 Consider the following single-machine scheduling problems. Job j has processing time p_j, due date d_j and weight w_j. If job j is completed at time C_j, then the tardiness T_j of job j is $\max(C_j - d_j, 0)$. The objective is to minimize the sum of the weighted tardinesses $\Sigma w_j T_j$.

Jobs	1	2	3	4
p_j	10	10	13	4
d_j	4	2	1	12
w_j	14	12	1	12

The neighborhood of a schedule contains all schedules that can be obtained through adjacent pairwise interchanges. The tabu-list is a list of pairs of jobs (j,k) that were swapped within the last two moves and cannot be swapped again. Initially, the tabu-list is empty.

As a first schedule the sequence $S_1 = 2,1,4,3$ is chosen; the corresponding value of the objective function is $\Sigma w_j T_j(2,1,4,3) = 500$. The aspiration criterion is therefore 500. There are three schedules in the neighborhood of S_1, namely 1,2,4,3; 2,4,1,3; and 2,1,3,4. The respective values of the objective function are 480, 436, and 652. Selection of the best non-tabu sequence results in $S_2 = 2,4,1,3$. The aspiration criterion is changed to 436. The tabu-list is updated and contains now the single pair $(1,4)$. The values of the objective functions of the neighbors of S_2 are

Sequence	4,2,1,3	2,1,4,3	2,4,3,1
$\Sigma w_j T_j$	460	500	608

Note that the second move is tabu. However, the first move is better than the second anyhow. The first move results in a schedule worse than the best one so far (which is the current one, which therefore is a local minimum). Nevertheless, $S_3 = 4,2,1,3$ and the tabu-list is updated and now contains $\{(2,4),(1,4)\}$. Neighbors of S_3 with the corresponding values of the objective functions are

Sequence	2,4,1,3	4,1,2,3	4,2,3,1
$\Sigma w_j T_j$	436	440	632

Now, although the best move is to $2,4,1,3$ (S_2), this move is tabu. Therefore S_4 is chosen to be $4,1,2,3$. Updating the tabu-list results in $\{(1,2),(2,4)\}$. The pair (1,4) drops from the tabu-list as the length of the list is kept to 2. Neighbors of S_4 with the corresponding values of the objective functions are

Sequence	1,4,2,3	4,2,1,3	4,1,3,2
$\Sigma w_j T_j$	408	460	586

The schedule 4,2,1,3 is tabu, but the best move is to the schedule 1,4,2,3. So $S_5 = 1,4,2,3$. The corresponding value of the objective is better than the aspiration criterion. So the aspiration criterion becomes 408. The tabu-list is updated by adding (1,4) and dropping (2,4). Actually, S_5 is a global minimum, but tabu-search, being unaware of this fact, continues. □

For an application of tabu-search to routing, see Glover and Laguna (1990) or Gendreau et al. (1994) and for an application to scheduling see Dell'Amico and Trubian (1991).

Genetic algorithms constitute a class of techniques that are more general and abstract than both simulated annealing and tabu-search. In what follows the use of genetic algorithms is illustrated through a scheduling problem. Genetic algorithms, when applied to scheduling, view sequences or schedules as *individuals* that are members of a *population*. Each individual is characterized by its *fitness*. The fitness of an individual is measured by the associated value of the objective function. The procedure works iteratively and each iteration is referred to as a *generation*. A generation of

Heuristic Methods

individuals consists of surviving individuals of the previous generation and new solutions or *children* from the previous generation. The population size usually remains constant from one generation to the next. The children are generated through reproduction and mutation of individuals that were part of the previous generation (the *parents*). Individuals are at times also referred to as *chromosomes*. A chromosome may consist of subchromosomes, each one containing the information regarding the job sequence on a machine. A mutation in a parent chromosome may be equivalent to an adjacent pairwise interchange in the corresponding sequence. In each generation the fittest individuals reproduce and the least fit die. The birth, death, and reproduction process that determine the composition of the next generation can be a complicated process that is usually a function of the fitness levels of the individuals of the current generation. A very simplified version of a genetic algorithm is described below.

Algorithm 2.2.4

Step 1
Set $k = 1$.
Select q initial sequences $S_{1,1}, \ldots, S_{1,q}$ through some heuristic.
Step 2
Select the best schedule among $S_{k,1}, \ldots, S_{k,q}$ and call it S_k^*.
Select the worst schedule among $S_{k,1}, \ldots, S_{k,q}$ and call it S_k^{**}.
Select a neighbor S from the neighborhood of S_k^*.

Replace S_k^{**} with S;
Keep all other schedules the same and go to step 3.
Step 3
Increment k by 1.
If $k = K$ then STOP,
otherwise go to step 2.

The following example illustrates the technique.

Example 2.2.5 Consider the scheduling problem of Example 2.2.3. In the first generation three individuals are selected at random.

Individual	2,1,3,4	3,4,1,2	4,1,3,2
Fitness	652	814	586

The individual 4,1,3,2 is allowed to reproduce. Its offspring, generated through a random swap, is 4,3,1,2 with fitness 758. The individual 3,4,1,2 dies. This results in a second generation.

Individual	2,1,3,4	4,3,1,2	4,1,3,2
Fitness	652	758	586

The individual 4,1,3,2 is allowed to reproduce again. Its offspring is 1,4,3,2 with cost 554. This results in a third generation.

Individual	2,1,3,4	1,4,3,2	4,1,3,2
Fitness	652	554	586

Now, 1,4,3,2 is allowed to reproduce with offspring 1,4,2,3 and cost 408, resulting in a fourth generation.

Individual	1,4,2,3	1,4,3,2	4,1,3,2
Fitness	408	554	586

So at the fourth generation the optimal sequence is found. Of course, the method is not aware of this and continues while retaining the best sequence obtained so far. □

The use of simulated annealing, tabu-search, and genetic algorithms has its advantages and disadvantages. One advantage is that they can be applied to a problem without having to know much about the structural properties of the problem. They can be coded very easily and provide solutions that are usually fairly good. However, the amount of computation time needed to obtain such a solution tends to be relatively long in comparison with the more rigorous problem specific approaches.

2.3. Filtered Beam Search

Filtered beam search is based on the ideas of branch and bound. Enumerative branch and bound methods are currently the most widely used methods for obtaining optimal solutions to NP-hard scheduling problems. The main disadvantage of branch and bound is that it usually is extremely time consuming, as the number of nodes one must consider is very large.

Consider, for example, a single-machine problem with n jobs. Assume that for each node at level k jobs have been selected for the first k positions. There is a single node at level 0, with n branches emanating from it to n nodes at level 1. Each node at level 1 branches out into $n - 1$ nodes at level 2, resulting in a total of $n(n - 1)$ nodes at level 2. At level k there are $n!/(n - k)!$ nodes. At the bottom level, level n, there are $n!$ nodes.

Heuristic Methods

Branch and bound attempts to eliminate a node by determining a lower bound on the objective for all partial schedules that sprout out of that node. If the lower bound is higher than the value of the objective under a known schedule, the node may be eliminated and its offspring disregarded. If one could obtain a reasonably good schedule through some clever heuristic before going through the branch and bound procedure, it may be possible to eliminate many nodes. Other elimination criteria may also reduce the number of nodes to be investigated. However, even after these eliminations there are usually still too many nodes to be evaluated. For example, it may require several weeks on a RISC workstation to find an optimal schedule for an instance of the problem considered in Example 2.2.3 with 20 jobs. The main advantage of branch and bound is that, after evaluating all nodes, the final solution is known with certainty to be optimal.

Filtered beam search is an adaptation of branch and bound in which not all nodes at any given level are evaluated. Only the most promising nodes at level k are selected as nodes to branch from. The remaining nodes at that level are discarded *permanently*. The number of nodes retained is called the *beam width* of the search. The evaluation process that determines which nodes are the promising ones is a crucial component of this method. Evaluating each node carefully, in order to obtain an estimate for the potential of its offspring, is time-consuming. There is a trade-off here: a crude prediction is quick, but may lead to discarding good solutions, but a more thorough evaluation may be prohibitively time-consuming. Here is where the filter comes in. For all the nodes generated at level k, a crude prediction is done. Based on the outcome of these crude predictions a number of nodes are selected for a thorough evaluation, while the remaining nodes are discarded permanently. The number of nodes selected for a thorough evaluation is referred to as the *filter width*. Based on the outcome of the careful evaluation of all nodes that passed the filter, a subset of these nodes (the number being equal to the beam width, which therefore cannot be greater than the filter width) is selected from which further branches will be generated.

A simple example of a crude prediction in scheduling is the following. The contribution of the partial schedule to the objective and the due date tightness or some other statistic of the jobs remaining to be scheduled are computed. Based on these values, the nodes at a given level can be compared with one another and an overall assessment can be made.

Every time a node has to undergo a thorough evaluation, all the jobs not yet scheduled are scheduled using a composite dispatching rule. Such a schedule can still be generated reasonably fast as it only requires sorting. The result of such a schedule is an indication of the promise of the node. If a large number of jobs is involved, nodes may be filtered out by examining the partial schedule obtained by scheduling only a limited number of the remaining jobs with a dispatching rule. This extended partial schedule may be evaluated, and based on its value a node may be discarded or retained. If a node is retained, it may be analyzed more thoroughly by having all its remaining jobs scheduled with the composite dispatching rule. The value of this schedule's objective then represents an upper bound on the best schedule among the offspring of that node. The following example illustrates a simplified version of beam search.

Example 2.3.1 Consider the instance of the scheduling problem in Example 2.2.3. As the number of jobs is rather small, only one type of prediction is made for the nodes at any particular level. No filtering mechanism is used. The beam width is chosen to be 2, which implies that at each level only two nodes are retained. The prediction at a node is made by scheduling the unscheduled jobs according to the WSPT rule (i.e., the remaining jobs are ordered in decreasing order of w_j/p_j).

A branch and bound tree is constructed assuming the sequence is developed starting out from $t = 0$. So, at the jth level of the tree jobs are put into the jth position. At level 1 of the tree there are four nodes: (1,*,*,*), (2,*,*,*), (3,*,*,*), and (4,*,*,*) (see Figure 1). Applying the WSPT rule to the three remaining jobs at each one of the four nodes results in the four sequences (1,4,2,3), (2,4,1,3), (3,4,1,2), and (4,1,2,3) with objective values 408, 436, 771, and 440. As the beam width is 2, only the first two nodes are retained.

Each of these two nodes leads to three nodes at level 2. Node (1,*,*,*) leads to nodes (1,2,*,*), (1,3,*,*), and (1,4,*,*) and node (2,*,*,*) leads to nodes (2,1,*,*), (2,3,*,*), and (2,4,*,*). Applying the WSPT rule to the remaining two jobs in each one of the six nodes at level 2 results in nodes (1,4,*,*) and (2,4,*,*) being retained and the remaining four being discarded.

The two nodes at level 2 lead to four nodes at level 3 (the last level), namely nodes (1,4,2,3), (1,4,3,2), (2,4,1,3), and (2,4,3,1). Of these four sequences sequence (1,4,2,3) is the best with a total weighted tardiness equal to 408. As observed in Examples 2.2.3, this sequence is optimal. □

Ow and Morton (1988) were the first to apply filtered beam search to scheduling.

2.4. Multiphase Approaches

For many combinatorial problems one may design procedures that combine elements of several of the techniques presented in this chapter.

For example, the following three-step approach has proved fairly use-

Heuristic Methods

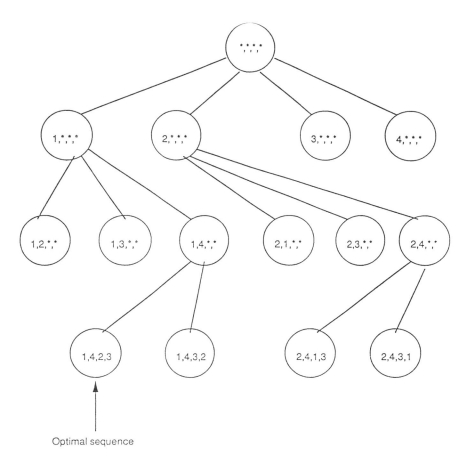

Figure 1 Application of beam search.

ful for solving scheduling problems in practice. It combines myopic rules with simulated annealing or tabu-search.

Algorithm 2.4.1

Step 1
Values of a number of statistics are computed, such as due date tightness, setup time severity, and so on.

Step 2
Based on the outcome of step 1 a myopic rule is selected and applied on the scheduling instance

Step 3

The schedule developed in step 2 is used as an initial solution for a tabu-search or simulated annealing procedure, which attempts to improve on the schedule for a limited amount of computer time.

Another important class of techniques is based on *machine pricing* and *time pricing* principles. Complicated job shop scheduling problems can be formulated through mathematical programs. These mathematical programs usually have several sets of constraints. For example, one set of constraints may have to enforce the fact that two jobs may not be assigned to the same machine at the same point in time. Such a set may be regarded as machine capacity constraints. Another set of constraints may have as its purpose that certain precedence constraints are enforced. Disregarding or relaxing one or more sets of constraints may make the solution of the scheduling problem significantly easier. It is possible to incorporate such a set of constraints in the objective function by multiplying it with a so-called *Lagrangian* multiplier. This Lagrangian multiplier is in effect the penalty one pays for violating the constraints. If the solution of the modified mathematical program violates any one of the relaxed constraints, the value of the Lagrangian multiplier has to be changed at the next iteration of the scheduling process in order to increase the penalty of violation and encourage the search for a solution that does not violate the relaxed constraints; see Fisher (1981).

3. ROUTING APPLICATIONS

Consider a *distribution system* with a single *depot* (e.g., a warehouse, production center, or school) and n geographically dispersed *demand points* or *customers* (e.g., retailers or bus stops). The demand points are numbered arbitrarily from 1 to n. At each demand point, there are a number of *items* (e.g., products or students), which are referred to as the *demand* and which must be brought to the depot using a fleet of vehicles.

3.1. Overview of Vehicle Routing Problems

Typically, three types of constraints are present in vehicle routing problems.

1. *Capacity constraints:* an upper bound on the number of items that can be carried by a vehicle. This upper bound is, of course, a result of the limited vehicle capacity.
2. *Distance (or travel time) constraints:* a limit on the total distance (or time) traveled by each vehicle and/or a limit on the amount of time an item can be in transit.

3. *Time window constraints:* a prespecified earliest and latest pickup or delivery time for each demand point and/or a prespecified time window in which vehicles must reach their final destination.

The problem is to design a set of routes for the vehicles such that each route starts and ends at the depot, each item is brought to the depot, no constraint is violated, and total distance traveled is as small as possible. This problem is a generic vehicle routing problem (VRP). It is clear that this problem is extremely hard to analyze, since the traveling salesman problem is a special case.

The vehicle routing problem appears in a large variety of applications such as the distribution of soft drinks, beer, gasoline, and pharmaceuticals, or the pickup and delivery of students by school buses. The wide applications of the vehicle routing problem have motivated academic researchers and private and public organizations to consider optimization techniques to improve the efficiency of distribution or transportation systems. Consequently, the problem has been analyzed extensively in the literature; an excellent survey of the literature may be found in Fisher (1992).

A special case of the vehicle routing problem is the capacitated vehicle routing problem. In the capacitated vehicle routing problem only the capacitated constraints are present. This problem can be described as follows: a set of customers dispersed in a given area must be served by vehicles of limited capacity, denoted by Q, initially located at a central depot. Associated with each customer i located at point x_i is a demand w_i, which represents the amount that must be delivered to that customer. The objective is to design efficient routes, starting and ending at the central depot, such that each customer receives its demand, the total load delivered by a vehicle does not exceed its capacity, and the total distance traveled is as small as possible.

The heuristics for this problem may be classified into the following three categories:

Constructive methods
Route first–cluster second methods
Cluster first–route second methods

In what follows the main characteristics of these three classes are described.

3.2. Constructive Methods

An example of a constructive method is the *savings algorithm* suggested by Clarke and Wright (1964). This is one of the earliest heuristics designed for this problem and, without a doubt, the most widely known. The idea of the savings algorithm is very simple: consider the depot and n demand points.

Suppose that initially a separate vehicle is assigned to each demand point. The total distance traveled by a vehicle that visits demand point i is $2d_i$, where d_i is the distance from the depot to demand point i. Therefore, the total distance traveled in this solution is

$$2d_1 + 2d_2 + \cdots + 2d_n$$

If two routes are combined, say points i and j are served on a single trip by the same vehicle, the total distance traveled by this vehicle is $d_i + d_{ij} + d_j$, where d_{ij} is the distance between demand points i and j. Thus, the *savings* obtained from combining demand points i and j, denoted by s_{ij}, is

$$s_{ij} = 2d_i + 2d_j - (d_i + d_j + d_{ij}) = d_i + d_j - d_{ij}$$

The larger the savings s_{ij} the more desirable it is to combine demand points i and j. Based on this idea, Clarke and Wright suggest the following algorithm, which is often referred to as the savings algorithm.

Algorithm 3.2.1

Step 1
Start with the solution for the capacitated vehicle routing problem in which a separate vehicle serves each demand point.

Step 2
For every pair of demand points i and j calculate the savings $s_{ij} = d_i + d_j - d_{ij}$.
Order the savings in decreasing order. This ordered list is called the savings list.

Step 3
Take the pair of demand points with the largest savings, say i, j, and eliminate it from the savings list.
Combine the tour containing i with the tour containing j if:
(i) the combined load of the tours containing i and j does not exceed the vehicle capacity.
(ii) both i and j are either the first or last demand point served on their route.

Step 4
Continue with step 3 until the savings list is exhausted.

3.3. Route First–Cluster Second Methods

The second class of heuristics are the route first–cluster second methods. This class contains all the heuristics that first order the customers according to their locations and then partition this ordering to produce feasible clusters. These clusters consist of sets of customers that are consecutive in the

Heuristic Methods

initial order. Customers are then routed within their cluster depending on the specific heuristic.

One such heuristic for the capacitated vehicle routing problem is the following *tour partioning heuristic* suggested by Beasley (1983). In this heuristic, called *optimal partitioning* heuristic, one constructs a traveling salesman tour through the customers and the warehouse. The tour is then *optimally* partitioned into segments, each containing a total load of at most Q items, by formulating an appropriate shortest-path problem; see Chapter 3 for a discussion of the shortest-path problem.

This is done as follows: given a traveling salesman tour through the customers and the depot, the points (i.e., customers) are numbered $x^{(0)}$, $x^{(1)}, \ldots, x^{(n)}$ in order of appearance on the tour, where $x^{(0)}$ is the depot. Let C_{jk} be the distance traveled by a vehicle that starts at the depot, visits customers $x^{(j+1)}, \ldots, x^{(k)}$ in this order, and returns to the depot if $\Sigma_{i=j+1}^{k} w_i \leq Q$; otherwise we let $C_{jk} = \infty$ (recall, w_i is the demand of customer i). If we find the shortest path from $x^{(0)}$ to $x^{(n)}$ in the directed graph with distance cost C_{jk}, we will have an optimal partition of the traveling salesman tour chosen.

The sweep algorithm suggested by Gillett and Miller (1974) can also be viewed as a route first–cluster second type of heuristic. In this algorithm, an arbitrary customer is selected as a starting customer. The other customers are ordered according to the angle between them, the depot, and the starting customer. Customers are then assigned to vehicles following this initial ordering and efficient routes are designed for each vehicle.

3.4. Cluster First–Route Second Methods

The third class of methods are the cluster first–route second methods. In this class customers are first clustered into groups to be served by the same vehicle (cluster first) and then efficient routes are designed for each cluster (route second). Heuristics of this class are usually more sophisticated than those of the previous class, since determining the clusters is often based on a mathematical programming approach.

To illustrate this class we describe here the heuristic suggested by Bramel and Simchi-Levi (1992). This heuristic is based on the following observation. A feasible solution that has attractive properties when the number of customers is very large can be described as follows: in this solution the length of every tour that visits a set of customers S consists of two parts. The first is the length of the *tour* that starts at the depot, visits the subregion (where its customers are located), and then goes back to the depot. The second is the additional distance obtained by *inserting* all the customers in S into this tour. It is clear, therefore, that if we can construct a heuristic that assigns customers to vehicles so as to minimize the sum of the lengths of all simple tours plus the total insertion costs of customers to each simple tour, then the heuristic will have the same structure as the above solution.

To construct such a heuristic we formulate the routing problem as a standard combinatorial problem commonly called (see, e.g., Pirkul, 1987) the *capacitated concentrator location problem*. This problem can be described as follows: given m possible sites for concentrators of fixed capacity Q, we would like to locate concentrators at a subset of these m sites and connect n terminals, where terminal i uses w_i units of a concentrator's capacity, in such a way that each terminal is connected to *exactly one* concentrator, the concentrator capacity is not exceeded, and the total cost is minimized. A site-dependent cost is incurred for locating each concentrator; that is, if a concentrator is located at site j, the *setup* cost is v_j, for $j = 1, 2, \ldots, m$. The cost of connecting terminal i to concentrator j is c_{ij} (the *connection* cost), for $i = 1, 2, \ldots, n$ and $j = 1, 2 \ldots, m$.

The capacitated concentrator location problem can be formulated as the following integer linear program. Let $y_j = 1$ if a concentrator is located at site j and 0 otherwise. Also, let $x_{ij} = 1$ if terminal i is connected to a concentrator at site j and 0 otherwise.

$$\text{Minimize} \quad \sum_{i=1}^{n} \sum_{j=1}^{m} c_{ij} x_{ij} + \sum_{j=1}^{m} v_j y_j$$

subject to

$$\sum_{j=1}^{m} x_{ij} = 1 \qquad \text{for } i = 1, \ldots, n$$

$$\sum_{i=1}^{n} w_i x_{ij} \leq Q \qquad \text{for } j = 1, \ldots, m$$

$$x_{ij} \leq y_j \qquad \text{for } i = 1, \ldots, n, j = 1, \ldots, m$$

$$x_{ij} \in \{0,1\} \qquad \text{for } i = 1, \ldots, n, j = 1, \ldots, m$$

$$y_j \in \{0,1\} \qquad \text{for } j = 1, \ldots, m$$

The first set of constraints ensures that each terminal is connected to exactly one concentrator, and the second ensures that the concentrator's capacity constraint is not violated. The third set of constraints guarantees that if a terminal is connected to site j, then a concentrator is located at that site. The last two sets of constraints ensure the integrality of the variables.

In formulating the capacitated VRP with unsplit demands as the capacitated concentrator location problem, every customer x_j in the VRP is a possible concentrator site in the location problem. The length of the simple

Heuristic Methods

tour in the VRP that starts at the depot, visits customer x_j, and then goes back to the depot is the setup cost in the location problem (i.e., $v_j = 2d_j$). Finally, the cost of inserting a customer into a simple tour in the VRP is the connection cost in the location problem (i.e., $c_{ij} = d_i + d_{ij} - d_j$). Hence, a solution for the capacitated VRP is obtained by solving a facility location problem with the data as described above. The solution obtained for the capacitated concentrator location problem is transformed (in an obvious way) to a solution for the capacitated VRP.

The capacitated concentrator location problem, although NP-hard, can be efficiently solved by the familiar Lagrangian relaxation technique, as described in Pirkul (1987) or Bramel and Simchi-Levi (1992), or by a cutting-plane algorithm, as described in Deng and Simchi-Levi (1992).

We can now describe the new algorithm for the capacitated VRP, called the location-based heuristic (LBH):

Algorithm 3.4.1

Step 1
Formulate the capacitated VRP as a capacitated concentrator location problem.
Step 2
Solve the capacitated concentrator location problem.
Step 3
Transform the solution obtained in step 2 into a solution for the VRP.

4. SCHEDULING APPLICATIONS

This section deals with heuristics for scheduling problems. One of the most important scheduling problems is the job shop problem.

Consider m machines and n jobs. A job has to be processed on each one of the m machines. The route a job has to follow is predetermined and fixed and the routes of the different jobs are not necessarily the same. The processing time of job j's operation on machine i is p_{ij}. Preemptions are not allowed. The objective is the minimization of the makespan.

4.1. Disjunctive Programming Formulation

Minimizing the makespan in a job shop can be represented elegantly by a disjunctive graph. Consider a directed graph $G = (N,A,B)$. The nodes N correspond to all of the processing operations performed on the n jobs. The so-called *conjunctive* (solid) arcs A represent the precedence relationships between the processing operations of a single job. Two operations, belonging to two different jobs, that are to be processed on the same machine are connected to one another by two *disjunctive* (broken) arcs going in opposite directions. The disjunctive arcs B form m cliques of double arcs, one clique for each machine. (A clique is a term in graph theory that refers to a graph in which all pairs of nodes are connected with one another.) All operations (nodes) that are connected to one another in one clique have to be done on the same machine. All arcs emanating from a node, conjunctive as well as disjunctive, have as length the processing time of the operation performed at that node. In addition, there are a source and a sink, which are dummy nodes. The source has conjunctive arcs emanating to all the first operations of the jobs and the sink has conjunctive arcs coming from all the last operations. The arcs emanating from the source have length zero (see Figure 2). A feasible schedule corresponds to a *selection* of at most one disjunctive arc from each pair such that the resulting directed graph is acyclic. This implies, then, that each selection from arcs within a clique must be acyclic. Such a selection determines the sequence in which the operations are to be performed on that machine.

That a selection from a clique has to be acyclic can be argued as follows. If there was a cycle within any clique, a feasible sequence of the operations on the corresponding machine would not have been possible. It is not immediately obvious that there may not be any cycle that is formed by disjunctive arcs from different cliques and conjunctive arcs. Such a cycle would correspond to a situation that in practice is often referred to as *deadlock*. For example, let (h,j) and (i,j) denote two consecutive operations that belong to job j and let (i,k) and (h,k) denote two consecutive operations that belong to job k. If under a schedule operation (i,j) precedes operation (i,k) on machine i and operation (h,k) precedes operation (h,j) on machine h, then the graph contains a four-arc cycle, two being conjunctive arcs and two being disjunctive arcs from different cliques. It is clear

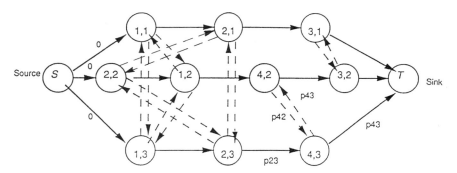

Figure 2 Directed graph for job shop with three jobs (Example 4.1.1.).

Heuristic Methods

that such a schedule is physically impossible. Summarizing, if D denotes the subset of the selected disjunctive arcs and the graph $G(D)$ is defined by the set of conjunctive arcs and the subset D, then D corresponds to a feasible schedule if and only if $G(D)$ contains no directed cycles.

The makespan of a feasible schedule that corresponds to an acyclic selection of disjunctive arcs is determined by the longest path (i.e., the *critical path*) from source to sink. The problem of minimizing the makespan reduces to finding a selection of disjunctive arcs that minimizes the length of this longest path.

There are several mathematical programming formulations for the job shop with no recirculation, including a number of integer programming formulations. However, the formulation most often used is a disjunctive programming formulation (see Example 1.4.1). This disjunctive programming formulation is closely related to the disjunctive graph model representation of the job shop.

In this formulation p_{ij} denotes the processing time of job j on machine i and the variable y_{ij} denotes the starting time of this operation. Set N denotes the set of all operations (i,j), corresponding to the nodes in the directed graph. Set A denotes the set of all precedence (routing) constraints $(i,j) \rightarrow (k,j)$ that require job j to be processed on machine i before it is processed on machine k; i.e., operation (i,j) precedes operation (k,j). This corresponds to the set of conjunctive arcs in the directed graph. The following mathematical program minimizes the makespan:

Minimize C_{max}

subject to

$$y_{kj} - y_{ij} \geq p_{ij} \quad \text{for all } (i,j) \rightarrow (k,j) \in A$$
$$C_{max} - y_{ij} \geq p_{ij} \quad \text{for all } (i,j) \in N$$
$$y_{ij} - y_{il} \geq p_{il} \text{ or } y_{il} - y_{ij} \geq p_{ij} \quad \text{for all } (i,l) \text{ and } (i,j),$$
$$i = 1, \ldots, m$$
$$y_{ij} \geq 0 \quad \text{for all } (i,j) \in N$$

The third set of constraints are often called the disjunctive constraints and represent the fact that some ordering must exist among operations of different jobs that are processed on the same machine. Because of these constraints, this formulation is referred to as the disjunctive programming formulation.

Example 4.1.1 Consider the following example with four machines and three jobs. The route, i.e., the machine sequence, and the processing times are given in the following table:

Jobs	Machine sequence	Processing times
1	1,2,3	$p_{11} = 10, p_{21} = 8, p_{31} = 4$
2	2,1,4,3	$p_{22} = 8, p_{12} = 3, p_{42} = 5,$ $p_{32} = 6$
3	1,2,4	$p_{13} = 4, p_{23} = 7, p_{43} = 3$

The objective consists of the single variable C_{max}. The first set of constraints consists of seven constraints: two for job 1, three for job 2, and two for job 3. For example, one of these is

$$y_{21} - y_{11} \geq 10 \ (=p_{11})$$

The second set consists of 10 constraints, one for each operation. An example is

$$C_{max} - y_{11} \geq 10 \ (=p_{11})$$

The set of disjunctive constraints contains eight constraints: three each for machines 1 and 2 and one each for machines 3 and 4 (there are three operations to be performed on machines 1 and 2 and two operations on machines 3 and 4). An example of a disjunctive constraint is

$$y_{11} - y_{12} \geq 3 \ (=p_{12}) \quad \text{or} \quad y_{12} - y_{11} \geq 10 \ (=p_{11})$$

The last set includes 10 nonnegativity constraints, one for each starting time. □

That a scheduling problem can be formulated as a disjunctive programming problem does not imply that there is a standard solution procedure available, which will work satisfactorily. The job shop problem is strongly NP-hard and solution procedures are based either on enumeration or on heuristics.

4.2. A Branch and Bound Approach

A technique often used is branch and bound. The enumeration (branching) procedure and the bounding procedure are somewhat special. In order to describe the branching procedure a definition is required.

Definition 4.2.1 *A feasible schedule is called active if no operation can be completed earlier by altering processing sequences on machines without delaying any other operation.*

So, in an active schedule it is impossible to reduce the makespan without increasing the starting time of some operation. Of course, there are

Heuristic Methods 603

many different active schedules. The branching scheme most often used is based on the generation of all active schedules. All active schedules can be generated by a simple algorithm. In the following algorithm Ω denotes the set of all operations all of whose predecessors have already been scheduled (i.e., Ω denotes the set of all schedulable operations) and r_{ij} denotes the earliest possible starting time of operation (i,j) in Ω.

Algorithm 4.2.2

Step 1
Let Ω contain the first operation of each job;
Let $r_{ij} = 0$, for all $(i,j) \in \Omega$.

Step 2
Compute

$$t(\Omega) = \min_{(i,j) \in \Omega} \{r_{ij} + p_{ij}\}$$

and let i^* denote the machine on which the minimum is achieved.

Step 3
Let Ω' denote the set of all operations (i^*,j) on machine i^* such that

$$r_{i^*j} < t(\Omega)$$

Put successively each operation in Ω' as the next one on machine i^*. For each such choice delete the operation from Ω; include its immediate follower in Ω and return to step 2.

Each node $\mathcal{V}$ in the branch and bound tree defines a selection of disjunctive arcs that corresponds to the order in which all the predecessors of set Ω have been scheduled. A branch out of node $\mathcal{V}$ corresponds to the selection of an operation (i^*,j) ($\in \Omega'$) as the next to be assigned on machine i^*. The disjunctive arcs $(i^*,j) \rightarrow (i^*,k)$ then have to be added to machine i^* for all operations (i^*,k) still to be processed on machine i^*. This implies that the newly created node, say node $\mathcal{V}'$, which corresponds to a partial schedule with only one additional operation scheduled, contains a number of additional selected disjunctive arcs. Let D' denote the set of disjunctive arcs selected at the newly created node. Call the graph with all the conjunctive arcs and set D' graph $G(D')$. The number of branches sprouting out of node $\mathcal{V}$ is equal to the number of operations in Ω'.

To find a lower bound on the makespan at node $\mathcal{V}'$ consider graph $G(D')$ and compute first all earliest possible starting times r_{ij} for all operations (i,j). The length of the critical path in this graph is a lower bound on the makespan at node $\mathcal{V}'$. Actually, even better (higher) lower bounds can be obtained relatively easily. Consider machine i and assume that all *other* machines are allowed to process more than one job at any point in time [as not all disjunctive arcs are selected yet in $G(D')$, more than one operation may be processed on a machine at one point in time]. For each operation (i,j) on machine i, compute now the minimum amount of time needed between the completion of (i,j) and the makespan by considering the critical path from node (i,j) to the sink in $G(D')$. This minimum amount of time needed, together with the lower bound on the makespan, then translates into a due date for operation (i,j). Sequencing all operations on machine i is now equivalent to a single-machine sequencing problem. In this single-machine sequencing problem the n jobs have release dates and due dates. In order to determine a sequence on this single machine that affects the makespan of the original job shop problem the least, the maximum lateness of the n jobs has to be minimized. This particular single-machine sequencing problem is in the scheduling literature often referred to as the $1|r_j|L_{max}$ problem. Although this $1|r_j|L_{max}$ problem is strongly NP-hard, relatively fast algorithms are available that lead to acceptable solutions. The optimal sequence obtained for this problem leads to a selection of additional disjunctive arcs to be added to D'. This may lead to a longer critical path, a larger makespan, and a better (higher) lower bound for node $\mathcal{V}'$. At node $\mathcal{V}'$ this can be done for each of the m machines separately. The largest makespan obtained this way can be used as a lower bound at node $\mathcal{V}'$. Although, at first sight it appears somewhat exaggerated to have to solve m strongly NP-hard scheduling problems to obtain one lower bound for another NP-hard problem, this type of bounding procedure has performed reasonably well in computational experiments.

Example 4.2.3 Consider the instance described in Example 4.1.1. The initial graph contains only conjunctive arcs and is depicted in Figure 3a. The makespan corresponding to this graph is 22. Applying the branch and bound procedure to this instance results in the following branch and bound tree.

Level 1: Applying Algorithm 4.2.2 yields

$$\Omega = \{(1,1), (2,2), (1,3)\}$$
$$t(\Omega) = \min(0 + 10, 0 + 8, 0 + 4) = 4$$
$$i^* = 1$$
$$\Omega' = \{(1,1),(1,3)\}$$

So there are two nodes of interest at level 1, one corresponding to operation (1,1) being processed first and the other to operation (1,3) being processed first.

If operation (1,1) is scheduled first, then the two disjunctive arcs depicted in Figure 3b are added to the graph. The node is characterized by the two disjunctive arcs

Heuristic Methods

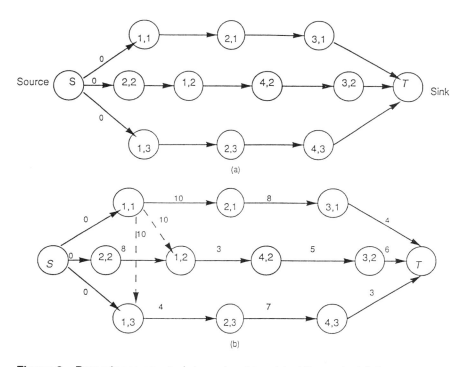

Figure 3 Precedence constraint graphs at level 1 of Example 4.2.3.

$(1,1) \rightarrow (1,2)$
$(1,1) \rightarrow (1,3)$

These two disjunctive arcs immediately increase the lower bound on the makespan to 24. An instance of the $1|r_j|L_{max}$ problem corresponding to machine 1, with the following release dates and due dates, can now be generated in order to improve the lower bound (see Figure 3b).

Jobs	1	2	3
p_j	10	3	4
r_j	0	10	10
d_j	10	13	14

The optimal sequence is 1,2,3 and the $L_{max} = 3$. This implies that a lower bound for the makespan at the corresponding node is $24 + 3 = 27$. An instance of $1|r_j|L_{max}$ corresponding to machine 2 can be generated in the same way. The release dates and due dates follow immediately from Figure 3b (assuming a makespan of 24).

Jobs	1	2	3
p_j	8	8	7
r_j	10	0	14
d_j	20	10	21

The optimal sequence is 2,1,3 with $L_{\max} = 4$. This leads to a better lower bound for the makespan at the node corresponding to operation (1,1) scheduled first, i.e., $24 + 4 = 28$. Analyzing machines 3 and 4 in the same way does not improve the lower bound.

The second node at level 1 corresponds to operation (1,3) scheduled first. If operation (1,3) is scheduled to go first, two different disjunctive arcs are added to the original graph. This leads to a lower bound for the makespan of 26. The associated instance of the $1|r_j|L_{\max}$ corresponding to machine 1 leads to an optimal sequence 3,1,2 with an $L_{\max} = 2$. This implies that the lower bound for the makespan at this node, corresponding to operation (1,3) scheduled first, is also equal to 28. Analyzing machines 2, 3, and 4 does not result in a better lower bound.

The next step is to branch out from node (1,1) at level 1 and generate nodes at the next level.

Level 2: Applying Algorithm 4.2.2 now yields

$$\Omega = \{(2,2), (2,1), (1,3)\}$$
$$t(\Omega) = \min(0 + 8, 10 + 8, 10 + 4) = 8$$
$$i^* = 2$$
$$\Omega' = \{(2,2)\}$$

There is one node of interest at this part of level 2, namely the node corresponding to operation (2,2) being processed first on machine 2. Two disjunctive arcs are added to the graph, namely (2,2) → (2,1) and (2,2) → (2,3). So this node is characterized by a total of four disjunctive arcs:

(1,1) → (1,2)
(1,1) → (1,3)
(2,2) → (2,1)
(2,2) → (2,3)

This leads to an instance of the $1|r_j|L_{\max}$ problem corresponding to machine 1 with release dates and due dates that can be computed easily (assuming a makespan of 28).

Heuristic Methods

Jobs	1	2	3
p_j	10	3	4
r_j	0	10	10
d_j	14	17	18

The optimal job sequence is 1,3,2 and the $L_{\max} = 0$. This implies that the lower bound for the makespan at the corresponding node is $28 + 0 = 28$. Analyzing machines 2, 3, and 4 in the same way does not increase the lower bound.

Continuing the branch and bound procedure results in the following job sequences for the four machines.

Machine	Job sequence
1	1,3,2 (or 1,2,3)
2	2,1,3
3	1,2
4	2,3

The makespan under this optimal schedule is 28 (see Figure 4). □

Many researchers have worked on the disjunctive programming formulation of the job shop and have developed branch and bound techniques; see, for example, Applegate and Cook (1991), Carlier and Pinson (1989), Lomnicki (1965), and Roy and Sussmann (1964).

4.3. The Shifting Bottleneck Heuristic

One of the most successful heuristic procedures developed for the job shop problem is the *shifting bottleneck* heuristic.

In the subsequent description of the shifting bottleneck heuristic M denotes the set of all m machines. In the description of an iteration of the heuristic it is assumed that in previous iterations a selection of disjunctive arcs has been determined for a subset M_0 of machines. That is, a job sequence for each one of the machines in M_0 has already been specified.

An iteration results in the selection of a machine from $M - M_0$ for inclusion in set M_0. The sequence in which the operations are to be processed on this machine has to be specified within the iteration as well. To determine which machine should be included next in M_0, an attempt is made to determine which unscheduled machine causes in one sense or another the severest disruption. To determine this, the original directed graph is modified by deleting *all* disjunctive arcs of the machines still to be scheduled

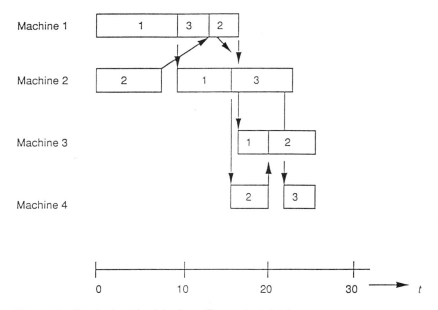

Figure 4 Gantt chart for job shop (Example 4.2.3.).

(i.e., the machines in set $M - M_0$) and keeping only the relevant disjunctive arcs of the machines in set M_0 (one from every pair). Call this graph G'. Deleting all disjunctive arcs of a specific machine implies that all associated operations, which originally were supposed to be done on the machine one after another, can now be done at any point in time in parallel (as if each one of these operations has a machine for itself or, equivalently, the machine can handle multiple jobs at the same time). The graph G' has one or more critical paths that determine the corresponding makespan. Call this makespan $C_{\max}(M_0)$.

Suppose that operation (i,j), $i \in M - M_0$, has to be processed in a time window in which the release data and due date are determined by the critical (longest) paths in G'. Consider each of the machines in $M - M_0$ as a separate $1|r_j|L_{\max}$ problem with the maximum lateness to be minimized. After solving all these single-machine problems, the single machine with the *largest* maximum lateness is chosen. This machine is, in a sense, the "bottleneck" among the remaining machines to be scheduled and therefore the one to be added next to M_0. Label this machine k, call its maximum lateness $L_{\max}(k)$, and schedule it according to the optimal solution obtained to the $1|r_j|L_{\max}$ problem for this machine. If the corresponding disjunctive

Heuristic Methods

arcs, which specify the sequence of operations on machine k, are inserted in graph G', then the makespan of the current partial schedule increases by at least $L_{max}(k)$, that is,

$$C_{max}(M_0 \cup k) \geq C_{max}(M_0) + L_{max}(k)$$

Before repeating the procedure and determining the next machine to be included, an additional step in which all machines in the original set M_0 are resequenced may be performed within the current iteration. That is, a machine, say machine l, is taken out of the set M_0 and a graph G'' is constructed by modifying graph G' through the inclusion of the disjunctive arcs that specify the sequence of operations on machine k and the exclusion of the disjunctive arcs associated with machine l. Machine l is resequenced by solving the corresponding $1|r_j|L_{max}$ problem with the release and due dates determined by the critical paths in graph G''. Resequencing each of the machines in the original set M_0 completes the iteration.

In the subsequent iteration the entire procedure is repeated and an additional machine is added to the current set $M_0 \cup k$.

The structure of this heuristic shows the relationship between the bottleneck concept and the more combinatorial concepts of critical (longest) path and maximum lateness. A critical path indicates the location and the timing of a bottleneck. The maximum lateness indicates the minimum amount by which the makespan increases if a machine is added to the set of machines already scheduled.

Example 4.3.1 Consider again the instance with four machines and three jobs described in Example 4.1.1. The route, i.e., the machine sequence, and the processing times are given in the following table:

Jobs	Machine sequence	Processing times
1	1,2,3	$p_{11} = 10, p_{21} = 8, p_{31} = 4$
2	2,1,4,3	$p_{22} = 8, p_{12} = 3, p_{42} = 5, p_{32} = 6$
3	1,2,4	$p_{13} = 4, p_{23} = 7, p_{43} = 3$

Iteration 1: Initially, set M_0 is empty and graph G' contains only conjunctive arcs and no disjunctive arcs. The critical path and the makespan $C_{max}(\emptyset)$ can be determined easily: this makespan is equal to the maximum total processing time required for any job. The maximum of 22 is achieved in this case by both job 1 and job 2. To determine which machine to schedule first, each machine is considered as a $1|r_j|L_{max}$ problem with the release

dates and due dates determined by the longest paths in G' (assuming a makespan of 22).

The data for the $1|r_j|L_{max}$ problem corresponding to machine 1 can be determined:

Jobs	1	2	3
p_j	10	3	4
r_j	0	8	0
d_j	10	11	12

This problem can be solved and the optimal sequences turns out to be sequence 1,2,3 with an $L_{max}(1) = 5$.

The data for the problem concerning machine 2 are:

Jobs	1	2	3
p_j	8	8	7
r_j	10	0	4
d_j	18	8	19

The optimal sequence for this problem is 2,3,1 with $L_{max}(2) = 5$. In a similar way it can be shown that

$$L_{max}(3) = 4$$

and

$$L_{max}(4) = 0$$

Hence, either machine 1 or machine 2 may be considered as a bottleneck. Breaking the tie arbitrarily, machine 1 is selected to be included in M_0. The graph G'' is obtained by adding the disjunctive arcs corresponding to the sequence of the jobs on machine 1 (see Figure 5). It is clear that

$$C_{max}(\{1\}) = C_{max}(\emptyset) + L_{max}(1) = 22 + 5 = 27$$

Iteration 2: Given that the makespan corresponding to G'' is 27, the critical paths in the graph can be determined. The three remaining machines have to be analyzed separately as $1|r_j|L_{max}$ problems. The data for the problem concerning machine 2 are now

Heuristic Methods

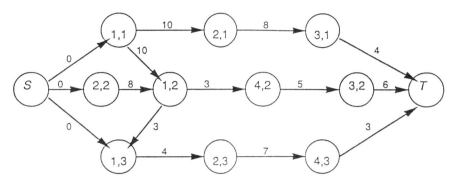

Figure 5 Iteration 1 of shifting bottleneck heuristic (Example 4.3.1.).

Jobs	1	2	3
p_j	8	8	7
r_j	10	0	17
d_j	23	10	24

The optimal schedule is 2,1,3 and the resulting $L_{max}(2) = 1$. The data for the problem corresponding to machine 3 are:

Jobs	1	2
p_j	4	6
r_j	18	18
d_j	27	27

Both sequences are optimal and $L_{max}(3) = 1$. Machine 4 can be analyzed in the same way and the resulting $L_{max}(4) = 0$. Again, there is a tie and machine 2 is selected to be included in M_0. So $M_0 = \{1,2\}$ and

$$C_{max}(\{1,2\}) = C_{max}(\{1\}) + L_{max}(2) = 27 + 1 = 28$$

The disjunctive arcs corresponding to the job sequence on machine 2 are added to G'' and graph G''' is obtained. At this point, still as a part of iteration 2, an attempt may be made to decrease $C_{max}(\{1,2\})$ by resequencing machine 1. It can be checked that resequencing machine 1 does not give any improvement.

Iteration 3: The critical path in G''' can be determined and machines 3 and 4 remain to be analyzed. These two problems turn out to be very simple with both having zero maximum lateness. Neither of the machines constitutes a bottleneck in any way.

The final schedule is determined by the following machine sequences: the job sequence 1,2,3 on machine 1; the job sequence 2,1,3 on machine 2; the job sequence 2,1 on machine 3; and the job sequence 2,3 on machine 4. The makespan is 28. □

Extensive numerical research has shown that the *shifting bottleneck* heuristic is extremely effective. Applied on a particular test problem with 10 machines and 10 jobs, which had remained unsolved for more than 20 years, the heuristic obtained a very good solution after only a couple of minutes of CPU time. This solution turned out to be optimal after a branch and bound approach, applied to the problem, obtained the same result and verified its optimality. The branch and bound approach, in contrast with the heuristic, needed many hours of CPU time. The disadvantage of the heuristic is, of course, that there is no guarantee that the solution it generates is optimal. The shifting bottleneck procedure is due to Adams et al. (1988).

The disjunctive graph formulation for the job shop problem described above can also be applied to the more general job shop problem in which a job, on its route through a shop, may have to visit a machine more than once. These job shops are said to be subject to recirculation. In this case, the set of disjunctive arcs for a machine may not be a clique. If two operations of the same job have to be performed on the same machine, then a precedence relationship has to exist between the two operations. These two operations are not connected by a pair of disjunctive arcs, because they are already connected by conjunctive arcs. The branch and bound approach decribed above still applies. However, the bounding mechanism is now not based on the solution of a $1\,|\,r_j|\,L_{max}$ problem as described above, but rather on the solution of a more general single-machine problem with the jobs subject to precedence constraints. The precedence constraints are the routing constraints on the different operations of the same job to be processed on the machine.

The shifting bottleneck heuristic can be adapted in order to be applied to more general job shop problems with recirculation as well as job shop problems with multiple machines at every stage; see Ovacik and Uzsoy (1992).

One variation of the shifting bottleneck heuristic is based on decomposition principles. This variation is especially suitable for the scheduling of flexible job shops; see Ovacik and Uzsoy (1992) and Pinedo (1995). The following five-step approach can be utilized for the flexible job shop.

Heuristic Methods

Algorithm 4.3.2

Step 1
The shop is divided into a number of workcenters that have to be scheduled.
A workcenter may consist of a single machine or a bank of machines in parallel.

Step 2
The entire job shop is represented through a disjunctive graph.

Step 3
A performance measure is computed in order to rank the workcenters in order of criticality. The schedule of the most critical workcenter, among the workcenters of which the sequences still have to be determined, is fixed.

Step 4
The disjunctive graph representation is used to capture the interactions between the workcenters already scheduled and those not yet scheduled.

Step 5
The workcenters that already have been sequenced are rescheduled using the new information obtained in step 4.
If all workcenters have been scheduled, stop. Otherwise go to step 3.

5. LOCATION AND LAYOUT APPLICATIONS

The quadratic assignment problem formulated in the introduction is an important model for facility layout problems. Unfortunately, solving large-scale quadratic assignment problems to optimality is very difficult. Again, efficient heuristics have been developed. One such heuristic is the *pairwise interchange heuristic*, which is based on the 2-opt algorithm introduced in Section 2; see Francis et al. (1992). To describe this approach in the context of facility layout let $s = (s_1, s_2 \ldots, s_i, \ldots, s_j, \ldots, s_n)$ be a vector representing the assignment of facilities to sites. In that case, s_i is the site to which facility i is assigned. Thus, given an assignment of facilities to sites, the material handling cost can be written as

$$f(s) = \sum_{i,j \mid i<j} w_{ij} d_{s_i s_j}$$

Given a vector of assignment of facilities to sites, does there exist another assignment vector with smaller material handling cost? In general, it is not easy to answer this question. However, if we can identify two facilities such that interchanging their locations (sites) reduces the total cost, then we can improve the solution. This is clearly an easier task to perform, and in fact the interchange heuristic does exactly that. At each step it finds the best two facilities to interchange, i.e., those whose interchange will reduce total cost by the largest amount. After interchanging these two facilities, the process continues until no other two facilities that reduce material handling cost are found.

Given an assignment vector s, we need an efficient algorithm that identifies a pair of facilities whose interchange reduces material handling cost by the largest amount. This can be done by calculating for every pair of facilities u, v the difference between total material handling cost in $s = (s_1, s_2 \ldots, s_u, \ldots, s_v, \ldots, s_n)$, $f(s)$, and the total material handling cost associated with $s' = (s_1, s_2 \ldots, s_v, \ldots, s_u, \ldots, s_n)$, denoted by $f(s')$. We then chose the pair for which the difference is the largest and it is positive.

To find the difference between the material handling cost associated with s and s', observe that in $f(s)$ the sum of all terms involving facilities u and v equals

$$\sum_{i \neq v} w_{iu} d_{s_i s_u} + \sum_{i \neq u} w_{iv} d_{s_i s_v} + w_{uv} d_{s_u s_v} = \sum_{i} w_{iu} d_{s_i s_u} + \sum_{i} w_{iv} d_{s_i s_v} - w_{uv} d_{s_u s_v}$$

On the other hand, the sum of all terms involving facilities u and v in $f(s')$ is equal to

$$\sum_{i} w_{iu} d_{s_i s_v} + \sum_{i} w_{iv} d_{s_i s_u} + w_{uv} d_{s_u s_v}$$

This is true because if $i = v$, $d_{s_v s_v} = 0$ and similarly if $i = u$, $d_{s_u s_u} = 0$. Thus,

$$f(s) - f(s') = \sum_{i} (w_{iu} - w_{iv})(d_{s_i s_u} - d_{s_i s_v}) - 2 w_{uv} d_{s_u s_v}$$

which can be efficiently calculated. If the right-hand side is positive, then $f(s) > f(s')$ and therefore the vector s' provides a better assignment of facilities to site.

Formulating the facility layout problem as a quadratic assignment problem has one major disadvantage. We must specify possible locations for all facilities, thus discretizing the problem. In practice, however, the facility layout problem is a continuous problem, because we can locate each facility anywhere within the available space. To overcome this difficulty a number of computer-aided methods have been developed in the past three decades. In all cases, the computer-aided facility layout methods use heuristics that are either *improvement*-type heuristics or *construction*-type heuristics. For a detailed review of the different methods we refer to the book by Francis et al. (1992).

6. DISCUSSION

Because most industrial problems tend to be NP-hard, the development of suitable heuristics is important. It appears that the heuristics in the future will often be combinations of classical approaches (e.g., branch and bound applied to an integer programming formulation of the problem) and more modern techniques (e.g., neighborhood search). A combined approach may be designed in such a way that the classical approach is applied first and the solution obtained is then fed into a neighborhood search technique.

The sophistication of the various phases of a composite heuristic is a function of the time allowed for the development of the procedure. If sufficient time is available for the development process, a very sophisticated integer programming technique may be appropriate. The development time of such an approach is considerable. However, the resulting code tends to give high-quality solutions in a relatively short time. If not much time is available for the development of the procedure, a local search technique may be used that obtains its initial solution from a simple myopic rule. Such a procedure can be developed in a relatively short time.

The sophistication of the procedure may also depend on the type of use of the procedure, i.e., whether it has to operate in real time or not. If it has to operate frequently in real time, then a relatively sophisticated procedure may be desirable. If it has to operate only on an overnight basis, then the level of sophistication may not necessarily be that high.

ACKNOWLEDGMENTS

Research supported in part by ONR contract N00014-90-J-1649 and NSF contracts DDM-8922712 and DDM-9322828

REFERENCES

Adams, J., Balas, E., Zawack, D., "The Shifting Bottleneck Procedure for Job Shop Scheduling," *Management Science*, Vol. 34, pp. 391–401, 1988.

Applegate, D., Cook, W., "A Computational Study of the Job-Shop Scheduling Problem," *ORSA Journal on Computing*, Vol. 3, pp. 149–156, 1991.

Barker, J. R., McMahon G. B., "Scheduling the General Job-Shop," *Management Science*, Vol. 31, pp. 594–598, 1985.

Beasley, J., "Route First-Cluster Second Method for Vehicle Routing," *Omega*, Vol. 11, pp. 403–408, 1983.

Bhaskaran, K., Pinedo, M., "Dispatching," in *Handbook of Industrial Engineering*, ed. G. Salvendy, pp. 2184–2198. Wiley, New York, 1992.

Bienstock, D., Bramel, J., Simchi-Levi, D., "A Probabilistic Analysis of Tour Partitioning Heuristics for the Capacitated Vehicle Routing Problems with Unsplit Demands," *Mathematics of Operations Research*, in press.

Bramel, J., Simchi-Levi, D., "A Location Based Heuristic for General Routing Problems," Working Paper, Department of IE&OR, Columbia University, New York, 1992.

Carlier, J., Pinson, E., "An Algorithm for Solving the Job Shop Problem," *Management Science*, Vol. 35, pp. 164–176, 1989.

Christofides, N., "Vehicle Routing," in *The Traveling Salesman Problem*, eds. E. L. Lawler et al., pp. 431–448. Wiley, New York, 1985.

Clarke, G., Wright, J. W., "Scheduling of Vehicles from a Central Depot to a Number of Delivery Points," *Operations Research*, Vol. 12, pp. 568–581, 1964.

Dell'Amico, M., Trubian, M., "Applying Tabu-Search to the Job Shop Scheduling Problem," *Annals of Operations Research*, Vol. 41, pp. 231–252, 1991.

Deng, Q., Simchi-Levi, D., "Valid Inequalities, Facets and Computational Results for the Capacitated Concentrator Location Problem," Working Paper, Department of IE&OR, Columbia University, New York, 1992.

Fisher, M. L., "The Lagrangian Relaxation Method for Solving Integer Programming Problems," *Management Science*, Vol. 27, pp. 1–18, 1981.

Fisher, M. L., "Vehicle Routing," in *Handbooks in Operations Research and Management Science*, volume on Networks and Distribution, eds. M. Ball, T. Magnanti, C. Monma, and G. Nemhauser, in press.

Francis, R. L., McGinnis, F., Jr., White, J. A., *Facility Layout and Location: An Analytical Approach*. Prentice-Hall, Englewood Cliffs, NJ, 1992.

Gendreau, M., Hertz, A., Laporte, G., "A Tabu Search Heuristic for the Vehicle Routing Problem," *Management Science*, Vol. 40, pp. 1276–1290, 1994.

Gillett, B. E., Miller, L. R., "A Heuristic Algorithm for the Vehicle Dispatch Problem," *Operations Research*, Vol. 22, pp. 340–349, 1974.

Glover, F., Laguna, M., "Tabu Search," in *Modern Heuristics Techniques for Combinatorial Problems*, ed. C. Reeves, pp. 70–141. Blackwell Scientific Publishing, Boston, 1993.

Glover, F., "Tabu Search: A Tutorial," *Interfaces*, Vol. 20, No. 4, pp. 74–94, 1990.

Golden, B. L., Stewart, W. R., "Empirical Analysis of Heuristics," in *The Traveling Salesman Problem*, eds. E. L. Lawler et al., pp 207–249. Wiley, New York, 1985.

Kirkpatrick, S., Gelatt, C. D., Vecchi, M. P., "Optimization by Simulated Annealing," *Science*, Vol. 220, pp. 671–680, 1983.

Lomnicki, Z. A., "A Branch and Bound Algorithm for the Exact Solution of the Three-Machine Scheduling Problem," *Operational Research Quarterly*, Vol. 16, pp. 89–100, 1965.

Matsuo, H., Suh, C. J., Sullivan, R. S., "A Controlled Search Simulated Annealing Method for the General Job Shop Scheduling Problem," Working Paper 03-44-88, Graduate School of Business, University of Texas, Austin, 1988.

Morton, T. E., Pentico, D. W., *Heuristic Scheduling Systems: With Applications to Production Systems and Project Management*. Wiley, New York, 1994.

Ovacik, I. M., Uzsoy, R., "A Shifting Bottleneck Algorithm for Scheduling Semiconductor Testing Operations," *Journal of Electronic Manufacturing*, Vol. 2, pp. 119–134, 1992.

Ow, P. S., Morton, T. E., "Filtered Beam Search in Scheduling," *International Journal of Production Research*, Vol. 26, pp. 297–307, 1988.

Heuristic Methods

Panwalkar, S. S., Iskander, W., "A Survey of Scheduling Rules," *Operations Research*, Vol. 25, pp. 45–61, 1977.

Pinedo, M. L., *Scheduling: Theory, Algorithms and Systems*. Prentice-Hall, Englewood Cliffs, NJ, 1995.

Pirkul, H., "Efficient Algorithms for the Capacitated Concentrated Location Problem," *Computers in Operations Research*, Vol. 14, pp. 197–208, 1987.

Roy, B., Sussmann, B., "Les Problèmes d'Ordonnancement avec Constraintes Disjonctives," Note DS No. 9 bis, SEMA, Montrouge, 1964.

van Laarhoven, P. J. M., Aarts, E. H. L., Lenstra, J. K., "Job Shop Scheduling by Simulated Annealing," *Operations Research*, Vol. 40, pp. 113–125, 1992.

Wilhelm, M. R., Ward, T. L., "Solving Quadratic Assignment Problems by 'Simulated Annealing,'" *IEE Transactions*, pp. 107-119, 1987.

Author Index

Abadie, J., 415, 483
Adams, J., 612, 615
Adulbhan, P., 518, 542
Afentakis, P., 259, 261
Aggarwal, A., 337, 382
Ahuja, R. K., 322, 382
Allais, M., 360, 382
Applegate, D., 259, 261, 607, 616
Arntzen, B. C., 258, 261
Avramovich, D., 116
Avriel, M., 1

Bailey, J., 120
Baker, K. R., 258-259, 261, 525, 542
Baker, T. E., 435, 483
Balas, E., 153, 157, 166, 169, 262, 612, 615
Balbirer, S. D., 116
Balinski, M. L., 147, 262
Barany, I., 258-259, 262
Bare, B., 118
Barnhart, C., 147, 225, 259, 262
Baybars, I., 258, 262
Bazaraa, M. S., 116
Beale, E. M. L., 200, 209, 262, 433, 483
Beasley, J. E., 257, 262, 597, 616
Bechtold, S. E., 259, 262
Beck, M. P., 554, 573
Beckwith, R. E., 23, 116

Belenson, S. M., 498, 541
Bell, D. E., 361, 382
Bellman, R. E., 322, 381-382
Benayoun, R., 497, 541
Bender, P. S., 258, 262
Benders, J. F., 555, 570
Benichou, M., 209, 262
Bennington, G. E., 303, 306
Berger, A. J., 559, 570
Berna, T. J., 437, 483
Bhaskaran, K., 583, 615
Birge, J. R., 543-544, 550, 555, 557-558, 562, 568, 570
Bitran, G. R., 258, 263, 568, 571
Blackwell, D., 325, 372, 382
Blais, J., 116
Bland, R. G., 64, 116
Blazewicz, J., 259, 263
Boddington, C. E., 433, 483
Bogetoft, P., 492, 541
Bookbinder, J. H., 116
Boyd, E. A., 166, 230, 263
Bradley, G. H., 118, 146, 263, 289, 306
Bramel, J., 597, 599, 616
Brooke, A., 5, 9
Brosch, L., 116
Brown, G. G., 306
Buck, R., 116
Bullock, W., 117

Burch, E., 120
Burkard, R. E., 4, 9
Buzby, B. R., 433, 483

Cammell, S., 121
Carino, D. R., 551, 571
Carlier, J., 607, 616
Carpenter, T. J., 559, 571
Carpentier, J., 415, 483
Cavalier, T. M., 118
Cenna, A. A., 524, 526, 542
Censor, Y., 547, 571
Chand, S., 522, 542
Chang, L., 258, 263
Chankong, V., 492
Charnes, A., 21–22, 64, 116, 494, 541, 564, 571
Chatto, L., 24, 118
Chen, C. D., 337, 342, 382
Christofides, N., 260, 263
Chvatal, V., 150, 165, 263
Cinlar, E., 342, 382, 550, 571
Ciriani, T. A., 4, 10
Clarke, G., 595–596, 616
Cleef, H. W., 568, 571
Conn, A. R., 433, 483
Conway, R. W., 259, 263
Cook, T. M., 116, 607, 616
Cook, W., 259, 261
Cooper, W. W., 21–22, 116, 494, 541, 564, 571
Cornuejols, G., 257, 263
Cox, C. B., 24, 118
Crainic, T. G., 209, 264
Crandall, H. W., 117
Crowder, H., 209, 263, 554, 573

Dantzig, G. B., 20–21, 64, 116–117, 552, 554, 558, 571
Daskin, M. S., 117
Deckro, R. F., 258, 263
Dell'Amico, M., 588, 616
Dempster, M. A. H., 558, 571
Denardo, E. V., 307, 322, 325–326, 337, 364–365, 369, 372–373, 376, 382–383
Deng, Q., 599, 616

Dennis, J. E., 404–405, 408, 484
De Porter, E. L., 118
Diaby, M., 258, 263
Dietrich, B. L., 166, 263
Dijkstra, E. W., 283, 306
Dikin, I. I., 83, 117
Donohue, C. J., 557–558, 562, 568, 570
Drud, A., 440, 484
Duckstein, L., 492, 541

Eaton, D. J., 117
Edmonds, J., 279, 293–294, 296, 306
Edmunson, H. P., 552, 571
Edwards, J. R., 117
Ellsberg, D., 360, 383
Elster, J., 552, 572
Emerson, H. P., 2, 102–103, 117
Engquist, M., 433, 484
Eppen, G. D., 259, 263, 550, 560, 563, 571
Erickson, V., 23, 117–118
Ermoliev, Y., 558, 571
Escudero, L. F., 166, 263, 568, 572
Euler, L., 292
Evans, J. R., 271, 286, 297, 306
Everett, H., 209, 263

Fan, Y., 440, 484
Fang, S. C., 117
Farley, A. A., 117
Farr, D., 21
Federgruen, A., 337, 376, 383–384
Feinberg, E., 383
Field, R. C., 118
Fisher, M. L., 257, 263, 594–595, 616
Fletcher, R., 407, 484
Flood, M. M., 21, 26, 117
Flynn, B. B., 300, 306
Forrest, J. J. H., 192, 204, 209, 264
Foulds, L. R., 298, 306
Fourer, R., 94, 117, 443, 484, 560
Francis, R. L., 42, 117, 613–614, 616
Franz, L. S., 112, 117, 259, 264
Fraundorfer, K., 552–553, 571
French, S., 523, 542
Fry, T. D., 519, 542

Author Index

Garey, M., 147, 230, 264
Garfinkel, R. S., 209, 264
Garg, U., 117
Gartska, S., 558, 571
Garvin, W. W., 117
Gassman, H. I., 557, 571
Gates, R. S., 118
Gaul, W., 568, 571
Gauthier, J. M., 209, 264
Gavish, B., 259, 261
Gay, D. M., 117, 560, 571
Gelatt, C. D., 586, 616
Gelders, L. F., 522–525, 542
Gendreau, M., 588, 616
Gendron, B., 209, 264
Geoffrion, A. M., 233, 235, 245, 257–258, 264
Gillet, B. E., 597, 616
Gilmore, P. C., 38, 117
Glassey, C. R., 555, 572
Glover, F., 123, 131, 147, 153, 157, 166, 169, 207, 209, 250, 253, 257, 264–265, 569, 572, 588, 616
Glynn, P., 552, 554, 558, 571–572
Golany, B., 1
Golden, B. L., 305–306, 583, 616
Gomory, R. E., 38, 117, 150, 153, 157, 265
Graves, G. W., 258, 264, 306
Graves, S. C., 258, 265, 337, 383
Greenberg, H. J., 102, 105, 108–109, 117, 251, 257, 265
Greene, J. H., 24, 118
Griffith, R. E., 430, 484
Grossmann, I. E., 257, 268
Gryna, F. M., 24, 118
Guignard, M., 250, 253, 257, 265
Gutterman, M. M., 112, 119

Haas, E. A., 568, 571
Haessler, R. W., 118
Haimes, Y. Y., 492, 541
Hall, N. G., 260, 265
Han, S. P., 433, 484
Hannan, E. L., 496, 542
Harbrauer, E., 121
Harpell, J. L., 22, 118

Harrison, H., 29, 118
Haverly, C. A., 461, 484
Hax, A. C., 258, 265
Held, M., 225, 236, 257, 265
Hertz, A., 588, 616
Heyden, Van der L., 343
Hibbard, W. R., 118
Hicks, C. R., 118
Higle, J., 552, 554, 558, 572
Hilal, S. S., 118
Hiller, R. S., 257, 265
Hillier, F. S., 4, 10
Hirst, J. P. H., 192, 204, 209, 264
Ho, J. K., 555, 572
Hoffman, A. J., 118, 383
Hoffman, K., 209, 266
Holcomb, M. C., 118
Holloran, T. J., 118
Holmes, D. F., 557–558, 562, 568, 570
Holzman, A. G., 4, 10
Hooker, J. N., 118
Hosel, Van S., 383–384
Huang, C. C., 552, 572
Huberman, G. R., 383
Hurlimann, T., 118
Huxley, S. J., 120
Hwang, C. L., 492, 541

Ibaraki, T., 4, 9
Ignizio, J. P., 118, 492, 494, 496, 541
Infanger, G., 554, 572
Iskander, W., 583, 617

Jacobs, F. R., 300, 306
Jacobs, L. W., 259, 262
Jansma, G., 117
Jarvis, J. J., 116
Jensen, J. L., 552, 572
Jeroslow, R. G., 157, 166, 169, 259, 262, 266
Johnson, D., 147, 230, 264
Johnson, E. L., 166, 209, 266, 293–294, 296, 306
Johnson, L. A., 302, 306
Johnston, B., 121
Jones, D., 569, 572

Kahneman, D., 360, 383
Kall, P., 552–553, 571–572
Kallberg, J. G., 544, 572
Kamesam, P. V., 568, 571
Kao, E., 563, 566, 572
Kapur, K. C., 498, 541
Karmakar, N., 83–85, 118
Karney, D., 569, 572
Karwan, M. H., 257, 266
Katz, P., 258, 266
Kella, O., 383
Kelly, J. P., 207, 266
Kelton, D., 550, 572
Kendrick, D., 5, 9
Kent, B., 118
Kent, T., 551, 571
Kernighan, B. W., 117, 560, 571
Khomenynk, V. V., 493
Kim, N., 433, 485
Kim, S., 250, 253, 257, 265
King, A. J., 568, 571
King, R. H., 119
Kirkpatrick, S., 586, 616
Klee, V., 83, 119
Klingman, D., 29, 33, 36, 119, 250, 253, 257, 264, 569, 572
Kolen, A., 384
Koopman, T., 21
Kortanek, K. O., 119
Krishnan, R., 119
Kruskal, J. B., 281, 306
Kuby, M., 257, 266

Laguna, M., 207, 229, 264, 266, 588, 616
Lamar, B. W., 258, 265
Lamont, J., 116
Land, A., 209, 266
Lane, M. S., 22, 118
Langston, G. D., 116
Lanzenauer Von, C. H., 121
Laporte, G., 588, 616
Larichev, O., 497, 541
Lasdon, L. S., 258–259, 261, 266, 385, 415, 425, 433, 435, 440, 485
Law, A., 550, 572

Leachman, R. C., 4, 10
Lee, C. Y., 337, 342, 382–383
Lee, S. M., 492, 494, 496, 541
Lee, Y., 166, 268
Lehrer, R. N., 24, 119
Leitmann, G., 493, 541
Lenstra, J. K., 118
Lent, A., 547, 571
Leong, G. K., 519, 542
Leung, J. M. Y., 257, 259, 266
Liberatore, M. J., 119
Lieberman, E. R., 492–493, 540
Lieberman, G. J., 4, 10
Liebman, J. F., 440, 443, 454, 484
Locke, M. H., 437, 483
Lockett, G., 113, 119
Loewenstein, G., 552, 572
Løkketangen, A., 207, 264
Lomnicki, Z. A., 607, 616
Louveaux, F. V., 555, 568, 572–573
Love, R. F., 42, 119, 257, 266
Luenberger, D. G., 406, 409, 412, 414, 484
Luss, H., 257, 266, 337, 342, 383
Lustig, I. J., 559, 571

MacKenzie, T., 383
Madansky, A., 549, 552, 573
Magee, J. F., 21, 119
Magee, T. M., 123
Magnanti, T. L., 257–259, 266, 382
Malcolm, D. G., 23, 119
Mangasarian, O. L., 385, 433
Manley, B., 119
Manne, A. S., 110, 119, 337, 383, 555, 572
Mansour, A. H., 22, 118
Markowitz, H., 547, 573
Marsten, R. E., 259, 267
Martin, R. K., 166, 259, 263, 267, 560, 563, 571
Masd, A. S. M., 492, 541
Matsuo, H., 568, 571, 586, 616
Mazolla, J. B., 257, 260
McAuley, P. T., 116
McGarrah, R. E., 23, 119

Author Index

McGinnis, L. F., Jr., 117, 613–614, 616
Meal, H. C., 258, 265
Meeraus, A., 5, 9
Mehring, J. S., 112, 119
Mehrotra, S., 119
Mei-Ko, K., 292, 296, 306
Mellon, B., 22, 116
Merten, A. G., 525, 542
Metzger, R. W., 24, 119
Might, R. J., 35, 119
Miller, J. L., 112, 117, 259, 264
Miller, L. J., 597, 616
Miller, T., 119
Minieka, E., 286, 297, 306
Minty, G., 83, 119
Mitten, L. G., 324, 383
Montgolfier De, J., 497, 541
Montgomery, D. C., 302, 306
More, J. J., 439, 441, 484
Morgenstern, O., 358, 384, 547, 574
Morris, J. G., 119
Morse, P. M., 120
Morton, T. E., 582, 592, 616
Mote, J., 119
Mukyangkoon, S., 501, 542
Mulvey, J. M., 543–544, 547, 551–552, 554, 558–559, 563, 570–571, 573
Murphy, F. H., 102, 105, 117, 120
Murtagh, B. A., 421, 441, 484
Murty, K. G., 146, 267
Myers, D. H., 551, 571

Nash, S. G., 5, 10, 439, 484
Neebe, A. W., 257, 260
Nemhauser, G. L., 147, 153, 209, 267
Neuman, Von J., 358, 384, 547, 574
Ng, Y. K., 259, 267
Nielsen, S. S., 547, 573
Nocedal, J., 406, 484
Nod, M. C., 555, 573

Oliff, M., 120
Orden, A., 22, 120
Orlin, J. B., 382

Osman, I., 207, 266
Osyczka, A., 492, 541
Ovacik, I. M., 612, 616
Ow, P. S., 592, 616
Ozkarahan, I., 120

Padberg, M. W., 153, 166, 209, 266–267
Palacios-Gomez, F. E., 433, 484
Panwalkar, S. S., 583, 617
Papadimitriou, C. H., 147, 267
Park, J. K., 337, 382
Park, K., 166, 267
Park, S., 166, 267
Parker, G., 147, 267
Pauker, S. G., 350, 383
Pauker, S. P., 350, 383
Pentico, D. W., 582, 616
Perlis, J. H., 496, 541
Peterson, R., 454, 484
Phillips, N., 119, 569, 572
Pierskalla, W. P., 251, 257, 265
Pinedo, M., 575, 583, 612, 617
Pinson, E., 607, 616
Pirkul, H., 598–599, 617
Plummer, J., 385
Pochet, Y., 258, 267
Potts, C. N., 259, 267
Powell, S., 209, 266
Price, W. L., 120
Prim, R. C., 282, 306
Pruzan, P., 492, 541
Pulleyblank, W. R., 383
Puthenpura, S., 117

Queyranne, M., 4, 9, 563, 566, 572

Ramakrishnan, K. G., 118
Ramsey, F. P., 358, 383
Randolph, P. H., 23, 117
Rapoport, L. A., 120
Rardin, R. L., 147, 257, 267
Rautman, C. A., 120
Reeves, C. M., 407, 484
Reid, R. A., 120
Reklaitis, G. V., 437, 484

Rema, P., 119
Ribeiro, C. C., 225, 258, 267
Ribiere, G., 209, 264
Rinaldi, G., 153, 267
Rinnooy Kan, A. H. G., 118
Robinson, E. P., 258, 268
Rockafellar, R. T., 547, 554, 573
Ross, S. M., 376, 383
Rothblum, U. G., 372, 383
Roundy, R. O., 337, 383
Rousseau, J., 116
Roy, B., 607, 617
Roy, Van T. J., 166, 268
Rudin, W., 371, 383
Ruszczynski, A., 558–559, 570, 573
Rutenberg, D., 558, 571
Ryder, E. E., 120

Sahinidis, N. V., 257, 268
Saltzman, M. J., 182, 209, 268
Salukvadze, M. E., 493, 541
Salvesos, M. E., 2, 10, 22–23, 120
Sarkar, S., 440, 484
Sarma, P. V. L. N., 437, 484
Saunders, M. A., 421, 441, 484
Savage, J., 358, 383
Savelsberg, M. W. P., 209, 268
Scarf, H., 376, 384
Schnabel, R. F., 404–405, 408, 484
Schneeberger, H., 522, 542
Schrage, L., 28, 120, 166, 267, 560, 563, 571
Schrijver, A., 118
Schulte, J., 116
Schwarzbek, R., 24, 119
Sear, T. N., 120
Sen, S., 552, 554, 558, 572
Shanno, D. F., 120, 559, 571
Shapiro, J. F., 146, 225, 257–259, 265, 268, 369, 384
Shapiro, M., 120
Shapley, L. S., 325, 384
Shaw, D., 116
Sherali, H. D., 116, 118, 166, 268
Shim, J., 492, 496, 541
Shuttleworth, D., 121

Silver, E. A., 454, 484
Simchi-Levi, D., 575, 597, 599, 616
Simmons, D., 117
Singleton, F. D., 121
Slyke Van, R., 555–556, 573
Small, J. B., 117
Smeers, Y., 55, 573
Smith, B. A., 121
Smith, S., 440, 485
Smith, W. E., 522, 524, 542
Smith, W. L., 22, 120
Sodaro, D., 119
Sommer, D. C., 147, 264
Sorenson, D. C., 434, 485
Soumis, F., 225, 258, 267
Sousa, L. J., 118
Soyster, A. L., 11, 118–120
Sparrow, W., 116
Spellman, R. A., 117
Speranza, M. K., 259, 268
Spielberg, K., 147, 262
Stacy, C., 551, 571
Steiger, D., 119, 569, 572
Steiglitz, K., 147, 267
Steuer, R. E., 492, 500, 540
Stewart, R. A., 430, 484
Stewart, W. R., 583, 616
Stoyan, D., 552, 572
Stroup, J. S., 120
Suh, C. J., 586, 616
Sullivan, R. S., 586, 616
Sussmann, B., 607, 617
Sutherland, F., 116
Svintsitski, O. G., 557–558, 562, 568, 570
Sweeney, P. E., 118
Sylvanus, M., 551, 571

Tabucanon, M. T., 487, 492, 501, 506, 518, 524, 526, 540, 542
Taha, H. A., 120
Tangedahl, L., 209, 264
Tapia, R. A., 120
Taylor, P. E., 120
Tergny, J., 497, 541
Terjung, R. C., 258, 266

Author Index

Thisse, J. F., 568, 573
Thizy, J. M., 257–258, 263, 268
Thompson, J. W., 23, 121
Thompson, R. G., 121
Thrall, R. M., 121
Threadgill, J., 119
Todd, M. J., 121
Tomlin, J. A., 192, 200, 204, 209, 262, 264
Tran, H. V., 298, 306
Triegeiro, W. W., 259, 268
Trubian, M., 588, 616
Tsai, C., 117
Turcotte, M., 120
Turner, A., 551, 571
Tversky, A., 360, 383
Tzur, M., 337, 383

Ukovich, W., 259, 268
Unger, V. E., 257, 267
Uzsoy, R., 612, 616

Vachani, R., 259, 266
Vanderbei, R. J., 559, 573
Van Hosel, S., 383
Van Roy, T. R., 166, 268
Van Wassenhove, L. N., 258–259, 267–268
Vasani, R., 23
Vecchi, M. P., 586, 616
Veinott, A. F., 383
Ventura, J., 11
Vickson, R. G., 544, 574
Vladimirou, H., 544, 547, 558–559, 573
Von Lanzenauer, C. H., 121
Von Neuman, J., 384, 547, 574

Wagelmans, A., 337, 384
Wagner, H. M., 117, 331, 384
Ward, T. L., 586, 617
Waren, A., 385, 415, 425, 440, 484
Wassenhove, Van L. N., 258–259, 267–268, 522–525, 542

Watanabe, K., 551, 571
Weintraub, A., 257, 269
Welch, J., 105, 121
Wesolowsky, G. O., 119
Westerberg, A. W., 437, 483
Wets, R., 547, 555–556, 558, 568, 571, 573–574
Whitaker, D., 121
White, J. A., 117, 613–614, 616
Whitin, T. M., 331, 384
Wierzbicki, A. P., 552, 574
Wiig, K. M., 111, 121
Wilhelm, M. R., 586, 617
Williams, H. P., 4, 10, 137, 147, 269
Willis, R. J., 121
Wilson, E. J. G., 121
Wirth, R., 119, 569, 572
Within, T. M., 331
Wittrock, R. J., 557, 574
Wollmer, R. D., 120
Wolsey, L. A., 147, 166, 267–268
Wong, R. T., 305–306
Wonnacott, R. J., 550, 552, 574
Wonnacott, T. H., 550, 552, 574
Wood, W. P., 117
Woolsey, R. E. D., 113, 121, 166, 269
Wright, J. W., 595–596, 616
Wright, S. J., 439, 441, 484

Yanasse, H., 568, 571
Ye, Y., 121
Young, D., 119, 569, 572
Yu, P. L., 493, 541

Zangwill, W. I., 337, 384
Zawack, D., 612, 615
Zeleny, M., 492, 541
Zenios, S. A., 547, 573
Zhang, J., 433, 485
Zhang, Y., 120
Zheng, Y. S., 376, 384
Ziemba, W. T., 544, 551–552, 571–572, 574
Zionts, S. C., 146, 269

Subject Index

Abelian theorem, 370–371, 374
Absolute value, 40–41
Acid dilution factor, 472
Activity analysis, 21
Adaptive strategies, 206
Adjacent, 272, 284
Affine scaling algorithm, 82–85, 88–90, 92
Affine transformation, 85–87
Aggregate employment, 23
Aggregate plan, 8, 327–331, 381
Aircraft loading problem, 259
Algebraic modelling systems, 399, 442
Alkylation process optimization, 470
Alternate optima, 46, 66–69
Amortization factor, 562
AMPL, 443, 560–561, 564–565, 567
Analytic hierarchy process, 491
Arc, 271, 273, 275, 282–287, 289–291, 316–318, 320–321, 337, 343, 354, 365, 381, 599–600, 604–612
  capacities, 276
  routing problems, 275–298, 306
Artificial variable, 65–67, 69
Ascent extreme direction, 47, 49
Assembly line balancing, 1, 123, 257–258
Assignment problem, 23, 25–26, 31, 229, 576
  generalized, 229, 246

Augmented Lagrangian, 441, 547, 558–559, 570
Automated guided vehicles, 291
Automatic storage and retrieval system, 5

Backtracking, 202, 336
Balancing storage, 454
Beam search, 592–593
  filtered, 590–591
Benders decomposition, 258, 555
  nested, 555–556, 562
Best bound rule, 176, 204
Best projection method, 204
BFGS update, 405
Bicriterion performance measures, 524
Bill of materials, 501
Binary variables, 6, 128, 137
Blending problem, 8, 22, 25, 35, 37, 461
Bottleneck, 608–610
Boundary point, 388
Branch, 273, 277, 279
Branch and bound, 124, 169–210, 231, 241–242, 246, 260, 578–579, 581, 590–592, 602–604, 607, 612
Branch and cut, 153
Bridges of Koenigsberg puzzle, 148
Bunching, 558

627

Candidature rules, 206
Capacitated concentrator location problem, 598–599
Capacity, 8, 33–34, 97–98, 102–105, 111, 273–275, 305, 329–330, 338, 340–341
  expansion, 257, 308, 337–342
  planning, 511, 559–561, 563, 570
Capital budgeting, 123, 129
Cardinal information, 492
Cargo loading, 129
Cellular manufacturing, 8, 300, 302, 306
Central differences, 444
Chance constraint, 564
Chemical process, 437
Chessie System, 33
Chinese postman problem, 147, 276, 291–297, 305
Chord, 273
Chvatal-Gomory rounding method, 150, 165–166
Clique, 600
Cluster analysis, 554–555
Cluster first-route second heuristic, 597
Cholesky factor, 405
Column generation, 225, 242, 252, 256, 259
Combinatorial optimization, 193, 204
Companies, agencies and organizations
  AFFCO Exports Ltd., 38
  Agrico Chemical Company, 569
  American Airlines, 26
  American Hospital Supply Corp., 35
  Amoco U.K. Ltd., 29, 112
  Argus Camera, 23
  Ballyclough Cooperative Creamery, 29
  Barber Ellis Fine Papers, 26
  Blue Bell, 38
  British Petroleum, 29
  Bureau of Labor Statistics, 24, 114
  Central Carolina Bank & Trust Co., 35

[Companies, agencies and organizations]
  Central Foundry of General Motors, 36
  Citgo, 24, 29, 36–37, 113
  Citgo Petroleum Corporation, 569
  Financial Services Group of Canada, 32
  Forest Research Institute of New Zealand, 35
  Frontline Systems, 442
  General Electric, 22
  General Motors, 24
  Institute of Management Science, 21–22
  Kelly-Springfield Tire and Rubber, 33
  McDonnell Douglas, 26
  North American Van Lines, 35, 112
  Olean Tile Company, 26
  O.R. Society of America, 21
  Owens-Corning, 33
  Red Cross Blood Donor Clinic, 42
  Sandia National Laboratories, 26
  Stauffer Chemical Company, 109, 111
  Tatung Wire and Cable, 33
  Transport Company of Montreal, 31
  U.S. Air Force, 21, 35
  U.S. Bureau of Labor, 19
  U.S. Department of Agriculture, 35
  United Airlines, 32, 112
  United Rescue Mission, 37
  W. R. Grace, 33, 569
Complementary slackness, 220, 288
Completion time, 522
Complexity, 82, 124, 282
Compulsory assignments, 189, 191
Computer-aided model building, 20, 93, 449
Concave function, 393
Cone, 244, 409
Conjugate gradient methods, 406, 417
CONOPT, 440
Conservation of flow, 274, 304

Subject Index

Constructive methods, 595
Continuity, 359
Continuous-time model, 369
Control, 382
Convex combination, 53–55, 69
Convex hull, 53–54, 134–137, 153–155, 222–223, 256
Convexity, 232, 409
Convex optimization, 234, 334, 391
Cost coefficient, 199
Cost goal, 501
Crashing costs, 518
Criterion cone, 500
Critical fractile equation, 330
Critical path, 28, 320–322, 601, 604, 609–612
Cubic interpolation, 407
Cubic polynomial, 407
Curse of dimensionality, 381–382
Curve fitting, 499
Customer's downtime, 530
Cuts, 150, 159
 all-integer, 160
 fractional, 160
 Gomory, 153, 249
Cutting plane, 124, 149, 155, 169, 209, 252, 554, 578, 599
 dual procedure, 156, 169
 fractional algorithms, 249
Cutting stock problem, 8, 25, 37–38
Cycle, 272–273, 276, 279–280, 287, 289–294, 296, 316, 321, 365, 600

Dantzig-Wolfe decomposition, 225–226
Data management, 20, 93
Deadheading, 292–293
Decision analysis, 348–361
 consistency checks, 357–358
 critique of, 360
 disutility in, 353
 indifference probability, 351
 postulates, 358–359
 sensitivity analysis, 349, 356–357
 utility, 353, 361

Decision tree, 342–348, 354–355, 358, 360–361, 365
 chance node, 343–344, 354
 choice node, 343
 rolling back, 347
 sensitivity analysis, 347
Decomposition, 40
Degeneracy, 56–57, 59, 61, 64, 469–470
De Morgan's laws, 139–140
Depot, 594–595, 597, 599
Descent algorithm, 421
Detached coefficient form, 27
Diagonal quadratic approximation, 559
Diffusion, 550
Dijkstra's algorithm, 283–286, 297
Directional derivative, 400
Discount factor, 366, 368, 370
Discounted infinite-horizon model, 370
Discounted Markov renewal model, 367–368
Discrete time model, 366, 369–370
Disjoint intervals, 145
Disjoint sets, 49
Disjunctive constraints, 137
Disjunctive programming, 169, 579–581, 599, 601–602, 607
Dispatching rule, 582–583, 591–592
Disposing of nuclear wastes, 26
Distribution, 2, 11, 29, 33, 96, 98, 100, 115, 258, 276, 595
 centers, 12, 41–42
 costs, 26
 depots, 13–14, 16
 efficiency, 29
 plan, 16
 system, 594
Domain violations, 446
Dual ascent, 242
Duality, 19–20, 70
Duality gap, 207–208, 241
Dual price, 215–218
Dual simplex method, 156, 163–164, 183, 190, 208–209

Dynamic lot size model, 331–337
Dynamic programming, 3, 7, 9, 307–384
  backtracking optimizers, 336
  monotonicity condition, 324–325, 381
  policy, 323
  principle of optimality, 323–324
  states, 322

Earliest due date, 522
Econometrics, 550
Economic order quantity, 454
Economy of scale, 307, 332, 337–338, 381, 398
Edge, 271–273, 275–276, 279–282, 291–298, 300–302
Efficiency, 2, 4, 33, 82, 97, 213, 241, 249, 272, 283, 286, 296, 306
Efficient points, 500
Efficient solution, 490
Either-or decisions, 125, 127–129
ELECTRE method, 491
Emergency medical service, 42
Euler's constant, 316
Euler tour, 147, 292–295, 298
Exact penalty problem, 433
Excel, 399, 440–442
Expected mean value solution, 548
Expected recourse problem solution, 548
Expected utility, 547–549
Expected value of perfect information, 543, 549
Extreme direction, 46–47, 49, 222
Extreme point, 46–47, 49, 54–58, 64, 69, 82, 88, 222, 388, 563

Facets, 151
Facility layout, 8, 23, 298–299, 300, 306, 568, 575, 579, 586, 614
Facility location, 3, 8, 40, 42, 123, 197, 257, 568, 575, 599, 613
  multifacility, 25, 39–41
Fast forward-fast back, 557
Fathoming, 187, 211

Feasible, 275, 304
  direction, 410
  region, 386
  halfspace, 44
  solution, 279, 287–288, 300
Fenchel duality, 166
Financial planning, 8, 544, 547, 555
Finite difference derivatives, 444
Finite discrete-time Markov decision, 373
Finite horizon, 36
Finite-horizon Markov decision model, 364–369
Fitting models to data, 447
Fixed charge problem, 130, 166
Fletcher-Reeves method, 406
Flexible manufacturing systems (FMS), 5
Forecasting, 8, 503
Forest, 273, 280
Forest planning, 35, 111, 129, 257
Forward difference approximation, 444
Functional equation, 308, 316, 318–319, 322–326, 329, 332, 335–336, 339, 341, 345, 347, 363–364, 368, 381

Game theory, 498
GAMS, 5, 102–104, 441–443, 463, 473
Gantt chart, 519, 608
Gauss-Newton method, 408
Generalized reduced gradient method (GRG), 403, 415
Genetic algorithms, 207, 584, 588
GINO, 440, 443
Global criterion method, 493, 532
Global optimum, 386, 390, 393, 465
Goal programming, 32, 434, 491, 494
Gradient, 234, 396, 411
Graph, 9, 271, 604, 608, 610
  connected, 273
  directed, 271–272, 303, 599–600, 607
  disjunctive, 613
  Euler, 292, 295, 298

Subject Index

[Graph]
  planar, 298
  undirected, 271–272
Graphical solution, 42–44, 49, 496

Halfspace, 43, 49, 51–53
Harmonic series, 316
Hessian matrix, 403, 436
Heuristics, 3, 7, 9, 31, 202, 210, 242, 261, 581
  cluster first-route second, 597
  construction-type, 614
  improvement-type, 613
  interchange, 614
  location-based (LBH), 599
  nearest neighbor (NNH), 582
  pairwise interchange, 613
  route first-cluster second, 596–597
  shifting bottleneck, 607, 611
  shortest processing time, 522
  shortest setup time first (SST), 582
  tour partitioning, 597
Hierarchical production planning, 3, 258
Hyperplane, 17, 51–53, 55, 244

Ideal distance minimization method, 493
Idle time, 518
Immediate strategies, 202
Implicit enumeration, 169
Implicit preference information, 492
Importance sampling, 552, 554, 558
IMSL, 441
Income stream, 366–369, 370–371
  Abelian theorem, 370
  continuously compounded, 367–369
  continuous time, 369
  discounted, 367
  discrete time, 366
  present value, 367
  undiscounted, 367
Incumbent solution, 170
Indicator variables, 129

Indifference probability, 351–353, 355–358, 360–361
Infeasible solution, 48, 66, 69, 76–77, 106, 109
Infinite-horizon model, 374
Initial basic feasible, 291
Inner linearization, 555
Integer infeasibility, 192, 203
Integer programming, 3, 9, 17, 31, 123, 277–278, 562, 566, 578, 601
  linear, 6, 577, 598
  mixed, 125, 130, 136, 149, 169, 302
  pure, 125
Integrality property, 30, 245
Integral quantities, 126
Interchange heuristic, 614
Interior, 52
Interior point methods, 19, 20, 50, 82, 84, 559, 570
Intermediate destinations, 27
Inventory, 2–3, 23, 35, 38, 93, 111–112, 382, 501
  control, 8, 375–381, 559, 566, 570
  holding costs, 32, 454
  management, 8, 32
  models, 452
  ordering costs, 454
  planning, 8, 109
Investment planning, 544, 547

Jensen's inequality, 553
Just in time, 258, 518

Kanban, 258, 518
Knapsack problem, 128–129, 148, 150–153, 166, 230
Kruskal's algorithm, 281–282
Kuhn-Tucker conditions, 409, 413, 435, 463

Labor grades, 326, 381
Lagrangean
  decomposition, 219, 250, 253, 255–257
  dual, 221–223, 232–235, 245, 251, 253

[Lagrangean]
  relaxation, 209–210, 215–220, 225–231, 245–248, 250, 253, 255–257, 259–260, 599
  saddle point, 558
Lagrange multiplier, 409, 413, 435, 457, 474
Largest processing time, 522
Least absolute error, 452
Least absolute value, 448
Least squares, 447, 408
Least squares fitting, 399
Levenberg-Marquardt method, 408
LIFO branch rule, 172, 188, 195, 203
Linear combination, 53
Linear equation, 45, 55, 79
Linear fractional program, 374
Linear function, 14
Linear inequality, 16
Linearly dependent equations, 53–55, 287
Linear programming, 3, 5, 9, 11–121, 134, 137, 149, 153–157, 277, 282–287, 331, 374, 307, 330, 334, 364–366, 381, 386, 434, 548, 554–555, 575–576
  basic feasible solution, 55, 57–62, 64, 66, 90, 92
  basic matrix, 54
  basic solution, 54–56, 80
  big-M method, 66, 92
  coefficient matrix, 27–28
  complementary slackness, 75
  dual, 73–78
  feasible region, 17, 43–49, 56–58, 69, 77, 82, 84
  feasible solution, 75, 91–92
  optimal simplex tableau, 79
  pivot, 163–165, 183
  primal, 228
  primal problem, 73–74, 77–78
  reduced costs, 60, 62–63, 66–69, 74, 88, 199, 288–289
  relaxation, 165, 197, 210, 219, 578
  right-hand side, 50, 62–64, 74, 78–80, 105, 108, 110

[Linear programming]
  sensitivity analysis, 19, 70, 78–81
  simplex, 20–22, 49–50, 54, 58, 61, 64, 66, 69, 81–84, 93–95, 102–103, 563
  simplex tableau, 59, 61–62, 68, 72, 81, 158–164, 183, 189, 191, 208
  standard form, 49–51, 57, 59, 61, 65, 67, 69, 82, 92
  two-phase method, 65–66, 69, 92
  unbounded solution, 46–47, 66, 69–71, 76–77, 88–89, 106, 108
Linear regression, 400
Linesearch, 401, 405–406, 417, 427
Local optimum, 386, 390, 393, 463, 465
Local search procedures, 581, 583
Location-based heuristic (LBH), 599
Locus, 16
Logical conditions, 138
Logistics, 29, 123, 257–258
Longest path, 28–29, 308, 322, 365, 601
Look-ahead strategies, 183, 190
Lottery, 358–359
Lotus 1-2-3, 442

Maintenance, 8
Man-machine chart, 1
Machine, 43, 104, 110, 115
  capacity constraint, 229
  grouping, 301
  loading, 22
  maintenance, 526
  pricing, 594
  sequencing, 518
Machine-part incidence matrix, 300–301
Maintenance capacity, 530
Makespan, 582, 599, 601–602, 604–606, 609–610
Manufacturing, 2–3, 24, 34
  cells, 300
  plants, 42
Manufacturing resources planning, 501

Subject Index

Marginal utility approach, 547
Marginal values, 475
Markov chains, 312–313, 361
Markov decision model, 361–375
  bias, optimal policy in, 372
  discrete time, 366
  finite, 362
  finite horizon, 364
  functional equation for, 363
  gain, optimal policy in, 372
  infinite horizon, 369
  optimal policy in, 362
  randomized policy in, 372–374
  renewal model, 368–369, 374
  return function, 362–363
  stationary policy in, 372–374
  turnpike theorem, 369–370
Material, 5, 11, 72
  availability, 511
  balances, 462, 503
  flows, 258
  handling, 24
Material requirement planning, 3, 501
Matrix-vector form, 2, 50
Maximum absolute error, 449
Maximum effectiveness method, 493
Maximum error, 452
Maximum tardiness, 518, 522
Mean flow time, 522
Method of approximation programming, 430
Midpoint approximation, 552
Minimax fit, 449
Minimum-cost flow, 275–276, 286–287, 290, 306
Minimum slack time, 522
MINOS, 441
MINTO, 209
Mixed-integer program (MIP), 577
Mixed-integer, stochastic linear programming, 560
Mixed master problems, 247
Mixing, 8, 25, 35
Model management, 19
Mononicity, 324–325, 359, 381

MPS file, 441
Multiattribute utility theory, 491
Multiobjective programming, 3, 7, 9
Multiple-attribute decision making, 489
Multiple-objective decision making, 487, 489, 551–552
Multiple optima, 46
Multiple products, 33
Multiplier adjustment method, 242
Multistage stochastic programming, 552, 556
Myopic rules, 581–582, 593

NAG Fortran Library, 441
Nearest neighbor (NN) heuristic, 582
Negative gradient direction, 416
Negative reduced gradient direction, 429
Network, 3, 9, 27–28, 32, 130, 153, 252, 271, 316, 320, 337, 365, 381, 558, 560
  acyclic, 321
  directed, 365
  flow, 273, 340
  flow problem, 274, 333–334, 339
  generalized, 547
  multistage stochastic, 559
  planar, 321
  simplex algorithm, 286–290
  stochastic, 547
Newsboy problem, 330
Newton's method, 403–404, 408, 425, 427, 436
NLPQL, 441
Node, 173–177, 180, 198, 200, 271, 274–276, 283, 285–287, 289, 294–296, 316–317, 319, 333, 337, 339, 366, 381, 599–600, 603–604, 606
  evaluation method, 203
  root, 287
  sink, 273–275, 282
  source, 273–275, 282
Node-arc incident matrix, 275, 287, 304
Nonanticipativity, 545–547, 553, 558

Nonconvex problems, 391
Nonconvex regions, 133-135, 137
Noncoplanar, 54
Nonlinear optimization, 3, 6, 9, 18, 385, 552
 feasibility tolerance, 445
 optimality tolerance, 445
 sensitivity analysis, 409, 414, 458, 534
 slack variables, 420, 424
 termination, 445
 tuning, 445
Nonlinear 0-1 polynomials, 141
Nonlinear regression, 479
Nonpolynomial-time algorithm, 83
Nonseparable functions, 145
Nonsingular, 54
Nonsmooth problems, 397
Norm estimate, 205
NP-hard problems, 575, 581
Null space, 87-88

Objective degradation, 194, 196
OMNI, 102
On-line process control, 437
Optimality conditions, 409
Operations research, 2, 8, 19, 22-24, 33
Optimal product, 43
Ordinal information, 492
Ordinal preference, 358
Ornstein-Ulenbeck equation, 551
OSL, 558, 562-563
Outer linearization, 555
Outranking/paired comparison methods, 491
Overachievement, 495

Pairwise interchange heuristics, 613
Parallel processors, 524
Parametric programming, 81
Part families, 300
Pascal's triangle, 310
Path, 272-273, 289, 291, 296, 303, 314-317, 319-321, 324, 381

Path length, 319-321
 longest arc, 319
 sum of arc, 317
 sum of node, 320
Payoff matrix, 490, 498
Penalties, 203
 analysis, 183
 methods, 217
 parameter, 446
 sequential linear programming, 433
Percentage error criterion, 205
Personnel staffing, 30
PERT/CPM network, 28-29
Petroleum industry, 461, 470
Piecewise linear approximation, 142-143
Piecewise linear optimization problem, 232
Planning horizon, 32
Plant, 96, 98, 100-104, 110, 115, 124, 559, 562
Policy, 322-323, 361-364, 370, 372, 374-375, 379, 381-382
Polyhedral annexation, 169
Polyhedral convex cone, 244
Polyhedral set, 43, 52-53, 55-57
Polyhedral theory, 257
Polynomial complexity, 83
Polynomial interpolation, 406
Polynomial-time algorithm, 83
Polytope, 52, 54
Pooling problem, 461
Predecessor function, 284-285
Preference information, 492
Prenodes, 209
Present value, 367
Prim's algorithm, 282
Priority ordering, 197
Probability tree, 311
Process planning, 257
Process selection, 21, 25, 34
Product, 5, 11-13, 16, 25, 27, 29, 36, 42-43, 71-72, 74, 78, 80, 93, 96, 100-101, 104, 110, 327, 342, 375, 559-560, 562, 566

Subject Index

Production, 2, 25, 33, 96, 99, 109, 111, 115, 275, 300, 303, 327, 331, 334–336, 382, 559, 566, 570
  capacity, 337
  data, 7, 42
  distribution, 97–98, 103, 105–106
  lot sizing, 2, 257–258, 302–303, 306
  multiperiod, 32
  planning, 6, 8, 23, 26, 33, 308, 501, 550, 568
  rates, 1, 5, 33
  requirements, 43
  scheduling, 21, 110, 123, 259, 276, 381
Productivity, 2–3
Product mix, 8, 21, 24–25, 34–35, 42, 44, 46, 62, 71–74, 76, 79, 81, 109, 506
Progressive hedging, 547–548, 558–559
Projective transformation, 85–86
Project management, 8, 259–260, 308, 381, 568
PROMETHEE, 491
Proportionality, 17
Proximal point algorithm, 558
Pseudocosts, 194, 204
Pull system, 518
Purchasing, 258
Purification scheme, 90

Quadratic assignment problem, 579, 613–614
Quadratic approximation, 435
Quadratic convergence, 433
Quadratic form, 396
Quadratic function, 391, 405
Quadratic interpolation, 408
Quadratic polynomial, 408
Quadratic program, 386, 435–436, 579
Quasi-convex, 375, 379
Quasi-Newton methods, 405–406, 417
Quattro Pro, 440
Queueing problems, 583

R&D project selection, 536
Random walk, 551

Rank, 54, 287
Rectilinear distance, 39–40
Recursion, 308–314, 318, 321, 323, 325–327, 329, 333, 336, 363, 371
  backward, 319, 332, 336
  forward, 319, 332, 336, 338, 341
Reduced gradient, 419, 423
Reduced objective, 419
Reduced problem, 419, 421
Refinery production planning, 435
Regeneration point, 325, 335, 339–341, 376, 381
Regression analysis, 472
Relative importance, 500
Reoptimizing strategy, 202
Resource allocation, 123
Resources, 566
Resource space tour, 200
Return function, 362
Returns to scale, 17
Right-hand side, 274
Risk aversion, 547, 558
Robustness, 356, 360
Robust optimization, 127, 551, 554
Rolling horizon, 566
Rounding, 126, 129
Route first-cluster second heuristic, 596–597
Routing, 586, 588, 594–598
Row-action algorithm, 547

Saddle point, 402
Sampling, 552, 558
Satisficing, 488
Savings algorithm, 595–596
Scaling, 240, 446, 546–548, 550–559, 562, 565–566
Scheduling, 2–3, 5, 8, 26, 32, 109, 112, 257, 259, 382, 518, 599, 602
  crew, 259
  employee, 7, 303–305, 570
  machine, 259, 575, 581, 584
  parallel-machines, 521
  personnel, 259, 559, 563

[Scheduling]
  problems, 576–578, 580, 582–583, 588–594
  single-machine, 587, 604
Search direction, 403, 429
Separable nonlinear functions, 142
Separable subproblems, 229
Separation algorithms, 151
Sequential decision process, 307, 496
Service-level constraints, 568
Setup time and setup cost, 5, 124, 259
Shifting bottleneck procedure, 612
Shortest path, 25, 27, 275–276, 282–285, 296, 298, 303, 306–308, 316–322, 324–326, 337, 365–366
  problem, 597
  tree, 286
Shortest processing time procedure, 522
Shortest setup time first (SST) heuristic, 582
Sifting, 558
Simulated annealing, 207, 584–586, 593–594
Simulation, 4, 550–551, 554, 584
Simultaneous linear equations, 2, 23
Single-machine sequencing problem, 518–519, 604
Single-pivot look-ahead, 184
Slack variable, 51
Smooth problems, 397
Software
  AMPL, 443, 560–561, 564–565, 567
  CONOPT, 440
  Excel, 399, 440–442
  GAMS, 102–104, 441–443, 463, 473
  GINO, 440, 443
  IMSL, 441
  Lotus 1-2-3, 442
  MathPro, 102, 105
  MINOS, 441
  MINTO, 209
  MPS, MPSX, 103, 441
  NAG Fortran Library, 441
  NLPQL, 441
  OMNI, 102

[Software]
  OSL, 558, 562–563
  QuattroPro, 440
Spanning tree, 273, 275, 277–282, 287–290, 301–302, 306
Sparse matrix representations, 440
Specially ordered sets, 142, 200
Specificity, 349
Splines, 449
Spreadsheet optimizers, 399, 441
SQP algorithm, 440
Squared error, 452
s-S policy, 376
Staffing, 3, 30–32, 93
Starting over strategy, 202
Stationary, 372
Stationary point, 402
Steepest ascent direction, 87
Steepest descent, 400
Step method, 497, 517
Step size, 401, 417
Stochastic decomposition, 552, 554, 558
Stochastic differential equations, 551
Stochastic processes, 550
Stochastic programming, 3, 7, 9, 18, 543–545, 547–551, 553–555
Stochastic programming approximation, 552
Stochastic quasi-gradient methods, 558
Stockless production systems, 518
Stopping criteria, 208
Strictly concave function, 393
Strictly convex function, 393
Strong duality theorem, 75–77
Structural/system constraints, 494
Subdifferential, 235
Subgradient, 235, 243
Subgradient optimization, 225, 232, 236, 238, 241–242, 252, 256–257
Substitution, 359
Successive linear programming, 430
Successive quadratic programming, 435
Superbasic variable, 423–424

Subject Index

Superposition theorem, 342
Surplus, 51
Surrogate constraint relaxations, 153
Surrogate dual, 219, 250, 252
Surrogate relaxation, 250
Sweep algorithm, 597

Tabu-list, 587
Tabu search, 207
Tabu-search, 584, 586, 593–594
Taylor series approximation, 425
Taylor's series expansion, 430
Telecommunications, 258
Termination condition, 241
Time pricing, 594
Time series statistical methods, 550
Tolerance bounds, 434
Total flow time, 518
Total tardiness, 522
Tour partitioning heuristic, 597
Transportation, 3, 8, 25, 27, 33, 93–96, 114
  cost, 12, 16
  model, 21, 24
  network, 26, 257
  problem, 16, 22
Transshipment, 8, 25, 27–29, 138
Trapezoidal approximation, 552
Traveling salesman problem (TSP), 147, 577, 582–583, 595, 597
Tree, 273, 284
Trust region, 434
Two-person-zero-sum method, 498
Two stage L-shaped method, 555

Unbalanced workload, 258
Unconstrained minimum, 389
Unconstrained problems, 399
Underachievement, 495
Undiscounted Markov decision, 367–369
Utility, 549
Utility function, 544

Valid inequalities, 151
Value of the stochastic solution, 543–544, 548
Variance reduction method, 552
Vector maximum, 500
Vehicle dispatching, 575
Vehicle routing, 258, 292, 581
Vertex, 244, 271–273, 275, 282–283, 293–299, 301
Vertex solution, 469

Wait-and-see solution, 549
Warehouse, 5, 11, 12, 16, 24, 138
Weak duality theorem, 75
Weighted flow time, 518
Weighted shortest processing time first rule, 583
Work factor analysis, 2
Workforce planning, 501
Worst-case analysis, 82
Worst-case bound, 210

Zero inventory, 518
Zero-one programming, 128
Zero-one variables, 128